Table of Atomic Weights

Element	Symbol	Atomic Number	Atomic Weight	Element	Symbol	Atomic Number	Atomic Weight
Actinium	Ac	89	(227)[a]	Manganese	Mn	25	54.9380
Aluminum	Al	13	26.98154	Mendelevium	Md	101	(258)
Americium	Am	95	(243)	Mercury	Hg	80	200.59
Antimony	Sb	51	121.75	Molybdenum	Mo	42	95.94
Argon	Ar	18	39.948	Neodymium	Nd	60	144.24
Arsenic	As	33	74.9216	Neon	Ne	10	20.179
Astatine	At	85	(210)	Neptunium	Np	93	237.0482
Barium	Ba	56	137.34	Nickel	Ni	28	58.71
Berkelium	Bk	97	(247)	Niobium	Nb	41	92.9064
Beryllium	Be	4	9.01218	Nitrogen	N	7	14.0067
Bismuth	Bi	83	208.9804	Nobelium	No	102	(255)
Boron	B	5	10.81	Osmium	Os	76	190.2
Bromine	Br	35	79.904	Oxygen	O	8	15.9994
Cadmium	Cd	48	112.40	Palladium	Pd	46	106.4
Calcium	Ca	20	40.08	Phosphorus	P	15	30.97376
Californium	Cf	98	(251)	Platinum	Pt	78	195.09
Carbon	C	6	12.01115	Plutonium	Pu	94	(244)
Cerium	Ce	58	140.12	Polonium	Po	84	(210)
Cesium	Cs	55	132.9054	Potassium	K	19	39.098
Chlorine	Cl	17	35.453	Praseodymium	Pr	59	140.9077
Chromium	Cr	24	51.996	Promethium	Pm	61	(147)
Cobalt	Co	27	58.9332	Protactinium	Pa	91	231.0359
Copper	Cu	29	63.546	Radium	Ra	88	226.0254
Curium	Cm	96	(247)	Radon	Rn	86	(222)
Dysprosium	Dy	66	162.50	Rhenium	Re	75	186.2
Einsteinium	Es	99	(254)	Rhodium	Rh	45	102.9055
Erbium	Er	68	167.26	Rubidium	Rb	37	85.4678
Europium	Eu	63	151.96	Ruthenium	Ru	44	101.07
Fermium	Fm	100	(257)	Samarium	Sm	62	150.4
Fluorine	F	9	18.99840	Scandium	Sc	21	44.9559
Francium	Fr	87	(223)	Selenium	Se	34	78.96
Gadolinium	Gd	64	157.25	Silicon	Si	14	28.086
Gallium	Ga	31	69.72	Silver	Ag	47	107.868
Germanium	Ge	32	72.59	Sodium	Na	11	22.98977
Gold	Au	79	196.9665	Strontium	Sr	38	87.62
Hafnium	Hf	72	178.49	Sulfur	S	16	32.06
Hahnium	Ha	105	(260)[b]	Tantalum	Ta	73	180.9479
Helium	He	2	4.00260	Technetium	Tc	43	98.9062
Holmium	Ho	67	164.9304	Tellurium	Te	52	127.60
Hydrogen	H	1	1.00797	Terbium	Tb	65	158.9254
Indium	In	49	114.82	Thallium	Tl	81	204.37
Iodine	I	53	126.9045	Thorium	Th	90	232.0381
Iridium	Ir	77	192.22	Thulium	Tm	69	168.9342
Iron	Fe	26	55.847	Tin	Sn	50	118.69
Krypton	Kr	36	83.80	Titanium	Ti	22	47.90
Kurchatovium	Ku	104	(260)[e]	Tungsten	W	74	183.85
Lanthanum	La	57	138.9055	Uranium	U	92	238.029
Lawrencium	Lr	103	(256)	Vanadium	V	23	50.9414
Lead	Pb	82	207.19	Xenon	Xe	54	131.30
Lithium	Li	3	6.941	Ytterbium	Yb	70	173.04
Lutetium	Lu	71	174.97	Yttrium	Y	39	88.9059
Magnesium	Mg	12	24.305	Zinc	Zn	30	65.38
				Zirconium	Zr	40	91.22

[a] Value in parentheses is the mass number of the most stable or best-known isotope.

[b] Suggested by American workers but not yet accepted internationally.

[c] Suggested by Russian workers. American workers have suggested the name Rutherfordium.

INTRODUCTION TO CHEMISTRY

INTRODUCTION TO CHEMISTRY

MARTHA J. GILLELAND

California State College,
Bakersfield

West Publishing Company

St. Paul   New York

Los Angeles   San Francisco

Student Study Guide

A study guide has been developed to assist students in mastering the concepts presented in this text. It reinforces chapter material presenting it in a concise format with review questions. An examination copy is available to instructors by contacting West Publishing Company. Students can purchase the study guide from the local bookstore under the title *Study Guide to Accompany Introduction to Chemistry*, prepared by Rebecca Williams, Richland College.

Copyediting Pamela McMurry
Design Janet Bollow
Artwork Vantage Art
Composition Progressive Typographers
Cover photo   color Schlieren photo of bunsen burner by Dagmar Hailer-Hamann, Peter Arnold
Cover design Janet Bollow

COPYRIGHT © 1986 By WEST PUBLISHING COMPANY
50 West Kellogg Boulevard
P.O. Box 64526
St. Paul, MN   55164-1003

Library of Congress Cataloging-in-Publication Data

Gilleland, Martha J.
   Introduction to chemistry.

   Includes index.
   1. Chemistry.   I. Title.
QD31.2.G5238   1986   540   85–22573
ISBN 0–314–93180–5
1st Reprint—1986

Photo Credits

1 assorted glassware by Werner H. Müeller, Peter Arnold; **2** The Bettmann Archive; **3** The Bettmann Archive; **6** Ron Church, Photo Researchers; **19** graduated cylinders by Herb Levart, Science Source/Photo Researchers; **33** Richard Wood, Taurus Photos; **47** images of graphite atoms by IBM Watson Research Center, Peter Arnold; **54** courtesy Michael Isaacson, Cornell University, and M. Ohtsuki, The University of Chicago; **73** abstract neon gas by Carol Kitman, Peter Arnold; **color insert page 1** (top) courtesy Sargent-Welch Scientific Company; (bottom left) Will McIntyre, Photo Researchers; (bottom right) Rip Griffith, Photo Researchers; **color insert page 2** (top left) Manfred Kage, Peter Arnold; (top right) Werner Müeller, Peter Arnold; (bottom left) Malcolm S. Kirk, Peter Arnold; (bottom right) John Zoiner, Peter Arnold; **97** sulfur crystallized by Manfred Kage, Peter Arnold; **123** sodium chloride and potassium chloride—crystals made from water solution by Manfred Kage, Peter Arnold; **150** The Bettmann Archive; **157** sterile reagent ampules: potassium chloride by Stan Levy, Science Source/Photo Researchers; **167** The Metropolitan Museum of Art; **173** sugar crystals by Science Photo Library, Science Source/Photo Researchers; **179** courtesy Perkin-Elmer Corporation; **189** lit candles by Warner H. Müller, Peter Arnold; **192** Burndy Library; **219** soap bubbles by Philip Jon Bailey, Taurus Photos; **220** David S. Strickler, Monkmeyer Press Photo Service; **221** The Bettmann Archive; **222** Martin M. Rotker, Taurus Photos; **253** table salt by Science Photo Library, Science Source/Photo Researchers; **263** Charles Farrow; **277** Westinghouse glassware by Dick Luria, Science Source/Photo Researchers; **309** perfume research beakers by Dick Luria, Science Source/Photo Researchers; **331** maldon sea salt by Dr. Jeremy Burgess, Science Photo Library, Science Source/Photo Researchers; **346** Paul Conklin, Monkmeyer Press Photo Service; **365** crystals of KCl by Omikron, Science

(continued following index)

CONTENTS

Preface   xv

CHEMISTRY: ORIGINS AND SCOPE   1

CHAPTER 1

1.1 The Origins of Chemistry	2
1.2 The Nature of Chemistry	2
1.3 Studying Chemistry	3
1.4 States of Matter	5
1.5 Classification of Matter	7
1.6 Subclassification of Matter	8
1.7 Physical and Chemical Changes	9
1.8 Properties of Matter	10
1.9 Conservation of Mass in Chemical Reactions	11
1.10 Energy	12
1.11 Energy Changes in Chemical Reactions	13
1.12 Energy and Matter	13
Summary	14
Study Questions and Problems	15

PERSPECTIVE
ACHIEVEMENTS OF ALCHEMY   3

SCIENTIFIC MEASUREMENTS   19

CHAPTER 2

2.1 Precision and Accuracy	20
2.2 Significant Figures	21
2.3 Calculations and Significant Figures	23
2.4 Scientific Notation	24
2.5 Systems of Measurement	27
2.6 Conversion of Units	29
2.7 Mass and Weight	32
2.8 Density	33
2.9 Specific Gravity	36
2.10 Units of Energy	36
2.11 Temperature	36
2.12 Specific Heat	39
Summary	41
Study Questions and Problems	42

PERSPECTIVE
DEVELOPMENT OF THE
THERMOMETER   39

CHAPTER **3**

ELEMENTS, ATOMS, AND COMPOUNDS — 47

3.1	Elements	49
3.2	The Periodic Table	52
3.3	Atoms	53
3.4	Subatomic Particles	54
3.5	Atomic Number	55
3.6	Mass Number	55
3.7	Isotopes	57
3.8	Atomic Mass Units	59
3.9	Atomic Weight	60
3.10	Compounds	61
3.11	Composition of Compounds	62
3.12	Chemical Formulas	62
3.13	Formula Weights of Compounds	63
3.14	Percentage Composition	65
	Summary	67
	Study Questions and Problems	68

PERSPECTIVE
MAKING NEW ELEMENTS   59

PROBLEM-SOLVING SKILLS
THE IMPORTANCE OF UNITS AND
LABELS   64

CHAPTER **4**

ELECTRON ARRANGEMENTS IN ATOMS — 73

4.1	Early Models of the Atom	74
4.2	The Nature of Light	76
4.3	Emission Spectra of the Elements	77
4.4	The Bohr Atom	78
4.5	Modern Ideas of Atomic Structure	80
4.6	Energy Levels of Electrons	81
4.7	Energy Sublevels	81
4.8	Electron Configurations	85
4.9	Quantum Numbers (optional)	89
	Summary	92
	Study Questions and Problems	93

PERSPECTIVE
RUTHERFORD'S GOLD FOIL
EXPERIMENT   75

CHAPTER **5**

CHEMICAL PERIODICITY — 97

5.1	The Periodic Law	98
5.2	The Modern Periodic Table	99
5.3	Periods of Elements	101
5.4	Groups of Elements	104
5.5	A Survey of the Representative Elements	104
5.6	The Transition Elements	109
5.7	Atomic Size	111
5.8	Ionization Energy	112
5.9	Electron Affinity	114
5.10	The Chemical Behavior of Metals and Nonmetals	115
	Summary	116
	Study Questions and Problems	118

PERSPECTIVE
IS THERE AN END TO THE PERIODIC
TABLE?   110

CHAPTER 6

CHEMICAL BONDS   **123**

6.1 Types of Chemical Bonds   124
6.2 Ionic Bonds   125
6.3 Electron Dot Formulas of Atoms and Monatomic Ions   126
6.4 Ionic Compounds   129
6.5 An Introduction to Oxidation and Reduction   131
6.6 Covalent Bonds   133
6.7 Molecular Compounds   135
6.8 Multiple Covalent Bonds   135
6.9 Polyatomic Ions and Coordinate Covalent Bonds   136
6.10 Electron Dot Formulas of Molecular Compounds and Polyatomic Ions   138
6.11 Molecular Shapes   141
6.12 Electronegativity   146
6.13 Polar Covalent Bonds   147
6.14 Polarity of Molecules   149
       Summary   151
       Study Questions and Problems   152

PERSPECTIVE
LINUS PAULING, CRUSADING SCIENTIST 150

CHAPTER 7

NAMING INORGANIC COMPOUNDS   **157**

7.1 Oxidation Numbers   159
7.2 Binary Ionic Compounds   161
7.3 Ionic Compounds Containing Polyatomic Ions   163
7.4 Binary Molecular Compounds   164
7.5 Acids   166
7.6 Inorganic Hydrates   167
       Summary   169
       Study Questions and Problems   169

PERSPECTIVE
THE ORIGIN OF SCIENTIFIC NOMENCLATURE IN CHEMISTRY 167

CHAPTER 8

CALCULATIONS BASED ON CHEMICAL FORMULAS   **173**

8.1 The Chemical Mole   174
8.2 Molar Masses   174
8.3 Calculation of Molar Masses   176
8.4 Additional Calculations Based on the Mole Concept   176
8.5 Empirical and True Formulas   178
8.6 Calculation of the Empirical Formula   181
8.7 Calculation of the True Formula from the Empirical Formula   184
       Summary   186
       Study Questions and Problems   186

PERSPECTIVE
A SIMPLE EXPERIMENT TO DETERMINE PERCENTAGE COMPOSITION 180

PROBLEM-SOLVING SKILLS
MANIPULATING RATIOS 182

CHAPTER **9**

PERSPECTIVE
ANTOINE LAVOISIER AND AN
EARLY DEMONSTRATION OF THE
CONSERVATION OF MASS IN
CHEMICAL REACTIONS   192

PROBLEM-SOLVING SKILLS
COMBINING STEPS IN THE
SOLUTIONS TO PROBLEMS   207

CHEMICAL EQUATIONS   189

9.1	The Language of Chemical Equations	190
9.2	Conservation of Mass in Chemical Reactions	191
9.3	Balancing Chemical Equations	191
9.4	Types of Chemical Reactions	197
	Decomposition	197
	Combination	197
	Single Replacement	197
	Double Replacement	198
9.5	Stoichiometry	199
9.6	Stoichiometric Calculations	201
	Mole-Mole Calculations	201
	Mole-Weight Calculations	202
	Weight-Mole Calculations	203
	Weight-Weight Calculations	205
9.7	Percentage Yield	208
9.8	Limiting Reactant	210
	Summary	213
	Study Questions and Problems	213

CHAPTER **10**

PERSPECTIVE
THE EARTH'S ATMOSPHERE   229

GASES   219

10.1	Properties of Gases	220
10.2	Volume and Pressure Relationships	223
10.3	Volume and Temperature Relationships	226
10.4	Pressure and Temperature Relationships	230
10.5	Combining the Gas Laws	231
10.6	Avogadro's Hypothesis	235
10.7	The Ideal Gas Equation	236
10.8	The Kinetic Molecular Theory	239
10.9	Gas Mixtures	241
10.10	Gas Density	242
10.11	Stoichiometric Calculations Involving Gas Volumes (optional)	243
	Mole-Volume Calculations	243
	Weight-Volume Calculations	244
	Volume-Volume Calculations	245
	Summary	247
	Study Questions and Problems	248

CHAPTER **11**

LIQUIDS AND SOLIDS   253

11.1	Intermolecular Forces	254
	Dipolar Attractions	255
	Hydrogen Bonds	255
	London Dispersion Forces	256

11.2 General Classes of Liquids   258
11.3 Densities of Liquids   258
11.4 Surface Tension   259
11.5 Viscosity   259
11.6 Vaporization   260
11.7 Boiling Point   262
11.8 Heat of Vaporization   263
11.9 The Solid State   265
         Crystalline Solids   265
         Amorphous Solids   267
11.10 The Melting Process   268
11.11 Heat of Fusion   268
11.12 Melting Point   269
11.13 Calculations Based on Phase Changes (optional)   270
          Summary   272
          Study Questions and Problems   273

PERSPECTIVE
LIQUID CRYSTALS   270

CHAPTER **12**   **WATER AND AQUEOUS SOLUTIONS**   **277**

12.1 Water: Structure and Properties   278
12.2 Properties of Solutions   280
12.3 Solution Formation   282
12.4 Solubility   284
12.5 Factors Affecting Solubility   285
         Pressure   285
         Temperature   286
         Chemical Structure   286
12.6 Concentrations of Solutions: Molarity   286
12.7 Concentrations of Solutions: Percent Concentration   290
12.8 Dilutions   291
12.9 Stoichiometric Calculations Involving Solutions   294
12.10 Colligative Properties of Solutions (optional)   295
          Vapor-Pressure Lowering   295
          Boiling-Point Elevation   296
          Freezing-Point Lowering   296
          Osmotic Pressure   298
12.11 Colloids and Suspensions (optional)   299
          Summary   301
          Study Questions and Problems   302

PERSPECTIVE
PREPARING SOLUTIONS OF
SPECIFIED MOLARITY   289

CHAPTER **13**   **CHEMICAL EQUILIBRIUM AND REACTION RATES**   **309**

13.1 Reversible Reactions   310
13.2 Rates of Chemical Reactions   311
13.3 Activation Energy   311

13.4 Factors Affecting Reaction Rates   312
        Temperature   312
        Concentration   313
        The Nature of the Reactants   314
        Catalysis   314
13.5 Equilibrium Constants   315
13.6 Application of Equilibrium Constants   318
13.7 Le Chatelier's Principle   319
        Change in Concentration   319
        Change in Temperature   322
13.8 Energy and Chemical Change   322
13.9 The Relationship between Chemical Equilibrium
        and Reaction Rates   324
        Summary   325
        Study Questions and Problems   326

PERSPECTIVE
ENERGY USE IN THE UNITED STATES
— PAST, PRESENT, AND
FUTURE   323

CHAPTER **14**   **ACIDS, BASES, AND SALTS**   **331**

14.1 Concepts of Acids and Bases   332
14.2 Properties of Acids and Bases   336
14.3 Common Acids and Bases   337
14.4 Acid Strength   338
14.5 Polyprotic Acids   341
14.6 Base Strength   343
14.7 Acid and Basic Anhydrides   345
14.8 Acid-Base Titrations   347
14.9 Normality (optional)   349
14.10 Salts   352
14.11 Solubilities of Salts and Other Ionic Compounds   354
14.12 Solubility Product Constants (optional)   356
        Summary   359
        Study Questions and Problems   359

PERSPECTIVE
ACID RAIN   346

CHAPTER **15**   **ELECTROLYTES, pH, AND BUFFERS**   **365**

15.1 Electrolytes   366
15.2 The Ionization of Water   367
15.3 pH   368
15.4 The pH Scale   371
15.5 Measurement of pH   373
15.6 Additional pH Calculations (optional)   374
15.7 Hydrolysis (optional)   377
15.8 Buffers   378
        Summary   381
        Study Questions and Problems   382

PERSPECTIVE
THE ACID-BASE BALANCE OF
BLOOD   380

CHAPTER **16**

OXIDATION AND REDUCTION   **385**

16.1 Concepts of Oxidation and Reduction   386
16.2 Balancing Redox Equations: The Oxidation State Method   387
16.3 Balancing Redox Equations: The Ion-Electron Method   389
  Reactions in Acidic Solution   389
  Reactions in Basic Solution   392
16.4 The Activity Series of Metals   393
16.5 Voltaic Cells   396
16.6 Batteries   397
  Lead Storage Battery   397
  Dry Cell   398
  Nickel-Cadmium Battery   399
16.7 Electrolytic Cells   399

PERSPECTIVE
CHARLES MARTIN HALL AND THE
PRODUCTION OF ALUMINUM 400

  Summary   401
  Study Questions and Problems   402

CHAPTER **17**

RADIOACTIVITY AND NUCLEAR PROCESSES   **407**

17.1 Radioactivity   408
17.2 Alpha Decay   409
17.3 Beta Decay   410
  Electron Emission   410
  Positron Emission   410
  Electron Capture   410
17.4 Gamma Decay   411
17.5 Nuclear Reactions   411

PERSPECTIVE
MADAME CURIE 411

17.6 Natural and Artificial Radioactivity   413
17.7 Detection and Measurement of Nuclear Radiation   415
  Detecting Nuclear Radiation   415
  Measurements of Nuclear Radiation   416
17.8 Radiation Safety   418
17.9 Applications of Radiochemistry   421
  Diagnostic X Rays   421
  Archeological Dating   422
  Isotopic Tracers   423
  Radiation Therapy   423
  Medical Diagnosis   424
17.10 Nuclear Energy   424
17.11 Nuclear Fission   426
  Nuclear Power Plants   426
  The Atomic Bomb   429

PERSPECTIVE
THE DEVELOPMENT OF THE ATOMIC
BOMB 430

17.12 Nuclear Fusion   430
  Summary   431
  Study Questions and Problems   432

CHAPTER **18**

INTRODUCTION TO ORGANIC CHEMISTRY   **437**

18.1 The Nature of Organic Chemistry   438
18.2 The Hydrocarbons   438
18.3 Alkanes   439
18.4 Naming Alkanes   442
18.5 Alkenes and Alkynes   447
18.6 Naming Alkenes and Alkynes   449
18.7 Cyclic Aliphatic Hydrocarbons   451
18.8 Benzene and Aromatic Hydrocarbons   452
18.9 Classification of Organic Compounds by Functional Group   454
18.10 Orbital Hybridization in Hydrocarbon Molecules (optional)   457
         Bonding in Methane   457
         Bonding in Ethene   458
         Bonding in Benzene   460
         Bonding in Ethyne   460
         Summary   463
         Study Questions and Problems   464

PERSPECTIVE
PETROLEUM REFINING 446

APPENDIX **A**

REVIEW OF BASIC MATHEMATICS   **471**

A.1 Diagnostic Test for Basic Calculations   471
A.2 Review of Basic Calculations   472
         Positive and Negative Numbers   472
         Addition   473
         Subtraction   475
         Multiplication   476
         Roots and Exponents   477
         Division   479
         Fractions   481
         Decimal Numbers and Percent   484
A.3 Diagnostic Test for Basic Algebra   485
A.4 Review of Basic Algebra   486
         Solving Algebraic Equations   486
         Word Problems   495
         Proportionalities   499
         Logarithms   502

APPENDIX **B**

DECIMAL pH VALUES   **507**

APPENDIX **C**

**SOLUTIONS TO EXERCISES AND ANSWERS TO SELECTED
STUDY QUESTIONS AND PROBLEMS**   **509**

Glossary   553
Index   563

PREFACE

Introduction to Chemistry is designed for students who have no formal background in chemistry but who need to prepare for a course in general chemistry or who need a working understanding of chemistry to achieve their career goals. Because the students using this textbook are not likely to have extensive mathematical skills, methods for solving mathematical problems are presented in detail and are based on dimensional analysis and logical reasoning. Examples of problems requiring mathematical skills and/or logical reasoning are abundant, and detailed solutions are shown. Immediately following each worked example is an exercise to reinforce the concept illustrated by the example. Detailed solutions to the exercises appear in appendix C.

Each chapter begins with a brief introduction and an outline of the topics covered in the chapter. Margin notes contain definitions of new terms, pronunciation of technical terms, and supplementary information. The margin notes called *Math Tips* explain or clarify the mathematical operations referred to in the text at that point. The boxed inserts called *Perspectives* describe historical developments, practical applications, current issues, or human involvement in chemistry. As the name implies, these are designed to help each student develop an individual perspective of chemistry. Where appropriate, boxed inserts entitled *Problem-Solving Skills* provide insight into strategies used in solving mathematical problems in chemistry. At the conclusion of each chapter are a summary and an extensive set of study questions and problems. Answers to selected study questions and problems appear in appendix C. Additional appendices contain a review of basic mathematics, including diagnostic tests, and explain how to calculate decimal pH values. The textbook concludes with a comprehensive glossary and an extensive index.

CHAPTER SEQUENCE

The text begins with the historical basis of chemistry and the fundamental nature of matter and energy. The first five chapters deal with scientific measurements; elements, atoms, and compounds; electron configurations; and chemical periodicity. These chapters provide a basis for laboratory work early in the course; in particular, classes of matter, specific heat, atomic weights, formula weights, and percentage composition are introduced early to establish a background for the typical laboratory experiments that might accompany this text. Chapters 6–9 contain additional basic material: chemical bonds, naming inorganic compounds, calculations based on chemical formulas, and chemical equations (including stoichiometric calculations).

Chapters 10–14 cover slightly more advanced topics: gases, liquids, and solids; water and aqueous solutions; chemical equilibrium and rates of reactions; and acids, bases, and salts. These chapters would probably be included in a one-semester course, although the chapter on gases could be omitted if this topic is not considered necessary for a particular audience. The last four chapters (15–18) contain advanced topics: electrolytes, pH, and buffers; oxidation and reduction; radioactivity and nuclear processes; and an introduction to organic chemistry. One or more of these chapters could be selected for inclusion in a one-semester course.

Several chapter sections are labeled *optional;* they are not essential to material that follows them in the textbook, and they may be included or omitted at the discretion of the instructor; an example is section 4.9, "Quantum Numbers."

TEXTBOOK SUPPLEMENTS

Excellent supplements are available for use with this text. The student study guide, prepared by Rebecca Williams of Richland College, contains an outstanding collection of study aids. The accompanying laboratory manual, by Thomas I. Pynadath of Kent State University, provides a rich assortment of student-tested experiments. Instructor's manuals are available for both the textbook and the laboratory manual.

ACKNOWLEDGMENTS

While working on this textbook, I was fortunate to receive thoughtful and perceptive comments from reviewers from a variety of academic locations. In particular, I wish to thank Rebecca Williams, Wayne Svoboda, and Robert C. Pfaff for their especially comprehensive reviews and generous suggestions. I am deeply grateful to them and all of the following reviewers:

Stan Ashbaugh, Orange Coast College; Dorothy S. Barnes, University of Massachusetts; Thomas M. Barnett, Johnson County Community College, Kansas; Wayne Boring, Stephen F. Austin University; Thomas B. Brill, University of Delaware; Thomas C. DeVore, James Madison University, Virginia; Lawrence P. Greis, Queens College of CUNY; Arthur H. Hayes, Santa Ana College, California; Charles Howard, University of Texas, San Antonio; Delores B. Lamb, Greenville Technical College, South Carolina; Pat Milliken, Triton College, Illinois; Gordon Parker, University of Toledo; Robert C. Pfaff, University of Nebraska; Michael J. Pizanis, Diablo Valley College, California; Lynn G. Savedoff, California State University, Northridge; Martha W. Sellers, Northern Virginia Community College; Ruth Sherman, Los Angeles City College; Wayne Svoboda, Northeastern Illinois University; Joseph Thomas, California State University, Fullerton; Richard C. Thompson, University of Missouri; Kathleen M. Trahanovsky, Iowa State University; Rebecca Williams, Richland College, Texas.

In addition, I would like to express my appreciation to my editors, Denise Simon and Barbara Fuller, for coordinating the project, to Pamela McMurry for editing the manuscript and checking solutions to the problems, and to Theresa O'Dell for her help with the text supplements.

INTRODUCTION TO CHEMISTRY

CHEMISTRY: ORIGINS AND SCOPE

CHAPTER 1

OUTLINE

1.1 The Origins of Chemistry
Perspective: Achievements of Alchemy
1.2 The Nature of Chemistry
1.3 Studying Chemistry
1.4 States of Matter
1.5 Classification of Matter
1.6 Subclassification of Matter
1.7 Physical and Chemical Changes
1.8 Properties of Matter
1.9 Conservation of Mass in Chemical Reactions
1.10 Energy
1.11 Energy Changes in Chemical Reactions
1.12 Energy and Matter
Summary
Study Questions and Problems

Chemistry is one of the physical sciences, those sciences that deal primarily with the study of inanimate objects. Astronomy, geology, and physics are also physical sciences. The physical sciences are closely interrelated, and chemistry is considered one of the more fundamental of the group. Chemistry involves the study of matter and its transformations, and non-living matter is the basis of all the physical sciences.

Because chemistry is a physical science, you might think that it is not a part of the living world; however, the principles of chemical behavior of living matter are the same as those of inanimate matter. In fact, chemistry is just as important to the study of the life sciences as it is to the study of the physical sciences. A good understanding of chemistry is essential in fields such as agriculture, biology, immunology, medical technology, medicine, microbiology, nursing, nutrition, and pharmacy, to name just a few of the life sciences and closely related applied sciences.

Matter (anything that takes up space and has mass) is all around us; even our bodies are made of matter. This means that chemistry pervades our lives and greatly influences our life styles and quality of life. Some examples of the vital roles that chemistry plays in the world are its applications in solving worldwide problems of nutrition and food availability, manufacturing commercial products, advancing the frontiers of medicine, and helping to alleviate the worldwide energy shortage.

1.1 THE ORIGINS OF CHEMISTRY

Alchemy
(AL-keh-me)

Historians believe that chemistry began with the ancient art of **alchemy,** a philosophy that attempted to understand the relationship between human beings and their surroundings here on earth. Alchemy probably originated in China during the fourth century B.C. At that time, the goal of alchemy was to prolong life. The Chinese alchemists believed that a person could achieve immortality by drinking a magic "elixir of life." Unfortunately, this potion, a solution of gold, existed only in the imagination of the alchemists. From its Chinese beginnings, alchemy spread through the Hindu, Greek, Arabic, and Latin cultures, and in the twelfth century A.D. it was brought to the Western world by Christian scholars returning from eastern travels. The European alchemists (see Figure 1-1), bothered by the scarcity of gold and hence the scarcity of the elixir of life, developed a second goal for alchemy—**transmutation,** the conversion of common, inexpensive metals such as lead and iron into gold. After searching for the secret of transmutation for three hundred years, the alchemists finally grew weary of their quest and devoted themselves to religious studies. As the mystical influence of alchemy diminished, the study of matter took on a more scientific approach, and chemistry emerged.

1.2 THE NATURE OF CHEMISTRY

Chemistry is the study of matter and its transformations. In your introduction to chemistry, you will study the fundamental units of matter, the

FIGURE 1-1 An Alchemist

⊙ PERSPECTIVE ACHIEVEMENTS OF ALCHEMY

Although alchemy did not create the elixir of life or discover the secret of transmutation, it did make important contributions. The work of fourteenth century alchemists began the characterization of the corrosive substances we now call acids. The use of these substances in concocting elixirs and in attempts at transmutation demonstrated many of the properties of acids. By allowing acids to react with other materials, the alchemists enlarged the list of known chemical substances, some of which are shown in Figure a. In addition, the alchemists developed

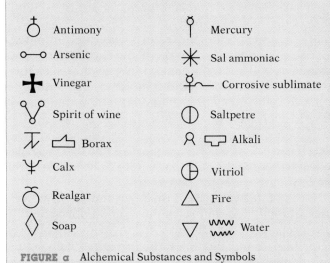

FIGURE a   Alchemical Substances and Symbols

FIGURE b   Paracelsus

laboratory techniques and equipment that are still in use today. These accomplishments provided a foundation of knowledge from which chemistry could develop. Furthermore, the elixirs made up by the alchemists became medicines for specific ailments, and the practice of medical chemistry was begun by Paracelsus (1494–1541) (Figure b), the most famous of the European alchemists. Paracelsus' work marked the beginning of pharmacy.

structure of matter, and how matter is transformed by chemical reactions and physical changes. As you study chemistry, you will discover a new way of looking at the world about you and the matter it contains. You should thus develop better insight into modern society and technology and equip yourself to function as a well-informed and responsible citizen.

1.3 STUDYING CHEMISTRY

Now that you are about to begin your study of chemistry, you may be wondering how to go about it. Chemistry is not a trivial subject, and to be successful at studying it requires a commitment, just as all important things in life do. Then just how should you approach chemistry? The answer is that you should approach it enthusiastically, actively, and responsibly. The key to the learning process is self-discipline. Think of learning chemistry just as you think of other goals: you must work at it. If you want to become an accomplished musician, lose weight, run the Boston marathon, or play professional football, you must exert a certain amount of self-discipline. The same is true of mastering chemistry. When you reach your goal, you will be rewarded with a sense of accomplishment, the good feeling that comes from achieving what you set out to do.

Now that you are ready to study chemistry with a sense of commitment, responsibility, and self-discipline, what activities are actually needed to learn chemistry? (Notice that *activities* are necessary; learning chemistry is an *active* process, not a passive one. You must *do* things to learn chemistry. It will not reach out and take hold of you; you must do work to master it.) Some tips for success are:

1. Attend *all* class meetings. Very few people can learn chemistry without benefit of an instructor and laboratory experience. It is very important that you arrive at your classes on time and prepared to listen, think along with the instructor, and take good notes.

2. Read through each assignment *before* the lecture. Class time is limited, and no instructor can (or would want to) recite the entire textbook to you. Read ahead and you will know the general subject of each lecture before you get to class. You may not understand every detail the first time you read it, but you will recognize the important terms (and know approximately how to spell them in your notes) and know which concepts are the difficult ones and which information is available in illustrations and tables. If you follow this advice, you will find that you have more time in class to listen and think, because it won't be necessary to try to write down every word the instructor says.

3. Take *good* notes in class. Listen to the instructor and jot down key words and phrases that will help you recall ideas and information. If you come to class prepared, you will not need to write down every word the instructor says, but in general, if the instructor feels something is important enough to write on the board, you should take it down in your notes. Also, be sure to take down practice problems so that you can work through them again later. Keep your notes neat, orderly, and legible, and they will encourage clear thinking.

4. Ask questions. Form the habit of asking questions of yourself as you read assignments. Make a list of the questions that you can't answer by reasoning or by referring to the textbook. If the answers do not become clear to you during lecture, ask the instructor in class or at a later time. The sooner you get answers to your questions, the faster and more completely you will learn.

5. Study chemistry *every day*. You will find that learning chemistry is a bit like building a house. First you install the foundation, and then you construct the rest of the house on the foundation. In learning chemistry, the first few weeks of study are critical, because they form the foundation of what is to come. Each chapter will depend on principles and concepts from all the preceding chapters, so you cannot skip over material for any period of time and expect to understand much of what follows. Budget your time carefully, so that you study chemistry at least an hour each day, preferably right after class while the subject is still fresh in your mind. During this study time you can review lecture notes

(outlining or rewriting them is an excellent way to study), work through practice problems given in class or in the textbook, reread the material in the textbook, answer assigned questions and problems (with and then without the aid of your textbook and class notes), and read the next assignment. You should also keep an organized notebook that contains all homework, handouts, returned exams and laboratory reports, and lecture notes. Always correct any questions missed on returned work immediately. These are only a few suggestions; don't be afraid to invent other activities. You will find that regular studying eliminates the need for cramming the night before an exam; you will need only to refresh yourself on the material you have already learned, and you can get a good night's rest.

6. Do not mistake memorizing facts for true learning. There will be factual information that you will want to commit to memory, but keep in mind that true learning involves *application* of factual information by reasoning and by solving problems. Memorization is not an end in itself but only a means to an end; the final goal is developing the ability to reason and to solve problems. To achieve this goal you must *practice* reasoning and problem-solving.

7. If at first you don't succeed . . . Not everyone is fortunate enough to do as well as he or she would like on the first exam in a course. If this happens to you, rework the exam (referring to your notes and the textbook) immediately. Master that material, because what comes next depends on it. Seek your instructor's assistance, especially if you cannot rework the exam correctly, and ask him or her to help you analyze your study habits. You may be able to find ways of improving them. If you think you need additional help, don't hesitate to make an appointment with a tutor. Above all, do everything you can to improve your study habits immediately.

These suggestions are intended not to discourage you but rather to encourage you to develop good study habits so that you will approach chemistry with confidence. It is a subject that can be mastered by the average college student, but it does require active involvement. In fact, if you enjoy thinking and learning, you just might find chemistry to be fun.

1.4 STATES OF MATTER

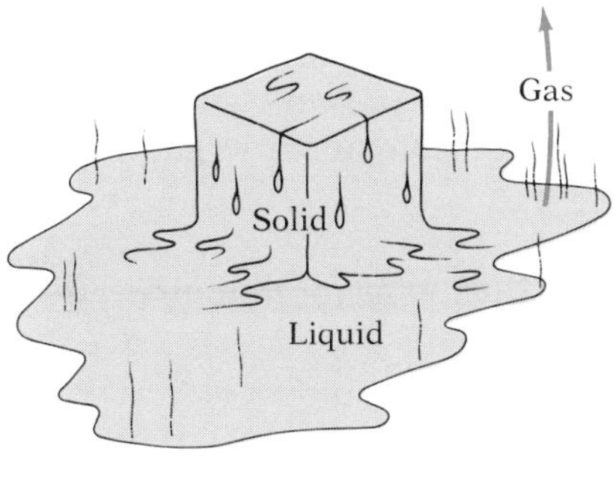

Since the study of chemistry involves the study of matter, let's review the definition of matter: Matter is anything that occupies space and has mass. It exists in three forms, commonly called the **states of matter** or **physical states:** solid, liquid, and gas. The three physical states of water are illustrated in Figure 1-2.

Solid matter is rigid, with a fixed volume. Baseballs, screwdrivers, and frying pans are examples of solid matter. In contrast, liquid matter such as water has no definite shape, and it will flow to take on the shape of its container. Remember, however, that a particular liquid sample has a fixed volume. This means, for example, that a gallon of water at a given tempera-

FIGURE 1-2 Water in Its Three Physical States: Solid (ice), Liquid, and Gaseous (water vapor)

FIGURE 1-3 Scuba Divers with Compressed Air Cylinders

Fluid

A substance that changes its shape easily and is capable of flowing.

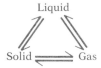

FIGURE 1-4 Changes in State Are Reversible

Identity

The unique nature of a particular person, object, or substance.

TABLE 1-1	Characteristics of the Three States of Matter		
Physical State	Shape	Volume	Examples
Solid	Fixed	Fixed	Iron, wood, table salt
Liquid	Variable; takes on the shape of its container	Fixed	Water, milk, gasoline
Gas	Variable; takes on the shape of its container	Variable; fills its container	Air, steam, helium

ture always occupies the same volume—one gallon—regardless of the size and shape of its container.

Like liquid matter, gaseous matter also has the ability to flow, so both liquids and gases can be classified as *fluids*. Unlike liquids, however, gases do not have fixed volumes; they can contract and expand to fill the volumes of their containers completely. Thus a gas has neither fixed shape nor fixed volume. The most familiar example of gaseous matter is the air we breathe. It can be compressed in cylinders to allow divers to breathe under water; as small amounts are released from the cylinder, the air expands to fill the diver's lungs (Figure 1-3).

The characteristics of the three states of matter are given in Table 1-1.

When matter changes from one physical state to another, the change is called a **change of state.** For example, ice (solid) melts to form water (liquid), and water can form water vapor (gas). (The temperature at which a substance melts is called its **melting point;** this is also the **freezing point** for the corresponding liquid. The temperature at which a substance boils is called its **boiling point.**) The reverse process also occurs: water vapor can condense to form liquid water, and liquid water can be frozen to make ice. Thus, changes in state are reversible, as shown in Figure 1-4. Such changes of state occur when water evaporates from the earth's surface and then precipitates as rain and snow. This cycle of events is known as the hydrologic cycle (Figure 1-5).

Although the changes of state of water are most familiar to us, other types of matter undergo similar changes. In every case, however, a change of state is only a *change in form; the identity of the substance remains the same.* We will discuss why and how changes of state occur in later chapters.

EXAMPLE 1.1

What is the physical state of each of the following at room temperature (72° F)?

a. Milk   **b.** Butter   **c.** Air   **d.** Mercury

Solution   **a.** Liquid
b. Butter melts to a liquid at room temperature.
c. Gas

Transpiration
The emission of water vapor by vegetation.

Precipitation
Falling products of moisture condensation in the atmosphere, such as rain, snow, or hail.

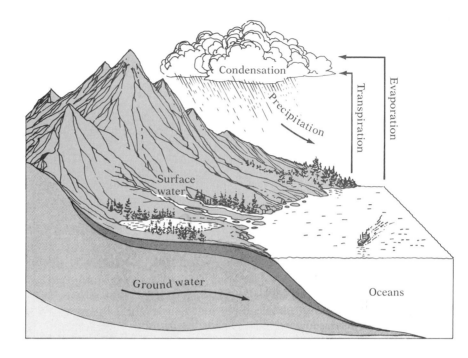

FIGURE 1-5 The Hydrologic Cycle

d. Liquid (Mercury is one of only two elements that are liquids at this temperature.)

EXERCISE 1.1 What is the physical state of each of the following at the temperature of a hot oven (450° F)?

a. Water **b.** Stainless steel **c.** Candle wax **d.** Cheese ▪

1.5 CLASSIFICATION OF MATTER

Homogeneous
(hoe-moe-GEE-nee-us): uniform throughout.

Matter can be classified as homogeneous or heterogeneous. **Homogeneous** matter has the same properties (identifying features and characteristics) and composition throughout. Since one state of matter has different properties than another state, homogeneous matter can consist of only a single physical state. Examples of homogeneous matter are water, water with sugar dissolved in it, air, stainless steel, and pure diamond.

Before we consider heterogeneous matter, we must understand the meaning of the term *phase*. A **phase** is a state of matter having clearly defined and distinguishable boundaries. Thus, when oil droplets are suspended in water, there are two phases—oil and water. In this example, both phases are liquids, but the boundaries that separate the two phases are easily distinguished. Homogeneous matter consists of only one phase. **Heterogeneous matter** is a mixture of two or more phases. Thus, the properties and composition of heterogeneous matter are not the same throughout. Examples of heterogeneous matter are oil suspended in water, mud suspended in water, dust suspended in air, and water with ice

Heterogeneous
(HEH-teh-row-gee-nee-us): nonuniform throughout.

Mixture
Two or more intermingled substances or phases.

cubes in it. Some substances do not appear heterogeneous by casual inspection: milk seems uniform, and so does fog. But if we look closely, perhaps with the aid of a magnifying glass, we see that milk is actually a suspension of tiny fat droplets in a liquid, and fog similarly contains droplets of water suspended in air (Figure 1-6).

EXAMPLE 1.2 Classify each of the following (at room temperature, 72° F) as homogeneous or heterogeneous.

a. Ice cream   **b.** Gold   **c.** Blood   **d.** Mercury

Solution **a.** Ice cream is heterogeneous because it contains milk, a heterogeneous substance.
b. Gold is homogeneous because it consists of only one substance in one phase.
c. Blood is heterogeneous because it contains suspended cells.
d. Mercury is homogeneous because it consists of only one substance in only one phase.

EXERCISE 1.2 Classify each of the following (at room temperature, 72° F) as homogeneous or heterogeneous.

a. Orange juice (with pulp in it)   **c.** Vinegar
b. Sea sand   **d.** Whipped cream   ■

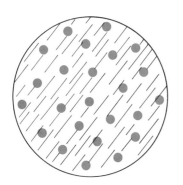

FIGURE 1-6 Suspended Droplets in Milk and Fog

1.6 SUBCLASSIFICATION OF MATTER

By definition, heterogeneous matter is a mixture, either of different substances in separate phases (as in a mixture of oil and water) or of different phases of the same substance (as in water containing ice cubes). All samples of heterogeneous matter can therefore be subclassified as one of these two types of mixtures.

Although it may seem contradictory, samples of homogeneous matter may also be mixtures, in this case, homogeneous mixtures. **Homogeneous mixtures** are a subclassification of homogeneous matter; they are mixtures having the same composition and properties throughout and consisting of only one phase. Thus, they qualify as homogeneous matter. Water with sugar dissolved in it, as mentioned earlier, is an example of a homogeneous mixture. So is air, a mixture of different gases evenly distributed through each other.

A second category of homogeneous matter also exists: **pure substances.** Pure substances contain only one substance and only one phase; they have the same composition and hence the same properties throughout. Examples of pure substances are aluminum foil, diamond, and distilled water. The classification and subclassification of matter are illustrated in Figure 1-7.

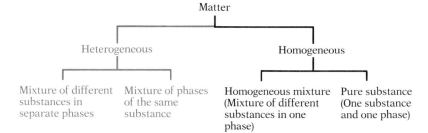

FIGURE 1-7 Classification and Subclassification of Matter

EXAMPLE 1.3

Classify each of the following as a heterogeneous mixture, homogeneous mixture, or pure substance.

a. Copper metal   **c.** Meringue (whipped egg white)
b. Salt dissolved in water   **d.** Creamy salad dressing

Solution

a. Copper metal is a pure substance because it contains only one substance in only one phase.
b. Salt dissolved in water forms a homogeneous mixture because the composition and properties of the mixture are the same throughout.
c. Meringue is a heterogeneous mixture because it contains pockets of air distributed through egg white.
d. Creamy salad dressing is a heterogeneous mixture because it contains oil droplets suspended in another liquid.

EXERCISE 1.3

Classify each of the following as a heterogeneous mixture, homogeneous mixture, or pure substance.

a. Smoke   **b.** Gasoline   **c.** Swiss cheese   **d.** Oxygen   ■

1.7 PHYSICAL AND CHEMICAL CHANGES

You have seen that matter easily undergoes changes of state; however, a change of state is not a change in identity. Ice is simply solid water; when it melts, it becomes liquid water. Evaporation converts liquid water to gaseous water, but regardless of its physical state, water is water. The same is true for any particular type of matter. A change that does not alter the identity of a substance is called a **physical change.** Changes of state are not the only kinds of physical changes that matter can undergo. Others are changes in size or shape. Breaking up ice cubes to make crushed ice reduces the sizes of the pieces of ice and changes their shapes, but the identity of the ice does not change. Mixing two different substances, such as sand and salt, together can be a physical change, provided that the identity of each substance is the same before and after mixing.

In contrast, any change that alters the identity of a substance is called a **chemical change.** In this kind of change, a substance is converted to one or more entirely *different* substances. Chemical changes are also called **chemical reactions.** Burning gasoline is a chemical change in which the liquid gasoline is converted to two new substances, carbon dioxide (a gas) and water vapor. The fact that the new substances are in a different physical state than the original substance is only coincidental; the important feature is the change in identity of the original substance. Other examples of chemical change are the tarnishing of silver, the rusting of iron, and the decaying of leaves.

EXAMPLE 1.4   Classify each of the following as a physical or chemical change.

a. The evaporation of alcohol   **c.** The breaking of glass
b. The burning of charcoal   **d.** The souring of milk

Solution   **a.** Physical change   **c.** Physical change
b. Chemical change   **d.** Chemical change

EXERCISE 1.4   Classify each of the following as a physical or chemical change.

a. The fermentation of grape juice to make wine
b. The mixing of cake batter
c. Wood being sawed
d. The formation of yogurt from milk

1.8 PROPERTIES OF MATTER

The **properties** of a sample of matter are those characteristics and features that distinguish it from all other kinds of matter. Thus, properties can be used to identify matter. There are two classes of properties: physical properties and chemical properties.

Physical properties can be observed and measured without changing the identity of the substance under consideration. Many physical properties can be noted by observing the appearance of a substance. Size, color, physical state, and shininess of surfaces are such physical properties. For other physical properties, one must observe how a substance interacts with its surroundings, provided there is no change in identity. Examples of these are taste, odor, conduction of heat or electricity, flexibility, and brittleness. Still other physical properties are melting point, boiling point, and density (to be discussed in chapter 2).

Chemical properties can be observed and measured only by changing the identity of the substance. Thus, chemical properties describe a substance's ability to undergo chemical change. One such chemical property is **flammability,** the ability to burn. This property helps to distinguish substances that burn, like gasoline and coal, from those that do not, such as iron and table salt. Rusting is a chemical property that distinguishes iron from stainless steel. A variety of chemical properties can be used to characterize a substance: examples are explosiveness, corrosiveness, toxicity, and susceptibility to tarnishing. Every chemical change that a substance can undergo describes a chemical property for that substance. Table 1-2 lists selected physical and chemical properties of a variety of common substances.

TABLE 1-2 Some Properties of Common Substances

Substance	Physical Properties			Chemical Properties
	Physical state	*Color*	*Melting point, °C*	
Acetic acid (component of vinegar)	Liquid	Colorless	16.6	Neutralizes lye
Iron	Solid	Gray	1530	Reacts with oxygen
Mercury	Liquid	Gray	−38.9	Reacts with oxygen
Oxygen	Gas	Colorless	−218.8	Reacts with hydrogen
Sodium chloride (table salt)	Solid	White	808	Can be decomposed by electricity
Wood	Solid	Brown	Decomposes	Flammable

EXAMPLE 1.5 Classify each of the following as a physical or chemical property.

a. Dynamite is explosive.
b. Water freezes at 32° F.
c. Chlorine is a bleaching agent.
d. Mercury is a liquid at room temperature.

Solution
a. Explosiveness is a chemical property.
b. The freezing point is a physical property.
c. Bleaching represents a change in identity of both the substance being bleached and the bleaching agent; thus, the ability to bleach is a chemical property.
d. The physical state of a substance is a physical property.

EXERCISE 1.5 Classify each of the following as a physical or chemical property.

a. Aluminum is easily corroded.
b. Iron is a hard, shiny metal.
c. Helium is a gas at room temperature.
d. Vinegar has a sour taste. ■

1.9 CONSERVATION OF MASS IN CHEMICAL REACTIONS

Throughout history, people have been intrigued by fire. For a long time it was thought that when a substance burned, most of the mass of the substance just disappeared, emitting heat and light and leaving behind a few ashes. But in the eighteenth century, the French chemist Antoine-Laurent Lavoisier gathered experimental evidence that suggested otherwise. He found that burning results from the chemical reaction of a substance with

FIGURE 1-8 Chemical Reactions Obey the Law of Conservation of Mass

Wood + oxygen = ashes + gases
Mass of wood + mass of oxygen = mass of ashes + mass of gases
Total mass before chemical change = total mass after chemical change

oxygen to produce one or more new substances. We have since come to realize that fire transforms a substance into one or more other substances, but no mass is destroyed. Burning, like all other chemical reactions, follows the principle called the **law of conservation of mass,** which states that matter is neither created nor destroyed during a chemical reaction. (The term **mass** refers to the amount of matter contained in a substance.) Thus, in burning and in all other chemical reactions, the total mass of the substances initially present is equal to the total mass of new substance or substances produced (Figure 1-8). Mass is always conserved during chemical change.

1.10 ENERGY

Energy is usually defined as the capacity to do work, and **work** means moving an object over a distance. In this sense, gasoline contains energy because burning gasoline can cause an automobile to move. Animals possess energy because they are able to move themselves or their body parts. In both cases, work is performed.

Energy exists in many forms, but all can be classified as either kinetic or potential energy. **Kinetic energy** is energy possessed by an object if it is moving. A moving car and a thrown baseball are examples of objects having kinetic energy; both have the capacity for doing work if they strike other objects. The car can displace a light post, a tree, or almost anything else in its path, and an errant throw of a baseball can move glass from a windowpane. **Potential energy** is energy that is stored in an object because of the object's position, condition, or composition. Because of its position, a car parked on top of a hill has stored energy that represents a potential for doing work. If the car begins to roll downhill, some of its potential energy is transformed into kinetic energy. A tightly drawn bowstring has potential energy because of its condition; when released, the bowstring imparts kinetic energy to an arrow. Dynamite contains potential energy due to its composition; it is a mixture of substances that explodes with great force when detonated. Its potential energy is transformed to kinetic energy as the dynamite explodes.

Kinetic
From the Greek *kinein*, meaning "to move."

EXAMPLE 1.6

Classify each of the following as having kinetic or potential energy.

a. Water at the top of a waterfall   **c.** A runner in a track race
b. A person hitting a ping pong ball   **d.** A sugar cube

Solution

a. If the water is flowing, it will have kinetic energy; regardless of that, it will certainly have potential energy due to its position.
b. Kinetic energy
c. Kinetic energy
d. A sugar cube has potential energy due to its stored chemical energy.

EXERCISE 1.6 Classify each of the following as having kinetic or potential energy.

a. A gopher digging a hole c. A ripe apple in a tree
b. A stretched rubber band   d. A dormant volcano   ■

There are many forms of energy: heat energy, mechanical energy, electrical energy, chemical energy, and light energy. Energy can undergo changes of form just as matter can. For example, heat energy is converted to mechanical energy when steam drives a piston. Heat energy is converted to electrical energy when steam turns a turbine that drives an electrical generator. Light energy is converted to heat energy when sunlight strikes and heats a dark surface, and the chemical energy of food is converted to mechanical energy by the metabolic processes that enable us to perform work. Although energy can be changed from one form to another, *energy cannot be created nor destroyed;* this principle is called the **law of conservation of energy.**

1.11 ENERGY CHANGES IN CHEMICAL REACTIONS

Combustion, or burning, is a chemical reaction in which an obvious energy change occurs: heat and light are released by the reaction. Energy changes accompany all chemical reactions, although some of the energy changes are not as evident as those in combustion. Some reactions release energy, and others absorb energy from their surroundings.

A combustion reaction can be thought of as a process in which the fuel (the substance that is burned) reacts with oxygen. In general, such reactions release energy. As you might expect, reactions that involve the reverse process, synthesis of a substance, usually require energy. An example of an energy-requiring reaction is the synthesis of fat in our bodies; in this case, energy released by the utilization of food is used to synthesize fat (Figure 1-9).

The energy changes of chemical reactions and physical changes may take many different forms, but specific terms are used when heat is involved. If heat is released by a process, the process is said to be **exothermic.** Conversely, if heat is absorbed by a process, the process is said to be **endothermic.**

Exothermic
Heat-releasing; from the Greek *exo-*, meaning ''outside,'' and *therm*, meaning ''heat.''

Endothermic
Heat-absorbing; from the Greek *endo-*, meaning ''within,'' and *therm*, meaning ''heat.''

1.12 ENERGY AND MATTER

The laws of conservation of mass and energy suggest that the amount of mass in the universe and the amount of energy in the universe remain separately constant. This idea was challenged by Albert Einstein in 1905,

FIGURE 1-9 The Energy Released by the Utilization of Food is Used to Synthesize Fat

Food + oxygen = decomposition products + energy
Energy + decomposition products = fat

when he suggested that mass and energy are interchangeable according to the following relationship,

$$E = mc^2$$

where E = energy, m = mass, and c = the speed of light (3×10^{10} cm/sec). The now famous Einstein equation states that energy is equal to mass multiplied by the square of the speed of light. In essence, energy and matter are related by the square of the speed of light. If, indeed, mass and energy are interchangeable, then the total mass of the universe can decrease, provided that the decrease is offset by an equivalent increase in the total amount of energy of the universe; the converse is also true. Einstein's ideas were far ahead of his time, and it was forty years before other scientists agreed with him.

Because the square of the speed of light is a very large number, a small amount of mass has the potential of producing an enormous amount of energy. However, chemical reactions involve energy changes of a rather small magnitude, and any mass changes are so very small that they are usually not detectable. Thus, the separate laws of conservation of mass and energy describe chemical reactions well enough to suit most purposes. In contrast, *nuclear reactions* result in significant mass changes and thus quite large energy changes (see chapter 17). In these cases, the Einstein equation describes the corresponding changes in energy and mass. Since mass and energy are interchangeable, we can go one step further and combine the two laws of conservation into a single principle: *The combined total of mass and energy in the universe remains constant.*

SUMMARY

Chemistry, the study of matter and its transformations, had its beginnings in the ancient art of alchemy. **Matter** is anything that occupies space and has mass. The three **states of matter** are solid, liquid, and gaseous matter. When matter changes from one physical state to another, the change is called a **change of state.**

Matter is classified as **homogeneous** or **heterogeneous.** Homogeneous matter has the same composition and properties throughout, while heterogeneous matter has variable composition and properties. Heterogeneous matter can be subclassified as a mixture of different substances in separate **phases** or as a mixture of different phases of the same substance. Homogeneous matter can be subclassified as a **homogeneous mixture** or as a **pure substance.**

Matter undergoes two kinds of changes: physical and chemical. A **physical change** is one that does not alter the identity of a substance, while a **chemical change** brings about a change in identity. Identifying features and characteristics of matter are called **properties. Physical properties** can be observed without changing the identity of a substance, but **chemical properties** describe a substance's ability to undergo chemical change. The **law of conservation of mass** states that matter is neither created nor destroyed during chemical change.

Energy, the capacity to do **work,** can be classified as **kinetic** or **poten-**

tial. Kinetic energy is the energy of a moving object, while potential energy is energy stored in an object due to position, condition, or composition. Energy exists in many forms, all of which are interchangeable; however, the **law of conservation of energy** states that energy can neither be created nor destroyed.

Energy changes accompany all chemical reactions and many physical changes. These energy changes take many different forms, but if heat is released, the process is **exothermic,** and if heat is absorbed, the process is **endothermic.**

The **Einstein equation,** $E = mc^2$, describes how energy and mass are related through the square of the speed of light. Given this relationship, we can formulate a new principle: The combined total of mass and energy in the universe remains constant.

STUDY QUESTIONS AND PROBLEMS

STATES OF MATTER

1. Define each of the following terms:
   a. Fluid   c. Freezing point   e. Change of state
   b. Melting point d. Boiling point
2. Describe the three states of matter and give two examples of each.
3. What is the physical state of each of the following?
   a. Ice cream at room temperature (72° F)
   b. Bread at room temperature (72° F)
   c. Water at 500° F
   d. Air at 200° F
   e. Candle wax at −10° F
4. Give a name to each of the following processes.
   a. The water from your breath forms a cloud on a cold day.
   b. Water is changed to ice.
   c. Alcohol spontaneously disappears from an open container.
   d. Milk gets hard at low temperature.
   e. Water bubbles vigorously when heated.
   f. Wet laundry hardens when hung outside on a cold day.

CLASSIFICATION OF MATTER

5. Define each of the following terms:
   a. Homogeneous matter   c. Phase
   b. Heterogeneous matter d. Mixture
6. Give two examples for each of the two classes of matter.
7. Classify each of the following as homogeneous or heterogeneous.
   a. Wood   d. A copper penney
   b. A glass windowpane e. A section of grapefruit
   c. Styrofoam   f. Gravel
8. How many phases of water are there in fog? Explain your answer.

SUBCLASSIFICATION OF MATTER

9. Define each of the following terms:
   a. Mixture   b. Pure substance

10. What are the two subclasses of heterogeneous matter?
11. What are the two subclasses of homogeneous matter?
12. Distinguish between the two types of heterogeneous mixtures, and give two examples of each.
13. Distinguish between homogeneous mixtures and pure substances; give two examples of each.
14. Classify each of the following as a heterogeneous mixture, homogeneous mixture, or pure substance.
    a. Alcohol dissolved in water
    b. Mist
    c. Fruit cake
    d. The solid residue left after evaporation of seawater
    e. Partially frozen pop
    f. Buttermilk
15. Vinegar is composed of water and acetic acid. Is vinegar a heterogeneous mixture, homogeneous mixture, or pure substance? Explain your answer.

PHYSICAL AND CHEMICAL CHANGES

16. Define each of the following terms:
    a. Physical change   b. Chemical change
17. Give three specific examples each of physical and chemical changes.
18. Classify each of the following as a physical or chemical change.
    a. The burning of leaves
    b. The crushing of grapes
    c. The sharpening of a knife
    d. The tarnishing of copper
    e. The expansion of water when it freezes
    f. The softening of lead at high temperature
19. When platinum wire is heated to extremely high temperatures, it glows red; when cooled, it returns to its original appearance. Do these observations describe a physical change or a chemical change? Explain your answer.
20. List each change that occurs in the following sequences, indicate whether it is a physical change or a chemical change, and explain your answers.
    a. Food is taken into the body by the mouth, where it is chewed, mixed with saliva, and swallowed. The food then enters the stomach, where it is mixed with acid and delivered to the small intestine, where it is digested.
    b. Iron ore is mined from the earth and crushed into small pieces. It is then heated with coal to produce iron.

PROPERTIES OF MATTER

21. Define each of the following terms:
    a. Properties (of a substance)   c. Chemical properties
    b. Physical properties   d. Flammability
22. Tell whether each of the following statements describes a physical or a chemical property. Explain your answers.

a. Sugar is a white solid at room temperature.
b. Copper conducts electricity.
c. Metals are corroded by acid.
d. Hydrogen is flammable.
e. Baking soda tastes salty.
f. Natural gas burns.

CONSERVATION OF MASS IN CHEMICAL REACTIONS

23. In your own words, summarize the law of conservation of mass.
24. Can the terms mass and matter be used interchangeably? Explain your answer.
25. When coal burns, what happens to it? Is any mass lost? Explain your answer.

ENERGY

26. Define each of the following terms:
    a. Energy   c. Kinetic energy
    b. Work   d. Potential energy
27. Classify each of the following as having kinetic or potential energy.
    a. A speeding bullet
    b. A pat of butter
    c. A dancing bear
    d. A hiker standing on the top of a mountain
    e. A flashlight battery
    f. Natural gas
28. In your own words, summarize the law of conservation of energy.

ENERGY CHANGES IN CHEMICAL REACTIONS

29. Define each of the following terms:
    a. Exothermic   b. Endothermic
30. Classify each of the following processes as exothermic or endothermic.
    a. The boiling of water   c. The formation of an ice crystal
    b. The burning of wood   d. The melting of wax

ENERGY AND MATTER

31. Write the Einstein equation and explain what it means.
32. Is it possible for mass to disappear from the universe? Explain your answer.
33. Is the conversion of mass to energy an important consideration in chemical reactions? Explain your answer.

SCIENTIFIC MEASUREMENTS

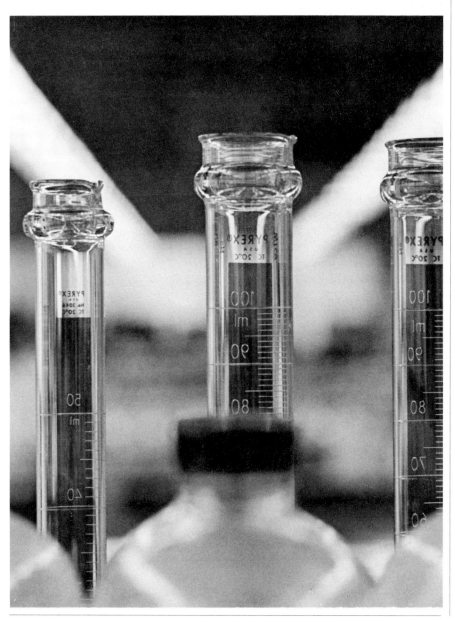

OUTLINE

2.1 Precision and Accuracy
2.2 Significant Figures
2.3 Calculations and Significant Figures
2.4 Scientific Notation
2.5 Systems of Measurement
2.6 Conversion of Units
2.7 Mass and Weight
2.8 Density
2.9 Specific Gravity
2.10 Units of Energy
2.11 Temperature
　　　Perspective: Development of the
　　　　Thermometer
2.12 Specific Heat
　　　Summary
　　　Study Questions and Problems

Measurements—where would we be without them? How could you tell someone how far away you live or where you found that terrific campsite last summer? How could you manage the 24 hours of each day, or for that matter, how could you manage your life? Just as measurements are essential to daily living, they are also essential to chemistry. Chemistry is an experimental science, and experiments involve making measurements.

Measurements always involve units, such as feet (for length), acres (for area), and gallons (for volume). Thus, units are the language of measurements. In this chapter you will learn about not only measurements but also their units and how to make conversions between various units.

Making unit conversions requires some basic math skills, most of which you learned in high school; however, if you need a review of basic math, now is the time to refer to appendix A. There you will find material designed to help you sharpen the math skills needed for this chapter and the following ones.

2.1 PRECISION AND ACCURACY

Precision and accuracy are terms that describe the quality of a measurement. Although they are often used interchangeably in casual conversation, they have distinctly different meanings when applied to scientific measurements. **Precision** refers to how closely repeated measurements of the same quantity agree with each other. If multiple measurements of the same quantity are in good agreement with each other, the measurements have a high degree of precision. In contrast, **accuracy** refers to how well a measurement or multiple measurements agree with the true value.

The distinction between precision and accuracy can be illustrated by considering the results of the archery contest shown in Figure 2-1. Archer A places five arrows at scattered positions on the target. Archer B places all five arrows in a small area on the target but not near the bull's-eye. Archer C is the best shot of all; all five arrows are clustered inside the bull's-eye. Let's imagine that the locations of the arrows correspond to measurements. The measurements of archer A are in poor agreement with each other and are not near the true value (the bull's-eye); hence these measurements have neither precision nor accuracy. Archer B's measurements are in good agreement with each other but not with the true value; thus they have good precision but poor accuracy. The best measurements are those of archer C; they are both precise and accurate.

Precision and accuracy are qualities to strive for in making scientific measurements. The precision of a measurement is often dictated by how finely the units of measurement are marked on the measuring device; in other words, precision often depends on the fineness of calibration of the measuring device. We would expect to make much more precise measurements with a stopwatch that is calibrated to the nearest tenth of a second than with an ordinary wristwatch that is calibrated only to the nearest second (see Figure 2-2). This is because the smallest difference that can be read on the stopwatch is about 0.01 seconds, whereas the smallest difference that can be read on the wristwatch is fifty times larger, about 0.5 seconds. Thus, more uncertainty of measurement is obtained

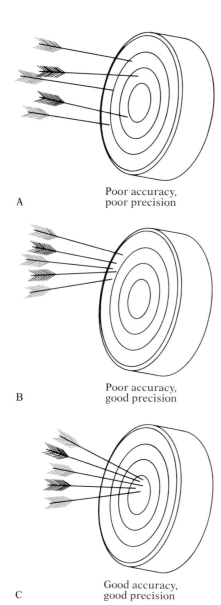

A   Poor accuracy,
      poor precision

B   Poor accuracy,
      good precision

C   Good accuracy,
      good precision

FIGURE 2-1   Precision and Accuracy in Archery

FIGURE 2-2 A Stopwatch Measures Time More Precisely Than Does a Wristwatch

with the less finely calibrated device, in this case the wristwatch. Thus, precision can also refer to how finely a measuring device is calibrated.

Accuracy also depends on the measuring device to some extent; the device must be carefully calibrated to a known standard. Accurate measurements have only small amounts of error; the ability to make almost error-free measurements depends on the skill of the person using the measuring device. Errors in measurement are typically caused by misreading the scale or choosing the wrong measuring device.

A high degree of precision usually indicates good accuracy, but this is not always true. If the same mistake is made throughout a series of measurements, the measured values will show a high degree of precision, but poor accuracy, like the measurements of archer B.

2.2 SIGNIFICANT FIGURES

In making and recording scientific measurements, we must be careful that the reported values indicate the precision of our measurements. Suppose we measure a sheet of paper with a ruler calibrated to 0.1-inch intervals and find the length, upon close inspection, to be about 20% of the way between 11.1 and 11.2 inches. We know with certainty that the length is closer to 11.1 inches than to 11.0 inches or 11.2 inches, and we might estimate it to be 11.12 inches. The last digit of our measurement is an estimate. Thus the first three digits are highly reliable, but the last digit is somewhat uncertain. In using scientific data, we always assume that the last digit is an estimate; the amount of uncertainty is frequently taken to be one unit of measure in the last digit. Thus our measurement of 11.12 inches has an uncertainty of 0.01 inches. If we were to use a more finely calibrated ruler, we might find the paper to be 11.123 inches long, and this measurement would have an uncertainty of 0.001 inches.

In our example, every digit in the number 11.12 was **significant;** that is, every digit had physical meaning, even the last digit (2) that was estimated. If we had measured with a ruler calibrated to only 1-inch intervals, a reported length of 11.12 inches would have no real meaning, as it would be physically impossible to estimate the last digit. The correct use of **significant figures** (those digits that have actual physical meaning) is essential to reporting scientific measurements validly. It is important to note that *all digits obtained by measurement are significant, and that the last digit to the right in such measurements is always an estimate.*

If there are zeros in a reported value, the number of digits may *not* correspond to the number of significant figures in the value. For example, the numbers 4 and 0.004 have different numbers of digits, but they both have only one significant figure, because the zeros in 0.004 are used only for showing where the decimal point belongs. The following rules will help you to handle the problem of zeros and significant figures easily.

1. *Zeros used to show where a decimal point belongs are not significant.* For example, each of the numbers below

    0.005, 0.05, 0.5, 50, 500

has only one significant figure. In such cases, the zeros serve only to locate the position of the decimal point.

2. *Zeros that are part of a measurement are significant.* For example, in the value 0.00705 inches, the zero between 7 and 5 is significant because it is a part of the measurement; however, the zeros before 7 are not significant because they serve only to locate the decimal point.

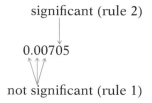

In another example, 1.240, the zero is also significant. It must be part of the measurement, because there is no other reason for its existence; it certainly is not needed to show where the decimal point belongs.

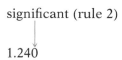

3. **Exact numbers,** numbers that have no uncertainty associated with them, are considered to have an infinite number of zeros following the decimal point. Thus, *exact numbers have an infinite number of significant figures.* One kind of exact number is a value used as a definition. For example, by definition, 1 gallon contains exactly 4 quarts:

$$1 \text{ gal} = 4 \text{ qt}$$

In the context of this definition, the numbers 1 and 4 are exact. Another kind of exact number is a value obtained by counting individual objects. A deck of playing cards always contains *exactly* 52 cards, not 51.7 or 52.3. A third kind of exact number is one that corresponds to the numbers found in reduced simple fractions. For example, in fractions like

$$\frac{1}{4}, \frac{2}{3}, \frac{8}{9},$$

the whole numbers in the numerators and denominators are exact numbers. Values such as these do not represent actual measurements.

EXAMPLE 2.1   How many significant figures are in each of the following measurements?

a. 751 inches   c. 400.2 miles
b. 600 seconds   d. 0.0086 gallons

Solution   a. 3   b. 1   c. 4   d. 2

EXERCISE 2.1   How many significant figures are in each of the following measurements?

a. 46.02 feet   c. 1.450 square inches
b. 1700 miles   d. 0.190 inches   ■

2.3 CALCULATIONS AND SIGNIFICANT FIGURES

So far, the emphasis in this chapter has been on making and recording measurements. What happens when those measurements are used in calculations? Suppose we want to find the area of a sheet of paper and we have measured the length to be 11.12 inches and the width to be 8.46 inches. Multiplying these dimensions together gives the area:

$$A = (\text{length})(\text{width})$$
$$= (11.12 \text{ in})(8.46 \text{ in})$$

Using a calculator, we find the answer to be 94.0752 square inches. Are we justified in reporting this six-digit value? The answer is no, because our measurements of length and width contained only four and three significant figures, respectively. Multiplying these values together gives a quantity whose value is only as good as the least precise measurement from which it is derived. Since the least precise figure in our multiplication contained only three significant figures, our answer can have only three significant figures. Thus, the correct value of the area, as determined by our measurements, is 94.1 square inches.

Since significant figures in calculations can be tricky, use the following rules to ensure that your answers always contain the correct number of significant figures:

1. *In multiplication and division, the answer must contain the same number of significant figures as the term with the least number of significant figures.* For example, in the calculation

$$(15.79)(13.6) = 214.744$$

   the answer should be reported as 215, because the term with the least number of significant figures (13.6) has only three significant figures. Therefore, the answer can contain only three significant figures.

2. *In addition and subtraction, the answer must contain the same number of decimal places as the term with the least number of decimal places.* For example, in the addition

$$\begin{array}{r} 17.01 \\ 13.2 \\ \underline{327.845} \\ 358.055 \end{array}$$

   the answer should be reported as 358.1, because the term with the least number of decimal places (13.2) has only one decimal place. Therefore, the answer can contain only one decimal place.

In these examples you probably noticed that we had to reduce the number of digits in our answers to comply with the rules. Whenever a calculated value contains more figures than are significant, we must *round off* the number. The following rules describe how to round off properly:

1. If the nonsignificant figure is *less* than 5 (or 50 if there are two nonsig-

---

MATH TIP

When multiplying measurements, first separate the numbers from the units

$$(11.12)(8.46)(\text{in})(\text{in})$$

Then multiply the numbers on a calculator. The units must also be multiplied. Since

$$(x)(x) = (x^1)(x^1) = x^{1+1} = x^2, \text{ then}$$
$$(\text{in})(\text{in}) = (\text{in}^1)(\text{in}^1) = \text{in}^{1+1} = \text{in}^2$$

When both the numbers and units have been multiplied, the answer has been obtained.

nificant figures, or 500 for three nonsignificant figures, . . .), simply *drop* the nonsignificant figure. Thus, 6.483 becomes 6.48 when rounded off to three significant figures.

2. If the nonsignificant figure is *more* than 5 (or 50 if there are two nonsignificant figures, or 500 for three nonsignificant figures, . . .), *drop the nonsignificant figure and increase the significant figure preceding it by 1*. Thus, 6.487 becomes 6.49 when rounded off to three significant figures.

3. If the nonsignificant figure is *exactly* 5 (or 50 if there are two nonsignificant figures, or 500 for three nonsignificant figures, . . .), *drop* the 5 (or the 50, 500, . . .) and *increase the preceding significant figure by 1 if it is odd or leave it the same if it is even*. For example, 9.35 becomes 9.4 when rounded off to two significant figures, and 9.45 also becomes 9.4 when rounded off to two significant figures.

EXAMPLE 2.2   Round off each number to three significant figures.

    a. 1.428   b. 0.44695   c. 0.000334772   d. 7.455

Solution   a. 1.43   b. 0.447   c. 0.000335   d. 7.46

EXERCISE 2.2   Round off each number to two significant figures.

    a. 75.992   b. 0.007795   c. 1.22500   d. 583,791

2.4 SCIENTIFIC NOTATION

In numbers containing one or more terminal zeros, as in 50 and 1400, it is difficult to tell how many significant figures are intended. We can eliminate this uncertainty by using scientific notation. A number written in scientific notation has the general form

$$N \times 10^{\text{exponent}}$$

where N is usually a number between 1 and 10, and the exponent is a whole number. (An **exponent** is the power to which a number is raised. For example, 1 million (1,000,000) can be expressed as 10^6, where 6 is the exponent and 10 (the number raised to a power) is called the **base**.) N must contain the correct number of significant figures as dictated by measurements or calculations. Table 2-1 illustrates the use of scientific notation to express numbers to two significant figures.

When you convert an ordinary number to scientific notation, you will probably have to move the decimal point. For example, the number 1 million (1,000,000) is written in scientific notation by moving the decimal point six places to the left and using the exponent 6. In this case, N is 1, and it is assigned the correct number of significant figures by placing zeros after the decimal point. For one significant figure, 1 million is expressed as 1×10^6.

$$1 \text{ million} = 1,000,000 = 1 \times 10^6$$

move decimal 6 places to left

TABLE 2-1 Scientific Notation	
Number	Scientific Notation
1,000,000	1.0×10^6
100,000	1.0×10^5
10,000	1.0×10^4
1,000	1.0×10^3
100	1.0×10^2
10	1.0×10^1
1	1.0×10^0
0.1	1.0×10^{-1}
0.01	1.0×10^{-2}
0.001	1.0×10^{-3}
0.0001	1.0×10^{-4}
0.00001	1.0×10^{-5}
0.000001	1.0×10^{-6}

Note that 10^0 is equal to 1.

A **positive exponent** indicates how many times a base must be multiplied by itself to give the original number, as illustrated below.

$$1 \text{ million} = 1,000,000 = (10)(10)(10)(10)(10)(10) = 1 \times 10^6$$

A positive exponent also signifies that the number is greater than 1.

The number 1/1,000,000 (one-millionth) can also be expressed in scientific notation. Here we must move the decimal point six places to the right and use the exponent −6.

$$\text{one-millionth} = \frac{1}{1,000,000} = 0.000001 = 1 \times 10^{-6}$$

move decimal 6 places to right

A **negative exponent** tells how many times 1 must be divided by the base to give the original number, as shown below.

$$\text{one-millionth} = \frac{1}{1,000,000} = \frac{1}{(10)(10)(10)(10)(10)(10)} = 1 \times 10^{-6}$$

A negative exponent also signifies that the number is less than 1.

To convert a number expressed in scientific notation to its ordinary form, simply reverse the operations just described. A positive exponent tells you that the decimal point must be moved to the right to give the ordinary number, and a negative exponent tells you to move the decimal point to the left to get the ordinary number.

EXAMPLE 2.3

Express 47,398 in scientific notation with three significant figures.

Solution

The goal is to express 47,398 in the form

$$N \times 10^{\text{exponent}},$$

where N is between 1 and 10. To do this, we must move the decimal point four places to the left and therefore use the exponent 4. We must also have three significant figures in our answer, so we round off N to 4.74. Thus, the correct answer is 4.74×10^4.

$$47{,}398 = 4.7398 \times 10^4$$

$$= 4.74 \times 10^4 \text{ (rounded off to three significant figures)}$$

EXERCISE 2.3   Express each of the following numbers in scientific notation with the number of significant figures indicated.

    a. 138,000 (three significant figures)
    b. 5,731 (two significant figures)
    c. 9,548,735 (four significant figures)
    d. 142,700 (two significant figures)   ■

EXAMPLE 2.4   Express 0.0000572 in scientific notation with two significant figures.

Solution   To give N a value between 1 and 10, we must move the decimal point five places to the right and use the exponent -5. This gives the expression 5.72×10^{-5}. Next we must round off N (5.72) to two significant figures, giving a final answer of 5.7×10^{-5}.

$$0.0000572 = 5.72 \times 10^{-5}$$

$$= 5.7 \times 10^{-5} \text{ (rounded off to two significant figures)}$$

EXERCISE 2.4   Express each of the following numbers in scientific notation with the number of significant figures indicated.

    a. 0.00598 (two significant figures)
    b. 0.0003714 (three significant figures)
    c. 0.0963 (three significant figures)
    d. 0.000000359 (two significant figures)   ■

MATH TIP

A scientific calculator can be used for calculations involving numbers in scientific notation. The correct key to use will be marked EE, Exp, or EEX. To enter the number 4.74×10^4, first enter 4.74, press the EE key, and then press 4. The calculator will display this as

    4.74   04

Then proceed with any calculations with this number on the calculator in the usual way.

    Now you can see how scientific notation clarifies the problem of significant figures for numbers ending in one or more zeros. For example, in the number 4,000, how many zeros are significant? We cannot tell just by looking at the number, but surely the person who reported the value as a measurement knows the number of significant figures intended. By changing 4,000 to scientific notation, it can be written in any one of the following ways, depending on how precisely the measurement was made.

$$4.000 \times 10^3 \text{ four significant figures}$$
$$4.00 \times 10^3 \text{ three significant figures}$$
$$4.0 \times 10^3 \text{ two significant figures}$$
$$4 \times 10^3 \text{ one significant figure}$$

2.5 SYSTEMS OF MEASUREMENT

There are two popular systems of measurement: the *British system* and the *metric system*. The British system, traditionally used in the United States, the United Kingdom, and Canada, has the familiar units of feet, pounds, and gallons. The metric system is used by scientists and by all industrialized nations of the world except the United States; however, the United States is now in the process of converting to the metric system. Recently a standard metric system called the *International System of Measurements*, or *SI* (from the French *Systeme International*), was adopted for worldwide use (see Table 2-2). The SI is not a new system of measurement but a modernization and simplification of the older metric system. The SI is composed of one unit in each type of measurement (or property). Both the metric system and the SI are based on the decimal system, and standard prefixes shown in Table 2-3 are used to denote powers of 10.

The basic unit of length in SI is the *meter* (also spelled metre), abbreviated as m. One meter is equivalent to 39.4 inches (or 1.09 yards). The

TABLE 2-2 Units in Measurement Systems (Boldface indicates SI standard units.)

Property	Metric	British	Unit Relationships
length	kilometer (km)	mile (mi)	1.61 km = 1 mi
	meter (m)	yard (yd)	1 m = 39.37 in
			1000 m = 1 km
	centimeter (cm)	foot (ft)	30.48 cm = 1 ft
		inch (in)	2.54 cm = 1 in
			100 cm = 1 m
	millimeter (mm)		1000 mm = 1 m
	micrometer (μm)		$10^6 \, \mu$m = 1 m
	nanometer (nm)		10^9 nm = 1 m
volume	**cubic meter** (m³)	quart (qt)	1 m³ = 1057 qt
	liter (L)		1 L = 1.057 qt
			1 L = 0.001 m³
	deciliter (dL)		10 dL = 1 L
	milliliter (mL)		1000 mL = 1 L
	cubic centimeter (cm³)		1 cm³ = 1 mL
	microliter (μL)		$10^6 \, \mu$L = 1 L
mass	**kilogram** (kg)	pound (lb)	1 kg = 2.2 lb
	gram (g)		453.6 g = 1 lb
			1000 g = 1 kg
	milligram (mg)		1000 mg = 1 g
	microgram (μg)		$10^6 \, \mu$g = 1 g
	nanogram (ng)		10^9 ng = 1 g
energy	**joule** (J)		1 J = 10^7 erg
	kilocalorie (kcal)		1 kcal = 4184 J
	calorie (cal)		1000 cal = 1 kcal
			1 cal = 4.184 J
	erg (erg)		4.184 × 10^7 erg = 1 cal
temperature	**Kelvin** (K)	Fahrenheit (°F)	K = °C + 273
	Celsius (°C)		°F = ⅘ °C + 32
			°C = (°F − 32)(⅚)

TABLE 2-3	Common Prefixes in the Metric System		
Prefix	Abbreviation	Meaning	Exponential Notation
nano-	n	one billionth (0.000000001)	10^{-9}
micro-	μ	one millionth (0.000001)	10^{-6}
milli-	m	one thousandth (0.001)	10^{-3}
centi-	c	one hundredth (0.01)	10^{-2}
deci-	d	one tenth (0.1)	10^{-1}
kilo-	k	one thousand (1,000)	10^{3}

metric prefixes given in Table 2-3 can be combined with the basic unit of length (the meter) as follows:

$$1 \text{ nanometer (nm)} = \frac{1}{1{,}000{,}000{,}000} \text{ meter (m)} = 10^{-9} \text{ meter (m)}$$

$$1 \text{ micrometer } (\mu\text{m}) = \frac{1}{1{,}000{,}000} \text{ meter (m)} = 10^{-6} \text{ meter (m)}$$

$$1 \text{ millimeter (mm)} = \frac{1}{1{,}000} \text{ meter (m)} = 10^{-3} \text{ meter (m)}$$

$$1 \text{ centimeter (cm)} = \frac{1}{100} \text{ meter (m)} = 10^{-2} \text{ meter (m)}$$

$$1 \text{ decimeter (dm)} = \frac{1}{10} \text{ meter (m)} = 10^{-1} \text{ meter (m)}$$

$$1 \text{ kilometer (km)} = 1{,}000 \text{ meter (m)} = 10^{3} \text{ meter (m)}$$

The volume of an object is the space occupied by the object. Since calculating the volume of an object of known dimensions requires cubing units of length, the SI standard unit of volume is the cubic meter (m^3). This is a large volume (about 274 gallons), and it has not yet become a popular unit. Most scientists use the unit of volume called the *liter* (L). This unit is defined as 0.001 cubic meter, and thus 1 L = 0.001 m^3, or 1000 L = 1 m^3. Subunits of the liter such as *milliliters* (mL) and *deciliters* (dL) are often used to measure small volumes. These subunits are defined in Table 2-2.

The milliliter (mL) is the unit of volume that you will use most frequently in making measurements in chemistry. A milliliter (mL) is the same as a cubic centimeter (cm^3). It takes about twenty drops from an eyedropper to give a volume of one milliliter. Figure 2-3 shows several pieces of equipment used in the chemistry laboratory for measuring liquid volume. All of these devices are calibrated in milliliters.

The SI unit of mass, the *kilogram* (kg), is also a large quantity, slightly more than two pounds, and subunits such as the *gram* (g), *milligram* (mg), and *microgram* (μg) are often used. These subunits are also defined in Table 2-2. We will discuss the topic of mass more completely later in this chapter.

MATH TIP
To perform calculations using metric measurements you will often need to multiply or divide by powers of ten. This can easily be done by applying the rules of exponents in multiplication and division:

$(a^m)(a^n) = a^{m+n}$ or $(10^2)(10^3) = 10^{2+3} = 10^5$

$\dfrac{a^m}{a^n} = a^{m-n}$ or $\dfrac{10^2}{10^3} = 10^{2-3} = 10^{-1}$

When performing these types of calculations on a calculator you must be careful. To enter 10^2, press 1 $\boxed{\text{EE}}$ 2, since $10^2 = 1 \times 10^2$. Do not enter 10 $\boxed{\text{EE}}$ 3, since the $\boxed{\text{EE}}$ key enters the 10.

Liter
(LEE-ter)

Kilogram
(KIH-lo-gram)

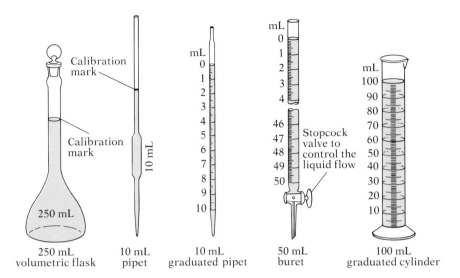

FIGURE 2-3 Laboratory Equipment Used for Measuring Liquid Volume

2.6 CONVERSION OF UNITS

Units may be converted from one to another by multiplying by appropriate conversion factors. In making unit conversions and working other problems in this text, we will rely on **dimensional analysis,** the use of dimensions (units) associated with a quantity as an aid in setting up the solution to a problem. Consider the relationship between feet and yards:

$$1 \text{ yd} = 3 \text{ ft}$$

(Note that 1 and 3 are exact numbers, because 1 yd is defined as 3 ft.) If we divide this relationship by 3 ft, we obtain

$$\frac{1 \text{ yd}}{3 \text{ ft}} = \frac{3 \text{ ft}}{3 \text{ ft}}$$

or

$$\frac{1 \text{ yd}}{3 \text{ ft}} = 1 \qquad (1)$$

Similarly, if we divide the relationship by 1 yd, we obtain another fraction that is equal to 1:

$$\frac{1 \text{ yd}}{1 \text{ yd}} = \frac{3 \text{ ft}}{1 \text{ yd}}$$

or

$$1 = \frac{3 \text{ ft}}{1 \text{ yd}} \qquad (2)$$

Now we have derived two fractions from the original unit relationship of 1 yd = 3 ft; both contain units and both are equal to 1. These fractions are

MATH TIP

Dividing both sides of an equation by the same quantity (in this case, 3 ft) does not change the meaning of the equation.

simply conversion factors that are called **unit factors;** since a unit factor is equal to 1, any expression can be multiplied by a unit factor without changing the value of the expression.

Now let's see how unit factors are used in making unit conversions. For example, if we wish to convert 2.00 yards to feet, we must choose the unit factor that can be multiplied by 2.00 yards and leave only feet as the final unit. We find that unit factor (2) meets these requirements:

$$(2.00 \text{ yd}) \left(\frac{? \text{ ft}}{? \text{ yd}} \right) = ? \text{ ft}$$

or

$$(2.00 \text{ yd}) \left(\frac{3 \text{ ft}}{1 \text{ yd}} \right) = 6.00 \text{ ft}$$

(Note that the number 2.00 dictates the number of significant figures in our answer, because 1 and 3 are exact numbers.)

We can use the same approach to convert feet to yards. Suppose we wish to calculate the number of yards corresponding to 10.0 feet. We choose unit factor (1) and multiply it by 10.0 feet:

$$(10.0 \text{ ft}) \left(\frac{? \text{ yd}}{? \text{ ft}} \right) = ? \text{ yd}$$

or

$$(10.0 \text{ ft}) \left(\frac{1 \text{ yd}}{3 \text{ ft}} \right) = 3.33 \text{ yd}$$

By using dimensional analysis to set up our problem, we have arrived at the **unit-factor method** for making conversions. Its three simple steps are as follows:

1. Write down the fundamental relationship between the two different units and derive the two unit factors. Referring to the last example, we would write

$$1 \text{ yd} = 3 \text{ ft}$$

$$\frac{1 \text{ yd}}{3 \text{ ft}} = 1$$

$$\frac{3 \text{ ft}}{1 \text{ yd}} = 1$$

2. Write down the quantity to be converted, followed by the unit factor that allows cancellation of the unwanted units, and then cancel the unwanted units. Again, in the last example, this would be

$$(10.0 \text{ ft}) \left(\frac{? \text{ yd}}{? \text{ ft}} \right) = ? \text{ yd}$$

or

$$(10.0 \text{ ft}) \left(\frac{1 \text{ yd}}{3 \text{ ft}} \right) = ? \text{ yd}$$

3. Complete the calculations.

$$(10.0 \text{ ft}) \left(\frac{1 \text{ yd}}{3 \text{ ft}} \right) = 3.33 \text{ yd}$$

This is a foolproof method for converting units from one system of measurements into those of another, and it can also be used for making conversions within one system.

EXAMPLE 2.5   Convert 4.2 yards to meters.

Solution   Step 1: Unit relationship (from Table 2-2): 1 m = 1.09 yd

$$\text{Unit factors:} \quad \frac{1 \text{ m}}{1.09 \text{ yd}} = 1$$

$$\frac{1.09 \text{ yd}}{1 \text{ m}} = 1$$

Step 2: Choose unit factor and cancel units:

$$(4.2 \text{ yd}) \left(\frac{? \text{ m}}{? \text{ yd}} \right) = ? \text{ m}$$

$$(4.2 \text{ yd}) \left(\frac{1 \text{ m}}{1.09 \text{ yd}} \right) = ? \text{ m}$$

Step 3: Complete calculations:

$$(4.2 \text{ yd}) \left(\frac{1 \text{ m}}{1.09 \text{ yd}} \right) = 3.9 \text{ m}$$

EXERCISE 2.5   Make the indicated conversions. (Consult Table 2-2 for unit relationships.)

a. 3.3 meters to millimeters
b. 16.5 cubic meters to quarts
c. 25.0 pounds to kilograms
d. 17 milliliters to liters

   The unit factor method is especially helpful when two or more conversion factors are needed (this occurs when there is not a direct relationship between initial units and final units). In these cases, unit factors are used in series; one unit factor converts initial units to a second set of units, and then one or more other unit factors are used to convert the second set of units to the desired final units. The following example illustrates the use of unit factors in series.

EXAMPLE 2.6   If 1 kilometer is equal to 0.621 miles and 1 mile contains 5,280 feet, convert 1.20 kilometers to feet.

Solution   Step 1: Unit relationships: 1 km = 0.621 mi

1 mi = 5,280 ft

$$\text{Unit factors:} \quad \frac{1 \text{ km}}{0.621 \text{ mi}} = 1$$

$$\frac{0.621 \text{ mi}}{1 \text{ km}} = 1$$

$$\frac{1 \text{ mi}}{5,280 \text{ ft}} = 1$$

$$\frac{5,280 \text{ ft}}{1 \text{ mi}} = 1$$

Step 2: Choose unit factors and cancel units:

$$(1.20 \text{ km}) \left(\frac{? \text{ mi}}{? \text{ km}} \right) \left(\frac{? \text{ ft}}{? \text{ mi}} \right) = ? \text{ ft}$$

or

$$(1.20 \text{ km}) \left(\frac{0.621 \text{ mi}}{1 \text{ km}} \right) \left(\frac{5,280 \text{ ft}}{1 \text{ mi}} \right) = ? \text{ ft}$$

Step 3: Complete calculations:

$$(1.20 \text{ km}) \left(\frac{0.621 \text{ mi}}{1 \text{ km}} \right) \left(\frac{5,280 \text{ ft}}{1 \text{ mi}} \right) = 3930 \text{ ft} = 3.93 \times 10^3 \text{ ft}$$

EXERCISE 2.6 Make the indicated conversions. (Refer to Table 2-2 for unit relationships.)

a. 1750 milligrams to kilograms
b. 4230 feet to kilometers
c. 3.65 kilograms to micrograms
d. 4.30 meters to millimeters

2.7 MASS AND WEIGHT

The amount of matter contained in an object is its **mass.** For example, the mass of 1 milliliter of water at 4° C is 1 gram. In this statement, *gram* is the unit of mass. The mass of an object (in contrast to its weight) is a constant value. Mass is measured on balances; several types of balances are shown in Figures 2-4 and 2-5. These balances find the mass of an object by balancing it with known masses.

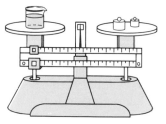 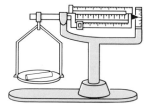 

FIGURE 2-4 Three Types of Laboratory Balances

Platform balance   Triple beam balance   A single pan analytical balance

The **weight** of an object is the force exerted on the object's mass by the earth's gravitational attraction. Weight is measured on a scale (your bathroom scale, for example), and it varies with distance from the center of the earth. The farther an object is from the center of the earth, the less its weight. Thus, you weigh less in the mile-high city of Denver than at sea level, even though your mass is the same at both altitudes. Although weight varies with altitude, the variation is very slight and we usually ignore it. The British units of ounces and pounds are used in many English-speaking countries for measuring weight. Even though mass and weight are clearly different properties of matter, you will find the terms used interchangeably. Just remember that when *weight* is mentioned in chemistry, the term *mass* is usually intended.

2.8 DENSITY

Since the mass and volume of a given material depend on the size of the sample, these quantities are considered to be variable quantities. However, they are used to define density, a property that does not depend on sample size. Density can be used to identify or characterize substances; for example, one way to identify pure gold is to determine its density.

Density is the amount of mass in a volume of one unit. In other words, density is mass divided by volume:

$$\text{density} = \frac{\text{mass}}{\text{volume}}$$

or

$$d = \frac{m}{v}$$

Thus, if we measure the mass and volume of a substance, we can calculate its density. Density is usually expressed in units of grams per milliliter (g/mL) or grams per cubic centimeter (g/cm^3). For very low density substances, such as a gas, density has units of grams per liter, g/L. Table 2-4 shows the densities of some common solids, liquids, and gases. For example, the density of lead is 11.4 g/mL; this means that 1 mL of lead has a

FIGURE 2-5 A Student Measures Mass with an Analytical Balance

TABLE 2-4   Densities of Common Substances (0° C, 1 atmosphere pressure)

Solids		Liquids		Gases	
Substance	Density (g/mL)	Substance	Density (g/mL)	Substance	Density (g/L)
Lead	11.4	Mercury	13.6	Air	1.3
Iron	7.9	Water	0.999	Oxygen	1.43
Cement	3	Butter	0.9	Nitrogen	1.25
Sand	2.3	Oil	0.8	Helium	0.18
Ice	0.917	Gasoline	0.7	Hydrogen	0.09
Balsa wood	0.11				

Balsa wood   Lead
0.11g/mL   11.4 g/mL

FIGURE 2-6 Comparison of the Densities of Balsa Wood and Lead at 0° C. Each substance has a volume of 1.00 mL.

mass of 11.4 g. Density varies with temperature because the volume of a substance, especially if it is a liquid or a gas, depends on the substance's temperature. Densities of liquids and gases also vary with pressure, as will be explained in chapters 10 and 11. For these reasons, the temperature and pressure at which a density is determined is usually reported, as in Table 2-4.

We generally consider substances of low density to be "light" and those of high density to be "heavy." What we really mean when we use these expressions is that a particular volume of one substance has a smaller or larger mass than the same volume of another substance. Thus, balsa wood, a "light" substance, is less dense than lead, a "heavy" substance. Figure 2-6 compares the densities of balsa wood and lead.

EXAMPLE 2.7

A sample of gold has a mass of 26.3 g and a volume of 1.36 mL. Find the density of the gold.

Solution

$$d = \frac{m}{v}$$

$$= \frac{26.3 \text{ g}}{1.36 \text{ mL}}$$

$$= 19.3 \text{ g/mL}$$

EXERCISE 2.7

A block of metal has a mass of 82.7 g and a volume of 10.7 mL. Find the density of the metal.   ▨

MATH TIP

To move a factor from one side of an equation to the other, simply reverse the mathematical operation involving that factor. For example, in the equation $d = m/v$, the right side is *divided* by v; moving v to the left side requires *multiplying* the entire equation by v.

$$d = \frac{m}{v} \qquad dv = \frac{m}{\cancel{v}} (\cancel{v}) \qquad dv = m$$

MATH TIP

In this case, d is a *multiplier*; to move it from one side of the equation to the other requires *dividing* the entire equation by d.

$$dv = m \qquad \frac{dv}{\cancel{d}} = \frac{m}{d} \qquad v = \frac{m}{d}$$

The expression for density involves three terms: density (d), mass (m), and volume (v). If any two of the terms are known, the third can be calculated. For example, density (d) can be calculated if mass (m) and volume (v) are known; the calculation consists of dividing mass (m) by volume (v). Similarly, if density (d) and mass (m) are known, the third term, volume (v), can be calculated. To make this calculation, we must first rearrange the expression for density by solving for the unknown quantity, volume (v). We make the rearrangement as follows (see appendix A for additional help with algebra):

$$d = \frac{m}{v}$$

$$dv = m$$

$$v = \frac{m}{d}$$

Now, mass can be divided by density to find volume.

EXAMPLE 2.8

A copper wire has a density of 8.92 g/mL and a mass of 4.83 g. Find the volume of the wire.

Solution

$$d = \frac{m}{v}$$

$$dv = m$$

$$v = \frac{m}{d}$$

$$= \frac{4.83 \text{ g}}{8.92 \text{ g/mL}}$$

$$= 0.541 \text{ mL}$$

MATH TIP

To divide g in the numerator by g/mL in the denominator, the fraction in the denominator must be inverted, and then multiplied by the fraction in the numerator:

$$\frac{g}{\frac{g}{mL}} = \frac{g}{1} \div \frac{g}{mL} = \frac{g}{1} \times \frac{mL}{g} = mL$$

EXERCISE 2.8 A piece of glass has a density of 2.32 g/mL and a mass of 11.23 g. What is the volume of the glass? ■

The expression for density can also be solved for mass (m), as shown below.

$$d = \frac{m}{v}$$

$$dv = m$$

or

$$m = dv$$

If the density (d) and volume (v) of a substance are known, the mass of the substance can be calculated by multiplying d by v.

EXAMPLE 2.9 A sample of mercury has a volume of 12.5 mL. If the density of mercury is 13.6 g/mL, what is the mass of the mercury sample?

Solution

$$d = \frac{m}{v}$$

$$dv = m$$

or

$$m = dv$$

$$= (13.6 \text{ g/mL})(12.5 \text{ mL})$$

$$= 170 \text{ g}$$

$$= 1.70 \times 10^2 \text{ g}$$

Note that the answer is converted to scientific notation to show that the final zero is significant.

EXERCISE 2.9 A lump of aluminum has a volume of 45 mL. If the density of aluminum is 2.70 g/mL, what is the mass of the aluminum? ■

2.9   SPECIFIC GRAVITY

A quantity closely related to density is **specific gravity**, defined as the density of a material relative to that of water at 4° C. (C stands for the Celsius temperature scale, which will be explained in section 2.11.) This standard density of water has a convenient value of 1.0000 g/mL, so that in the metric system, specific gravity turns out to be simply the density of a substance without units:

$$\text{specific gravity} = \frac{\text{density of substance in g/mL}}{\text{density of water at 4° C in g/mL}}$$

$$= \frac{\text{density of substance in g/mL}}{1.0000 \ g/mL}$$

$$= \text{density of substance (no units)}$$

As you can see, specific gravity is a quantity that shows how many times more dense the substance is than water at 4° C. Note once again that specific gravity has no units.

EXAMPLE 2.10 What is the specific gravity of a sample of ice whose density is 0.917 g/mL?

Solution
$$\text{specific gravity} = \frac{0.917 \ g/mL}{1.0000 \ g/mL}$$
$$= 0.917$$

EXERCISE 2.10 What is the specific gravity of a sample of seawater whose density is 1.02 g/mL?   ■

2.10   UNITS OF ENERGY

The traditional metric unit of energy is the **calorie** (cal), historically defined as the quantity of energy that will raise the temperature of 1 g of water from 14.5° C to 15.5° C. For measuring larger quantities of energy, *kilocalories* (kcal) are used; kilocalories are the same as the *Calories* (note the uppercase C) often used in nutrition. Recently, the SI unit **joule** (J), named for the English physicist James Prescott Joule (1818–89), has become more popular, and the calorie is now defined in terms of joules:

Joule
(jool)

$$1 \ cal = 4.184 \ J$$

(Note that 1 and 4.184 are exact numbers, because 1 cal is defined as 4.184 J.)

2.11   TEMPERATURE

Heat is a familiar form of energy that can be transferred between objects. If a warm object and a cool object are brought together, energy is transferred from the warm object to the cool one. The two objects had different temperatures initially, but the heat transfer will continue until both objects have the same temperature.

We use the word **temperature** to indicate the coldness or warmth of an object. We often measure body temperature in *Fahrenheit* degrees, the temperature scale used traditionally in medicine and meteorology. The Fahrenheit scale uses two reference points: the freezing point and the boiling point of pure water. In Fahrenheit, these values are 32° and 212°, respectively.

The *Celsius* scale, formerly called the *centigrade* scale, is used for scientific work, and as the United States converts to the metric system, use of the Celsius scale is becoming widespread in this country. The Celsius scale is also based on the freezing point and boiling point of pure water as reference points, but the Celsius scale sets these values at 0° and 100°, respectively.

A comparison of the Fahrenheit and Celsius scales (Figure 2-7) shows that there are 180 Fahrenheit degrees between the freezing and boiling points of water but only 100 Celsius degrees in the same interval. The ratio of the number of Celsius to Fahrenheit degrees in this interval is $^{100}/_{180}$, which reduces to $^{5}/_{9}$; this fraction will be used in converting readings on one temperature scale to readings on the other. We must also note that the interval of 0° – 32° on the Fahrenheit scale is below zero on the Celsius scale. Thus, when we convert Fahrenheit readings to Celsius readings, we must first subtract 32 degrees from the Fahrenheit reading. We then multiply the result by $^{5}/_{9}$, to reduce the number of Fahrenheit degrees to the right number of Celsius degrees. These steps give us the following relationship:

$$°C = (°F - 32)\left(\frac{5}{9}\right) \qquad (3)$$

Fahrenheit scale
Temperature scale named for its originator, Daniel Fahrenheit (1686–1736), a German physicist who spent most of his life in Holland.

Celsius scale
Temperature scale named for its originator, Anders Celsius, a Swedish astronomer.

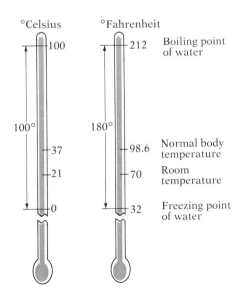

FIGURE 2-7 Comparison of the Fahrenheit and Celsius Temperature Scales

For conversion of Celsius temperature to Fahrenheit, we need only rearrange the equation as follows:

$$\frac{9}{5}\,°C = °F - 32$$

$$32 + \frac{9}{5}\,°C = °F$$

This relationship is usually written as

$$°F = \frac{9}{5}\,°C + 32 \tag{4}$$

You can keep the factors ⁵⁄₉ and ⁹⁄₅ straight by remembering that in our reference interval, there are more Fahrenheit degrees than Celsius degrees. Hence, when we want to convert to Celsius temperature, we must reduce the number of Fahrenheit degrees by multiplying by a factor whose value is less than one (⁵⁄₉). The opposite is true for conversion of Celsius to Fahrenheit.

If you prefer, the decimal number 1.8 can be used in place of the fraction ⁹⁄₅. Then equation (3) can be written as

$$°C = \frac{°F - 32}{1.8} \tag{5}$$

and equation (4) can be written as

$$°F = 1.8°\,C + 32 \tag{6}$$

EXAMPLE 2.11   Convert normal body temperature, 98.6° F, to its Celsius temperature.

Solution
$$°C = \frac{5}{9}\,(°F - 32)$$

$$= \frac{5}{9}\,(98.6 - 32)$$

$$= \frac{5}{9}\,(66.6)$$

$$= 37.0$$

Note that both ⁵⁄₉ and 32 are exact numbers. Thus, we round off our answer to three significant figures, since there are three significant figures in our limiting term, 98.6.

EXERCISE 2.11   Make the indicated conversions.

a. 72° F to °C   **b.** 54° C to °F   **c.** 0° F to °C   **d.** −45° C to °F

A third temperature scale, the *Kelvin* (or absolute) scale, is used in SI and has units of Kelvins (K). This temperature scale is derived from the theoretical behavior of gases at very low temperatures and will be discussed in chapter 10. Kelvin temperature is related to Celsius temperature as follows:

$$K = °C + 273$$

PERSPECTIVE DEVELOPMENT OF THE THERMOMETER

Although the difference between hot and cold has been distinguished since the days of cave men, it was not until 1592 that the first thermometer was invented. This milestone was achieved by the Italian scientist Galileo, who called his device a thermoscope. It was a very simple apparatus, consisting of a large glass bulb with a long, narrow, openmouthed neck inverted over a vessel of colored water or spirit of wine (alcohol) (see Figure a). Upon driving some air out of the bulb, Galileo found that the liquid rose a short distance into the neck and that subsequent changes of temperature caused changes of volume of the trapped air and thus changes in the level of the liquid in the neck. By 1611, a colleague of Galileo by the curious name of Sanctorius Sanctorius came up with the idea of providing a scale for the thermoscope, thus creating an instrument capable of measuring temperature. Twenty-one years later, the French physician Jean Rey created a simpler device (Figure b) that was not dependent on air as an indicator of temperature changes but instead contained a liquid that responded directly to temperature changes.

As time passed, scientists learned to seal their thermometers to prevent the sensitivity to barometric pressure that had earlier been observed, and in the period of 1657–67, Italian scientists began using mercury as the liquid in their thermometers. By the early eighteenth century, as many as 35 temperature scales had been devised. The most famous of all, the Fahrenheit scale, was devised in Holland by the German instrument maker Daniel Gabriel Fahrenheit in the period of 1700–30. It is a tribute to Fahrenheit's skill that the presently accepted value for normal human body temperature, 98.6° F, is so close to the value of 96° that he found in the eighteenth century.

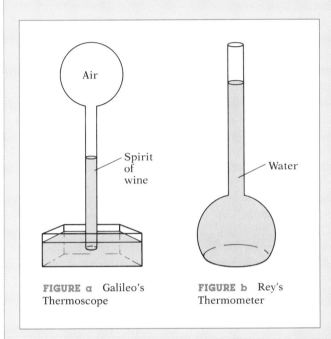

FIGURE a Galileo's Thermoscope

FIGURE b Rey's Thermometer

2.12 SPECIFIC HEAT

Adding heat to a substance usually causes the temperature of the substance to increase. However, each material has a different capacity for absorbing heat. Water must absorb 1.00 calorie per gram in order for its temperature to increase by 1° C; in contrast, it only takes 0.031 calories of heat to raise the temperature of one gram of gold by 1° C. Thus, water has a high **heat capacity** relative to gold. The term **specific heat** is used to describe the amount of heat, measured in calories, required to raise the temperature of one gram of a substance 1° C. Thus,

specific heat = calories per gram per °C of temperature change

or

$$\text{specific heat} = \frac{\text{cal}}{\text{g}\Delta T}$$

Heat capacity

The capacity of a substance for absorbing heat; heat capacity differs from specific heat in that heat capacity depends on the mass of the sample, whereas specific heat does not.

TABLE 2-5 Specific Heats of Selected Materials (cal/g °C)

Solids		Liquids	
ice	0.492	water	1.00
aluminum	0.215	ethyl alcohol	0.581
iron	0.108	glycerine	0.540
copper	0.092	ethyl ether	0.529
silver	0.057	mineral oil	0.50
lead	0.038	olive oil	0.471
gold	0.031		

where ΔT represents the temperature change from T_1 (initial temperature) to T_2 (final temperature), or $\Delta T = T_2 - T_1$. Specific heat has units of cal/g °C. Since the expression for specific heat contains three variable quantities, any one quantity can be calculated if the other two are known. Table 2-5 shows the specific heats of various solids and liquids.

The high specific heat of water is beneficial to life in many ways. Because it takes such a large amount of heat to change the temperature of water, our bodies, which are composed of about 70% water, are able to maintain a relatively constant temperature. Similarly, a variety of other organisms are able to control their temperatures and thus maintain a favorable internal environment for sustaining life. In addition, large bodies of water take a long time for both heating and cooling; this allows the warm waters of the Gulf Stream to keep much of western Europe warmer than the corresponding region of eastern North America. We shall return to the topic of specific heat in chapter 11.

EXAMPLE 2.12   It takes 166 calories of heat to raise the temperature of 500.0 grams of mercury from 25° C to 35° C. What is the specific heat of mercury?

Solution

$$\text{specific heat} = \frac{\text{cal}}{\text{g}\Delta T}$$

$$= \frac{166 \text{ cal}}{500.0 \text{ g} (35° \text{ C} - 25° \text{ C})}$$

$$= \frac{166 \text{ cal}}{500.0 \text{ g} (10° \text{ C})}$$

$$= 0.033 \text{ cal/g °C}$$

EXERCISE 2.12   It takes 2,178 calories of heat to raise the temperature of 125 grams of porcelain from 20° C to 87° C. What is the specific heat of porcelain? ■

EXAMPLE 2.13   The specific heat of water is 1.00 cal/g °C. How many calories of heat are required to raise the temperature of 125 grams of water from 30° C to 90° C?

Solution

$$\text{specific heat} = \frac{\text{cal}}{\text{g}\Delta T}$$

$$\text{cal} = (\text{specific heat})(\text{g})(\Delta T)$$
$$= (1.00 \text{ cal/g °C})(125 \text{ g})(90° \text{ C} - 30° \text{ C})$$
$$= (1.00 \text{ cal/g °C})(125 \text{ g})(60° \text{ C})$$
$$= 7500 \text{ cal}$$
$$= 7.5 \times 10^3 \text{ cal}$$

EXERCISE 2.13   The specific heat of lead is 0.038 cal/g °C. How many calories of heat are needed to raise the temperature of 15.0 grams of lead by 25° C?   ▧

EXAMPLE 2.14   A sample of water at 25° C was heated by adding 250 calories. If the mass of the water was 7.3 grams, what was the final temperature of the water? The specific heat of water is 1.00 cal/g °C.

Solution

$$\text{specific heat} = \frac{\text{cal}}{\text{g}\Delta T}$$

$$(\text{specific heat})(\Delta T) = \frac{\text{cal}}{\text{g}}$$

$$\Delta T = \frac{\text{cal}}{(\text{g})(\text{specific heat})}$$
$$= \frac{250 \text{ cal}}{(7.3 \text{ g})(1.00 \text{ cal/g °C})}$$
$$= 34° \text{ C}$$

MATH TIP
To solve for T_2, T_1 must be moved to the opposite side of the equation. This requires adding T_1 to both sides of the equation, which does not change the meaning of the equation.

$$\Delta T + T_1 = T_2 - T_1 + T_1$$
$$\Delta T + T_1 = T_2$$

Since

$$\Delta T = T_2 - T_1,$$

then

$$T_2 = \Delta T + T_1,$$

and

$$T_2 = 34° \text{ C} + 25° \text{ C}$$
$$= 59° \text{ C}$$

EXERCISE 2.14   A block of iron weighing 11.4 grams was heated by adding 1500 calories. If the initial temperature of the iron was 10° C, what was the final temperature of the iron? The specific heat of iron is 0.108 cal/g °C.   ▧

SUMMARY

Precision and accuracy are terms that describe quality of measurement. **Precision** means how closely multiple measurements of the same quantity agree; precision also refers to how finely a measuring device is calibrated. **Accuracy** indicates how well a measurement agrees with the true value. **Significant figures** are digits in a number that have physical meaning. Rules for determining significant figures are given on pp. 21-22. Rules for using significant figures in calculations are given on p. 23, and rules for rounding off are given on pp. 23-24. **Scientific notation,** a method of expressing numbers as small whole numbers multiplied by powers of 10, can be used to clarify the significance of terminal zeros in a number. (A **positive exponent** indicates how many times a base must be multiplied by itself to give the corresponding ordinary number, while a **negative exponent** tells how many times 1 must be divided by the base to give the ordinary number.)

Of the two popular measuring systems, the **metric system** is the one used by scientists. The metric system is based on the decimal system, and standard prefixes are used to denote powers of 10. The **SI** is a recently developed simplification of the metric system. Popular metric units of measure for length, volume, and mass are the meter, liter, and gram, respectively. Units can be converted from one to another by use of **unit factors;** two unit factors can be derived from any single unit relationship. To interconvert two units that do not have a direct relationship, unit factors are used in series.

The amount of matter contained in an object is its **mass. Weight** is the force exerted on an object's mass by the earth's gravitational attraction. The mass of an object is a fixed quantity, in contrast to its weight, which varies with altitude. **Density,** a fixed property, is the mass of an object divided by its volume ($d = m/v$). Since the expression for density contains three terms, any one quantity can be calculated if the other two are known. **Specific gravity** is the density of a material relative to that of water at 4° C.

Two metric units of energy, the **joule** and the **calorie,** are now used. The calorie, once defined as the amount of energy needed to raise the temperature of 1 g of water from 14.5° C to 15.5° C, is now defined in terms of joules (1 cal = 4.184 J). **Heat** is a familiar form of energy that can be transferred from a warmer object to a cooler one. **Temperature** indicates the coldness or warmth of an object. The two popular temperature scales, **Fahrenheit** and **Celsius,** use the freezing and boiling points of water as their reference points. Readings on the two scales are easily interconverted. **Specific heat** describes the amount of heat needed to raise the temperature of one gram of a substance 1° C. Compared to most substances, water has a high specific heat.

STUDY QUESTIONS AND PROBLEMS

(More difficult questions and problems are marked with an asterisk.)

PRECISION AND ACCURACY

1. Define each of the following terms:
   a. Precision   b. Accuracy

2. Is it possible to make precise but not accurate measurements? Explain your answer.
3. Is it possible to make accurate but not precise measurements? Explain your answer.
4. Which is more dependent on human error, precision or accuracy? Explain your answer.

SIGNIFICANT FIGURES

5. Define each of the following terms:
   a. Significant figures   b. Exact numbers
6. How many significant figures are in each of the following measurements?
   a. 370.5 feet   e. 0.150 kilograms
   b. 4,020 square yards   f. 6,167 miles
   c. 0.0010 grams   g. 700 minutes
   d. 1.203 inches   h. 300.0 milliliters
7. How many significant figures are in each of the following measurements?
   a. 1075 pounds   e. 3040 meters
   b. 650.2 hours   f. 75.02 centimeters
   c. 0.0072 milliliters   g. 12.00 seconds
   d. 87.3 years   h. 1500.2 millimeters

CALCULATIONS AND SIGNIFICANT FIGURES

8. Round off each of the following numbers to two significant figures.
   a. 4,735   c. 2.550   e. 551.0
   b. 0.003651   d. 0.1693   f. 35.50
9. Round off each of the following numbers to three significant figures.
   a. 3.6275×10^5 c. 200.5   e. 0.0048732
   b. 7,396   d. 4.8669×10^{-2} f. 1.975×10^{-7}
10. Perform the indicated operations and round off each answer to the correct number of significant figures.

   a. $(102.5)(11.17)$ c. $(8.1)(1.23)(0.47)$ e. $\dfrac{(19)(1.059)(75)}{(6.2)(1.86)}$

   b. $(216.5)(0.018)$ d. $75 - 0.76 - 1.49$ f. $\dfrac{(1.008)(32.6)}{149.1}$

11. Perform the indicated operations and round off each answer to the correct number of significant figures.

   a. $\dfrac{149 + 2.54}{376}$ c. $55.841 - 4.10$ e. $\dfrac{4.15}{2.077}$

   b. $39.0 + 42.6 + 1.39$ d. $\dfrac{67.9 - 12.73}{0.5152}$ f. $\dfrac{73.8}{65.21 - 47}$

SCIENTIFIC NOTATION

12. Define each of the following terms:
   a. Exponent   b. Base

13. Express each of the following numbers in scientific notation.
    a. 732.6   c. 7,000,000   e. 0.0000000000437
    b. 100.4   d. 0.0538   f. 10,573
14. Express each of the following numbers in scientific notation.
    a. 7,498,500   c. 458   e. 9,134,998,375
    b. 0.0003718   d. 0.01269   f. 0.00000000001
*15. Express each of the following as ordinary numbers without powers of 10.
    a. 5.689×10^2 c. 1.499×10^{-7} e. 4.761×10^6
    b. 2.367×10^4 d. 3.65×10^{-3} f. 7.2517×10^{-9}

SYSTEMS OF MEASUREMENT

16. Define each of the following terms:
    a. British system   b. Metric system   c. SI
17. Many people find the metric system more convenient to use than the British system. Why might this be true?

CONVERSION OF UNITS

18. Define each of the following terms:
    a. Dimensional analysis   c. Unit-factor method
    b. Unit factor
19. Make the indicated conversions. (Refer to Table 2-2 for unit relationships.)
    a. 67.3 nm to m   c. 0.457 mL to L   e. 0.059 g to ng
    b. 84.92 g to kg   d. 52.3 mm to cm   f. 0.394 m to mm
20. Make the indicated conversions. (Refer to Table 2-2 for unit relationships.)
    a. 1.759 L to mL   c. 20.6 mm to m   e. 143 g to mg
    b. 7.6 g to mg   d. 72 kg to g   f. 3.0 L to cm^3
21. Make the indicated conversions. (Refer to Table 2-2 for unit relationships.)
    a. 14.7 lb to g   c. 24.833 qt to L   e. 166 km to mi
    b. 17.32 in to m   d. 13.0 yd to m   f. 39 cm to in
22. Make the indicated conversions. (Refer to Table 2-2 for unit relationships.)
    a. 760 in to cm   c. 250 mL to qt   e. 12.6 in to cm
    b. 145 g to lb   d. 2.63 m to ft   f. 1.65 qt to L
23. The Empire State Building is 1,250 feet tall. Calculate its height in each of the following units:
    a. meters   c. centimeters
    b. kilometers   d. millimeters
24. Without using a source of magnification, the normal human eye can distinguish objects having a diameter of 0.1 mm. What is this diameter in inches?
*25. An ice cube has edges that are 1.5 inches long. What is the volume of the ice cube in cubic centimeters (cm^3)?
26. A quarter weighs about 0.25 ounces. What is the weight of the quarter in grams?

27. On the average, the moon is 238,857 miles from earth. How far is this in kilometers?
*28. If light travels at a speed of 3.0×10^{10} cm/sec, how long does it take light reflected from the moon to reach the earth? (The moon is 238,857 miles from earth.)
*29. An athlete ran a marathon in 2 hours and 58 minutes. If a marathon is 26.22 miles, what was the average speed of the athlete in kilometers per hour?
30. An automobile gas tank holds 14.7 gallons. What is the volume of the gas tank in liters?

MASS AND WEIGHT

31. Define each of the following terms:
    a. Mass   b. Weight
32. At which location does a 15-pound bowling ball weigh more, in Death Valley (elevation −280 ft) or on Mount Whitney (elevation 14,500 ft)? Explain your answer.

DENSITY

33. Define density in words and by use of a mathematical expression.
34. A block of ice at −10° C has a volume of 1.00186 L and a mass of 1.000 kg. Calculate its density.
35. A metal object has a volume of 3.0 mL and weighs 16.6 g. Calculate the density of the object.
36. A sample of the "heaviest" substance known, osmium, has a mass of 61.88 g and a volume of 2.75 mL. Calculate the density of osmium.
*37. A quarter-pound stick of butter has a volume of 7.8 cubic inches (in^3). What is the density of the butter?
38. Calculate the volume occupied by a 20.0-gram sample of mercury, density 13.5939 g/mL.
39. What volume is occupied by 736 g of water, density 1.00 g/mL?
40. An alcohol has a density of 0.79 g/mL. What is the mass of 109 mL of the alcohol?
41. A sample of marble, density 2.70 g/mL, has a volume of 7.4 mL. What is the mass of the marble?

SPECIFIC GRAVITY

42. Define specific gravity; how is it related to density?
43. Copper has a specific gravity of 8.96. What volume of copper has a mass of 74.3 g?
44. The specific gravity of air at 25° C is 1.1843×10^{-3}. At this temperature, how many grams of air can the human lungs hold if average lung capacity is 6,000 mL?
45. Aluminum and molybdenum have specific gravities of 2.70 and 10.2, respectively. Which occupies the greater volume, 25 g of aluminum or 75 g of molybdenum?

TEMPERATURE

46. Distinguish between heat and temperature.

47. Make the following temperature conversions.
    a. 1200° F to °C   c. 37° C to °F   e. 78° F to °C
    b. 32° F to °C   d. 45° C to °F   f. 590° C to °F

48. Make the following temperature conversions.
    a. 0° F to °C   c. −150° F to °C   e. 600° C to °F
    b. −21° C to °F   d. 220° F to °C   f. 14° C to °F

49. A person has a fever of 102.3° F. By how many degrees is his temperature elevated on the Celsius scale? (Normal body temperature is 98.6° F.)

SPECIFIC HEAT

50. Define specific heat.

51. How much heat, in calories, is needed to raise the temperature of 100.0 mL of water from 25.0° C to 100.0° C?

52. The specific heat of an alcohol is 0.581 cal/g °C. How many calories are needed to raise the temperature of 93.7 g of the alcohol from 29° C to 78° C?

53. Freon, a gas used in refrigeration systems, has a specific heat of 0.1297 cal/g °C. Calculate the amount of heat required to change the temperature of 472 g of freon from 0° C to 25° C.

54. A beaker contained 250.0 mL of water at a temperature of 78° F. The beaker was heated until the water temperature was 200.0° F. How many calories of heat were needed to bring about this temperature change?

55. How many calories of heat must be removed from 52.7 g of water to lower its temperature from 78° F to 32° F?

ELEMENTS, ATOMS, AND COMPOUNDS

CHAPTER 3

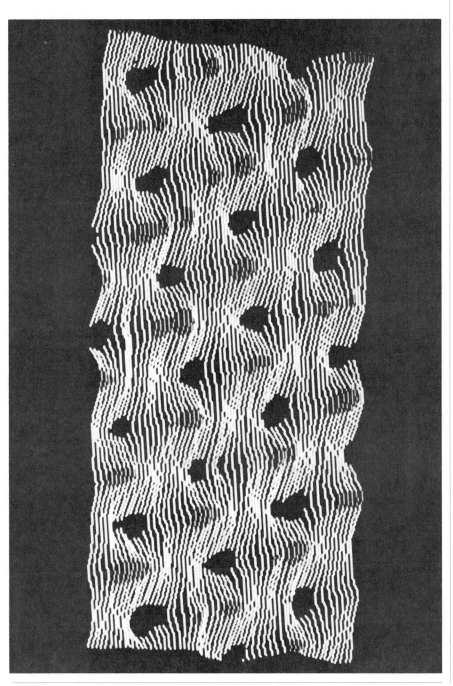

OUTLINE

3.1 Elements
3.2 The Periodic Table
3.3 Atoms
3.4 Subatomic Particles
3.5 Atomic Number
3.6 Mass Number
3.7 Isotopes
  Perspective: Making New Elements
3.8 Atomic Mass Units
3.9 Atomic Weight
3.10 Compounds
3.11 Composition of Compounds
3.12 Chemical Formulas
3.13 Formula Weights of Compounds
  Problem-Solving Skills: The
    Importance of Units and Labels
3.14 Percentage Composition
  Summary
  Study Questions and Problems

Robert Boyle, a modest and retiring English gentleman-scientist who wrote about elements as early as 1661, was the first person to recognize the unique nature of the chemical elements. His work influenced Antoine Lavoisier, a French chemist who published the first list of elements in 1789 (Figure 3-1).

As scientists learned more about elements, they also began to learn how elements combine to form chemical compounds. During the period 1803–1810, the English chemist John Dalton first explained the laws of chemical combination, and in doing so, he postulated the existence of

	Noms nouveaux.	Noms anciens correſpondans.
	Lumière........	Lumière.
	Calorique........	Chaleur. Principe de la chaleur. Fluide igné. Feu. Matière du feu & de là chaleur.
Subſtances ſimples qui appartiennent aux trois règnes & qu'on peut regarder comme les élémens des corps.	Oxygène........	Air déphlogiſtiqué. Air empiréal. Air vital. Baſe de l'air vital.
	Azote...........	Gaz phlogiſtiqué. Mofète. Baſe de la mofete.
	Hydrogène.......	Gaz inflammable. Baſe du gaz inflammable.
Subſtances ſimples non métalliques oxidables & acidifiables.	Soufre..........	Soufre.
	Phoſphore........	Phoſphore.
	Carbone,........	Charbon pur.
	Radical muriatique.	Inconnu.
	Radical fluorique .	Inconnu.
	Radical boracique,.	Inconnu.
Subſtances ſimples métalliques oxidables & acidifiables.	Antimoine.......	Antimoine.
	Argent..........	Argent.
	Arſenic..........	Arſenic.
	Biſmuth.........	Biſmuth.
	Cobolt..........	Cobolt.
	Cuivre..........	Cuivre.
	Etain...........	Etain.
	Fer.............	Fer.
	Manganèſe.......	Manganèſe.
	Mercure.........	Mercure.
	Molybdène.......	Molybdène.
	Nickel..........	Nickel.
	Or..............	Or.
	Platine..........	Platine.
	Plomb..........	Plomb.
	Tungſtène.......	Tungſtene.
	Zinc...........	Zinc.
Subſtances ſimples ſalifiables terreuſes.	Chaux........	Terre calcaire, chaux.
	Magnéſie........	Magnéſie, baſe du ſel d'Epſom.
	Baryte..........	Barote, terre peſante.
	Alumine.........	Argile , terre de l'alun, baſe de l'alun.
	Silice...........	Terre ſiliceuſe , terre vitrifiable.

LAVOISIER'S LIST OF THE ELEMENTS (1789).

FIGURE 3-1 The First List of Elements (Reproduction from A. L. Lavoisier, *Traite elementaire de chimie*, 1789)

atoms. In time, Dalton's atomic hypothesis became universally accepted and with a few minor modifications by modern scientists, it remains today one of the cornerstones of chemistry. In this chapter you will learn about the nature of elements, how they are composed of atoms, and a little about chemical compounds.

3.1 ELEMENTS

The Greek philosophers who lived about twenty-five hundred years ago were intrigued with the world about them, and they believed that all matter was composed of four fundamental elements: air, earth, fire, and water (Figure 3-2). Their ideas on the elements were accepted for almost two thousand years, but as chemistry developed into a science, more and more elements were recognized. Twenty-three had been identified by the end of the eighteenth century. Many more have been discovered since then, and at last count there were 109. We now define **elements** as pure substances that cannot be separated into simpler substances by ordinary processes. We also know that what the ancient Greeks thought were elements are actually mixtures or chemical combinations of the true elements. For example, air is a mixture of gases, mostly nitrogen and oxygen; both of these are true elements. Earth (soil) is composed of over twenty elements, both free and in chemical combination. Water is a compound of the elements hydrogen and oxygen. The fourth Greek element, fire, is simply a chemical reaction that occurs between oxygen and flammable substances and emits heat and light. Some familiar examples of true elements are the metals gold, silver, tin, and lead and the nonmetals sulfur, neon, chlorine, and phosphorus; you will meet many more examples as you continue your study of chemistry.

Space exploration has revealed exciting information about the distribution of elements in the solar system. Table 3-1 summarizes what we know at this time. As you can see, hydrogen is by far the most abundant element.

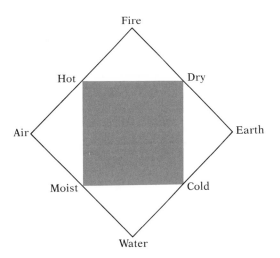

FIGURE 3-2 The Four Greek Elements

TABLE 3-1	The Most Abundant Elements in the Solar System
Element	Relative Abundance (% of Total Atoms)
Hydrogen	91
Helium	9
Oxygen	
Nitrogen	
Carbon	
Neon	
Iron	
Silicon	0.1
Magnesium	
Sulfur	
Nickel	
Aluminum	
Calcium	

It is thought to be the basic material from which all other elements in the universe are synthesized.

Although hydrogen is abundant in the solar system, it is found only to a small extent here on earth. Hydrogen and helium, the second most abundant element in the solar system, are gases of very low density. Even though they may once have occurred in large amounts on earth and in its atmosphere, they have long since escaped into space. Table 3-2 shows that iron is the most plentiful element on earth, followed by oxygen, magnesium, and silicon. Some iron is found in the crust that forms the outer portion of the earth, but most is in the earth's hot, molten core. Magnesium occurs primarily in minerals in the earth's crust and on its surface in silicate compounds such as sand. Oxygen is found as the uncombined element in the earth's atmosphere. Table 3-3 lists sources and uses of some familiar elements on earth.

TABLE 3-2	Elemental Composition of the Earth and Its Atmosphere (percent by weight)		
Element	%	Element	%
Iron	35.4	Manganese	0.1
Oxygen	27.8	Potassium	0.1
Magnesium	17.0	Titanium	
Silicon	12.6	Phosphorus	
Sulfur	2.7	Chromium	
Nickel	2.7	Hydrogen	0.3
Calcium	0.6	Chlorine	
Aluminum	0.4	Nitrogen	
Cobalt	0.2	Argon	
Sodium	0.1		

TABLE 3-3 Some Well-Known Elements

Element	Source	Uses
Hydrogen	Various compounds containing hydrogen	Manufacture of ammonia; conversion of oils to margarine and shortening; rocket fuel
Helium	Underground gas deposits	Inert component of welding gases; weather balloons; cryogenics
Boron	Underground deposits of ore (borax)	Hardening agent in steel; component of semiconductors, glass, and washing powders
Carbon	Underground deposits (diamond, graphite, coal)	Jewels (diamonds); lubricant (graphite); coloring agent (carbon black)
Nitrogen	Earth's atmosphere	Manufacture of ammonia; liquid coolant
Oxygen	Earth's atmosphere	Metallurgical processes; manufacture of chemicals; medical respirators
Neon	Earth's atmosphere	Electric signs
Aluminum	Underground deposits of ore (bauxite)	Aircraft construction; building materials; containers and wrapping materials
Sulfur	Underground deposits	Processing dried fruits; manufacture of gunpowder and rubber; fertilizers
Chromium	Underground deposits (chromite)	Manufacture of alloys (chrome, stainless steel)
Iron	Underground deposits of iron ore	Construction of all types; machines and parts; tools; stainless steel
Copper	Underground deposits	Electrical wires; manufacture of alloys (brass, bronze); coins
Silver	Underground deposits	Silver bullion and coins; silverware; jewelry; photographic industry
Tin	Natural deposits of tin oxide	Plating steel cans; metal bearings; alloys (soft solder, pewter, bronze)
Platinum	Underground deposits	Electrical apparatus; jewelry; dental alloys
Gold	Underground deposits	Gold bullion; jewelry; electrical devices
Mercury	Natural deposits of mercury sulfide (cinnabar)	Thermometers; barometers; electrical switches; dental fillings
Lead	Natural deposits of lead sulfide	Roofing materials; underwater coverings; linings for water pipes; storage batteries; ammunition; radiation shielding

3.2 THE PERIODIC TABLE

As scientists learned to recognize elements, they began to classify them, and in the late nineteenth century our current classification scheme was developed. This remarkably simple arrangement, called the **periodic table,** is shown in Figure 3-3 and inside the front cover of this book.

The elements are arranged in the periodic table in horizontal rows called **periods,** which are layered one on top of another. The vertical columns of elements are called **groups** or **families.** Members of a chemical group tend to have similar properties. Across any period in the table, properties change from one element to the next in a very regular way. If the elements and their properties were listed in the same order as they appear in the periodic table, similar properties would recur periodically. This predictable pattern of recurring similarities is the basis for the name *periodic table.*

Notice that elements in the periodic table are represented by **symbols.** This practice originated with the alchemists, and modern chemists continue it. Most names for elements are derived from Latin and Greek, and

PERIODS	IA	IIA	IIIB	IVB	VB	VIB	VIIB	VIIIB			IB	IIB	IIIA	IVA	VA	VIA	VIIA	NOBLE GASES 0
1	1 **H** 1.00797																	2 **He** 4.00260
2	3 **Li** 6.941	4 **Be** 9.01218											5 **B** 10.81	6 **C** 12.01115	7 **N** 14.0067	8 **O** 15.9994	9 **F** 18.99840	10 **Ne** 20.179
3	11 **Na** 22.98977	12 **Mg** 24.305											13 **Al** 26.98154	14 **Si** 28.086	15 **P** 30.97376	16 **S** 32.06	17 **Cl** 35.453	18 **Ar** 39.948
4	19 **K** 39.098	20 **Ca** 40.08	21 **Sc** 44.9559	22 **Ti** 47.90	23 **V** 50.9414	24 **Cr** 51.996	25 **Mn** 54.9380	26 **Fe** 55.847	27 **Co** 58.9332	28 **Ni** 58.71	29 **Cu** 63.546	30 **Zn** 65.38	31 **Ga** 69.72	32 **Ge** 72.59	33 **As** 74.9216	34 **Se** 78.96	35 **Br** 79.904	36 **Kr** 83.80
5	37 **Rb** 85.4678	38 **Sr** 87.62	39 **Y** 88.9059	40 **Zr** 91.22	41 **Nb** 92.9064	42 **Mo** 95.94	43 **Tc** 98.9062	44 **Ru** 101.07	45 **Rh** 102.9055	46 **Pd** 106.4	47 **Ag** 107.868	48 **Cd** 112.40	49 **In** 114.82	50 **Sn** 118.69	51 **Sb** 121.75	52 **Te** 127.60	53 **I** 126.9045	54 **Xe** 131.30
6	55 **Cs** 132.9054	56 **Ba** 137.34	57 ***La** 138.9055	72 **Hf** 178.49	73 **Ta** 180.9479	74 **W** 183.85	75 **Re** 186.2	76 **Os** 190.2	77 **Ir** 192.22	78 **Pt** 195.09	79 **Au** 196.9665	80 **Hg** 200.59	81 **Tl** 204.37	82 **Pb** 207.19	83 **Bi** 208.9804	84 **Po** (210)	85 **At** (210)	86 **Rn** (222)
7	87 **Fr** (223)	88 **Ra** 226.0254	89 **†Ac** (227)	104 **Ku** (261)	105 **Ha** (260)	106 (263)	107 (264)	108 (266)	109 (266)									

*	58 **Ce** 140.32	59 **Pr** 140.9077	60 **Nd** 144.24	61 **Pm** (147)	62 **Sm** 150.4	63 **Eu** 151.96	64 **Gd** 157.25	65 **Tb** 158.9254	66 **Dy** 162.50	67 **Ho** 164.9304	68 **Er** 166.264	69 **Tm** 168.9342	70 **Yb** 174.06	71 **Lu** 174.97
†	90 **Th** 232.0381	91 **Pa** 231.0359	92 **U** 238.029	93 **Np** 237.0482	94 **Pu** (244)	95 **Am** (243)	96 **Cm** (247)	97 **Bk** (247)	98 **Cf** (251)	99 **Es** (254)	100 **Fm** (257)	101 **Md** (258)	102 **No** (255)	103 **Lr** (256)

FIGURE 3-3 The Periodic Table

the symbols are abbreviations of these names. For example, the name *carbon* comes from the Latin word *carbo*, which means "coal"; the symbol for carbon is C. Some symbols do not resemble English names. One example is the symbol for potassium, K, which comes from the Latin word *kalium;* in another case, the symbol for lead, Pb, comes from the Latin word *plumbum*. For some man-made elements, pseudo-Latin names such as berkelium (Bk), californium (Cf), and einsteinium (Es) are used to indicate where they were created or to honor famous scientists. Notice also that some symbols are a single letter (for example, H, C, O, N) while others have two letters (Sn, Pb, Au, Cl, for example). When a chemical symbol is only a single letter, the letter is always uppercase; when two letters are used, the first letter is always uppercase and the second letter always lowercase. A complete list of the elements and their symbols appears inside the front cover of this book.

3.3 ATOMS

The existence of atoms was first proposed by the Greek philosopher Democritus in the fifth century B.C. Democritus believed that a sample of matter could be divided into smaller and smaller parts, but that eventually there would be a point at which basic, indivisible units of matter would be reached. Democritus' ideas were called the theory of discontinuous matter, and his fundamental units of matter came to be known as atoms. Democritus' theory was challenged by Aristotle, who believed that matter was infinitely divisible, with each part having identical properties regardless of size. Aristotle's concept, called the theory of continuous matter, became more popular than that of Democritus, and its popularity lasted until experiments in the sixteenth century discredited it. The return to the discontinuous theory of matter forced scientists to rethink their ideas on the composition of matter, and in the early nineteenth century the schoolmaster John Dalton proposed his **atomic theory.** Dalton's ideas are summarized as follows:

Democritus
(deh-MOCK-rih-tus)

Aristotle
(AIR-is-totl)

Atomic theory
The theory that the smallest unit representative of an element is an atom.

1. Elements are composed of tiny, indivisible particles called *atoms*, whose identities are maintained throughout physical and chemical changes.
2. Each element has identical atoms that are distinctly different from atoms of any other element.
3. Chemical compounds are composed of atoms of two or more elements.
4. Atoms of different elements are present in compounds in small, whole-number ratios. (An example is the compound carbon dioxide, in which one carbon atom is present for every two oxygen atoms; thus, in carbon dioxide, the ratio of carbon to oxygen is 1 : 2.)
5. Atoms of the same elements may form more than one compound by combining in different ratios. (This principle is illustrated by comparing the compounds carbon dioxide and carbon monoxide. Both are composed of the same elements, carbon and oxygen, but the ratio of

carbon to oxygen in carbon dioxide is 1 : 2, while the ratio of carbon to oxygen in carbon monoxide is 1 : 1.)

Dalton is considered one of the fathers of modern chemistry, and his atomic theory is fundamental to chemistry. However, a few modifications have been made to account for more recent evidence. For example, we now know that atoms are composed of even smaller units called *subatomic particles* (section 3.4) and, under certain conditions, atoms can be broken apart (chapter 17). In addition, all atoms of an element have the same size but not the same mass, as you will learn in section 3.7.

We now define an **atom** as the smallest unit of an element that has all of the properties of the element. An atom is very small indeed, and although atoms vary in size from one element to another, the approximate diameter of an atom is 10^{-8} cm. It would take about one million atoms to make a speck of 0.1 mm in diameter, about half the size of the period at the end of this sentence. Even powerful magnifying devices that produce photographic images of large atoms give only fuzzy pictures (Figure 3-4).

Such pictures have not helped us much in finding out what is inside atoms, but other evidence suggests that atoms are highly complex structures composed of smaller, more fundamental particles. Atoms are approximately spherical in shape, and at the center of each is a very small nucleus. The nucleus contains such a variety of exotic particles that some modern physicists refer to it as a "zoo." The nucleus takes up only a negligible amount of space within the atom; by far, most of an atom is empty space where tiny particles called electrons move about.

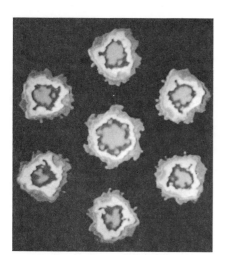

FIGURE 3-4 Images of Atoms. In this photograph of uranyl acetate clusters, individual uranium atoms are seen as roundish spots with bright edges and gray centers.

3.4 SUBATOMIC PARTICLES

Subatomic particles are those particles that exist within atoms. Subatomic particles inside the nucleus are called **nucleons.** Although a dozen or so nucleons are known to exist, the most important ones for our purposes are protons and neutrons. A **proton,** symbolized by a p, is a particle of matter having a positive electrical charge and a mass of 1.6726×10^{-24} grams. This is such a small mass that it would take about 100 million trillion protons to equal the mass of a single gnat! A **neutron,** symbolized by an n, has a similar mass, 1.6748×10^{-24} grams, but it has no electrical charge. Its electrical neutrality is the basis for its name. Protons and neutrons, along with other nucleons, are packed tightly together to form the dense core of the atom.

Although atoms contain the protons' positive charges, they do not have overall electrical charge. This means that the positive charges in an atom's nucleus are exactly balanced by the same number of negative charges elsewhere in the atom. The negative charges are possessed by **electrons,** symbolized by e^-, which move constantly about the nucleus. Hence, each atom has exactly the same number of electrons as protons. An electron has a mass of 9.1096×10^{-28} grams, about $1/_{2,000}$ the mass of a proton or a neutron.

Ion
(EYE-on)

Cation
(CAT-eye-on)

It is possible for an atom to lose or gain one or more electrons. When this happens, the atom is no longer electrically neutral, and we call it an ion. An **ion** is an atom or a group of atoms having a net electrical charge other than zero. If the ion has a net positive charge, it is called a **cation.** Cations have net positive charges because they have more protons than electrons. An example is the sodium cation, Na^+. In this symbol, the single + sign on the upper right indicates that the ion has lost an electron and thus has one net positive charge; therefore, its number of protons exceeds its number of electrons by 1. Another example of a cation is the magnesium ion, Mg^{2+}. In this ion, two electrons have been lost, resulting in two more protons than electrons. (Note that the number of the charge is written in the superscript before the sign of the charge, except for single charges, where only the sign of the charge is used.)

Anion
(AN-eye-on)

An ion with a net negative charge is called an **anion;** an anion has gained one or more electrons and thus has more electrons than protons. An example of an anion is the fluorine anion, F^-. In this anion, there is one more electron than there are protons. An example of an anion with two negative charges is the oxygen anion, O^{2-}. It contains two more electrons than protons.

Remember that in all ions, the charge arises because of a gain or loss of one or more electrons; the number of protons in an ion remains the same as in the atom or group of atoms from which the ion was formed.

3.5 ATOMIC NUMBER

Take another look at the periodic table and you will see a whole number above the symbol of each element. In fact, the elements are arranged in order of these numbers; they are called atomic numbers. The **atomic number** is simply the number of protons in an atom of the element. All atoms of any given element have the same number of protons, and thus all atoms of the element have the same atomic number. Elements are listed in the periodic table in order of increasing atomic number, and since protons cannot exist in fractional amounts, atomic numbers are always whole, exact numbers.

3.6 MASS NUMBER

Another quantity useful for characterizing an atom is its **mass number,** defined as the sum of the number of protons and neutrons in its nucleus:

$$\text{mass number} = \text{number of protons} + \text{number of neutrons}$$

or,

$$\text{mass number} = \text{atomic number} + \text{number of neutrons}$$

If any two of these quantities are known for an atom, the third can be calculated. Thus,

number of neutrons
           = mass number − atomic number (or number of protons)

and,

atomic number (or number of protons)
                    = mass number − number of neutrons

An atom's mass number thus is always equal to or larger than its atomic number.

EXAMPLE 3.1   A helium (He) atom has 2 protons and 2 neutrons. What is its mass number?

Solution   Since the mass number is the sum of protons and neutrons, the mass number for helium is (2 + 2) or 4.

EXERCISE 3.1   A potassium (K) atom has 19 protons and 20 neutrons. What is its mass number?   ▪

EXAMPLE 3.2   An atom having a mass number of 52 has 24 protons. How many neutrons does the atom have? Identify the element.

Solution   number of neutrons = mass number − number of protons
                                       = 52 − 24
                                       = 28

Since the atom contains 24 protons, 24 is also its atomic number. The name of element 24 is chromium, Cr.

EXERCISE 3.2   An atom having a mass number of 119 has 50 protons. How many neutrons does the atom have? Identify the element.   ▪

EXAMPLE 3.3   An atom having a mass number of 80 has 45 neutrons. How many protons are in the atom? Identify the element.

Solution   number of protons = mass number − number of neutrons
                                      = 80 − 45
                                      = 35

Since there are 35 protons in the atom, its atomic number must also be 35. Element 35 is Br, bromine.

EXERCISE 3.3   An atom having a mass number of 55 has 30 neutrons. How many protons are in the atom? Identify the element.
                                                                ▪

3.7 ISOTOPES

All atoms of a specific element have the same number of protons, but atoms of the same element can have different numbers of neutrons and thus different mass numbers. For example, there are three kinds of hydrogen atoms (atomic number 1); they differ only by the number of neutrons in each. Atoms of the same element that differ only in the number of neutrons they possess are called **isotopes.** Hydrogen is the only element whose isotopes have been given special names. The hydrogen isotope called *protium* has no neutrons; thus its mass number is 1 (since it contains one proton and no neutrons). Protium is the most abundant form of hydrogen; 99.9% of all naturally existing hydrogen is protium. A second kind of hydrogen, *deuterium*, is much less abundant, comprising only 0.1% of all hydrogen in nature. Deuterium contains one neutron in addition to its proton, so its mass number is 2. The third type of hydrogen, *tritium*, is radioactive and is found only in trace amounts in nature. Tritium contains two neutrons in addition to its proton, and its mass number is 3. The three hydrogen isotopes are diagrammed in Figure 3-5.

Isotopes of an element are represented by writing the element's symbol with subscripts and superscripts to the left of the symbol. The subscript is the atomic number of the element, and the superscript is the mass number.

$$^M_A E \qquad \begin{matrix} M = \text{mass number} \\ E = \text{symbol of the element} \\ A = \text{atomic number} \end{matrix}$$

For example, the three isotopes of hydrogen are represented as shown below:

protium $\quad ^1_1 H$
deuterium $\quad ^2_1 H$
tritium $\quad ^3_1 H$

All elements have isotopes, and a total of about fourteen hundred isotopes are known to exist. Interestingly, it is the number of protons in atoms that distinguishes one element from another. All isotopes of the same element have the same chemical properties but slightly different physical properties.

Isotope
(EYE-sah-tope)

Protium
(PRO-tih-um)

Deuterium
(doo-TEH-rih-um)

Tritium
(TRIH-tih-um)

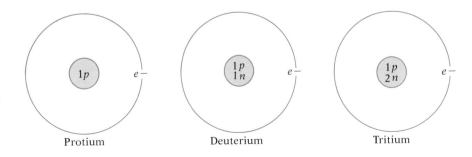

FIGURE 3-5 The Three Isotopes of Hydrogen. (The small inner circle represents the nucleus, and the larger outer circle represents the electron (e^-) moving about the nucleus. Protons and neutrons within nuclei are represented by *p* and *n*, respectively.)

EXAMPLE 3.4

How many protons, neutrons, and electrons are in $^{226}_{88}$Ra?

Solution

Since the atomic number of Ra, radium, is 88, an atom of radium contains 88 protons and the atom must also have 88 electrons. The difference between the mass number and atomic number, $226 - 88$, gives us the number of neutrons, 138.

EXERCISE 3.4

How many protons, neutrons, and electrons are in $^{112}_{48}$Cd? ■

Monatomic

Consisting of one atom; derived from the Greek prefix *mono-*, meaning "one," and the word atom.

Let us reconsider ions in light of the isotopic variations within elements. In determining the number of electrons in an ion derived from a single atom (a **monatomic ion**), you must remember that the ion has a different number of electrons than protons — this is why the ion has a charge. You must also remember that a monatomic ion has the same number of protons as the atom it was derived from. Let's look at the sodium cation, Na$^+$. A sodium atom has 11 protons (atomic number 11), and thus Na$^+$ must have 11 protons. The single positive charge on Na$^+$ tells us that the ion has one proton more than it has electrons. Hence, it must have 10 electrons. If the cation was formed from the sodium isotope of mass number 23, then $^{23}_{11}$Na$^+$ must have 12 neutrons, the same number as the atom it was derived from. Thus, a monatomic ion has the same number of both protons and neutrons as its parent atom. (Note: it is not necessary to show atomic numbers and mass numbers with the symbols for ions unless these quantities are needed to make a point in a discussion.) In the case of $^{24}_{12}$Mg^{2+}, the same reasoning tells us that there are 12 protons, 10 electrons, and 12 neutrons. Thus, regardless of their mass numbers, both Na$^+$ and Mg^{2+} have the same number of electrons, 10. Ions or atoms of different elements but with the same number of electrons are said to be **isoelectronic.**

Isoelectronic

Derived from the Greek prefix *iso-*, meaning "equal," and the word electron.

EXAMPLE 3.5

How many protons, neutrons, and electrons are in the iron cation $^{56}_{26}$Fe^{2+}?

Solution

The ion must have the same number of protons and neutrons as its parent atom; thus, $^{56}_{26}$Fe^{2+} contains 26 protons and 30 neutrons. Its two positive charges mean that it contains 2 more protons than electrons; thus it contains 24 electrons.

EXERCISE 3.5

How many protons, neutrons, and electrons are in the aluminum cation, $^{27}_{13}$Al^{3+}? ■

The amount of negative charge on an anion indicates how many more electrons than protons are present. For example, in the oxygen anion, O^{2-}, there are two more electrons than protons. If the anion was formed from $^{16}_{8}$O, then $^{16}_{8}$O^{2-} has 8 protons, 8 neutrons, and 10 electrons. Notice that O^{2-} happens to be isoelectronic with Na$^+$ and Mg^{2+}.

EXAMPLE 3.6

How many protons, neutrons, and electrons are in the chlorine anion, $^{35}_{17}$Cl$^-$?

Solution

$^{35}_{17}$Cl$^-$ has 17 protons, 18 neutrons, and 18 electrons (one more than the number of protons).

Since late 1982, two new elements have been synthesized, both by the same team of scientists in Darmstadt, West Germany. The group, led by Peter Armbruster and Gottfried Munzenberg, made the new elements with the Universal Linear Accelerator, a device that bombards target atoms with an accelerated stream of ions. In creating element 109 in late 1982, the research group used $^{209}_{83}$Bi as target atoms and bombarded them with $^{56}_{26}$Fe ions. The iron nucleus combined with the bismuth nucleus to create a new nucleus with 109 protons. It was calculated that the combination of the two nuclei should occur only once in every 10^{14} collisions, and it took a week of bombarding target atoms before element 109 was created. This achievement was especially remarkable in that element 108 had not yet been created, although element 107 had been made. Perhaps most astonishing of all is that the research group was able to identify the new element even though only one atom was created.

The identification of only a single atom of the new element was compared to identifying one grain of sand in a whole trainload. Adding to the difficulty of identifying the single atom was the fact that element 109 is radioactive and its single atom existed for only 5 milliseconds before it disintegrated.

The West German team achieved synthesis of element 108 in the spring of 1984. To make this element, $^{208}_{82}$Pb atoms were bombarded with $^{58}_{26}$Fe ions, creating a new nucleus with 108 protons. Surprisingly, element 108 is three times more stable than theoretically predicted, and the scientists succeeded in making three atoms of the new element.

The six newest elements, numbers 104–109, do not yet have official names but will be named by international agreement at some time in the future. Meanwhile, the Darmstadt group will attempt to make element 116, expected to be relatively stable.

EXERCISE 3.6 How many protons, neutrons, and electrons are in the bromine anion, $^{80}_{35}$Br⁻? ■

3.8 ATOMIC MASS UNITS

The term *mass number* is misleading because it tells us only the number of protons and neutrons in an isotope, not the actual mass of the isotope. The isotopic mass is an exceedingly small number of grams, usually expressed in scientific notation. For convenience, scientists have devised atomic mass units to refer to isotopic masses. An atomic mass unit is a relative unit created by comparison to a standard, in this case, the carbon isotope of mass number 12. Scientists have defined the **atomic mass unit (amu)** as $1/12$ the mass of the carbon isotope of mass number 12. This corresponds to a mass of 1.6604×10^{-24} grams for an atomic mass unit. It also means that, according to this standard, $^{12}_{6}$C has a mass of exactly 12 amu. Given this information, we can calculate the masses of subatomic particles in amu as follows:

$$\text{mass of proton} = (1.6726 \times 10^{-24}\ \text{g}) \left(\frac{1\ \text{amu}}{1.6604 \times 10^{-24}\ \text{g}} \right) = 1.007\ \text{amu}$$

$$\text{mass of neutron} = (1.6748 \times 10^{-24}\ \text{g}) \left(\frac{1\ \text{amu}}{1.6604 \times 10^{-24}\ \text{g}} \right) = 1.0087\ \text{amu}$$

$$\text{mass of electron} = (9.1096 \times 10^{-28}\ \text{g}) \left(\frac{1\ \text{amu}}{1.6604 \times 10^{-24}\ \text{g}} \right)$$

$$= 5.4864 \times 10^{-4}\ \text{amu}$$

MATH TIP

To clearly see the solution to this conversion, rewrite the problem separating the numbers, exponents, and units:

$$\left(\frac{9.1096}{1.6604} \right) \left(\frac{10^{-28}}{10^{-24}} \right) \left(\frac{\text{g·amu}}{\text{g}} \right)$$

$$= 5.4864 \times 10^{[-28-(-24)]}\ \text{amu}$$

$$= 5.4864 \times 10^{-4}\ \text{amu}$$

If this problem is solved using a calculator, it should be entered in the following way: Press 9.1096 EE 28 +/− ÷ 1.6604 EE 24 +/− = . The display reads 5.486389 −0.4. The answer should be rounded off to the correct number of significant figures and written as 5.4864 × 10^{-4} amu.

TABLE 3-4 Subatomic Particles

Particle	Symbol	Mass (amu)	Charge
proton	p	1	+1
neutron	n	1	0
electron	e^-	0	−1

By expressing these masses in amu, we no longer need to use scientific notation for masses of protons and neutrons. In fact, we often round off these quantities and say that protons and neutrons both have masses of 1 amu. Scientific notation is still required to express the mass of an electron, but this mass is so small compared to the masses of protons and neutrons that we usually assign it a value of 0 amu. Table 3-4 summarizes the main features of the proton, neutron, and electron.

3.9 ATOMIC WEIGHT

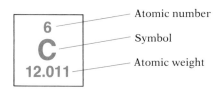

FIGURE 3-6 The Atomic Number and Atomic Weight as They Appear in the Periodic Table

In the periodic table the atomic number is shown just above the symbol for each element. The number just below each element's symbol is the **atomic weight** (see Figure 3-6). The **atomic weight** of an element is the average of the masses of all isotopes of the element found in nature. The atomic weight of an element can be calculated from isotopic masses and the percentages of each isotope as found in nature. Let's examine how the atomic weight of chlorine (Cl) is calculated. Chlorine, found in nature as a mixture of two isotopes, consists of 75.53% $^{35}_{17}$Cl and 24.47% $^{37}_{17}$Cl. $^{35}_{17}$Cl has a mass of 34.968 amu and $^{37}_{17}$Cl has a mass of 36.956 amu. We first find the contribution of each isotope toward the average mass of isotopes. This is done by multiplying the decimal equivalent of the percentage of each isotope by its isotopic mass:

MATH TIP

To find the decimal equivalent of the percentage of each isotope, divide the percentage by 100%. This can also be accomplished by moving the decimal point two places to the left.

$$\frac{75.53\%}{100\%} = 0.7553$$

The number 0.7553 is referred to as the "fraction of the $^{35}_{17}$Cl isotope."

contribution of each isotope
$$= \text{fraction of each isotope} \times \text{mass of each isotope}$$

Then, adding the individual contributions gives the average mass, or atomic weight.

$$^{35}_{17}\text{Cl: } (0.7553)(34.968 \text{ amu}) = 26.41 \text{ amu}$$
$$^{37}_{17}\text{Cl: } (0.2447)(36.956 \text{ amu}) = \underline{9.043 \text{ amu}}$$
$$\text{atomic weight of Cl: } 35.45 \text{ amu}$$

EXAMPLE 3.7

Copper exists in nature as a mixture of 69.16% $^{63}_{29}$Cu and 30.84% $^{65}_{29}$Cu. The mass of $^{63}_{29}$Cu is 62.930 amu and the mass of $^{65}_{29}$Cu is 64.928 amu. What is the atomic weight of copper?

Solution

$$^{63}_{29}\text{Cu: } (0.6916)(62.930 \text{ amu}) = 43.52 \text{ amu}$$
$$^{65}_{29}\text{Cu: } (0.3084)(64.928 \text{ amu}) = \underline{20.02 \text{ amu}}$$
$$\text{atomic weight of Cu: } 63.54 \text{ amu}$$

EXERCISE 3.7   Magnesium exists in nature as a mixture of 78.99% $^{24}_{12}$Mg, 10.00% $^{25}_{12}$Mg, and 11.01% $^{26}_{12}$Mg. What is the atomic weight of magnesium? (Assume a mass of 1.0 amu for each neutron and proton.)   ■

Most atomic weights are decimal numbers, but the periodic table shows whole-number atomic weights in parentheses for some elements. This indicates that these elements are radioactive. (Radioactivity will be discussed in chapter 17.) Their atoms undergo spontaneous disintegration, and the number in parentheses is the mass number of the most stable or best known isotope of the element. All elements past neptunium (atomic number 93) in the periodic table are man-made and not found in nature.

3.10 COMPOUNDS

Elements combine with each other to form a wide variety of more complex substances called compounds. A **compound** is a pure substance composed of two or more elements combined in definite proportions by weight. Many compounds form spontaneously in nature, while others are prepared in laboratories. Over 5 million compounds are known, and hundreds more are being discovered or prepared daily. When elements combine to form a compound, they lose their separate identities, and the compound has characteristics different from those of the constituent elements.

Compounds can be composed of either ions or molecules. Table salt, sodium chloride, is an example of an ionic compound; it contains sodium cations, Na^+, and chlorine anions, Cl^-, held together by the attraction of their opposite electrical charges. Water is an example of a compound composed of molecules. A **molecule** is a group of two or more atoms held together by the attraction of individual nuclei for electrons belonging to the other atoms of the molecule. Figure 3-7 illustrates differences between ionic and molecular compounds.

Let's take a closer look at these two compounds. Sodium chloride contains sodium cations and chloride anions combined in a ratio of one cation for each anion. For any sample of sodium chloride, 39.3% of the weight is contributed by sodium and 60.7% by chlorine. The compound is a white, crystalline solid at room temperature and has a very high melting point, 808° C. It is quite soluble in water; 36 g of sodium chloride will dissolve in 100 mL of water at 0° C. Although sodium chloride is not very reactive with other substances, it can be decomposed by electricity to its constituent elements.

Water is composed of molecules; each molecule contains two atoms of hydrogen and one atom of oxygen. By weight, it contains 11.2% hydrogen and 88.8% oxygen. Water is a colorless liquid at room temperature; it has a melting point of 0° C and a boiling point of 100° C. It dissolves many substances but is not very reactive. However, it too can be decomposed by electricity to its constituent elements. Both sodium chloride and water have unique sets of properties that distinguish them from all other sub-

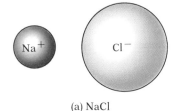

(a) NaCl

(b) H$_2$O

FIGURE 3-7 Ionic and Molecular Compounds. (a) Na$^+$ and Cl$^-$ are attracted to each other to form the ionic compound sodium chloride, NaCl. (b) Two hydrogen atoms share electrons with an oxygen atom to form a molecule of water, H$_2$O.

stances. Try to imagine, if you will, the enormous combinations of properties possessed by the over 5 million known elements and compounds.

3.11 COMPOSITION OF COMPOUNDS

Every chemical compound has its own unique composition that stays the same from one sample to another; that is, every sample of sodium chloride, regardless of its origin, contains 39.3% sodium and 60.7% chlorine by weight. Likewise, every drop of pure water, regardless of where it is collected, contains 11.2% hydrogen and 88.8% oxygen by weight. Countless experiments demonstrating this constancy in proportions have led to the formulation of the **law of definite proportions:** a given compound always contains the same elements combined in the same proportions by weight. In other words, the composition of any given compound is always the same, regardless of where the compound came from or how it was formed. This fundamental principle of chemistry allows us to distinguish any compound from the countless other compounds known to modern chemistry.

3.12 CHEMICAL FORMULAS

Chemical formulas are the shorthand of chemistry. You have probably heard water referred to as "H-two-O"; this simply means that the formula of water is H_2O. For sodium chloride, the formula is NaCl. **Formulas** represent compounds; formulas use element symbols and subscripts to indicate the proportions in which atoms or ions of the elements are combined. The formula for sodium chloride tells us that there is one ion of sodium for every ion of chlorine. In water, two atoms of hydrogen exist for every atom of oxygen in water molecules. Thus, the formula of a compound tells us not only which elements are present but also how many atoms or ions of each exist in the simplest unit of the compound. The term *simplest unit* means the simplest combination of the atoms or ions that has all of the properties of the compound.

The examples given so far are rather uncomplicated ones, but can you tell what is meant by $HClO_3$ or $Al_2(SO_4)_3$, for example? Chances are that these are somewhat baffling. It takes a bit of experience to get used to seeing chemical formulas and understanding what they mean, but the following paragraphs summarize the major features of chemical formulas.

1. The formula of a compound contains symbols of all of the elements in the compound with subscripted numbers indicating the proportions in which the atoms (or ions) exist in the simplest unit of the compound. If, for example, the symbol of an element carries a subscript of 2, that means that there are two atoms (or ions) of that element in the simplest unit of the compound. The only number never used as a subscript is 1. If the symbol of an element appears without a subscript, the symbol is understood to have a subscript of 1.

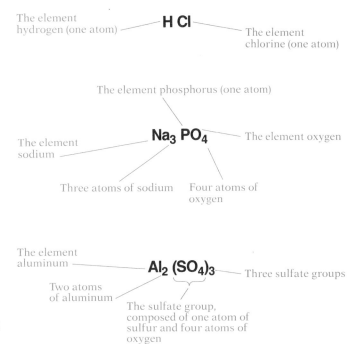

FIGURE 3-8 Notation Used in Chemical Formulas

2. If the formula of a compound contains symbols in parentheses, this means that these atoms are bound together as a group, and any subscript following the parentheses refers to the entire group of atoms. For example, in $Al_2(SO_4)_3$, Al_2 means that there are two atoms of aluminum in the simplest unit of the compound; $(SO_4)_3$ means that SO_4 (one atom of sulfur and four of oxygen) behaves as a group, and there are three of these groups in the simplest unit of the compound. In this compound, there is a total of three sulfur atoms and twelve oxygen atoms in the simplest unit of the compound, but they exist as three individual SO_4 groups. The formula of this compound is spoken as "A-L-two-S-O-4-taken three times."

3. Formulas do not indicate how the atoms (or ions) are arranged in the compound nor do they indicate the types of forces that hold the atoms (or ions) together.

Figure 3-8 illustrates the notation used in chemical formulas.

3.13 FORMULA WEIGHTS OF COMPOUNDS

We can calculate the mass of the simplest unit of a compound by using atomic weights of the constituent elements. Regardless of whether a compound is composed of ions or molecules, the formula describes the simplest unit of a compound; we call this the **formula unit** of the compound. If we multiply the atomic weight of each element in a formula by the number of atoms (or ions) of that element in the formula, then add the

▽ **PROBLEM-SOLVING SKILLS**

THE IMPORTANCE OF UNITS AND LABELS

The use of dimensional analysis greatly simplifies solving problems in chemistry. The key to dimensional analysis is to *always include units when setting up the solution to a problem*. In this way, you can cancel units before going on with the calculations; if, after cancelling units, you are left with only the units of the answer you are seeking, then you can be reasonably sure that the solution you have set up will solve the problem correctly.

In addition to always including units in setting up solutions to problems, *using labels is also a big help*. In the example where we found the formula weight of H_2O, notice how the atoms of each element were labelled: 2 atoms H and 1 atom O. With these labels, numerical values don't get misplaced and it is clear that "2 atoms H" should be multiplied by "1.0 amu/atom H" to obtain the contribution of hydrogen to the formula weight of water, and the contribution of oxygen is equally clear. Using units and labels may seem a bit time-consuming at first, but the time is well spent. Doing so helps you to organize your thinking and thus solve the problem correctly on the first attempt.

result for each element, we get what is called the **formula weight (FW)** for that compound. Formula weights carry units of amu, just as atomic weights do.

Let's see how the formula weight of water, H_2O, is calculated. We first find the contribution of each element toward the total mass of the formula weight. This is done by multiplying the number of atoms of the element in the formula unit by the atomic weight of the element.

contribution of element
= number of atoms of element × atomic weight of element

Then we add the individual contributions to get the formula weight of the compound.

$$\text{FW of } H_2O = (2 \text{ atoms H})(\text{atomic weight of H})$$
$$+ (1 \text{ atom O})(\text{atomic weight O})$$

$$= (2 \text{ atoms H}) \left(\frac{1.0 \text{ amu}}{\text{atom H}} \right) + (1 \text{ atom O}) \left(\frac{16.0 \text{ amu}}{\text{atom O}} \right)$$

$$= 2.0 \text{ amu} + 16.0 \text{ amu}$$

$$= 18.0 \text{ amu}$$

(Note that the number of atoms of each element in a formula unit is an exact number and thus does not influence the number of significant figures in the answer.)

As another example, we can calculate the formula weight of magnesium chloride, $MgCl_2$, as follows.

$$\text{FW of MgCl}_2 = (1 \text{ atom Mg})(\text{atomic weight Mg})$$
$$+ (2 \text{ atoms Cl})(\text{atomic weight Cl})$$

$$= (1 \text{ atom Mg}) \left(\frac{24.3 \text{ amu}}{\text{atom Mg}} \right) + (2 \text{ atoms Cl}) \left(\frac{35.5 \text{ amu}}{\text{atom Cl}} \right)$$

$$= 24.3 \text{ amu} + 71.0 \text{ amu}$$

$$= 95.3 \text{ amu}$$

EXAMPLE 3.8 Calculate the formula weight of carbon dioxide, CO_2.

Solution

$$\text{FW of CO}_2 = (1 \text{ atom C})(\text{atomic weight of C}) + (2 \text{ atoms O})$$
$$(\text{atomic weight of O})$$

$$= (1 \text{ atom C}) \left(\frac{12.0 \text{ amu}}{\text{atom C}} \right) + (2 \text{ atoms O}) \left(\frac{16.0 \text{ amu}}{\text{atom O}} \right)$$

$$= 12.0 \text{ amu} + 32.0 \text{ amu}$$

$$= 44.0 \text{ amu}$$

EXERCISE 3.8 Calculate the formula weight of each of the following compounds.

a. Methane, CH_4 c. Sodium sulfate, Na_2SO_4
b. Glucose, $C_6H_{12}O_6$

3.14 PERCENTAGE COMPOSITION

Formula weights can be used for many practical purposes; one is to calculate the weight percentage of an element in a compound. For example, an important concern today is the amount of sodium in our diets. Ordinary table salt, sodium chloride (NaCl), is usually the major dietary source of sodium. We can calculate the percentage by weight of sodium in table salt by using the formula weight of NaCl, 58.5 amu. We calculate the percentage of sodium just as we would calculate any other percentage: divide the part (mass of Na in NaCl) by the whole (mass of NaCl; this is simply the formula weight) and multiply by 100% (an exact quantity). The calculation is as follows:

$$\%\text{Na} = \left(\frac{\text{mass of Na in NaCl}}{\text{FW of NaCl}} \right) (100\%)$$

$$= \left(\frac{23.0 \text{ amu}}{58.5 \text{ amu}} \right) (100\%)$$

$$= 39.3\%$$

Thus, our calculation shows that sodium makes up 39.3% of the table salt in our diet.

Now that you see how to calculate the weight percentage of a single element in a compound, let's proceed to calculate the weight percentage of every element in a compound; this gives us the **percentage composition** of the compound. We will calculate the percentage composition of magnesium hydroxide, $Mg(OH)_2$, the main ingredient in milk of magnesia. To do this, we simply calculate the individual weight percentages of Mg, O, and H in $Mg(OH)_2$:

$$\%Mg = \left(\frac{\text{mass of Mg in } Mg(OH)_2}{\text{FW of } Mg(OH)_2}\right)(100\%)$$

$$= \left(\frac{24.3 \text{ amu}}{58.3 \text{ amu}}\right)(100\%)$$

$$= 41.7\%$$

$$\%O = \left(\frac{\text{mass of O in } Mg(OH)_2}{\text{FW of } Mg(OH)_2}\right)(100\%)$$

$$= \left(\frac{32.0 \text{ amu}}{58.3 \text{ amu}}\right)(100\%)$$

$$= 54.9\%$$

$$\%H = \left(\frac{\text{mass of H in } Mg(OH)_2}{\text{FW of } Mg(OH)_2}\right)(100\%)$$

$$= \left(\frac{2.0 \text{ amu}}{58.3 \text{ amu}}\right)(100\%)$$

$$= 3.4\%$$

We can check our calculations by making sure that the percentages add up to 100%:

$$41.7\% + 54.9\% + 3.4\% = 100.0\%$$

EXAMPLE 3.9   Calculate the percentage composition of glucose, $C_6H_{12}O_6$.

Solution

$$\text{FW of } C_6H_{12}O_6 = (6 \text{ atoms C})\left(\frac{12.0 \text{ amu}}{\text{atom C}}\right) + (12 \text{ atoms H})\left(\frac{1.0 \text{ amu}}{\text{atom H}}\right)$$

$$+ (6 \text{ atoms O})\left(\frac{16.0 \text{ amu}}{\text{atom O}}\right)$$

$$= 72.0 \text{ amu} + 12 \text{ amu} + 96.0 \text{ amu}$$

$$= 180 \text{ amu}$$

Then we calculate the percentage of each element in glucose:

$$\%C = \left(\frac{\text{mass of C in glucose}}{\text{FW of glucose}}\right)(100\%)$$

$$= \left(\frac{72.0 \text{ amu}}{180 \text{ amu}}\right)(100\%)$$

$$= 40.0\%$$

$$\%H = \left(\frac{\text{mass of H in glucose}}{\text{FW of glucose}}\right)(100\%)$$

$$= \left(\frac{12 \text{ amu}}{180 \text{ amu}}\right)(100\%)$$

$$= 6.7\%$$

$$\%O = \left(\frac{\text{mass of O in glucose}}{\text{FW of glucose}}\right) (100\%)$$
$$= \left(\frac{96.0 \text{ amu}}{180 \text{ amu}}\right) (100\%)$$
$$= 53.3\%$$

Checking:

$$40.0\% + 6.7\% + 53.3\% = 100.0\%$$

EXERCISE 3.9 Calculate the percentage composition of sodium bicarbonate, $NaHCO_3$, also known as baking soda. ■

SUMMARY

Elements are pure substances that cannot be separated into simpler substances by ordinary processes. The elements are represented by their **symbols** in the **periodic table,** where they are arranged in horizontal rows called **periods** and in vertical columns called **groups** or **families.**

An **atom** is the smallest unit of an element that has all the properties of the element. An atom has a core called the **nucleus** that contains protons, neutrons, and other particles. Outside of the nucleus are one or more electrons. A **proton** has a mass of 1 amu and a single positive charge; a **neutron** also has a mass of 1 amu, but it has no charge. An **electron** has negligible mass and a negative charge.

Atoms are electrically neutral but may form **ions** by gain or loss of one or more electrons. A positively charged ion is a **cation;** a negatively charged ion is an **anion.** The magnitude of the charge on a monoatomic ion indicates how many electrons were gained or lost by the parent atom.

The **atomic number** is the number of protons in an atom. The **mass number** is the sum of protons and neutrons in an atom. Atoms of the same element that differ only by the number of neutrons they possess are called **isotopes.** Isotopes of an element are represented by the symbol of the element with the atomic number subscripted and the mass number superscripted to the left of the symbol.

An **atomic mass unit (amu)** is $1/12$ the mass of the carbon isotope of mass number 12. The **atomic weight** of an element is the weighted average of masses of all isotopes of the element; it is calculated from isotopic mass and percentage of each isotope as found in nature.

A **compound** is a pure substance composed of two or more elements combined in definite proportions by weight. Compounds can be composed of either ions or molecules. A **molecule** is a group of two or more atoms held together by attraction of individual nuclei for electrons belonging to the other atoms of the molecule. The **law of definite proportions** states that a given compound always contains the same elements combined in the same proportions by weight. **Formulas** are representations of compounds; formulas use element symbols and subscripts to indicate the proportions in which atoms or ions of the elements are combined. A **formula unit** is the simplest unit of a compound as described by the formula. The **formula weight** of a compound is the sum of the atomic weights of the elements in the formula after each atomic weight has been multiplied by the number of atoms (or ions) of each element present. The **percentage composition** of a compound is the percentage, by weight, of every element in the compound.

STUDY QUESTIONS AND PROBLEMS

(More difficult questions and problems are marked with an asterisk.)

ELEMENTS

1. In your own words, define the term *element.*
2. Were the four elements of the ancient Greeks true elements? Explain your answer.
3. Hydrogen is thought to be the basic material from which all other elements are synthesized, yet there is very little elemental hydrogen on earth or in its atmosphere. Why?

THE PERIODIC TABLE

4. Define each of the following terms:
   a. Periodic table   c. Group (or family) of elements
   b. Period of elements   d. Element symbol
5. What is the basis for the name periodic table?
6. Using the list of elements inside the front cover, give the name of each of the following elements.
   a. He   c. F   e. Ag
   b. Au   d. Fe   f. Kr
7. Using the list of elements inside the front cover, give the symbol of each of the following elements.
   a. Polonium   c. Tungsten   e. Fermium
   b. Barium   d. Uranium   f. Chromium

ATOMS

8. Describe the discontinuous and continuous theories of matter. Which theory is accepted today?
9. In your own words, summarize Dalton's atomic hypothesis. How has it been modified to account for modern evidence?
10. What is an atom? In theory, can an atom be subdivided? Explain your answer.

SUBATOMIC PARTICLES

11. Define each of the following terms:
    a. Subatomic particles   d. Neutron   g. Cation
    b. Nucleon   e. Electron   h. Anion
    c. Proton   f. Ion
12. A silicon atom contains 14 protons, 14 neutrons, and 14 electrons. Draw a simplified diagram of this silicon atom.

13. Classify each of the following ions as a cation or an anion. Indicate whether each ion has an excess or deficiency of electrons relative to protons, and give the number of excessive or deficient electrons.
   a. I^- b. Ca^{2+} c. S^{2-} d. Al^{3+}

ATOMIC NUMBER

14. What is the atomic number of an element?
15. Using the periodic table or list of elements inside the front cover, give the atomic number of each of the following elements.
   a. Pb   c. Ho   e. Tc
   b. Np   d. Cs   f. Mo
16. Using the periodic table or list of elements inside the front cover, give the name of the element corresponding to each of the following atomic numbers.
   a. 74   c. 12   e. 32
   b. 96   d. 15   f. 49

MASS NUMBER

17. What is meant by mass number? How does it differ from atomic number?
18. Each of the following describes the nucleus of an atom of a particular element. Name the element.
   a. mass number 56; 30 neutrons
   b. mass number 201; 121 neutrons
   c. mass number 197; 118 neutrons
   d. mass number 238; 146 neutrons
   e. mass number 84; 48 neutrons
   f. mass number 207; 125 neutrons

ISOTOPES

19. Define each of the following terms:
   a. Isotope   c. Deuterium   e. Isoelectronic ions
   b. Protium   d. Tritium
20. Give the number of protons, neutrons, and electrons in each of the following atoms.
   a. $^{11}_{5}B$ c. $^{65}_{30}Zn$ e. $^{197}_{79}Au$
   b. $^{41}_{20}Ca$ d. $^{181}_{73}Ta$ f. $^{131}_{54}Xe$
21. Give the number of protons, neutrons, and electrons in each of the following ions.
   a. $^{137}_{56}Ba^{2+}$ c. $^{27}_{13}Al^{3+}$ e. $^{119}_{50}Sn^{2+}$
   b. $^{1}_{1}H^-$ d. $^{128}_{52}Te^{2-}$ f. $^{85}_{37}Rb^+$
22. Which of the following ions are isoelectronic?
   a. Br^- c. Sr^{2+} e. Ca^{2+}
   b. S^{2-} d. Li^+ f. K^+

ATOMIC MASS UNITS

23. Define isotopic mass. How does this quantity differ from mass number? From atomic number?

24. What is an atomic mass unit (amu)? Why were atomic mass units created?

25. List the three most important subatomic particles and give their masses in atomic mass units.

ATOMIC WEIGHT

26. What is the atomic weight of an element? How does it differ from isotopic mass?

27. Using the periodic table or the list of elements inside the front cover, give the atomic weight of each of the following elements.
    a. Y   **c.** Mn   **e.** Fr
    b. Tl   **d.** Ar   **f.** Rh

28. Naturally occurring neon, the gas used in brightly colored electric signs, is composed of 90.92% $^{20}_{10}$Ne, 0.257% $^{21}_{10}$Ne, and 8.82% $^{22}_{10}$Ne. Calculate the atomic weight of neon. (Assume a mass of 1.0 amu for each neutron and proton.)

29. Sulfur, the yellow, solid nonmetal, occurs naturally as 95.02% $^{32}_{16}$S, 0.76% $^{33}_{16}$S, and 4.22% $^{34}_{16}$S. Calculate the atomic weight of sulfur. (Assume a mass of 1.0 amu for each neutron and proton.)

COMPOUNDS

30. Define each of the following terms:
    a. Compound   **b.** Molecule

31. What are the two types of compounds? How do they differ from each other?

32. How are compounds and elements different from each other? How are they similar?

COMPOSITION OF COMPOUNDS

33. Does the law of definite proportions apply to mixtures? Explain your answer.

34. If sodium chloride contains 39.3% sodium by weight, how many grams of sodium are in 50.0 g of sodium chloride?

CHEMICAL FORMULAS

35. What is a chemical formula? How does it differ from a chemical symbol?

36. For each of the following compounds, name the elements present.
    a. Ammonia, NH_3
    b. Calcium carbonate, $CaCO_3$
    c. Sulfuric acid, H_2SO_4
    d. Sodium hydroxide, $NaOH$
    e. Magnesium sulfate, $MgSO_4$
    f. Potassium permanganate, $KMnO_4$

37. For each of the following compounds, give the number of atoms of each element in the simplest unit of the compound.
    a. Sodium carbonate, Na_2CO_3
    b. Magnesium phosphate, $Mg_3(PO_4)_2$
    c. Calcium hydrogen sulfate, $Ca(HSO_4)_2$

  d. Barium acetate, $Ba(C_2H_3O_2)_2$
  e. Aluminum bromide, $AlBr_3$
  f. Strontium nitrate, $Sr(NO_3)_2$
38. Given below is the composition of the simplest unit of some compounds; write the formula of each.
  a. Sodium oxide: two sodium atoms and one oxygen atom.
  b. Nitric acid: one hydrogen atom, one nitrogen atom, and three oxygen atoms.
  c. Sodium acetate: one sodium atom, two carbon atoms, three hydrogen atoms, and two oxygen atoms.
  d. Aluminum oxide: two aluminum atoms and three oxygen atoms.
  e. Carbon tetrachloride: one carbon atom and four chlorine atoms.
  f. Iron(III) sulfide: two iron atoms and three sulfur atoms.

FORMULA WEIGHTS OF COMPOUNDS

39. In your own words, give definitions for formula unit and formula weight.
40. Calculate the formula weight of each of the following compounds.
  a. Aluminum bromate, $Al(BrO_3)_3$
  b. Barium cyanide, $Ba(CN)_2$
  c. Calcium chromite, $CaCr_2O_4$
  d. Chromium(III) sulfate, $Cr_2(SO_4)_3$
41. Calculate the formula weight of each of the following compounds.
  a. Cobalt(III) acetate, $Co(C_2H_3O_2)_3$
  b. Copper(II) nitrate, $Cu(NO_3)_2$
  c. Iron(III) bromide, $FeBr_3$
  d. Lithium sulfate, Li_2SO_4

PERCENTAGE COMPOSITION

42. Rust, the reddish crust that forms on iron when it is exposed to the weather, is iron(III) oxide, Fe_2O_3. Calculate the percentage composition of rust.
43. Nitroglycerin, the liquid explosive, has the formula $C_3H_5(NO_3)_3$. Calculate the percentage composition of nitroglycerin.
44. Table sugar is sucrose, $C_{12}H_{22}O_{11}$. Calculate the percentage composition of sucrose.
45. Cryolite, a mineral used in the production of aluminum, has the formula Na_3AlF_6. Calculate the percentage composition of cryolite.

GENERAL PROBLEMS

46. How many protons would it take to equal the mass of a drop of water, 0.05 g? How many neutrons? How many electrons?
47. Iron, the strong, heavy metal, occurs naturally as 5.84% $^{54}_{26}Fe$, 91.68% $^{56}_{26}Fe$, 2.17% $^{57}_{26}Fe$, and 0.31% $^{58}_{26}Fe$. Calculate the atomic weight of iron. (Assume a mass of 1.0 amu for each neutron and proton.)
*48. Water contains 88.8% oxygen by weight. If electricity is used to decompose water, how many grams of water would be required to produce 17.39 g of oxygen?
*49. Chromium metal is produced from the ore chromite, $FeCr_2O_4$. How

many tons of chromite would be required to produce 1.00 ton of chromium metal?

*50. Which of the following has more atoms, 1.1 g of hydrogen atoms or 14.7 g of chromium atoms?

51. Phosphorus forms four compounds with sulfur: P_4S_3, P_4S_5, P_4S_7, and P_4S_{10}. Find the percentage composition of each compound.

ELECTRON ARRANGEMENTS IN ATOMS

CHAPTER
4

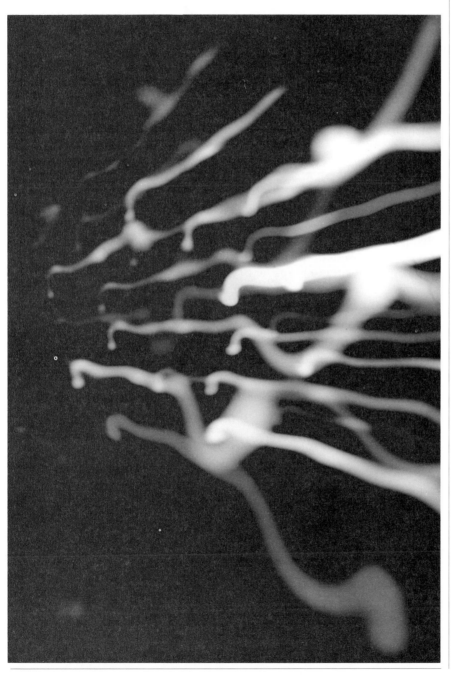

OUTLINE

4.1 Early Models of the Atom
  Perspective: Rutherford's Gold Foil
  Experiment
4.2 The Nature of Light
4.3 Emission Spectra of the Elements
4.4 The Bohr Atom
4.5 Modern Ideas of Atomic Structure
4.6 Energy Levels of Electrons
4.7 Energy Sublevels
4.8 Electron Configurations
4.9 Quantum Numbers (*optional*)
  Summary
  Study Questions and Problems

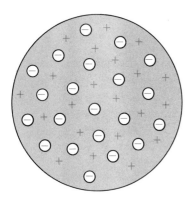

FIGURE 4-1 The Thomson Plum Pudding Model of the Atom

Scientific model

A mental image that describes the behavior of a scientific system.

The electron is considered the lightest stable elementary particle of matter. Each electron carries a single negative charge, the basic charge of electricity. Although electrical effects have been observed since the dawn of history, it was not until 1747 that Benjamin Franklin proposed that electricity consists of particles. In 1833, the English scientist Michael Faraday performed experiments that led him to believe that electric power was associated with each atom of matter. In 1897 the electron itself was discovered by J. J. Thomson, an English physicist who was awarded the Nobel Prize for Physics in 1906. In 1916 Robert Millikan, a physicist at the University of Chicago, measured the charge on an electron; for this accomplishment, he was honored with the Nobel Prize for Physics in 1923.

While these discoveries were taking place, scientists were convinced that electrons came from inside atoms, but they weren't sure why electrons were there in the first place. Since then we have learned that the arrangement of electrons in an atom largely determines the physical and chemical behavior of the atom, and that will be the subject of this chapter.

4.1 EARLY MODELS OF THE ATOM

The earliest **scientific model** of the atom was proposed in 1902 by an English physicist, Lord Kelvin, but Kelvin's model was supported so strongly by J. J. Thomson that it became known as the Thomson model. According to Kelvin and Thomson, an atom was a sphere of uniformly distributed positive charge with electrons stuck in it like raisins in a pudding; because of its resemblance to the English favorite, plum pudding, this model is often called the **Thomson plum pudding model** (see Figure 4-1).

In 1910, Ernest Rutherford, an English physicist, performed some novel experiments with gold foil (see the following Perspective section). The

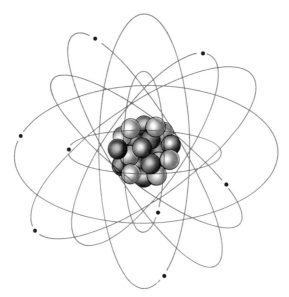

FIGURE 4-2 The Solar System Model of the Atom. The nucleus is in the center of the atom, with electrons orbiting about it.

PERSPECTIVE RUTHERFORD'S GOLD FOIL EXPERIMENT

When Rutherford conducted his gold foil experiment in 1910, he relied heavily on his earlier findings regarding the nature of radioactivity. About a year earlier he had learned that one type of radiation emitted by radioactive substances was composed of streams of helium nuclei. These particles are known as *alpha* (α) *particles*, and each bears a 2+ charge. While characterizing alpha particles, Rutherford studied how a narrow beam of alpha particles scattered as they passed through thin gold foil. A diagram of his experiment is given in Figure a.

Rutherford found that most of the alpha particles passed straight through the foil, but a few were deflected, and some even bounced straight back in the direction from which they had come. An undergraduate student working in Rutherford's laboratory said it was almost as if a 15-inch shell fired into a piece of tissue paper bounced back.

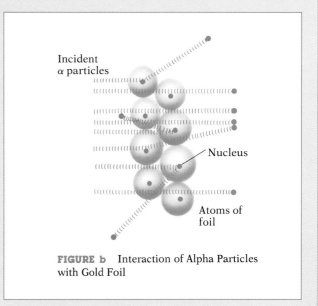

FIGURE b Interaction of Alpha Particles with Gold Foil

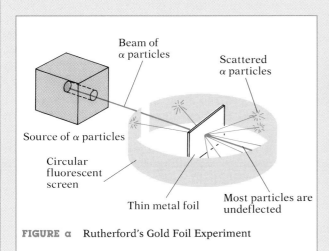

FIGURE a Rutherford's Gold Foil Experiment

Rutherford explained his observations by proposing a new atomic model. He suggested that most of the alpha particles passed directly through the foil because the gold atoms contained mostly empty space (see Figure b). However, a few of the alpha particles were deflected because they came close to the small nuclei of the gold atoms, where the positive charges of the atoms were located. Since each gold atom contained 79 protons, the repulsion between the positive charges of the gold atom and those of an alpha particle coming near it was strong enough to deflect the particle. Furthermore, an alpha particle headed directly toward a gold atom was so strongly repelled that the alpha particle reversed its direction, thus being deflected backward.

results suggested that the atom's positive charge and mass were concentrated in a very small region at the center of the atom. On this basis, Rutherford proposed a new atomic model in 1911. He proposed that most of the mass and all of the positive charge are concentrated in a very small volume of the atom, and he called this central core the *nucleus*. According to the **Rutherford model,** the negatively charged electrons have almost no mass but occupy most of the volume of the atom, traveling about the nucleus in orbits much like the planetary orbits of our solar system. Rutherford's model is often referred to as the **solar system model** of the atom (see Figure 4-2).

Rutherford's atomic model was a step in the right direction, but it did not explain one important feature of atoms: they emit certain wavelengths of light when they are heated to high temperatures. We will return to this shortcoming of the Rutherford model after a brief discussion of light and its properties.

4.2 THE NATURE OF LIGHT

Light is a form of radiant energy, that is, energy given off by an object (Figure 4-3). Light, like all radiant energy, travels at a speed of 3.0×10^8 m/sec. Radiant energy moves in waves, similar to those generated on the water's surface when a rock is thrown into a pond. Thus light has a wave nature. A distinguishing characteristic of radiant energy is its **wavelength,** defined as the distance the wave moves before it starts to repeat itself. Or, in other words, wavelength is the distance between two equivalent points in the path traveled by the wave. Figure 4-4 illustrates the distance of a wavelength, which is represented by the Greek symbol λ (*lambda*).

Visible light stimulates nerve receptors in our eyes and is thus a visible form of energy. Light, in a more general sense, can also be a form of radiant energy that we cannot see, such as ultraviolet and infrared radiation. Light energy is defined as radiant energy having wavelengths in the range of $10^{-8} - 10^{-6}$ *m* (see Figure 4-5); this is equivalent to 10 – 1000 nm, where 1 nm (nanometer) $= 10^{-9}$ m. Visible light occupies the range of wavelengths between 380 nm and 780 nm. This wavelength range is referred to as the **visible spectrum.** Each specific color of visible light has its own specific wavelength.

Red light has the longest visible wavelength, and violet light has the shortest. Thus, light with a longer wavelength than red light is called *infrared* (beneath red) light, while light with a shorter wavelength than violet light is called *ultraviolet* (beyond violet) light. There are other kinds of radiant energy, such as radio waves, microwaves, and X rays, that are not classified as light because their wavelengths are not in the range of light.

White light is composed of all wavelengths in the visible range; metals give off white light when heated to temperatures high enough to make

FIGURE 4-3 A Light Bulb Gives Off Visible Radiant Energy

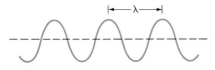

FIGURE 4-4 A Wavelength (λ) Is the Distance Between Two Equivalent Points in the Path Traveled by a Wave

Spectrum

An array of light waves ordered by wavelength; the plural of spectrum is **spectra.**

Infra-

Latin prefix meaning "below."

Ultra-

Latin prefix meaning "beyond."

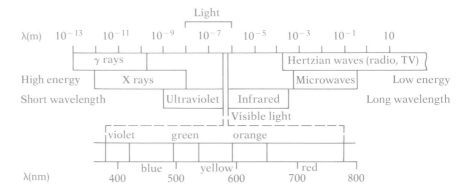

FIGURE 4-5 Wavelengths of Forms of Radiant Energy

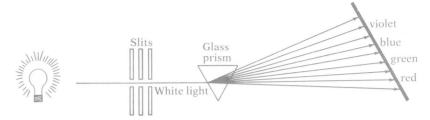

FIGURE 4-6 The Separation of White Light into Its Component Colors by a Prism

them glow. When white light passes through a prism, it separates into its component colors, as shown in Figure 4-6. Rain droplets serve as many small prisms that can separate sunlight into its component colors and form a rainbow.

Each wavelength of light has a specific amount of energy associated with it, and the energy of the light is inversely proportional to the light's wavelength (see appendix A for a review of proportionalities). Thus,

$$E \propto \frac{1}{\lambda}$$

Inserting a constant of proportionality (k) gives

$$E = k \left(\frac{1}{\lambda} \right)$$

MATH TIP

A proportionality can be converted to an equation by replacing the proportionality sign with an equals sign followed by a constant, k.

or,

$$E = \frac{k}{\lambda}$$

The constant k is actually composed of two constants,

$$k = hc,$$

where h is Planck's constant (named for Max Planck, the German physicist who first recognized the relationship between energy and wavelength) and c is the speed of light. The final relationship is

$$E = \frac{hc}{\lambda}$$

The inverse relationship between the energy of light and its wavelength tells us that the larger the value of λ, the smaller the energy of the light, and vice versa. Thus, ultraviolet light has much more energy than visible light and can harm living organisms. X rays and *gamma* radiation have even shorter wavelengths and hence possess even more energy; these forms of radiation have the potential to cause extreme damage to living tissues.

4.3 **EMISSION SPECTRA OF THE ELEMENTS**

The colored light that passes out of a prism is a continuous display of all the wavelengths of the entire range of visible light. Such a display is called a **continuous spectrum.** When individual atoms of an element are heated

FIGURE 4-7 The Atomic Emission Spectrum of Hydrogen. Each line corresponds to a wavelength of light emitted by hydrogen atoms.

to extremely high temperatures, light is given off by the atoms, but only one or a few colors of light are produced, and the colors are characteristic of the element. The reddish-orange color of neon lights is an example, and so are the colors produced by fireworks. The light emitted by a hot, gaseous element is called the **atomic emission spectrum** of the element.

Atomic emission spectra do not contain a continuous series of wavelengths but rather only a few, discrete wavelengths. For this reason, they are referred to as *discrete spectra* or *line spectra*. The number of wavelengths of light in atomic emission spectra increases with the complexity of atoms. Thus, elements of high atomic number have more wavelengths in their emission spectra than do atoms of low atomic number. Hydrogen, the simplest atom, has the simplest emission spectrum, shown in Figure 4-7.

4.4 THE BOHR ATOM

In the line spectra of elements, each line represents a discrete wavelength of light and hence a specific energy. Emission spectra were known when Rutherford proposed his solar system model, yet the model failed to explain them. Indeed, Rutherford's model suggested that electrons, being attracted to the protons in atomic nuclei, could easily be disturbed from their orbits, thus quickly losing energy and spiraling into the nucleus. In such a situation, the energy lost would produce a continuous spectrum of light rather than the characteristic line spectrum. Thus, Rutherford's model predicted very short lives for atoms and emission of continuous spectra; neither prediction fit the facts.

To resolve the difficulties with the Rutherford model, the Danish physicist Neils Bohr put forth a new atomic model in 1913. The **Bohr model** described the atom as having a nucleus at the center and definite paths (orbits) about the nucleus in which the electrons move (Figure 4-8). Bohr proposed that the electrons orbit on the surfaces of imaginary spherical shells of increasing radius layered one on top of another. Since each shell is a specific distance from the nucleus, each electron in a particular shell moves at a certain distance from the nucleus. Because an electron's energy depends on its distance from the nucleus, only certain well-defined energies are allowed for electrons.

Bohr's electron shells are designated as K, L, M, N, O, and so forth, starting with the one closest to the nucleus. An electron in any specific shell is closer to the nucleus and possesses less energy than an electron in

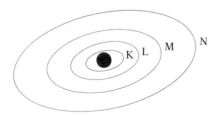

FIGURE 4-8 The Bohr Model of the Atom. The electron orbits are pictured as concentric rings about the nucleus.

SPECTRUM CHART

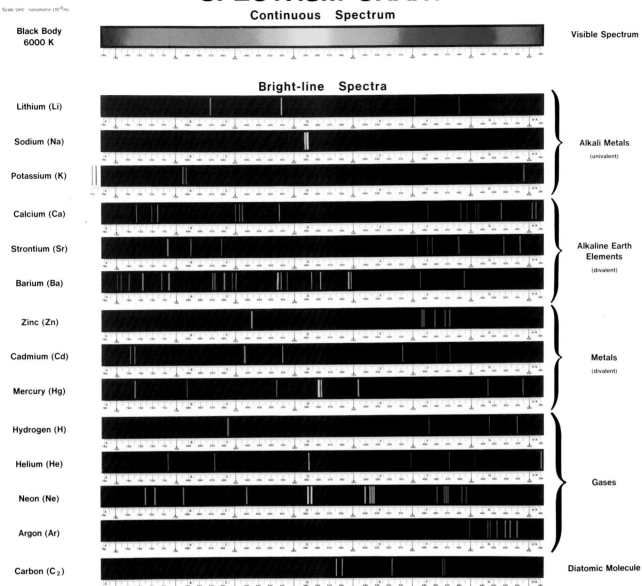

A Continuous Spectrum (top) and Emission Spectra of Some Elements (Courtesy of Sargent-Welch Scientific Co.)

Neon signs produce light because an electric current passes through a gas-filled tube, producing an emission spectrum. Different colors are produced by using different gases inside the tubes.

The brilliant colors of fireworks are produced when metal salts mixed with burning chemicals emit light.

Table salt (sodium chloride, NaCl) forms cubic crystals.

Particle accelerators such as this one in Geneva, Switzerland, are used to create trans-uranium elements. A particle accelerator speeds up positive monatomic ions to extremely high velocities and then directs them at target atomic nuclei. The particles collide with the target nuclei and transform them into nuclei of new elements.

A Synfuel Plant in Washington. Synfuel plants convert coal to gaseous or liquid fuels that produce less air pollution than does coal.

a higher shell. The Bohr model stipulated that an electron in an atom remains indefinitely in a specific shell; however, the atom can absorb radiant energy. When this happens, the electron moves to a higher shell, and the amount of energy absorbed corresponds exactly to the difference in energy between the two shells. The energy difference, ΔE, is given as follows:

$$\Delta E = E_2 - E_1,$$

where

ΔE = energy absorbed by the atom

E_2 = energy of higher shell

E_1 = energy of lower shell

According to the Bohr model, in the absence of radiant energy, the atom is stable and contains some minimum amount of energy; it is said to be in its **ground state.** After absorption of energy and elevation of one or more electrons to higher shells, the atom is said to be in an **excited state.**

An atom in an excited state is unstable; it achieves stability when its electrons return to their original shells and emit energy. The return of the electrons results in emission of exactly the same amount of energy the atom previously absorbed. Thus, an atom absorbs and re-emits energy in definite quantities, or **quanta.**

Let's see how this works with the hydrogen atom (see Figure 4-9). In its ground state, the only electron of the hydrogen atom occupies the K shell.

Quanta
Plural of **quantum,** Latin word meaning "quantity."

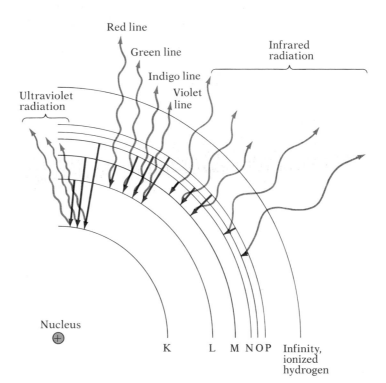

FIGURE 4-9 Diagram of Electron Transitions Leading to the Emission Spectrum of Hydrogen

If radiant energy is supplied to the atom, by an electric discharge for example, the electron moves to a shell farther away from the nucleus, perhaps M or higher. At this point, the hydrogen atom is in an excited state. Since excited states are unstable and have short lifetimes, the electron will return to the K shell in a short time. In returning to the lowest level, it may make short stops in the intervening shells. Each transition to a lower shell results in emission of a quantum of energy in the form of light, and each different quantum of energy emitted corresponds to a different wavelength of light, resulting in the line spectrum of hydrogen. In suggesting that atoms in their ground states are stable indefinitely and that specific wavelengths of light are emitted by excited atoms, the Bohr model proved to be an acceptable successor to the Rutherford model.

4.5 MODERN IDEAS OF ATOMIC STRUCTURE

The Bohr model was an important milestone because it introduced the idea that electrons exist in specific energy states in atoms and that a transition between any two states involves a definite amount, or quantum, of energy. However, this model, like its predecessors, proved to have its shortcomings. It was adequate for the hydrogen atom and for monatomic ions having only one electron, but it provided only a crude account for lines in the emission spectra of more complex atoms and ions. Its major deficiency was that it did not address the issue of repulsion between the electrons in an atom. As you shall see, modern ideas of atomic structure accommodate this repulsion of one electron for another.

de Broglie
(dee BROE-lee)

Eleven years after the Bohr model was proposed, a young French graduate student of physics, Louis de Broglie, made a radical hypothesis. It was known at the time that, in certain ways, energy possesses particlelike properties; for example, in certain situations, X-rays behave like a stream of particles. De Broglie argued that if energy could exist in packets, or quanta, and behave like particles at times, then particles of matter should behave like energy at times and should possess wave properties.

Heisenberg
(HIGH-sin-burg)

De Broglie's way of looking at matter and energy opened new avenues for many scientists, one of whom was the German physicist Werner Heisenberg. Heisenberg accepted the fact that a subatomic particle can exhibit properties of a wave, and he asked the question: Is it possible to specify the location of the particle exactly? The answer is no, no more than it is possible to speak of the precise location of a wave. A wave extends into space, and although its wavelength can be measured, the wave travels on indefinitely. This line of reasoning led to the **Heisenberg uncertainty principle** in 1927. When applied to electrons, it states that it is simply not possible to know both the exact energy of an electron and its location in space.

Heisenberg's uncertainty principle marked the beginning of the development of our modern atomic model. Since the idea of an electron existing at a specified energy level is still valid, then the uncertainty principle tells us that we cannot know its location with certainty. Instead, we must describe electron locations on the basis of statistics. In the modern model

81

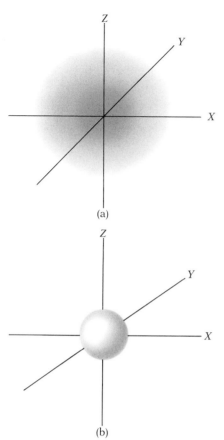

(a)

(b)

FIGURE 4-10 (a) Cloud Density Map of the Hydrogen Atom in Its Ground State. (b) Boundary Surface Diagram of the Hydrogen Atom in Its Ground State, Illustrating the 90% Probability Region for the Electron.

MATH TIP
Since $2n^2 = 2(n)(n)$, if $n = 1$, then $2n^2 = 2(1)(1) = 2$; and if $n = 2$, then $2n^2 = 2(2)(2) = 8$.

---

of the atom, we speak of the probability of finding an electron in a certain region of space at a given instant. The region of space near an atomic nucleus where an electron is most likely to be found is called an **atomic orbital.** In contrast to an orbit, an orbital does not define a specific path in which an electron moves. Instead, an orbital defines a region of high probability for finding an electron. Any given electron is free to move about as it chooses, but the volume of space it occupies most of the time is its atomic orbital. Since an electron moves at the speed of light, 3.0×10^8 m/sec, such rapid motion spreads out the charge of the electron into a cloud of charge. The cloud is denser in some regions than in others, and thus the cloud density is proportional to the probability of finding an electron. **Cloud density maps** are now used to illustrate atomic orbitals. The cloud density map of the hydrogen atom in its ground state is shown in Figure 4-10(a). Figure 4-10(b), called a **boundary surface diagram,** illustrates the region of space around the hydrogen nucleus where there is a 90% chance of finding the ground-state electron.

4.6 ENERGY LEVELS OF ELECTRONS

Like the Bohr model, modern atomic theory considers the electron distribution of an atom containing a number of electrons to be divided into **shells.** A shell is a region where the probability of finding an electron is high. Each shell is known as a **principal energy level** and is designated by the symbol n. The value of n is a positive whole number: 1, 2, 3, 4, . . . ; these values correspond to the letter designations K, L, M, N, . . . , of the Bohr model. The higher the value of n, the higher the energy of the electrons in the shell and the farther the shell is from the nucleus. Each shell can accommodate only a certain number of electrons; this number is $2n^2$. Thus, when $n = 1$, the shell capacity is 2; when $n = 2$, the shell capacity is 8, and so on. Shell capacities are summarized in Table 4-1.

4.7 ENERGY SUBLEVELS

Each shell consists of one or more **subshells** or **energy sublevels,** and the number of energy sublevels in a principal energy level is the same as the value of n. In other words, when $n = 1$, there is one energy sublevel; when $n = 2$, there are two, and so on. Each sublevel consists of one or more atomic orbitals. Atomic orbitals are designated by the lowercase letters s,

TABLE 4-1 The Electron Capacities of Shells

Principal energy level (n)	1	2	3	4	5
Letter designation	K	L	M	N	O
Capacity ($2n^2$)	2	8	18	32	50

TABLE 4-2   Subshells and Types of Orbitals in the First Five Principal Energy Levels

Principal Energy Level (n)	Number of Subshells (n)	Types of Orbitals
1	1	$1s$
2	2	$2s$
		$2p$
3	3	$3s$
		$3p$
		$3d$
4	4	$4s$
		$4p$
		$4d$
		$4f$
5	5	$5s$
		$5p$
		$5d$
		$5f$
		$5g$

p, d, f, g, h, and so forth. (The first four letters — s, p, d, and f — were first used as abbreviations of descriptions of lines in emission spectra: *s*harp, *p*rincipal, *d*iffuse, and *f*undamental.) In any one principal energy level, an s orbital has lower energy than a p orbital, which in turn has lower energy than a d orbital, and so on. Thus, for any principal energy level, the energy of the atomic orbitals increases in the order s, p, d, f, ... The principal energy level and orbital type must both be specified when referring to a specific orbital; for example, $1s$, $2p$, $3d$. Table 4-2 summarizes the subshells and types of orbitals for the first five principal energy levels.

Each type of atomic orbital has its own characteristic shape. s orbitals (Figure 4-11) are spherical, while p orbitals are shaped like dumbbells. The dumbbell-shaped orbitals can point into any one of three directions in space, so p orbitals exist in sets of three: one for the x-axis (p_x), one for the y-axis (p_y), and one for the z-axis (p_z) (see Figure 4-12). The three p orbitals in any one principal energy level are of equal energy content. Each of the two pear-shaped sections of a p orbital is called a *lobe;* an electron in a p orbital is equally likely to be found in either lobe.

The shapes of d and f orbitals are more complex than those of s and p orbitals. Their geometry is such that d orbitals exist in sets of five, while f orbitals exist in sets of seven. In any specific principal energy level, all d orbitals are of equal energy, as are all f orbitals. The shapes of d orbitals are shown in Figure 4-13. We will not be concerned with shapes of f orbitals.

Each atomic orbital can hold no more than two electrons. Electrons, with their like charges, have a built-in repulsion for each other, and it may seem surprising that two electrons can occupy the same atomic orbital. However, electrons behave as if they were spinning about their axes, much like the earth spins about its axis. This behavior generates a mag-

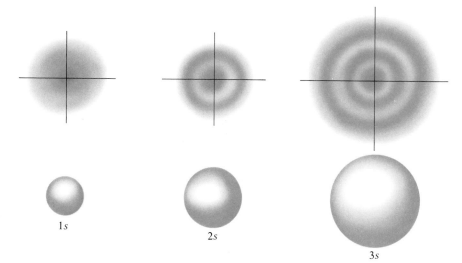

FIGURE 4-11 Shapes of *s* Orbitals. The upper figures are cloud density diagrams, while the lower figures are boundary surface diagrams for 90% probability regions.

netic field, and thus electrons behave like tiny magnets. There are only two directions of spin, clockwise and counterclockwise. Two electrons spinning in opposite directions are similar to two opposite magnetic poles (see Figure 4-14); they attract each other enough to be able to occupy the same atomic orbital despite their charge repulsions. Such oppositely spinning electrons in the same atomic orbital are said to be **paired,** and each atomic orbital can accommodate two paired electrons.

We noted earlier that the electron capacity of a principal energy level is $2n^2$; this is based on the number of orbitals and the fact that each orbital can contain a maximum of two electrons (see Table 4-3). Thus, for $n = 1$, there is one atomic orbital, with a capacity of 2 electrons; the first principal energy level therefore has a total electron capacity of 2. For $n = 2$, there are four atomic orbitals, each with a capacity of 2 electrons; hence, the second principal energy level has a total electron capacity of 8. Electron capacities of higher principal energy levels are calculated similarly.

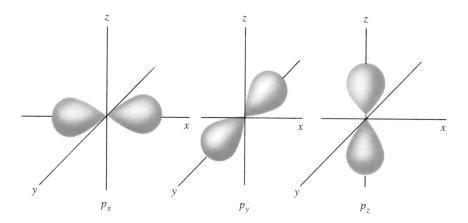

FIGURE 4-12 Boundary Surface Diagrams for 90% Probability Regions of the Three *p* Orbitals

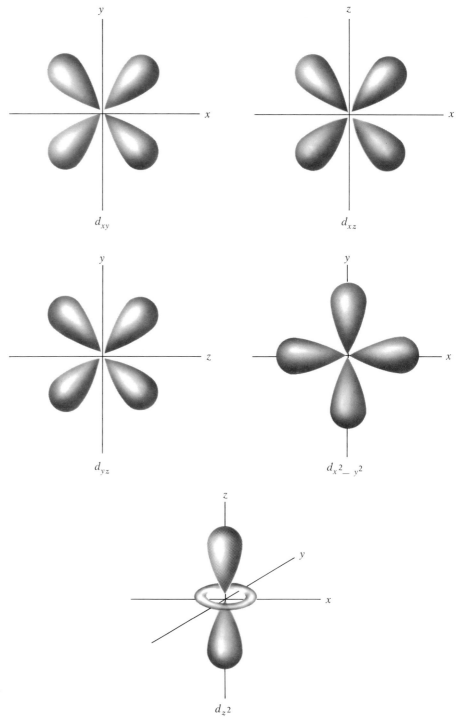

FIGURE 4-13 Boundary Surface Diagrams for 90% Probability Regions of the Five *d* Orbitals

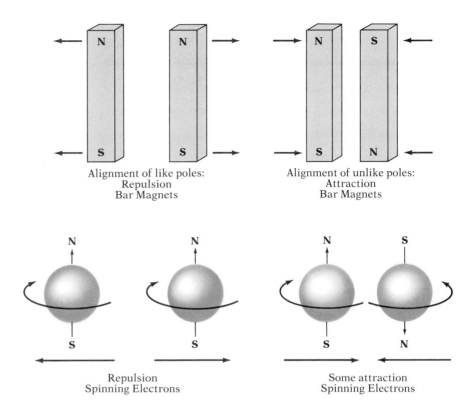

FIGURE 4-14 Magnetic Fields of Spinning Electrons

4.8 ELECTRON CONFIGURATIONS

The arrangement of electrons in the atomic orbitals of an atom is called the **electron configuration** of the atom. We can determine the ground-state electron configurations of the first 18 elements by assuming that the electrons occupy the shells and atomic orbitals within each shell in order of increasing energy content.

TABLE 4-3	Electron Capacities of the First Three Principal Energy Levels	
Principal Energy Level (n)	Subshells and Orbitals	Electron Capacity
1	$1s$	2
2	$2s$	2
	$2p_x 2p_y 2p_z$	$\underline{6}$
		8
3	$3s$	2
	$3p_x 3p_y 3p_z$	6
	$3d_{xy} 3d_{xz} 3d_{yz} 3d_{x^2-y^2} 3d_{z^2}$	$\underline{10}$
		18

For the simplest atom, hydrogen, there is one electron; it occupies the lowest orbital of the lowest principal energy level. Therefore, the single electron of the ground-state hydrogen atom is in the $1s$ orbital. We write its electron configuration as $1s^1$.

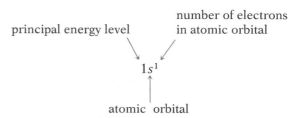

The helium atom has two electrons, which can both be accommodated by the $1s$ orbital; therefore, the electron configuration of helium is $1s^2$. Lithium has three electrons, but there is only enough space in the first principal energy level for two electrons; the third must go into the orbital of lowest energy in the second principal energy level, so the electron configuration of Li is $1s^2 2s^1$. As we consider atoms of increasing atomic number (and thus increasing number of electrons), we can make our way through the first 10 elements, at which point the second principal energy level becomes filled, as shown in Figure 4-15. (Note that we will not distinguish among the three equivalent p orbitals, the five equivalent d orbitals, or the seven equivalent f orbitals in writing electron configurations.) The element at the end of the first period on the periodic table, He, concludes the filling of the first principal energy level. Similarly, the element at the end of the second period, Ne, concludes the filling of the second principal energy level.

The developing pattern suggests that the first element in a period (H in period 1 and Li in period 2) marks the beginning of filling a new shell of electrons and the last element in the period signals completion of the shell. At first glance, the third period appears to follow this pattern. Thus, the electron configuration for Na, at the beginning of period 3, is $1s^2 2s^2 2p^6 3s^1$, while that for Ar, at the end of the period, is $1s^2 2s^2 2p^6 3s^2 3p^6$. However, there are only 8 electrons in the third principal energy level of Ar, and the third principal energy level has a capacity of 18 electrons, according to the formula $2n^2$. Thus, the last element in the third period does not conclude the filling of the third principal energy level. However, the first element in period 4, K, has the electron configuration $1s^2 2s^2 2p^6 3s^2 3p^6 4s^1$; it has started to fill the fourth principal energy level before the third level has been filled. What does this mean? It means simply that the $4s$ orbital is at a lower energy state than the $3d$ orbitals, even though the average energy of the fourth principal energy level is greater than that of the third principal energy level. That is, these two principal energy levels overlap each other in energy content. Once the $4s$ orbital is filled, filling of the $3d$ orbitals begins; hence, the electron configuration for Ca, the second element in period 4, is $1s^2 2s^2 2p^6 3s^2 3p^6 4s^2$ and that for Sc, the third element in the period, is $1s^2 2s^2 2p^6 3s^2 3p^6 4s^2 3d^1$.

The $4s$ and $3d$ orbitals are not the only ones that seem out of order due to the overlap in energy content of principal energy levels; the $5s$ and $4d$

Period 1

    H $1s^1$

    He $1s^2$

Period 2

    Li $1s^2 2s^1$

    Be $1s^2 2s^2$

    B $1s^2 2s^2 2p^1$

    C $1s^2 2s^2 2p^2$

    N $1s^2 2s^2 2p^3$

    O $1s^2 2s^2 2p^4$

    F $1s^2 2s^2 2p^5$

    Ne $1s^2 2s^2 2p^6$

Period 3

    Na $1s^2 2s^2 2p^6 3s^1$

    Mg $1s^2 2s^2 2p^6 3s^2$

    Al $1s^2 2s^2 2p^6 3s^2 3p^1$

    Si $1s^2 2s^2 2p^6 3s^2 3p^2$

    P $1s^2 2s^2 2p^6 3s^2 3p^3$

    S $1s^2 2s^2 2p^6 3s^2 3p^4$

    Cl $1s^2 2s^2 2p^6 3s^2 3p^5$

    Ar $1s^2 2s^2 2p^6 3s^2 3p^6$

FIGURE 4-15 Electron Configurations of the First Eighteen Elements

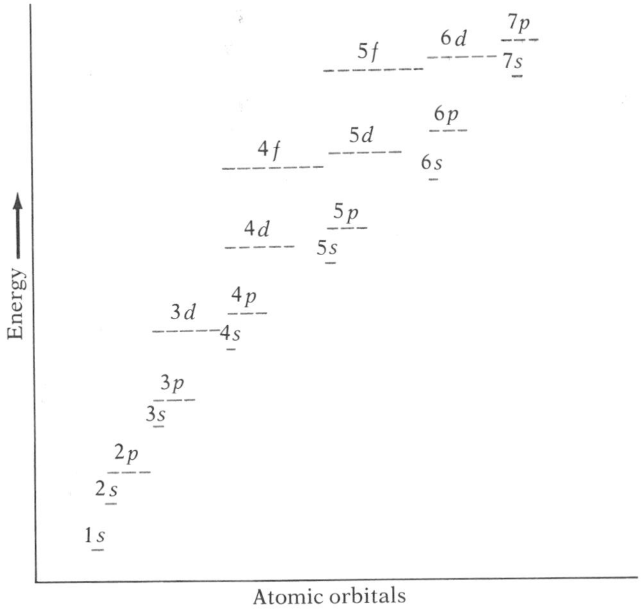

$$7p$$
$$6d \quad \overline{}$$
$$5f \quad \overline{---} \quad 7s$$

$$6p$$
$$5d \quad \overline{--}$$
$$4f \quad \overline{----} \quad 6s$$

$$5p$$
$$4d \quad \overline{---} \quad 5s$$

$$3d \quad 4p$$
$$\overline{----} \quad 4s$$

$$3p$$
$$3s$$

$$2p$$
$$2s$$

$$1s$$

Energy →

Atomic orbitals

FIGURE 4-16 Relative Energies of Atomic Orbitals

orbitals are also out of order, as are the $5p$ and $4f$ orbitals and several others. These energy overlaps are illustrated in Figure 4-16.

Rather than trying to remember those particular atomic orbitals that add electrons out of order, it is much simpler to rely on the diagram in Figure 4-17. You can easily construct this diagram at any time you need it. First, write all of the atomic orbitals of each principal energy level in horizontal rows, starting at the bottom with $1s$. Align all of the s orbitals vertically, all of the p orbitals vertically, and so on. Then draw a stairstep up the right side where the rows end, and draw a straight vertical line just to the left of the column of s orbitals. Finally, draw a diagonal arrow from each corner of the stairstep through the orbital notations to the vertical line. The lowest arrow will pass through the first orbital to be filled, the next arrow through the second orbital to be filled, and so on. Practice drawing this diagram so that you can reproduce it at will. (It is especially helpful during exams.) It is an easy method for deriving electron configurations without really having to remember the order of filling of the atomic orbitals.

A convenient short form reduces the amount of space needed to write most electron configurations. In this short form, the atomic symbol of the element at the end of the immediately preceding period is shown in brackets, followed by symbols for orbitals present beyond those of the element in brackets. For example, the electron configuration of Li, $1s^2 2s^1$, can be written as $[\text{He}]2s^1$, and that of C, $1s^2 2s^2 2p^2$, can be written as $[\text{He}]2s^2 2p^2$. Other examples are

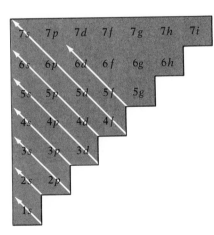

FIGURE 4-17 The Order of Filling of Atomic Orbitals

O $1s^2 2s^2 2p^4$ or $[\text{He}]2s^2 2p^4$

Na $1s^2 2s^2 2p^6 3s^1$ or $[\text{Ne}]3s^1$

Si $1s^2 2s^2 2p^6 3s^2 3p^2$ or $[\text{Ne}]3s^2 3p^2$

A complete list of electron configurations is given inside the back cover of this book. Note that there are some deviations from predicted configurations, especially for elements where d and f orbitals are half-filled. For example, Cr (atomic number 24) has the argon core of electrons and then $4s^1 3d^5$ instead of the expected $4s^2 3d^4$. Other exceptions are Cu (atomic number 29) and La (atomic number 57). Although the exceptional configurations have some significance, we will not dwell on them. Use the table of electron configurations for reference and for checking yourself in writing electron configurations, but use it only for these purposes. It is important to practice deriving electron configurations for yourself.

EXAMPLE 4.1   Write the electron configuration of tin, Sn.

Solution   From the list of elements inside the front cover, we find that Sn has an atomic number of 50; thus, we must place 50 electrons in atomic orbitals, using a diagram like the one in Figure 4-17 to determine the order of filling. We must also remember the total electron capacities of orbital types: s, 2; p, 6; d, 10; f, 14; and so forth. The electron configuration of Sn then is as follows:

$$1s^2 2s^2 2p^6 3s^2 3p^6 4s^2 3d^{10} 4p^6 5s^2 4d^{10} 5p^2 \quad \text{or} \quad [\text{Kr}]5s^2 4d^{10} 5p^2$$

As a check, we count electrons in the electron configuration to see that we have accounted for all 50 electrons of the Sn atom.

EXERCISE 4.1   Write the electron configuration of each of the following elements.

a. Rb   **b.** Cl   **c.** Ba   **d.** Xe   ■

Electron configurations of monatomic ions can be written by using the same approach used for atoms. The only difference is that a monatomic ion has an imbalance of electrons and protons, as indicated by its charge. Thus, K^+ has one fewer electron than its parent atom. The electron configuration for K (atomic number 19) is $1s^2 2s^2 2p^6 3s^2 3p^6 4s^1$ (or $[\text{Ar}]4s^1$), with a total of 19 electrons, while the electron configuration of K^+ is $1s^2 2s^2 2p^6 3s^2 3p^6$ (or $[\text{Ar}]$), with a total of only 18 electrons.

EXAMPLE 4.2   Write the electron configuration of Mg^{2+}.

Solution   Using the list of elements, we find that Mg has an atomic number of 12; therefore, Mg^{2+} has a total of 10 electrons ($12e^- - 2e^-$ lost) to be accounted for in its electron configuration. Using a diagram like that in Figure 4-17 to determine the order of filling and keeping in mind the total electron capacities of orbitals, we determine the following configuration for Mg^{2+}:

$$1s^2 2s^2 2p^6 \quad \text{or} \quad [\text{Ne}]$$

We can check our answer by counting electrons to see that they add up to 10.

EXERCISE 4.2   Write the electron configuration of each of the following ions.

a. Ca^{2+} **b.** Br^- **c.** Sr^{2+} **d.** S^{2-} ■

4.9 QUANTUM NUMBERS *(optional)*

Max Planck's work on the relationship of energy and wavelength of light (section 4.2) caused scientists to begin thinking of energy as existing in definite quantities; Planck called such a quantity of energy a **quantum.** When it was found that an atom absorbs energy and emits it in definite quantities, or quanta, it became possible to think of atomic structure in terms of energy quanta. This way of thinking about atomic structure, combined with de Broglie's ideas about the wave nature of electrons (section 4.5), led to a mathematical description of the atom called **wave mechanics** or **quantum mechanics.** In quantum mechanics, the location of each electron of an atom is identified by four **quantum numbers,** which indicate principal energy level, orbital type, orbital orientation, and electron spin.

The principal quantum number, n, corresponds to a principal energy level of an atom. The value of n is a positive whole number:

$$n = 1, 2, 3, \ldots$$

Each atomic orbital in a shell has an **orbital type quantum number, l.** The values of l are determined by the value of the principal quantum number of the energy level to which the orbital belongs. The value of l is also a positive whole number, and it is related to n as follows:

$$l = 0, 1, 2, 3, \ldots, n - 1$$

Thus, when n is 1, the only value of l is 0, and there is only one type of atomic orbital at that energy level, an s orbital. When $n = 2$, there are two types of orbitals, having l values of 0 and 1, in order of increasing energies; these are the s and p orbitals. When $n = 3$, the three types of orbitals have values of 0, 1, and 2, corresponding to s, p, and d orbitals. Thus, the number of values allowed for l determines the number of orbital types in the energy sublevel. (Note that the maximum value l can have is $n - 1$.)

EXAMPLE 4.3 What are the values of l and types of orbitals when $n = 4$?

Solution Since $l = 0, 1, 2, 3, \ldots, n - 1$,
then the values allowed for l are

$$l = 0$$
$$l = 1$$
$$l = 2$$
$$l = 3$$

The four values for l signify four atomic orbitals of increasing energy; therefore, the four orbital types must be s, p, d, and f.

EXERCISE 4.3 What are the values of l and types of orbitals when $n = 5$? ∎

Each orbital within a sublevel is also identified by a **magnetic orbital quantum number,** m_l, which describes the number of spatial orientations a particular type of orbital may have. (This quantum number is sometimes

TABLE 4-4 Relationships among Values of n, l, and m_l, through $n = 4$

n	l	Orbital Designation	m_l	Number of Orbitals
1	0	$1s$	0	1
2	0	$2s$	0	1
	1	$2p$	1, 0, −1	3
3	0	$3s$	0	1
	1	$3p$	1, 0, −1	3
	2	$3d$	2, 1, 0, −1, −2	5
4	0	$4s$	0	1
	1	$4p$	1, 0, −1	3
	2	$4d$	2, 1, 0, −1, −2	5
	3	$4f$	3, 2, 1, 0, −1, −2, −3	7

referred to as the **orientation quantum number**.) For any sublevel, the values of m_l are as follows:

$$m_l = +l, +(l-1), +(l-2), \ldots, 0, \ldots -(l-2), -(l-1), -l$$

Thus, for $l = 0$, there is only a single orientation for the s orbital in the sublevel. When $l = 1$, there are three values for m_l: +1, 0, and −1; this means that there are three orientations, one for each of the three p orbitals. When $l = 2$, m_l can be +2, +1, 0, −1, −2, corresponding to five orientations, one for each of the five d orbitals. Notice that values of m_l are derived from values of l, which in turn are derived from values of n. Each orbital in an atom is thus identified by a set of values for n, l, and m_l. The relationships among values of n, l, and m_l are illustrated in Table 4-4.

EXAMPLE 4.4   What are the values of l and m_l when $n = 3$? What are the types of orbitals, and how many of each exist?

Solution   For $n = 3$,

$$l = 0$$
$$l = 1$$
$$l = 2$$

The three values of l indicate three orbital types: s, p, and d.

$$m_l = +l, +(l-1), +(l-2), \ldots, 0, \ldots, -(l-1), -(l-2), -l$$

For $l = 0$,

$$m_l = 0$$

For $l = 1$,

$$m_l = +l = 1$$
$$m_l = +(l-1) = +(1-1) = 0$$
$$m_l = -l = -1$$

For $l = 2$,

$$m_l = +l = 2$$
$$m_l = +(l - 1) = +(2 - 1) = 1$$
$$m_l = +(l - 2) = +(2 - 2) = 0$$
$$m_l = -(l - 1) = -(2 - 1) = -1$$
$$m_l = -l = -2$$

For $l = 0$, the single value for m_l indicates one s orbital. For $l = 1$, the three values for m_l indicate three p orbitals. For $l = 2$, the five values for m_l indicate five d orbitals.

EXERCISE 4.4 What are the values of l and m_l when $n = 4$? What are the types of orbitals, and how many of each exist? ∎

The fourth quantum number, the **spin quantum number**, m_s, indicates the spin of the electron. Since there are only two directions for an electron to spin, m_s can have only two values: $+\frac{1}{2}$ and $-\frac{1}{2}$.

By use of quantum numbers, each electron in an atom can be described by a unique set of four numbers:

1. n gives the principal energy level.

$$n = 1, 2, 3, \ldots$$

TABLE 4-5 Electron Configurations of the Ground States of the First Six Elements

	Quantum Numbers of Electrons				Electron Configuration
	n	l	m_l	m_s	
$_1$H	1	0	0	$+\frac{1}{2}$	$1s^1$
$_2$He	1	0	0	$+\frac{1}{2}$	$1s^2$
	1	0	0	$-\frac{1}{2}$	
$_3$Li	1	0	0	$+\frac{1}{2}$	$1s^2 2s^1$
	1	0	0	$-\frac{1}{2}$	
	2	0	0	$+\frac{1}{2}$	
$_4$Be	1	0	0	$+\frac{1}{2}$	$1s^2 2s^2$
	1	0	0	$-\frac{1}{2}$	
	2	0	0	$+\frac{1}{2}$	
	2	0	0	$-\frac{1}{2}$	
$_5$B	1	0	0	$+\frac{1}{2}$	$1s^2 2s^2 2p^1$
	1	0	0	$-\frac{1}{2}$	
	2	0	0	$+\frac{1}{2}$	
	2	0	0	$-\frac{1}{2}$	
	2	1	$+1$	$+\frac{1}{2}$	
$_6$C	1	0	0	$+\frac{1}{2}$	$1s^2 2s^2 2p^2$
	1	0	0	$-\frac{1}{2}$	
	2	0	0	$+\frac{1}{2}$	
	2	0	0	$-\frac{1}{2}$	
	2	1	$+1$	$+\frac{1}{2}$	
	2	1	0	$+\frac{1}{2}$	

2. l gives the type of the orbital.

$$l = 0, 1, 2, \ldots , n - 1$$

3. m_l designates the orientation of the orbital.

$$m_l = +l, +(l - 1), +(l - 2), \ldots , 0, \ldots , -(l - 2), -(l - 1), -l$$

4. m_s refers to the spin of the electron.

$$m_s = +\frac{1}{2} \quad \text{or} \quad -\frac{1}{2}$$

Table 4-5 illustrates how electron configurations can be described by use of quantum numbers.

SUMMARY

The earliest scientific model of the atom, the **Thomson plum pudding model,** described the atom as a sphere of uniformly distributed positive charge with electrons stuck in it. The next atomic model, the **Rutherford model,** proposed that electrons move in orbits about the nucleus, but it failed to explain the basis of atomic emission spectra.

Light is radiant energy in the wavelength range of $10^{-8} - 10^{-6}$ m. A **wavelength** is the distance between two equivalent points in the path traveled by the wave. Each specific color in visible light corresponds to a specific wavelength of light. The energy of light is related to its wavelength by the equation $E = hc/\lambda$. When atoms of an element are heated to extremely high temperatures, the light given off by the atoms produces a **line spectrum;** such spectra are called **atomic emission spectra.**

The **Bohr model,** the third atomic model, described the atom as having a nucleus at the center and electron orbits around the nucleus. The Bohr model explained atomic emission spectra on the basis of electron transitions between orbits of different energies. The **modern atomic model** is based on the fact that electrons have wave nature. The **Heisenberg uncertainty principle** states that it is not possible to know both the exact energy of an electron and its location in space. The region of space near an atomic nucleus where an electron is likely to be found is called an **atomic orbital.** Cloud density maps and boundary surface diagrams are used to illustrate atomic orbitals.

The electrons of an atom containing a number of electrons are distributed among **shells,** which are regions where the probability of finding an electron is high. Each shell is a **principal energy level** and consists of one or more energy sublevels. Each energy sublevel contains one or more **atomic orbitals,** which are designated by the letters s, p, d, f, g, h, and so forth. s orbitals are spherical, p orbitals are shaped like dumbbells, and other orbitals have more complex shapes. Each atomic orbital can hold no more than two electrons. The designation of electron distribution in the atomic orbitals of an atom is called the **electron configuration** of the atom. Electron configurations are derived on the basis of atomic number, principal energy levels, energy sublevels, orbitals, order of orbital filling, and electron capacities of orbitals.

Quantum numbers are numbers that describe electron locations in atoms. The four quantum numbers indicate principal energy level, orbital type, orbital orientation, and electron spin.

STUDY QUESTIONS AND PROBLEMS

(More difficult questions and problems are marked by an asterisk.)

EARLY MODELS OF THE ATOM

1. What is a scientific model?
2. Describe the Thomson plum pudding model of the atom. Did this model acknowledge that protons are particles of matter? Explain your answer.
3. Describe the Rutherford model of the atom. How was it an improvement over the Thomson model?
4. What was the major deficiency of the Rutherford model of the atom?

THE NATURE OF LIGHT

5. Define each of the following terms:
   a. Radiant energy
   b. Light
   c. Wavelength
   d. Visible spectrum
   e. Infrared light
   f. Ultraviolet light
   g. Planck's constant
6. How do rainbows occur?
7. Describe how the energy of light is related to its wavelength. Does infrared light have more energy than ultraviolet light? Explain your answer.
8. Why do ultraviolet light, X-rays, and *gamma* radiation have the potential of damaging living tissues?

EMISSION SPECTRA OF THE ELEMENTS

9. Define each of the following terms:
   a. Continuous spectrum
   b. Atomic emission spectrum
   c. Line spectrum
*10. Tungsten filaments in electric light bulbs give off a continuous spectrum. How do you think this happens?
11. Explain how a continuous spectrum differs from a line spectrum. Which kind of spectrum is emitted by neon lights? Explain your answer.
12. Describe the conditions that would cause copper metal to emit a line spectrum. How would you go about demonstrating this?

THE BOHR ATOM

13. Define each of the following terms:
   a. Bohr atomic model
   b. Electron shell
   c. Ground state of an atom
   d. Excited state of an atom
   e. Quantum of energy

14. According to the Rutherford model, what kind of spectrum should a hot, gaseous element emit? Explain your answer.
15. Describe the Bohr model of the atom. How was it an improvement over the Rutherford model?
16. How did the Bohr model account for atomic emission spectra?
17. How does an atom change from its ground state to an excited state? Does it remain in an excited state indefinitely? How does it revert to its ground state?
18. What do the letters K, L, M, N, . . . designate?

MODERN IDEAS OF ATOMIC STRUCTURE

19. Define each of the following terms:
    a. De Broglie hypothesis
    b. Heisenberg uncertainty principle
    c. Atomic orbital
    d. Cloud density map
    e. Boundary surface diagram
20. What was the major deficiency of the Bohr atomic model?
21. What was de Broglie's new way of considering matter?
22. Why is it not possible to know both the exact energy and location of an electron?
23. How does an orbital differ from an orbit?

ENERGY LEVELS OF ELECTRONS

24. In terms of modern atomic theory, what is a principal energy level?
25. How does one calculate the electron capacity of a shell? What is the electron capacity of each of the following shells?
    a. 1   b. 2   c. 3   d. 4

ENERGY SUBLEVELS

26. What is an energy sublevel in an atom? How does it differ from an orbital?
27. How many energy sublevels are in a particular principal energy level?
28. What are the letter designations of atomic orbitals? In what order of atomic orbitals does energy increase?
29. Sketch a boundary surface diagram of an s orbital. How is a $2s$ orbital different from a $1s$ orbital?
30. How many p orbitals can exist in a given principal energy level?
31. Sketch boundary surface diagrams of a set of p orbitals. How does a p_x orbital differ from a p_y orbital? From a p_z orbital?
32. How does a $3p$ orbital differ from a $2p$ orbital?
33. Does a $4p_x$ orbital contain more energy than a $4p_y$ orbital? Explain your answer.
34. What is the maximum number of electrons that an orbital can hold? Why?
35. What is the maximum number of s electrons that can exist in any one principal energy level? How many p electrons? How many d electrons? How many f electrons? Explain your answers.

ELECTRON CONFIGURATIONS

36. What is an electron configuration of an atom?
37. Using only the periodic table or list of elements, write the electron configuration of each of the following atoms.
    a. B   c. Ar   e. Cd
    b. S   d. V   f. Re
38. Using only the periodic table or list of elements, write the electron configuration of each of the following ions.
    a. I^- b. Ra^{2+} c. Se^{2-} d. Al^{3+}
39. What is the common feature of all the electron configurations of elements in Group IA? In Group VIIIA?

QUANTUM NUMBERS

40. What quantum number does each of the following symbols represent?
    a. n b. l c. m_l d. m_s
41. What does each quantum number symbolized in question 40 signify?
42. What are the values of m_l for a d orbital? How many orbitals are there?
43. When $l = 3$, how many values does m_l have? How many orbitals are there?
44. How many orbitals are there in the $n = 4$ principal energy level?
*45. Each of the following sets of quantum numbers (n, l, m_l, m_s) describes an electron in an atom. Give the principal energy level and orbital designated by each.
    a. $3, 2, 1, -\frac{1}{2}$ c. $5, 1, -1, +\frac{1}{2}$
    b. $4, 2, 2, +\frac{1}{2}$ d. $4, 2, -1, -\frac{1}{2}$
*46. Using quantum numbers, explain the absence of d orbitals in the second principal energy level.

CHEMICAL PERIODICITY

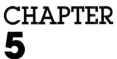

CHAPTER
5

OUTLINE

5.1 The Periodic Law
5.2 The Modern Periodic Table
5.3 Periods of Elements
5.4 Groups of Elements
5.5 A Survey of the Representative
      Elements
5.6 The Transition Elements
      Perspective: Is There an End to the
      Periodic Table?
5.7 Atomic Size
5.8 Ionization Energy
5.9 Electron Affinity
5.10 The Chemical Behavior of Metals
      and Nonmetals
      Summary
      Study Questions and Problems

As soon as scientists recognized the unique nature of the elements, they searched for ways of classifying these fundamental substances. One of the first to propose a classification system was Johann Wolfgang Dobereiner, a German chemist. Born in 1780 as the son of a coachman, Dobereiner had little opportunity for formal schooling as a boy, but when he was apprenticed to an apothecary, he developed a strong interest in science. He eventually received an advanced degree from a local university and became supervisor of science instruction there. In 1817, Dobereiner found that the atomic weight of strontium is midway between those of calcium and barium. As he continued to study the elements, he found several other such "triads"; among them were chlorine, bromine, and iodine, and lithium, sodium, and potassium. In each triad, the average of the largest and smallest atomic weights was approximately equal to the atomic weight of the middle element. Thus, he proposed that elements be classified as triads. Unfortunately, Dobereiner could not find enough triads to convince other scientists of his hypothesis, and his system of classification was regarded only as an interesting curiosity. However, the quest to classify the elements continued.

In 1864, John Alexander R. Newlands, an English industrial chemist, proposed classifying the elements in order of increasing atomic weight. He pointed out that every eighth element had similar properties, and by analogy to the musical scale, his observations were referred to as the "law of octaves." Newlands' classification system was at first ignored and then later ridiculed by his fellow scientists, but it was eventually acknowledged to be of great importance in the development of the periodic table.

Meanwhile, in St. Petersburg, Russia, a chemistry professor by the name of Dmitry Ivanovich Mendeleyev could not find a suitable textbook for the course he was teaching and began writing his own. While writing the book, which turned out to be a classic textbook, Mendeleyev searched for ways to classify the elements. The classification system he devised is the basis for the modern periodic table, and while classifying the elements, he formulated the periodic law.

Triad

A group of three closely related things; derived from the Greek words *tri-*, meaning "three," and *-ad*, meaning "related to."

5.1　THE PERIODIC LAW

Mendeleyev, like Newlands, had listed the elements in order of increasing atomic weight, and he found a periodic recurrence of similar properties of the elements; we now call this **chemical periodicity.** In 1869, Mendeleyev proposed the **periodic law,** by which the elements arranged by atomic weights show a periodic change in properties. Lother Meyer, a German chemist, had independently reached a similar conclusion, but since his publication appeared after Mendeleyev's, we usually associate Mendeleyev's name with the periodic table.

Mendeleyev's periodic table, illustrated in Figure 5-1, contained all of the 63 elements known at the time, arranged in 17 columns. Each of the columns contained elements that were similar to one another, and he boldly left gaps for elements that had not been discovered. He was so confident of his classification system that he even predicted properties of

Period	I a	I b	II a	II b	III a	III b	IV a	IV b	V a	V b	VI a	VI b	VII a	VII b	VIII a	VIII b		(0)
1	H 1.0																	He 4.0
2	Li 6.9		Be 9.0		B 10.8		C 12.0		N 14.0		O 16.0		F 19.0					Ne 20.2
3	Na 23.0		Mg 24.3		Al 27.0		Si 28.1		P 31.0		S 32.1		Cl 35.5					Ar 39.9
4	K 39.1		Ca 40.1		Sc 45.0		Ti 47.9		V 50.9		Cr 52.0		Mn 54.9		Fe 55.8	Co 58.9	Ni 58.7	
4		Cu 63.5		Zn 65.4		Ga 69.7		Ge 72.6		As 74.9		Se 79.0		Br 79.9				Kr 83.8
5	Rb 85.5		Sr 87.6		Y 88.9		Zr 91.2		Nb 92.9		Mo 95.9		Tc		Ru 101.1	Rh 102.9	Pd 106.4	
5		Ag 107.9		Cd 112.4		In 114.8		Sn 118.7		Sb 121.8		Te 127.6		I 126.9				Xe 131.3
6	Cs 132.9		Ba 137.3		La* 138.9		Hf 178.5		Ta 180.9		W 183.9		Re 186.2		Os 190.2	Ir 192.2	Pt 195.1	
6		Au 197.0		Hg 200.6		Tl 204.4		Pb 207.2		Bi 209.0		Po		At				Rn
7	Fr		Ra		Ac**													

*	Ce 140.1	Pr 140.9	Nd 144.2	Pm	Sm 150.4	Eu 152.0	Gd 157.3	Tb 158.9	Dy 162.5	Ho 164.9	Er 167.3	Tm 168.9	Yb 173.0	Lu 175.0
**	Th 232.0	Pa	U 238.0	Np	Pu	Am	Cm	Bk	Cf	Es	Fm	Md	No	Lr

FIGURE 5-1 An Adaptation of the Mendeleyev Periodic Table of 1871 (The colored squares indicate elements not known in 1871.)

three of the missing elements — scandium (Sc), gallium (Ga), and germanium (Ge) — and he lived to see his predictions come true, as the three elements were discovered during his lifetime and were found to have the properties he predicted.

In order to place elements of similar properties in the same columns, Mendeleyev sometimes made minor departures from the order of increasing atomic weights. For example, iodine (I, atomic weight 126.9 amu) comes *after* tellurium (Te, atomic weight 127.6 amu), and nickel (Ni, atomic weight 58.7 amu) comes *after* cobalt (Co, atomic weight 58.9 amu). Mendeleyev assumed that some of the atomic weights were in error, but they were actually all very close to presently accepted values. We now know that the atomic number is the true basis for ordering the elements in the periodic table, but the concept of atomic number was still unknown in Mendeleyev's day. However, the **periodic law** can be restated based on atomic number: *The physical and chemical properties of the elements are functions of their atomic numbers.*

5.2 THE MODERN PERIODIC TABLE

Since Mendeleyev's time, the form of the periodic table has changed somewhat. In addition, the table is longer because new elements have been discovered or created. Each of the seven horizontal rows is called a **period of elements.** The layered arrangement of the periods results in vertical columns of elements called **groups** or **families of elements.** All but one of the groups are labelled with a roman numeral followed by the letter A or B. Each period begins with an element in the IA column. For example, period 1 begins with hydrogen (H), period 2 begins with lithium (Li), and so on. Each element in the IA column marks the beginning of filling a new principal energy level with electrons. Each consecutive element in a period has one more electron in its atoms; thus, elements in each group have identical outer shell electron configurations except, of course, for the principal energy level designation. It is this outer shell electron configuration that accounts for the chemical properties of elements; thus, elements in any given group have similar chemical properties because of their identical outer shell electron configurations. The outer shell electrons of an atom are called **valence electrons,** and the outer shell is referred to as the **valence shell.** Each period ends with an element in Group O.

Members of a chemical group also tend to have similar physical properties, although there are many exceptions. For example, Figure 5-2 illustrates how boiling points vary with atomic number. Notice that each major peak on the graph corresponds to an element in Group IVA, while the lowest points correspond to elements in Group O. Thus elements in Group IVA have relatively high boiling points and those in Group O have rela-

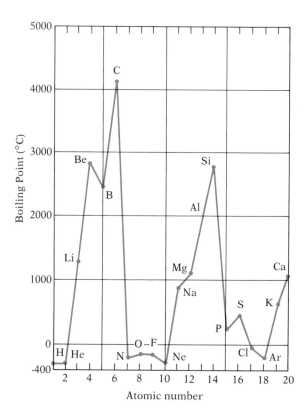

FIGURE 5-2   The Boiling Points of the Elements as a Function of Atomic Number

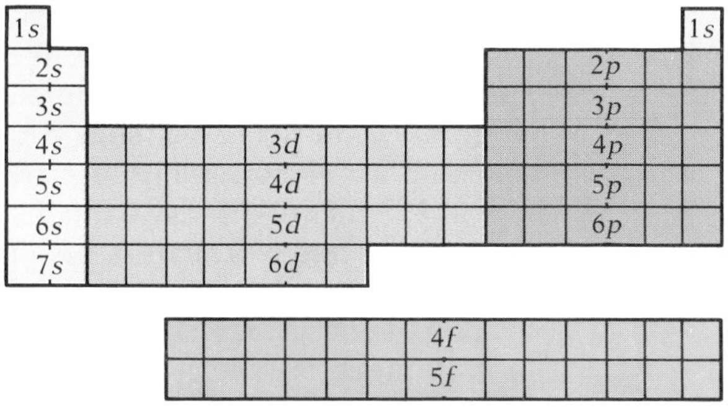

FIGURE 5-3 Blocks of Elements as Defined by the Type of Orbital Being Filled with Electrons

tively low ones. The boiling points of the elements are an example of periodicity.

The diagonal zigzag line that starts at boron (B) on the periodic table divides the metals from the nonmetals. **Metals** have shiny surfaces, can be pounded into sheets and drawn into wires, and are good conductors of heat and electricity. All metals except mercury are solids at room temperature. The **nonmetals** do not usually have shiny surfaces, are brittle if they are solids, and are poor conductors of heat and electricity. The nonmetals are gases, liquids, or solids at room temperature. Metals lie to the left of the zigzag line, while nonmetals are on the right. (Hydrogen is an exception; it is a nonmetal but is not usually shown on the right.) Bordering both sides of the zigzag line are the **metalloids,** which have both metallic and non-metallic properties. Although there are some exceptions, atoms of metals tend to have one, two, or three valence electrons, while those of nonmetals tend to have four, five, six, seven, or eight valence electrons.

The periodic table displays many patterns. One can be seen by dividing it into blocks of elements according to the type of orbital being filled with electrons, as illustrated in Figure 5-3. Atoms of elements in column IA have a single s electron in their valence shells, and those in column IIA have two s electrons in theirs. Helium (He) also contains two s electrons in its valence shell. These elements make up the s block of elements. The p block includes Groups IIIA, IVA, VA, VIA, VIIA, and Group O, except for helium. In these elements, p orbitals are being filled. In the block where d orbitals are being filled are Groups IB through VIIIB; these elements are called the transition elements and will be discussed later in this chapter. The f block contains elements whose f orbitals are being filled; it is composed of elements 58–71 and 90–103. These elements, sometimes called the inner transition elements, will also be discussed later in this chapter.

Metalloid

Metallike.

5.3 PERIODS OF ELEMENTS

Each period is numbered according to the principal energy level whose filling is begun by the first element in the period. Thus, in period 1, the first principal energy level starts to fill; in period 2, the second principal energy level starts to fill, and so on.

The number of elements in each period is determined by the electron capacities of the orbitals being filled. For example, period 1 contains 2 elements, corresponding to the electron capacity (2) of the $1s$ orbital being filled. Period 2 contains 8 elements because there are 8 spaces available for electrons in the s and p orbitals of the second principal energy level. Period 3 also contains 8 elements due to the 8 spaces available in the s and p orbitals of the third principal energy level.

Period 4 marks the point at which the principal energy levels begin to overlap in energy content. Thus, Ca has a full $4s$ orbital, but then the $3d$ orbitals fill before the $4p$ orbitals. The $3d$ orbitals are filled by elements 21–30, and gallium (Ga, atomic number 31) marks the beginning of filling $4p$ orbitals. There are 18 elements in this period, corresponding to the number of spaces for electrons in the $4s$, $3d$, and $4p$ orbitals. The energy overlap of principal energy levels continues through the remaining periods. In period 5, the $5s$, $4d$, and $5p$ orbitals are filling, and in period 6, the $6s$, $4f$, $5d$, and $6p$ orbitals are being filled. In the last series, period 7, the orbitals being filled are $7s$, $5f$, $6d$, and $7p$. The correlation between the number of elements in a period and the orbitals being filled is given in Table 5-1.

Because periods 1–3 contain fewer elements than the other periods, they are called the *short periods*. Periods 4–7 are called the *long periods* because they contain more elements than the other periods.

Period 1 contains only 2 elements, hydrogen (H) and helium (He). Both are colorless gases at room temperature and both are nonmetals, but their chemical properties are quite different. Hydrogen is rather reactive and exists on earth predominantly in the form of compounds. The pure element consists of **diatomic molecules,** that is, molecules that are each composed of two atoms. The formula for the hydrogen molecule is H_2. In contrast, helium is totally unreactive and does not form compounds. Thus, it exists on earth only as the uncombined element, which consists of single atoms rather than molecules. Helium concludes the filling of the first principal energy level, and its electron configuration is $1s^2$.

Period 2 is made up of 8 elements of widely varying properties. It begins with lithium (Li), a very reactive metal. The properties of elements in

Period 1
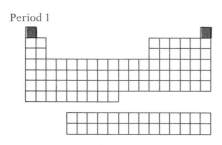

Period 2
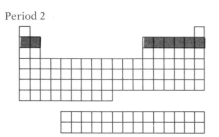

Period	Number of Elements	Orbitals Being Filled
		TABLE 5-1 The Number of Elements in Each Period
1	2	$1s$
2	8	$2s2p$
3	8	$3s3p$
4	18	$4s3d4p$
5	18	$5s4d5p$
6	32	$6s4f5d6p$
7	32*	$7s5f6d7p$

* Period 7 is incomplete at present.

Period 3

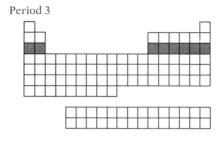

Period 4

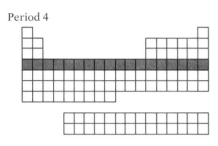

Period 5

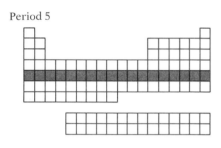

Period 6

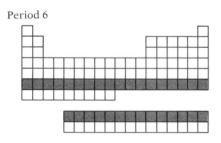

Period 7

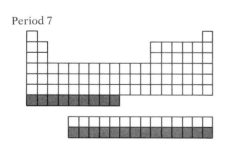

period 2 gradually change from metallic to nonmetallic, ending with the colorless gas neon (Ne), which is totally unreactive. Neon concludes the filling of the second principal energy level, with a valence shell configuration of $2s^2 2p^6$.

Period 3 also contains 8 elements, beginning with sodium (Na), a highly reactive metal, and ending with the colorless gas argon (Ar), a nonmetal. You can begin to see the general trend in properties across periods: except for period 1, the properties of the elements change from metallic at the beginning of a period to nonmetallic at the end of the period. In period 3, argon is an unreactive element, with a valence shell configuration of $3s^2 3p^6$. Notice that argon does not conclude the filling of the third principal energy level; the $3d$ orbitals are yet to be filled by period 4 elements.

Period 4, the first of the long periods, has 18 elements; it starts with potassium (K), a reactive metal, and ends with the colorless gas krypton (Kr), an unreactive nonmetal. Although its valence shell configuration is $4s^2 4p^6$, krypton does not conclude the filling of the fourth principal energy level, because the $4d$ orbitals will be filled by period 5 elements and the $4f$ orbitals will be filled by period 6 elements. Recall, however, that the $3d$ orbitals are filled in period 4. Scandium (Sc, atomic number 21) begins the filling of the $3d$ orbitals, and zinc (Zn, atomic number 30) completes their filling.

Period 5, the second long period, also has 18 elements; it continues the pattern of starting with a reactive metal, rubidium (Rb), and ending with an unreactive nonmetal, the colorless gas xenon (Xe). This period begins and completes the filling of the $4d$ orbitals, and the valence shell configuration of xenon is $5s^2 5p^6$.

Period 6 contains 32 elements, starting with cesium (Cs), a reactive metal, and ending with the colorless gas radon (Rn), an unreactive nonmetal. In this period, the $6s$, $4f$, $5d$, and $6p$ orbitals are being filled. Period 6 contains the **lanthanides.** These elements, extending from cerium (Ce, atomic number 58) to lutetium (Lu, atomic number 71), are placed at the bottom of the table for convenience; this allows a more compact table than if the lanthanides were put in the body of the table. Since they actually follow lanthanum (La, atomic number 57), these elements are given the name *lanthanides.*

Period 7 contains 23 elements but is incomplete at present. Elements in period 7 fill the $5f$ orbitals and begin filling the $6d$ orbitals. The newest elements in this group, beginning with plutonium (Pu, atomic number 94) have been created and do not exist in nature. The period begins with francium (Fr), and elements 108 and 109, recently created and not yet named, are the latest additions to the period. In theory, period 7 has space for 32 elements, ending with element 118, which should have properties like those of radon. In this period, the $7s$, $5f$, $6d$, and $7p$ orbitals are filled. The $5f$ orbitals are filled in the **actinides,** running from thorium (Th, atomic number 90) to lawrencium (Lr, atomic number 103); these elements are named for actinium (Ac, atomic number 89), which they follow; they are also placed at the bottom of the table for convenience.

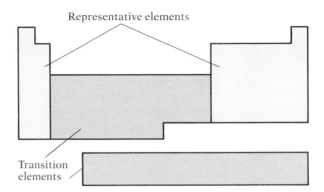

FIGURE 5-4 Representative and Transition Elements

5.4 GROUPS OF ELEMENTS

All but one of the groups of elements are labeled with A or B. Elements in the A groups and Group O are called the **representative elements.** Representative elements are filling s and p orbitals (see Figure 5-4). For all representative elements except those in Group O, the number of valence (outer shell) electrons is given by the group number. For example, the Group IA element lithium (Li) has the configuration $1s^2 2s^1$, and the Group VIA element oxygen (O) has the configuration $1s^2 2s^2 2p^4$. Except for helium, all elements in Group O have 8 valence electrons. Helium is exceptional in this group because its atoms have only 2 electrons, and both are valence electrons.

Elements in the B groups are called **transition elements.** These include the lanthanides and the actinides. In contrast to the representative elements, the transition elements are filling inner orbitals. Transition elements in the body of the table are filling inner d orbitals, and the lanthanides and actinides are filling inner f orbitals. The transition elements usually have 1 or 2 valence electrons, but the number of valence electrons varies considerably. Thus, group numbers for transition elements do not usually indicate the number of valence electrons.

5.5 A SURVEY OF THE REPRESENTATIVE ELEMENTS

Since most of the chemistry in this textbook involves the representative elements, this is a good time for you to get acquainted with them. The following paragraphs describe the groups of representative elements.

Group IA The Group IA elements are called **alkali metals;** the name stems from the fact that compounds of these elements form alkaline solutions. Each atom of the Group IA elements has a single s electron as its valence electron. We can write their valence shell configuration as ns^1, where n is the period of the individual element. All of the alkali metals have shiny, grayish surfaces and are quite soft, so soft that they can be cut with a butter knife. Their densities are relatively low for metals, ranging

Group I A

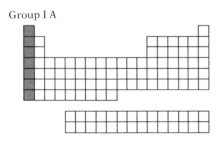

Alkaline solution

A solution that neutralizes acids.

from 0.535 g/mL (20° C) for lithium (Li), the lightest metallic element, to about 3 g/mL (20° C) for francium (Fr). Like most metals, the alkali metals consist of atoms rather than molecules. The alkali metals are quite reactive, and when cut, their shiny surfaces immediately take on a dull oxide coating due to their reaction with oxygen in the air. They are usually stored under an unreactive liquid, such as mineral oil, to prevent reaction with oxygen. The alkali metals react violently with water, and for that reason, special precautions are taken to prevent contact with water. Within the group, reactivity increases going down the column from lithium to francium. Because of their high reactivity, the alkali metals are found in their natural state only in compounds. Sodium (Na) and potassium (K) are the most abundant members.

Although hydrogen is often shown as a Group IA element (because its electron configuration is $1s^1$), it is not an alkali metal. It is actually a colorless, gaseous nonmetal and has little in common with the alkali metals. The greatest amount of hydrogen on earth is found in compounds. It forms compounds with almost all of the elements, and thus it forms more compounds than any other element. Its most important and most abundant compound is water, H_2O. Hydrogen is a unique element and cannot be classified with any group of elements; in many periodic tables, it is given a position by itself.

Group II A

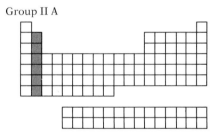

Alloy

A homogeneous blend of two or more metals.

Group III A

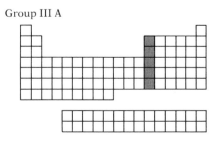

Group IIA   Before the nineteenth century, substances other than metals that were insoluble in water and unchanged by fire were called "earths." Since many compounds of the Group IIA elements fit this definition of earth and also resemble compounds of the alkali metals, the Group IIA elements were named the **alkaline earth metals.** These elements have a valence shell configuration of ns^2. All of the alkaline earth metals are hard, shiny, grayish metals with densities that are rather low, but not as low as those of the alkali metals. The alkaline earth metals are quite reactive, though not as reactive as the alkali metals. They react easily with oxygen in the air to form oxides, and they react slowly with water. Although not found naturally as free elements, they are widely distributed in minerals and in the sea. Calcium (Ca) is the second most abundant metal on earth. Beryllium (Be) and magnesium (Mg) are used for making strong, lightweight **alloys** for aircraft parts. Magnesium is also used in flashbulbs and fireworks because of the bright white light it gives off when burned.

Group IIIA   The Group IIIA elements have no special name, probably because their properties vary widely. All have ns^2np^1 as their valence shell configuration, giving a total of three valence electrons. The first member, boron (B), is a metalloid whose nonmetallic properties outweigh its metallic properties. It occurs in deposits of the compound borax in desert areas of California and India, but it is exceedingly difficult to produce the free element in a high state of purity. The small quantities of pure boron that have been prepared show it to have a black color with a metallic sheen. It is extremely hard, apparently second only to diamond in hardness among the elements. Although it is found only in compounds in nature, boron is chemically unreactive once isolated from its compounds.

The second member of the group, aluminum (Al), is also a metalloid, but it has such predominant metallic properties that it is often classified strictly as a metal. Aluminum is the most common metallic element in the earth's crust; it is mined as its ore, bauxite. Aluminum is a hard, strong, white metal with a relatively low density for a metal, 2.7 g/mL (20° C), and it is less reactive than the alkaline earth metals.

The other metals in Group IIIA—gallium (Ga), indium (In), and thallium (Tl)—are soft, white metals that are somewhat more reactive than aluminum. Gallium, for unexplained reasons, has an abnormally low melting point, 29.8° C, and will melt if held in the hand. Because its boiling point is 2070° C, it has the longest liquid range of any known substance, it is used as a liquid in thermometers that measure high temperatures.

Group IVA   The Group IVA elements also have varied properties and no special group name. There are four valence electrons in atoms of this group; their valence shell configuration is ns^2np^2. The first member, carbon (C), is a nonmetal that exists naturally in a variety of compounds. It is second only to hydrogen in the number of compounds it can form. As a free element carbon exists in nature in two different forms or **allotropes.** The two allotropes of carbon are graphite and diamond. Graphite, the more common of the two, is a black, flaky powder that is used as a lubricant, coloring agent, and pencil "lead." Diamond exists as colorless crystals and is one of the hardest solids known. Graphite can be converted to diamond by application of high pressures at high temperatures, and it is thought that naturally occurring diamond was formed under these conditions by geological processes. Neither form of carbon is very reactive, although each can be made to burn.

The second member of Group IVA, silicon (Si), is a metalloid with predominantly nonmetallic properties. It is second only to oxygen in abundance in the earth's crust, and it is especially plentiful in silicate compounds; sand is an example of a silicate. Although silicon is not found as the free element, it can be obtained from its compounds, and it has found extensive use in recent years as a semiconductor (a substance that conducts a limited amount of electricity) in the electronics industry. The third element of this group, germanium (Ge), is also a metalloid, but with more metallic than nonmetallic properties. Both silicon and germanium are hard, grayish solids at room temperature, and germanium is also a semiconductor. Both are components of computer "chips."

The other Group IVA elements, tin (Sn) and lead (Pb), are typical metals; they are obtained from their ores in the earth's crust. Tin is familiar to us from its use in "tin" cans, which are actually made of steel coated with tin. Lead is a very dense metal (11.34 g/mL at 20° C) and it is rather soft. Not being very reactive, it resists corrosion and finds wide use in equipment constructed for use in the chemical industry. Lead interferes with the function of the nervous system and is thus highly poisonous.

Group VA   Group VA is the first family of elements to include a gas (unless hydrogen is thought of as being in Group IA). The group shows

Group IV A

Allotrope

Derived from the Greek words *allo-*, meaning "other," and *-tropo*, meaning "change."

Group V A

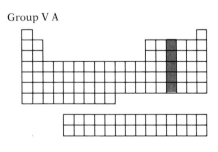

dramatic changes in properties, starting with the nonmetal nitrogen (N) and ending with the metal bismuth (Bi). The valence shell configuration for the group is ns^2np^3, giving a total of five valence electrons for each member.

Nitrogen, an element that is not very reactive, constitutes 80% of the earth's atmosphere and appears in a number of compounds on the earth's surface and in living organisms. It is a gas at room temperature and exists as diatomic molecules with the formula N_2. Nitrogen is used commercially for making ammonia (NH_3) and fertilizers.

The next member of the group, phosphorus (P), is a reactive nonmetal that is a solid at room temperature. It is not found free in nature, except in a few meteorites, but it occurs widely in compounds in rocks, minerals, plants, and animals. Phosphorus exists as molecules, each of which contains four atoms; thus, the formula for the phosphorus molecule is P_4. The largest commercial use of phosphorus is in making phosphoric acid (H_3PO_4), used to flavor soft drinks and make fertilizers.

The third member of the group, arsenic (As), is a metalloid with predominantly nonmetallic properties. It is widely distributed in nature, although not in large quantities, and it is occasionally found uncombined. Although not as reactive as phosphorus, arsenic also exists as molecules containing four atoms each (As_4). Arsenic is extremely toxic and is used in making insecticides and weed killers.

Antimony (Sb), the fourth member of the group, is also a metalloid, but is more metallic than nonmetallic. A bright, silvery white solid at room temperature, it is hard and brittle and occurs in nature chiefly as the mineral stibnite. It is about as reactive as arsenic. It is used for making strong, hard alloys for automobile batteries, bullets, and coverings for telephone cables. Like phosphorus and arsenic, antimony exists as molecules that contain four atoms each (Sb_4).

The last element in the group, bismuth (Bi), is a metal. It is the least abundant of the family and is found in nature only in compounds. It is a brittle, shiny white solid with a reddish tinge not found in any other metal. It is rather unreactive, and it is used for making alloys that melt at low temperatures for use in fire-control sprinkler systems.

Group VI A

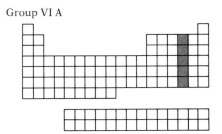

Group VIA   Elements of Group VIA show a large variation in properties, ranging from nonmetallic to metallic. They have six valence electrons and a valence shell configuration of ns^2np^4. The first member, oxygen (O), is a gaseous nonmetal that constitutes about 20% of the atmosphere. It is found in abundance in compounds on the surface of the earth. Oxygen exists as two allotropes. The common, stable allotrope is a colorless gas consisting of diatomic molecules, O_2. The second allotrope, ozone, is produced by the action of electrical discharge or ultraviolet light on O_2. Ozone is a colorless gas with a sour odor and taste. It consists of triatomic molecules, O_3, and is present in small amounts in polluted air and in the upper atmosphere, where it helps to screen out the harmful ultraviolet rays produced by the sun. Both allotropes of oxygen are quite reactive, but O_3 is more reactive than O_2.

The second element in the group is sulfur (S), a nonmetal. It occurs widely in nature as the uncombined element and in compounds. At room temperature, sulfur is a yellow, brittle solid composed of molecules, with each molecule containing eight sulfur atoms connected in a ring. Thus the formula for a sulfur molecule at room temperature is S_8. Sulfur is one of the most reactive and important of all elements. It is used in drying fruits and in producing gunpowder, rubber, fertilizers, matches, paper pulp, insecticides, fungicides, and fumigants.

The third element in the group, selenium (Se), is also a nonmetal. It is widely distributed in nature, but only in small quantities, and it can be found occasionally as the free element. The electrical conductivity of selenium increases when light strikes it, so it is used in photoelectric cells, solar cells, and photographic exposure meters. It also gives a clear, red color to glass and is used in making signal lights.

The next element in the group is tellurium (Te), a metalloid that is more nonmetallic than metallic. It occurs naturally in a variety of compounds, but not in great abundance, and it is rarely found as the free element. Tellurium is not very reactive, and it is used for making alloys.

The last member of the group, polonium (Po), is a metalloid with mostly metallic properties. Polonium was discovered in 1898 by Marie Curie, and she named it for her native Poland. A radioactive, silvery gray solid, polonium is a very rare element that is found in nature as a radioactive decomposition product of uranium, thorium, and actinium. Because it is so highly radioactive, there are very few practical uses for polonium.

Marie Curie
Distinguished Polish scientist who studied and worked in France; one of the few people to receive two Nobel Prizes, one in physics in 1903 and the other in chemistry in 1911.

Group VII A

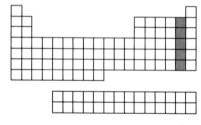

Group VIIA   The Group VIIA elements are known collectively as the **halogens.** Their group name comes from the Greek words *hal*, meaning "salt," and *gen*, meaning "to produce." Since members of Group VIIA all produce sodium salts of similar properties, the name halogen is appropriate. All of the halogens are nonmetals except astatine, which is a metalloid with predominantly nonmetallic properties.

The first two members, fluorine (F) and chlorine (Cl), are diatomic gases (F_2, Cl_2) at room temperature. Fluorine has a greenish color, while chlorine is a pale greenish yellow. Both have acrid, burning odors that are harmful to the lungs and nasal passages. Bromine (Br) is a dense, reddish brown liquid composed of Br_2 molecules, and iodine (I) is a shiny, dark purple solid, also composed of diatomic molecules (I_2). Bromine gives off reddish brown fumes and iodine gives off purple fumes, both of which are harmful to breathe. Astatine (As) is radioactive, and only very small amounts have been available for study. It is known, however, to be a solid at room temperature. All of the halogens are highly reactive and occur only in compounds in nature. Except for astatine, which occurs naturally only in trace amounts, the compounds of the halogens are widely distributed, especially in seawater. The halogens show great similarity to each other in chemical behavior; their valence shell configuration is ns^2np^5, all having seven valence electrons.

Chlorine, the best known halogen, is used for purifying water and in a number of chemical processes. Fluorine compounds are used to prevent tooth decay, as refrigeration fluids, and as lubricants. Iodine is used as an antiseptic, and bromine is used to prepare a gasoline additive that prevents deposits of lead in engines.

Hydrogen is sometimes shown to be a member of Group VIIA. Although it does have some resemblance to the halogens, it has more differences. For example, it is a gaseous, diatomic molecular nonmetal, but it is colorless and not nearly as reactive as the halogens. In addition, it shows more versatility in its chemical behavior than the halogens. Thus, hydrogen is not really a member of Group VIIA any more than it is a member of Group IA. It is truly a unique element.

Group 0

Octet

A group of eight.

Group O The Group O elements are known as the **noble gases.** All are gaseous nonmetals that consist of atoms rather than molecules. Until about twenty years ago, these gases were thought to be incapable of forming compounds. In fact, they were called the noble gases because of their aloofness from other elements. However, in 1962, two compounds of xenon (Xe) were prepared, and since then a number of krypton (Kr), xenon, and radon (Rn) compounds have been made. Due to their lack of reactivity and their scarcities, the noble gases are also called the *inert gases* and the *rare gases*. The group is sometimes labeled as Group VIIIA in the periodic table.

The valence shell configuration of helium (He), the first member of the group, is $1s^2$, but all other members have the valence shell configuration of ns^2np^6. Thus, all of the noble gases except helium have an octet of valence electrons. The **octet rule** is based on the lack of reactivity of the noble gases; it says that an atom or ion is least reactive when it contains eight electrons in its valence shell. The octet rule applies to many elements and compounds, but as you shall see in later chapters, there are some exceptions. The octet rule is of major importance in understanding chemical reactivity.

All of the noble gases are found in the earth's atmosphere, with argon (Ar) present to the extent of about 1% by volume; the others are present in only trace amounts. Helium has a very low density (0.18 g/L at 0° C and standard pressure), and it is used in balloons and dirigibles. Second only to hydrogen in lifting power, helium has the advantage of being nonflammable. (The infamous Hindenburg dirigible was filled with hydrogen.) Helium is also used in breathing mixtures for deep-sea divers and in welding. Neon (Ne) is the gas inside neon light tubes, and argon is used to provide an inert atmosphere inside ordinary electric light bulbs to lengthen the lifetime of the glowing filaments. Krypton is used in fluorescent lamps, and xenon is a filler gas in high intensity bulbs such as flashbulbs. Radon, the radioactive member of the group, has been used in radiation treatment of cancer.

5.6 THE TRANSITION ELEMENTS

All of the transition elements are metals and thus have similar properties. One chemical characteristic of metals is that they form cations by losing their valence electrons. An atom of a representative metal loses all of its valence electrons each time it forms a cation; thus, cations of a given representative metal always carry the same number of positive charges. Most transition elements, however, can lose electrons from their inner *d*

⊙ℛ **PERSPECTIVE**   IS THERE AN END TO THE PERIODIC TABLE?

Elements with atomic numbers higher than that of uranium (atomic number 92) are called transuranium elements. Except for neptunium (atomic number 93), all of these elements are man-made and do not exist naturally. All are radioactive, with some decomposing over tens of millions of years and others in fractions of a second. The synthesis of these elements involves bombarding atoms of lower atomic number with neutrons, protons, and even atoms or ions.

Glenn T. Seaborg, an American scientist, has been responsible for the creation of nine transuranium elements, and in 1951, he was honored with a Nobel Prize for his discoveries in the chemistry of the transuranium elements. Seaborg's studies led the way to making predictions about elements not yet known. Based on the number of protons and neutrons in their nuclei, certain elements not yet known are expected to be especially stable. The numbers of protons and neutrons in atoms of these elements are called *magic numbers*. These magic numbers predict that superheavy elements with atomic numbers from 110 to 126 should be rather stable and may even exist in nature. Efforts to find or create them are under way.

Naming new elements was once a fairly easy thing to do. The discoverer or creator simply chose a name and assigned it to the element. In recent years, however, American and Soviet scientists have disagreed about names for new elements. To reduce such possible conflicts in the future, a system has been proposed for naming elements past lawrencium, element number 103. The system uses specific syllables corresponding to the numbers 0–9, as shown at the top of the following column.

Number	Syllable	Abbreviation
0	nil	n
1	un	u
2	bi	b
3	tri	t
4	quad	q
5	pent	p
6	hex	h
7	sept	s
8	oct	o
9	en	e

Names would be constructed for elements by combining the syllables corresponding to the atomic numbers and then adding the ending *-ium*. Thus, the name proposed for element 104 is *un-nil-quad-ium*, or unnilquadium, and the symbol is Unq. Using this system, the name for element 105 is unnilpentium (Unp), that for element 106 is unnilhexium (Unh), and that for element 107 is unnilseptium (Uns).

The proposed naming system would apply to all elements past atomic number 103, but the question remains: how many more elements are there likely to be? It is predicted that there should be a limit of atomic stability at which point any additional electrons would be so far from the nucleus that they would not be attracted to it. On the basis of current knowledge, that limit apparently lies somewhere between elements 170 and 210. This means that someday there may be twice as many elements as we know now. And, if the last one is element 210, its name will probably be biunnil. Let's hope that a pronunciation guide is published for the new element names!

and f orbitals, as well as their valence electrons. This allows a transition metal to form cations having at least two different numbers of positive charges. For example, vanadium (V, atomic number 23) can form cations with charges of 2+, 3+, 4+, and 5+. Iron (Fe, atomic number 26) is capable of forming cations with two different charges, 2+ and 3+.

The transition elements include many of the most common metals, such as iron, nickel (Ni), copper (Cu), and mercury (Hg), as well as some of the most valued metals, platinum (Pt), gold (Au), and silver (Ag). In addition to the elements of the groups labelled B in the body of the periodic table, the transition elements also include the lanthanides and the actinides, located along the bottom of the table. Most of the transition metals form highly colored compounds.

5.7 ATOMIC SIZE

Atomic size is a striking example of chemical periodicity. The atomic radii of most of the representative elements are given in Figure 5-5. (Note that these values are given in nanometers (nm); 1 nm = 10^{-9} m.)

Notice that atomic size tends to decrease from left to right across a period of representative elements. The decrease across a period can be explained by considering two effects. The first is that the same principal energy level is being filled with electrons. This in itself would cause little change in size in going from one atom to the next. However, the second and overriding effect is that each successive nucleus has an additional proton. This increased positive charge draws the electrons closer and closer to the nucleus, and thus the atomic sizes decrease across a period.

Also notice that atomic size increases going down a group; this is due to one major effect. Each successive atom is starting to fill a new principal energy level, and each new principal energy level is farther from the nucleus than the last. Even though there is greater positive charge in each nucleus going down a group, the increased nuclear attraction does not offset the effect of additional principal energy levels.

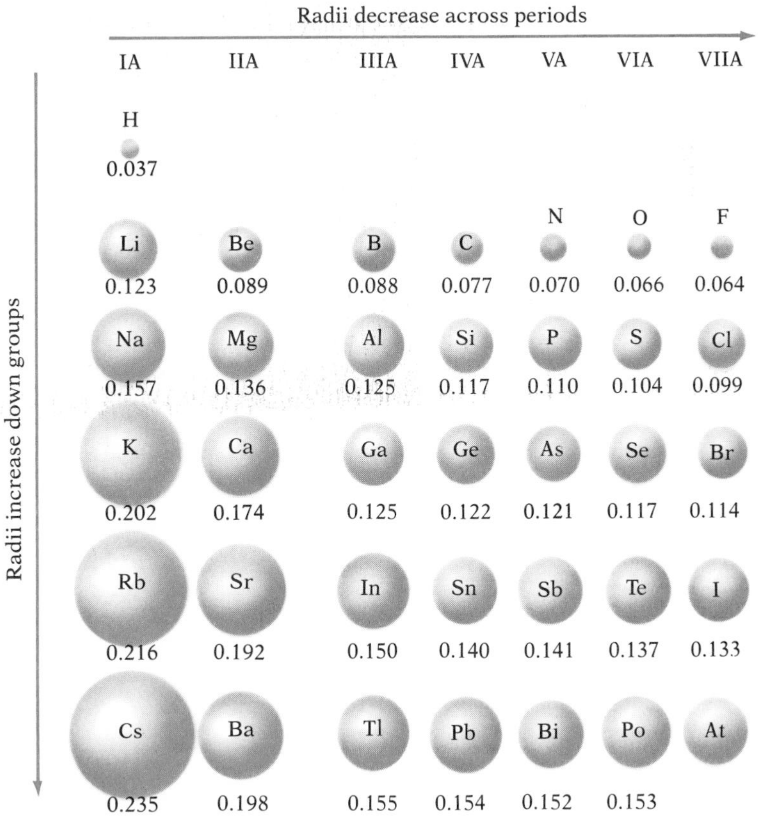

Radii decrease across periods

	IA	IIA	IIIA	IVA	VA	VIA	VIIA
	H 0.037						
	Li 0.123	Be 0.089	B 0.088	C 0.077	N 0.070	O 0.066	F 0.064
	Na 0.157	Mg 0.136	Al 0.125	Si 0.117	P 0.110	S 0.104	Cl 0.099
	K 0.202	Ca 0.174	Ga 0.125	Ge 0.122	As 0.121	Se 0.117	Br 0.114
	Rb 0.216	Sr 0.192	In 0.150	Sn 0.140	Sb 0.141	Te 0.137	I 0.133
	Cs 0.235	Ba 0.198	Tl 0.155	Pb 0.154	Bi 0.152	Po 0.153	At

Radii increase down groups

FIGURE 5-5 Atomic Radii (in nanometers) of Representative Elements

5.8 IONIZATION ENERGY

Another example of chemical periodicity is **ionization energy** (I.E.), the energy required to remove the most loosely held electron from a gaseous, ground-state atom or ion. Energy must be added to the ground-state atom or ion to completely remove the electron from the atom; thus, loss of an electron is an endothermic (heat-absorbing) process.

$$\text{atom} + \text{energy} \longrightarrow \text{cation}^+ + e^-$$

The energy required for removal of one electron is called the **first ionization energy,** that required for removal of a second electron is called the **second ionization energy,** and so on. Table 5-2 gives the first ionization energies of the representative elements in periods 2 and 3.

The ionization energies indicate how strongly each atom or ion holds on to its outermost electron. The values in Table 5-2 show that first ionization energies tend to increase going across a period. This happens because the atoms become smaller and smaller, and thus the outermost electron of each element becomes successively closer to its nucleus. The closer the outermost electron is to the nucleus and its positive charges, the more tightly the electron is held by the atom. Thus, atoms of the noble gas family are the smallest in any given period and require the most energy for removal of an electron. In contrast, the element with the lowest first ionization energy is the one at the beginning of a period. Since metals appear at the beginnings of periods and nonmetals near the ends of periods, this means that metals form cations more easily than do nonmetals.

The change in first ionization energies going down a group is different from what happens going across a period. In general, first ionization energies decrease within a group. In this case, as we go from one atom to the next, the atoms become larger and larger, and the electron to be removed is farther and farther away from the nucleus. Thus, removal of the most loosely held electron requires less and less energy as we go down the group. Another way of stating this is that the tendency to form cations increases within a group. These general trends of first ionization energies are summarized in Figure 5-6, and the periodicity of first ionization energies is illustrated in Figure 5-7.

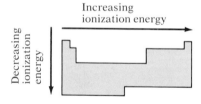

Increasing ionization energy →

Decreasing ionization energy ↓

FIGURE 5-6   General Trends of First Ionization Energies

TABLE 5-2	First Ionization Energies (I.E.) of Representative Elements in Periods Two and Three		
Element	I.E. (kJ/mol)	Element	I.E. (kJ/mol)
Li	520	Na	496
Be	900	Mg	738
B	801	Al	578
C	1086	Si	786
N	1402	P	1102
O	1314	S	1000
F	1681	Cl	1251
Ne	2081	Ar	1520

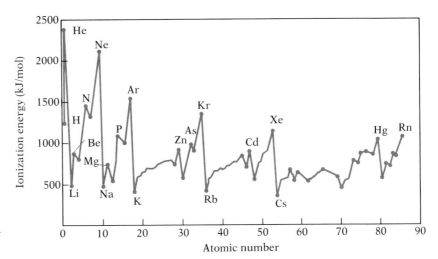

FIGURE 5-7 The First Ionization Energies of the Elements as a Function of Atomic Number

After one electron has been removed from an atom, it is harder to remove a second and even harder to remove a third. When the first electron is removed, the resulting cation becomes smaller, due to increased nuclear attraction for the remaining, smaller number of electrons. Then, the outermost electron is closer to the nucleus and is harder to remove than the first one was. The trends of the first four ionization energies of Na, Mg, and Al are shown in Table 5-3.

Notice the big difference between the first and second ionization energies for Na. The first electron to be removed from Na ($1s^2 2s^2 2p^6 \mathbf{3s^1}$) comes from the partly filled third principal energy level, but the second electron removed comes from the underlying filled principal energy level of Na$^+$ ($1s^2 \mathbf{2s^2 2p^6}$). In the case of Mg, the first two ionization energies are relatively low but there is a big difference between the second and third ionization energies. The first two electrons come from the partly filled third principal energy level ($1s^2 2s^2 2p^6 \mathbf{3s^2}$), but the third comes from the underlying filled principal energy level ($1s^2 \mathbf{2s^2 2p^6}$). For Al, the first three ionization energies are relatively low but a large difference appears between the third and fourth ionization energies. The first three electrons come from the partly filled third principal energy level ($1s^2 2s^2 2p^6 \mathbf{3s^2 3p^1}$), but the fourth comes from the underlying filled principal energy level ($1s^2 \mathbf{2s^2 2p^6}$). These three examples illustrate how very difficult it is to remove an electron from a filled octet; thus, ionization energies show conclusive support for the octet rule. For this reason, metallic representative

TABLE 5-3 The First Four Ionization Energies of Na, Mg, and Al

Element	First I.E. (kJ/mol)	Second I.E. (kJ/mol)	Third I.E. (kJ/mol)	Fourth I.E. (kJ/mol)
Na	496	4565	6912	9540
Mg	738	1450	7732	10,550
Al	577	1816	2744	11,580

elements tend to lose all of the electrons in their partly filled highest principal energy level, but none from their underlying filled shells. Or, we can say that *metallic representative elements tend to lose all of their valence electrons when they form cations and thus achieve an octet in the shell lying just beneath the valence shell.* Thus, the octet rule can be used to predict the charges on cations formed by representative elements.

EXAMPLE 5.1   Using only the periodic table, choose the element that would be expected to have the higher first ionization energy: Be or N? Why?

Solution   Be and N are in the same period (Period 2). Since first ionization energies tend to increase going across a period, the element farther to the right, in this case N, would be expected to have the higher first ionization energy.

EXERCISE 5.1   Using only the periodic table, choose the element that would be expected to have the higher first ionization energy: S or Te? Why?   ■

5.9 ELECTRON AFFINITY

The energy change that is brought about by adding an electron to a gaseous, ground-state atom is called the **electron affinity** of the atom.

$$\text{atom} + e^- \longrightarrow \text{anion}^- + \text{energy}$$

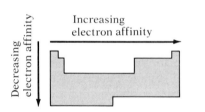

FIGURE 5-8   General Trends of Electron Affinities

In this case, the atom forms an anion. Energy is *usually* released by this process, but not always. Electron affinity trends are indicated in Figure 5-8. Thus, electron affinities increase across a period and decrease down a group, indicating that smaller atoms have a greater tendency to gain electrons than do larger atoms. Since the added electron is closer to the positively charged nucleus in the smaller atom than it is in the larger atom, it is attracted to the nucleus of the smaller atom with greater force. Thus nonmetals have a greater tendency to gain electrons than do metals. The tendency of metals to lose electrons is complemented by the tendency of nonmetals to gain electrons.

There is a limit to the tendency of nonmetals to gain electrons. Once a principal energy level becomes filled, much less energy is released (or energy may even be required) to add another electron, because the new electron must go into a higher principal energy level and will not be attracted as strongly to the nucleus. Thus, when nonmetals form anions, they tend to gain only enough electrons to fill their already partly filled valence shell, that is, they tend to gain enough electrons to have an octet in their valence shells.

EXAMPLE 5.2   Using only the periodic table, choose the element that would be expected to have the higher electron affinity: Ca or Ba? Why?

Solution   Ca and Ba are in the same group (Group IIA). Since electron affinity tends to decrease going down a group, the element higher in the group, in this case Ca, would be expected to have the higher electron affinity.

EXERCISE 5.2   Using only the periodic table, choose the element that would be expected to have the higher electron affinity: Na or S? Why?   ■

5.10   THE CHEMICAL BEHAVIOR OF METALS AND NONMETALS

The examples of chemical periodicity in the previous sections help us to understand the chemical behavior of metals and nonmetals. In general, metals have large sizes, low ionization energies, and low electron affinities. This means that metals, relative to nonmetals, have greater tendencies to lose electrons. In fact, the most metallic elements are those with the greatest tendency to lose electrons. Metals lie in the center and on the left side of the periodic table, and the most metallic elements appear in the lower left-hand corner of the periodic table. Thus francium (Fr) is the most metallic element. In contrast, nonmetals have small sizes, high ionization energies, and high electron affinities, so nonmetals have a greater tendency to gain electrons than do metals. The most nonmetallic elements, those with the greatest tendency to gain electrons, are located on the right side of the periodic table, next to the noble gases; those in the upper right-hand corner, next to the noble gases, are the most nonmetallic. Thus fluorine (F) is the most nonmetallic of all elements. (The noble gases are exceptionally stable and do not have tendencies to gain or lose electrons.) These trends are summarized in Figure 5-9.

The noble gases are especially stable because their atoms have the smallest size in each period and because they have very high ionization energies and very low electron affinities. Their filled outer s and p orbitals are the reason for their stability. To add an electron to a noble gas would require a great deal of energy because the added electron would have to go into an orbital of much higher energy than the previous ones. Removing an electron is also difficult because of the great attraction for electrons by the nuclei of these small atoms. Thus, the noble gases set the example for the octet rule, but their observed chemical behavior (or lack of it) can be explained by chemical periodicity.

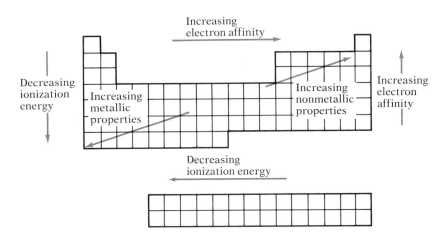

FIGURE 5-9   Chemical Periodicity and the Chemical Behavior of Metals and Nonmetals

EXAMPLE 5.3   Using only the periodic table, choose the element that would be expected to be more metallic: Ge or Na? Why?

Solution   Since metals have a greater tendency to lose electrons than do nonmetals, the more metallic element will be the one with the greater tendency to lose electrons. This means that the element closer to the left side of the periodic table will be the more metallic. In this case, the more metallic element is Na.

EXERCISE 5.3   Using only the periodic table, choose the element that would be expected to be more metallic: Cl or I? Why?   ■

SUMMARY

In his search for a classification system of the elements, Mendeleyev observed **chemical periodicity,** a periodic recurrence of elements with similar properties. Mendeleyev also proposed the **periodic law;** stated in modern terms, it says that the physical and chemical properties of the elements are functions of their atomic numbers. The **periodic table** is composed of horizontal rows **(periods)** and vertical columns **(groups** or **families).**

Each period begins with an element in column IA and ends with a noble gas. Each new period marks the beginning of filling a new principal energy level with electrons, and each consecutive element in the period adds one electron to its atoms. Elements in each group have identical outer shell configurations except for the principal energy level designation. Outer shell electrons are called **valence electrons,** and an outer shell is referred to as a **valence shell.** Members of a group tend to have similar properties, although there are many exceptions.

Metals are divided from **nonmetals** in the periodic table by a diagonal zigzag line. Bordering both sides of the zigzag line on the periodic table are the **metalloids,** which have both metallic and nonmetallic properties.

The periodic table can be divided into blocks of elements according to the type of orbital being filled with electrons. Each period is numbered according to the principal energy level whose filling is begun by that period. The number of elements in a period is determined by the electron capacities of the orbitals being filled. Period 4 marks the point at which the principal energy levels begin to overlap in energy content. Period 1 contains two elements, H and He; both are nonmetals. Periods 2–6 begin with metals and end with nonmetals. Period 7 is incomplete. Elements in the A groups and Group O (the **representative elements**) are filling s and p orbitals, and all underlying principal energy levels are filled. For all representative elements except those in Group O, the number of valence electrons is given by the group number. In Group O, He has 2 valence electrons but all other members have 8. Elements in the B groups (**transition**

elements) are filling inner d orbitals (B elements in the body of the periodic table) and f orbitals (the lanthanides and actinides). The group numbers of the transition elements do not usually indicate the number of valence electrons.

Group IA elements (the **alkali metals**) are all metals, and their valence shell configuration is ns^1. (Hydrogen is not an alkali metal and does not belong to any group.) Group IIA elements (the **alkaline earth metals**) are also all metals. Their valence shell configuration is ns^2. Group IIIA contains two metalloids; the rest of the group are nonmetals. The group valence shell configuration is ns^2np^1. Group IVA, with a group valence shell configuration of ns^2np^2, contains one nonmetal, two metalloids, and two metals. Group VA contains two nonmetals, two metalloids, and one metal and has a group valence shell configuration of ns^2np^3. Group VIA contains three nonmetals and two metalloids, with a group valence shell configuration of ns^2np^4. Group VIIA (the **halogens**) contains four nonmetals and a metalloid; the first four members consist of diatomic molecules. The valence shell configuration for the halogens is ns^2np^5. Group O elements (the **noble gases**) are very poorly reactive and consist of atoms rather than molecules. The group valence shell configuration is ns^2np^6. The lack of reactivity of the noble gases led to formulation of the **octet rule:** An atom or ion is least reactive when it contains eight electrons in its valence shell. The transition elements are all metals; the atoms of a transition element form cations having at least two different numbers of positive charges. The transition elements include the lanthanides and actinides.

Atomic radii decrease going across a period, as a result of successively adding one proton to each nucleus while adding one electron to the same principal energy level. Atomic sizes increase down a group, as a result of adding new principal energy levels. **Ionization energy** is the energy required to remove the most loosely held electron from a gaseous, ground-state atom or ion. First ionization energies increase across a period because the atoms are getting smaller and hold their electrons more tightly. First ionization energies decrease down a group because atomic sizes are getting larger and the atoms do not hold their electrons as tightly. After one electron has been removed from an atom, it gets progressively harder to remove additional electrons because of increased nuclear attraction for the remaining electrons. A very large ionization energy is observed for removal of an electron from a filled principal energy level.

Electron affinity is the energy change that occurs when an electron is added to a gaseous, ground-state atom. Electron affinities increase across a period because the atoms get smaller and the added electron is progressively closer to the nucleus and thus is attracted with greater force. Electron affinities decrease down a group because atoms get larger and the added electron is attracted with less and less force. Metals tend to lose electrons to achieve an octet of electrons in the next underlying principal energy level, while nonmetals tend to gain enough electrons to achieve an octet of valence electrons. The noble gases are especially stable because their atoms have very small sizes, very high ionization energies, and very low electron affinities.

STUDY QUESTIONS AND PROBLEMS

(More difficult questions and problems are marked with an asterisk.)

THE PERIODIC LAW

1. In your own words, explain what *chemical periodicity* means.
2. Why did Mendeleyev leave gaps in his periodic table? Was he right in doing so? Explain your answer.
3. Was Mendeleyev's periodic law correct? State the modern periodic law and explain how it differs from Mendeleyev's.

THE MODERN PERIODIC TABLE

4. Define each of the following terms:
   a. Period of elements
   b. Group or family of elements
   c. Valence electrons
   d. Valence shell
   e. Metal
   f. Nonmetal
   g. Metalloid
5. How does the modern periodic table differ from Mendeleyev's table?
6. Using the periodic table, give the number of elements in each of the following periods.
   a. 5   b. 7   c. 3   d. 1
7. Using the periodic table, give the number of elements in each of the following groups.
   a. 0   b. IIIA   c. IA   d. VIA
8. Using only the group number from the periodic table, give the number of valence electrons in each of the following atoms.
   a. Sn   c. In   e. K
   b. Mg   d. Ba   f. Sb
9. Describe how the boiling points of the elements vary with atomic number.

PERIODS OF ELEMENTS

10. How are the numbers of each period related to principal energy levels?
11. Using only the periodic table and Figure 5-3, give the valence shell electron configuration of each of the following atoms.
    a. C   c. Ar   e. Cs
    b. Ba   d. As   f. S
12. How is the number of elements in each period determined?
13. At which period do the principal energy levels begin to overlap in energy content? How is this overlap related to the transition elements?
14. Why is period 7 incomplete?
15. Hydrogen and helium are in period 1. How are they alike? How are they different?
16. Which elements begin each period? Which ones end each period?
17. What is the trend of element properties going across a period?

18. Which orbitals are being filled by the period 4 transition elements? The period 5 transition elements? The period 6 transition elements?
19. What are the lanthanides and actinides? In which periods are they?

GROUPS OF ELEMENTS

20. What do the letters A and B following group numbers signify?
21. What are the representative elements? The transition elements? How do they differ?
22. Using only the periodic table, classify each of the following as a representative element or a transition element.

    a. Br   d. Ti   g. Si   j. Sr
    b. W   e. Pb   h. In   k. Te
    c. Ra   f. F   i. Os   l. Ar

23. What do the group numbers of the representative elements signify?
24. How many valence electrons do the Group O elements have? Why is helium an exception?

A SURVEY OF THE REPRESENTATIVE ELEMENTS

25. Define each of the following terms:

    a. Alkali metals   e. Ozone
    b. Diatomic molecule   f. Halogens
    c. Alkaline earth metals   g. Noble gases
    d. Allotropes   h. Octet rule

26. Give the valence shell configuration for each group of representative elements.
27. To which group do the most reactive metals belong? The most reactive nonmetals?
28. Which elements exist as diatomic molecules? Write formulas for their molecules.
29. Why is hydrogen sometimes classified as a Group IA element?
30. Which groups of representative elements contain no metalloids? Which group contains only one metalloid?
31. Which group of elements contains only gases?
32. What useful purpose is served by the ozone in the upper atmosphere? How does ozone differ from oxygen?

THE TRANSITION ELEMENTS

33. In general, what are the physical properties of the transition elements?
34. How do cations of transition elements differ from those of representative elements?

ATOMIC SIZE

35. How do atomic sizes vary within a period? What is the basis for this variation?
36. How do atomic sizes vary within a group? What is the basis for this variation?
37. Which atom do you think has the smallest radius? Which would you expect to have the largest radius?

IONIZATION ENERGY

38. What does *ionization energy* mean? First ionization energy? Second ionization energy?

39. How do first ionization energies vary within a period? Why?

40. How do first ionization energies vary within a group? Why?

41. In each of the following pairs, select the element that would be expected to have the higher first ionization energy. Use only the periodic table. Give reasons for your selections.
  a. Cl or Al   **b.** As or Rb   **c.** B or Li   **d.** Fr or Cs

42. Which type of element forms cations more readily, metals or nonmetals? Why?

43. Why is the second ionization energy of lithium (Li) much higher than its first ionization energy?

44. Why is the third ionization energy of strontium (Sr) much higher than its second ionization energy?

45. How do values for ionization energies support the octet rule?

ELECTRON AFFINITY

46. What is meant by electron affinity?

47. How does electron affinity vary within a period? Why?

48. How does electron affinity vary within a group? Why?

49. In each of the following pairs, select the element that would be expected to have the higher electron affinity. Use only the periodic table. Give reasons for your selections.
  a. K or Fe   **b.** Ca or Se   **c.** As or N   **d.** Po or Cl

50. Which type of element forms anions more readily, metals or nonmetals? Why?

THE CHEMICAL BEHAVIOR OF METALS AND NONMETALS

51. Where are the most metallic elements located in the periodic table? Why are these most metallic?

52. Where are the most nonmetallic elements located in the periodic table? Why are these most nonmetallic?

53. In each of the following pairs, select the element that would be expected to be more metallic. Use only the periodic table. Give reasons for your selections.
  a. B and Ca   **b.** Ba and N   **c.** Li and Fr   **d.** Al and S

54. On the basis of atomic sizes, ionization energies, and electron affinities, explain why the noble gases are so unreactive.

GENERAL EXERCISES

55. Using only the periodic table, classify each of the following elements as a metal, nonmetal, or metalloid.
  a. Cl   **d.** Ce   **g.** Ge   **j.** Rn
  b. Am   **e.** Rb   **h.** Be   **k.** B
  c. Mo   **f.** P   **i.** Na   **l.** Cd

56. Why are periods 4–6 longer than periods 1–3?

57. A region of California where many computer components are made is referred to as "Silicon Valley." What do you think is the basis of this name?

58. Why is helium preferred to hydrogen as the gas in balloons and dirigibles?

*59. From your knowledge of trends in ionization energies, predict which of the following cations is likely to exist. Explain your answer.
   a. Al^{4+} c. Rb^+ e. Sr^+
   b. Mg^{3+} d. F^+ f. K^{2+}

*60. Compounds of Kr, Xe, and Rn have been made, but compounds of He, Ne, and Ar are not known to exist. How do these observations correlate with trends in atomic sizes, ionization energies, and electron affinities?

CHEMICAL BONDS

CHAPTER
6

OUTLINE

6.1 Types of Chemical Bonds
6.2 Ionic Bonds
6.3 Electron Dot Formulas of Atoms
and Monatomic Ions
6.4 Ionic Compounds
6.5 An Introduction to Oxidation and
Reduction
6.6 Covalent Bonds
6.7 Molecular Compounds
6.8 Multiple Covalent Bonds
6.9 Polyatomic Ions and Coordinate
Covalent Bonds
6.10 Electron Dot Formulas of Molecular
Compounds and Polyatomic Ions
6.11 Molecular Shapes
6.12 Electronegativity
6.13 Polar Covalent Bonds
Perspective: Linus Pauling,
Crusading Scientist
6.14 Polarity of Molecules
Summary
Study Questions and Problems

The word *bond* means something that binds, fastens, confines, or holds together. In marriage, two people are held together by the "bond of matrimony." When nations strike an agreement, their association can be referred to as a "bond between nations." In both cases, the bond is an abstraction; it is not a rope or a chain, but a force of attraction. The atoms or ions of chemical compounds are held together by similarly abstract forces of attraction; in this case, we call the forces **chemical bonds.** In the last chapter you learned that valence electrons are important to chemical reactivity. In this chapter you will learn how those valence electrons form chemical bonds.

6.1 TYPES OF CHEMICAL BONDS

In the early nineteenth century, it was found that chemical compounds could be divided into two major categories. Compounds in one category conducted electricity when dissolved in water, while those in the other category did not. It was soon discovered that compounds that conduct electricity produce ions in solution, and the ions are responsible for electrical conductivity. Thus, when sodium chloride, NaCl, dissolves in water, the ions Na^+ and Cl^- are the actual electrical conductors. From studies of sodium chloride and other compounds that conduct electricity, scientists concluded that most of these compounds are composed of ions; these compounds are now called **ionic compounds.** The forces that hold ions together in a compound are called **ionic bonds.** An ionic compound usually contains a metal and one or more nonmetals.

Scientists also found that the compounds whose solutions did not conduct electricity are composed of molecules instead of ions. When these compounds dissolve in water, their simplest units are molecules. (Some molecular compounds react with water to produce ions; however, the majority do not.) Since molecules do not have net charges, they are not capable of conducting electricity. Thus, we refer to compounds in the second category as **molecular compounds.** Molecular compounds are usually composed of nonmetals only. The forces that hold atoms together in a molecule are called **covalent bonds.**

Ionic and covalent bonds are the two major types of attractive forces in chemical compounds; these attractive forces are called **chemical bonds.**

EXAMPLE 6.1   Classify each of the following compounds as ionic or molecular.

a. CH_4 **b.** Na_2O **c.** $MgCl_2$ **d.** H_2S

Solution   We can recognize an ionic compound from its formula by the presence of both a metal and one or more nonmetals. Thus, compounds b and c are ionic. In contrast, molecular compounds usually contain nonmetals only; thus, a and d are molecular compounds.

EXERCISE 6.1   Classify each of the following compounds as ionic or molecular.

a. $CuCl_2$ **b.** NO_2 **c.** HgO **d.** SO_3

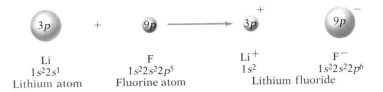

FIGURE 6-1 Formation of Lithium Fluoride from Lithium and Fluorine Atoms (The small p inside each nucleus refers to the protons in the nucleus.)

6.2 IONIC BONDS

In chapter 5, you learned that metals tend to lose their valence electrons while nonmetals (except for the noble gases) tend to gain enough electrons to achieve a valence shell octet. Thus, metals tend to form cations and most nonmetals tend to form anions. It is not surprising then that an ionic compound is formed when an atom of a metal loses its valence electrons to an atom of a nonmetal. Let's see how this happens when a lithium atom and a fluorine atom interact to form the ionic compound lithium fluoride, LiF (see Figure 6-1).

Notice that lithium loses its single valence electron to form the cation Li^+, which is isoelectronic with the noble gas helium. By losing its valence electron, Li achieves a noble gas configuration, and the resulting ion, Li^+, is stable. The electron lost by Li is gained by F, giving it a noble gas configuration also. The net result is that an electron is transferred from Li to F, and two ions are formed. Due to their noble gas configurations, both ions are less reactive than their parent atoms. Notice also that the ions are of different sizes than their parent atoms. The lithium cation is smaller than the atom for two reasons: (1) the ion has no electrons in its second principal energy level, and (2) the two remaining electrons of Li^+ are now attracted more strongly to the three protons of the nucleus. On the other hand, the fluorine anion is larger than the atom. Although it has the same number of principal energy levels as the atom, its ten electrons are not attracted as strongly to the nine protons of the nucleus.

The interaction between a sodium atom and a chlorine atom, shown in Figure 6-2, is similar. The sodium atom achieves an octet in its second principal energy level by losing its valence electron, and the chlorine atom achieves an octet in its valence shell by gaining the electron. The result of the electron transfer is the formation of the ionic compound sodium chloride, NaCl. In both LiF and NaCl, the positively charged cations are attracted to the negatively charged anions. Such electrical attractions between oppositely charged ions are called **ionic bonds.**

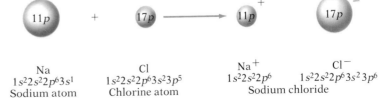

FIGURE 6-2 Formation of Sodium Chloride from Sodium and Chlorine Atoms (The small p inside each nucleus refers to the protons in the nucleus.)

IA	IIA	IIIA	IVA	VA	VIA	VIIA	O
H·							He:
Li·	Be:	:B	:C·	:N·	·O:	:F:	:Ne:
Na·	Mg:	:Al	:Si·	:P·	·S:	:Cl:	:Ar:
K·	Ca:	:Ga	:Ge·	·As·	·Se:	:Br:	:Kr:

FIGURE 6-3 Electron Dot Formulas for Atoms of the First Twenty-Six Elements

6.3 ELECTRON DOT FORMULAS OF ATOMS AND MONATOMIC IONS

The valence electrons of atoms and ions can be conveniently represented by electron dot formulas (also called electronic formulas or Lewis formulas). The **electron dot formula** of an atom is simply the symbol of the element surrounded by dots that represent valence electrons. Unpaired electrons are represented as single dots, and paired electrons are shown as paired dots. Thus, the electron dot formula for lithium ($1s^2 2s^1$) is **Li·**, and that for beryllium ($1s^2 2s^2$) is **Be:**. Figure 6-3 shows electron dot formulas for atoms of the first 26 elements.

For monatomic cations, electron dot formulas have no dots, emphasizing that the valence electrons have been lost. However, electron dot formulas for monatomic anions have eight dots, corresponding to their octets of valence electrons. Thus, the electron dot formula for Na^+ is simply **Na⁺**, while that for Cl^- is **:Cl:⁻**.

Atoms of hydrogen and Group IA elements can lose their single valence electron to form cations with single positive charges:

$$M \cdot \longrightarrow M^+ + e^-,$$

where M represents an atom of the element. The atoms of Group IIA elements lose both of their valence electrons when they form cations

$$M : \longrightarrow M^{2+} + 2e^-$$

and atoms of *metals* of Group IIIA (including the metallic metalloid aluminum) lose all three of their valence electrons when they form cations.

$$: \overset{\cdot}{M} \longrightarrow M^{3+} + 3e^-$$

Notice in each case that the cation has the same number of positive charges as its group number.

Atoms of Group IVA elements represent a balance between tendencies to lose and to gain electrons; thus, they seldom do either and are not often found as ions.

Atoms of *nonmetals* in Groups VA–VIIA tend to form anions. Those in Group VA achieve a noble gas configuration by gaining three electrons

$$: \overset{\cdot}{Z} \cdot + 3e^- \longrightarrow : \overset{\cdot\cdot}{Z} : ^{3-}$$

where Z represents the nonmetal. Group VIA nonmetals add two electrons to achieve a noble gas configuration,

The electron dot formula of an ion must show the valence electrons, if there are any, and the charge of the ion.

$$\cdot \ddot{\underset{\cdot}{Z}}: + 2e^- \longrightarrow :\ddot{\underset{\cdot\cdot}{Z}}:^{2-}$$

and those in Group VIIA add one electron.

$$:\ddot{\underset{\cdot}{Z}}: + e^- \longrightarrow :\ddot{\underset{\cdot\cdot}{Z}}:^-$$

EXAMPLE 6.2   Write the electron dot formula of

a. the cation of magnesium.
b. the anion of oxygen.

Solution   a. An atom of magnesium, a Group IIA element, has two valence electrons. A magnesium atom forms a cation by losing its two valence electrons and taking on a charge of 2+. Therefore, the electron dot formula of the magnesium cation is

$$Mg^{2+}$$

b. An atom of oxygen, a Group VIA element, has six valence electrons; it will form an anion by gaining enough electrons (two) to complete its valence shell octet. Therefore, the electron dot formula of the oxygen anion is

$$:\ddot{\underset{\cdot\cdot}{O}}:^{2-}$$

EXERCISE 6.2   Write the electron dot formula of each of the following ions.

a. The cation of gallium   c. The anion of bromine
b. The anion of sulfur

We can use electron dot formulas to illustrate the transfer of an electron from a lithium atom to a fluorine atom

$$Li\cdot + :\ddot{\underset{\cdot}{F}}: \longrightarrow Li^+ :\ddot{\underset{\cdot\cdot}{F}}:^-$$

formula: LiF

and from a sodium atom to a chlorine atom

$$Na\cdot + :\ddot{\underset{\cdot}{Cl}}: \longrightarrow Na^+ :\ddot{\underset{\cdot\cdot}{Cl}}:^-$$

formula: NaCl

In both of these examples, the single electron lost by the metal was enough to complete the valence shell octet of the nonmetal. What happens, though, when a single metal atom cannot provide enough electrons for the nonmetal atom to attain a noble gas configuration? In such cases the electrons lost by two or more metal atoms are required to satisfy the valence shell octet of the nonmetal. The following equations illustrate this.

$$2\,Li\cdot + \cdot\ddot{\underset{\cdot}{O}}: \longrightarrow (Li^+)_2 \;:\ddot{\underset{\cdot}{O}}:^{2-}$$

formula: Li$_2$O

$$3\,Na\cdot + :\overset{\cdot}{\underset{\cdot}{N}}\cdot \longrightarrow (Na^+)_3 \;:\ddot{\underset{\cdot}{N}}:^{3-}$$

formula: Na$_3$N

Some metal atoms have more than enough valence electrons for some nonmetal atoms; in these cases, two or more nonmetal atoms are needed to accommodate the electrons from the metal atom:

$$\text{Mg:} + 2 \; \ddot{\underset{\cdot}{\text{Cl}}}\text{:} \longrightarrow \text{Mg}^{2+} \; (\text{:}\ddot{\underset{\cdot\cdot}{\text{Cl}}}\text{:}^-)_2$$

formula: MgCl_2

$$\text{:}\dot{\text{Al}} + 3 \; \text{:}\ddot{\underset{\cdot}{\text{Br}}}\text{:} \longrightarrow \text{Al}^{3+} \; (\text{:}\ddot{\underset{\cdot\cdot}{\text{Br}}}\text{:}^-)_3$$

formula: AlBr_3

And finally, there are occasions when two or more of both the metal and the nonmetal atoms are needed for the electron transfer:

$$2\text{:}\dot{\text{Al}} + 3 \cdot \ddot{\underset{\cdot}{\text{O}}}\text{:} \longrightarrow (\text{Al}^{3+})_2 \; (\text{:}\ddot{\underset{\cdot\cdot}{\text{O}}}\text{:}^{2-})_3$$

formula: Al_2O_3

$$3\text{ Ca:} + 2 \; \text{:}\dot{\underset{\cdot}{\text{P}}}\cdot \longrightarrow (\text{Ca}^{2+})_3 \; (\text{:}\ddot{\underset{\cdot\cdot}{\text{P}}}\text{:}^{3-})_2$$

formula: Ca_3P_2

As you see, things can get complicated; however, remember that the total number of electrons lost by all atoms of the metal must be the same as the total number of electrons gained by all atoms of the nonmetal. The resulting ionic compound has equal positive and negative charges and is electrically neutral. To find the formula of an ionic compound that contains just two elements, write down the formulas of the cation and anion; then use the number of the charge on each ion as the subscript on the other ion:

$$\text{Li}^{①+} \rightleftarrows \text{O}^{②-} \qquad \text{Li}_2\text{O}$$

$$\text{Mg}^{②+} \rightleftarrows \text{Cl}^{①-} \qquad \text{MgCl}_2$$

$$\text{Al}^{③+} \rightleftarrows \text{O}^{②-} \qquad \text{Al}_2\text{O}_3$$

It is always necessary to write the *simplest formula,* that is, the formula that has the lowest possible whole-number ratios of ions. The examples above all give simplest formulas. However, if the subscripts are *identical* and have values of two or greater, the simplest formula is derived by dividing each subscript by itself, for example

$$\text{Mg}^{②+} \rightleftarrows \text{O}^{②-} = \text{Mg}_2\text{O}_2 = \text{Mg}_{2/2}\text{O}_{2/2} = \text{MgO}$$

If two nonidentical subscripts have values of two or greater and if the larger subscript is evenly divisible by the smaller, the simplest formula is derived by dividing both subscripts by the smaller, for example

$$\text{Sn}^{④+} \rightleftarrows \text{O}^{②-} = \text{Sn}_2\text{O}_4 = \text{Sn}_{2/2}\text{O}_{4/2} = \text{SnO}_2$$

This method automatically makes the total positive and negative charges in formula equal.

EXAMPLE 6.3   Write the electron dot formula of the ionic compound formed by calcium and iodine.

Solution First use the group number to determine the number of valence electrons for each atom. An atom of calcium, Group IIA, has two valence electrons. Since calcium is a metal, it will form a cation by losing its valence electrons. Thus, the electron dot formula of the cation is Ca^{2+}. An atom of iodine, Group VIIA, has seven valence electrons. Since iodine is a nonmetal, it will form an anion by gaining one electron to complete its valence shell octet. Therefore, the electron dot formula of the iodine anion is

$$:\!\overset{\cdot\cdot}{\underset{\cdot\cdot}{I}}\!:^{-}$$

Now, write the electron dot formulas of the cation and anion together, and choose the appropriate subscript(s) to make total positive charge equal to total negative charge.

$$Ca^{2+}\,(:\!\overset{\cdot\cdot}{\underset{\cdot\cdot}{I}}\!:^{-})_2$$

EXERCISE 6.3 Write the electron dot formula of the ionic compound formed by each pair of elements.

a. Potassium and sulfur   c. Aluminum and fluorine
b. Calcium and oxygen   ■

EXAMPLE 6.4 Write the formula of the ionic compound composed of rubidium and oxygen.

Solution Write down the formulas of the cation and anion; then use the number of the charge on each ion as a guide for choosing the subscript(s).

$$Rb^{①+} \overset{}{\underset{}{\times}} O^{②-} \quad \text{or} \quad Rb_2O$$

EXERCISE 6.4 Write the formula of the ionic compound composed of each pair of elements.

a. Barium and oxygen   c. Strontium and nitrogen
b. Gallium and bromine   ■

6.4 IONIC COMPOUNDS

Ionic compounds are always composed of cations and anions, and the total number of positive charges is always equal to the total number of negative charges. For an ionic compound such as sodium chloride, NaCl, the simplest unit, or formula unit, would contain one Na^+ and one Cl^-, and the size of this simplest unit would be much too small to see. On a more practical level, if you sprinkle some table salt on a dark background, you will see that sodium chloride is actually made up of cubic crystals. A sodium chloride crystal is diagrammed in Figure 6-4.

Crystals of sodium chloride are composed of sodium ions and chloride ions arranged in an orderly array called a **crystal lattice.** The six nearest neighbors of each cation are chlorine anions, and the six nearest neighbors of each anion are sodium cations. In theory, a crystal of sodium chloride could be composed of an infinitely large number of ions; how-

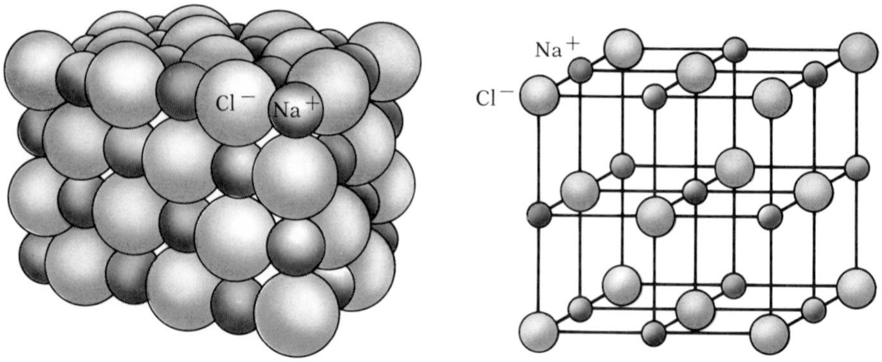

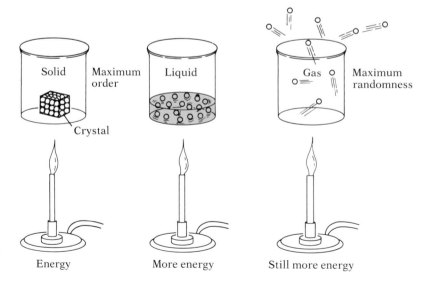

FIGURE 6-4 Two Representations of the Crystal Structure of Sodium Chloride. The diagram on the left shows the ions in their correct relative sizes, while the diagram on the right emphasizes the location of ions in the crystal lattice.

ever, crystals are easily fractured by handling, and small crystals are more common than large ones. Even in small crystals, there are trillions of ions, but regardless of how many ions are present, the ratio of sodium cations to chlorine anions is always 1 : 1. Each crystal is electrically neutral, containing the same number of positive charges as negative charges.

All ionic compounds form crystals, although the crystal shape of one compound may be different from that of another. Since a crystal can be of any size, the number of ions in a crystal is indefinite. The ionic bonds that hold ionic crystals together are not localized to specific pairs of ions but extend indefinitely through the crystal. For example, in the case of NaCl, each Na^+ is attracted most strongly to its six nearest Cl^- neighbors and less strongly to all other nearby Cl^- ions that lie beyond the six nearest neighbors. Similarly, each Cl^- is attracted most strongly to its six nearest Na^+ neighbors and less strongly to all other nearby Na^+ ions that lie beyond the six nearest neighbors. However, because an ionic compound has no net electrical charge, the formula of an ionic compound gives the simplest ratio of ions needed for electrical neutrality; this simplest ratio thus is referred to as a **formula unit.**

FIGURE 6-5 The Melting and Subsequent Vaporization of a Crystalline Solid

TABLE 6-1 Melting Points and Boiling Points of Some Ionic Compounds

Compound	Formula	Melting Point (°C)	Boiling Point (°C)
Ammonium chloride	NH_4Cl	340	520
Barium chloride	$BaCl_2$	963	1560
Barium oxide	BaO	1920	2000
Calcium fluoride	CaF_2	1418	2500
Calcium oxide	CaO	2590	2850
Lead chloride	$PbCl_2$	500	950
Lithium chloride	$LiCl$	610	1360
Magnesium fluoride	MgF_2	1263	2230
Magnesium oxide	MgO	2800	3580
Potassium chloride	KCl	772	1411
Sodium chloride	$NaCl$	808	1473
Sodium fluoride	NaF	1450	2490

Binary
Consisting of two.

The compounds discussed so far have been composed of ions of only two elements. Such compounds are referred to as **binary ionic compounds.** (You will study ionic compounds containing more than two elements later in this chapter.) Regardless of their complexity, ionic compounds are always composed of oppositely charged ions that are electrically attracted to each other. These attractions are quite strong, and ionic compounds tend to be solids at room temperature and have very high melting points. When a solid melts, its particles become much more mobile, moving randomly through the liquid. It requires much energy, hence high temperatures, to disrupt ionic bonds and cause the ions to move about. Much higher temperatures are required to separate the ions completely and thus form a gas. Figure 6-5 illustrates the melting and vaporizing processes for a crystalline solid. Examples of the high melting and boiling points of ionic compounds are given in Table 6-1.

6.5 AN INTRODUCTION TO OXIDATION AND REDUCTION

When a shiny surface of sodium metal is exposed to air, the dull coating that collects on the surface is sodium oxide, Na_2O. During the reaction between sodium and oxygen, electrons are transferred from sodium atoms to oxygen atoms. The electron dot formula of sodium oxide is

$$(Na^+)_2 \; : \overset{..}{\underset{..}{O}} : ^{2-}$$

formula: Na_2O

We can imagine that in the course of the reaction, every atom of sodium lost one electron (and every two sodium atoms lost two electrons)

$$2\,Na \cdot \longrightarrow 2\,Na^+ + 2\,e^- \qquad (1)$$

and each oxygen atom gained two electrons,

$$\cdot \overset{..}{\underset{.}{O}} : + 2\,e^- \longrightarrow : \overset{..}{\underset{..}{O}} : ^{2-} \qquad (2)$$

The sum of these two processes is

$$2\,\text{Na}\cdot + \cdot\ddot{\underset{\cdot}{\text{O}}}\!: \longrightarrow (\text{Na}^+)_2\ :\!\ddot{\underset{\cdot\cdot}{\text{O}}}\!:^{2-} \tag{3}$$

In reaction (1), sodium atoms changed to sodium cations by losing electrons. A reaction in which electrons are lost is called an **oxidation;** in this example, sodium was oxidized. Sodium atoms have no charge, so the charge on sodium went from 0 to 1+ during the reaction.

In oxidation reactions, we often refer to **oxidation numbers.** These are numbers assigned to atoms and ions according to some arbitrary rules that will be explained more fully in chapter 7. In this chapter, we will only be concerned with deriving oxidation numbers for monatomic ions and uncombined atoms, whose oxidation numbers are assigned on the basis of charge. Like charges, oxidation numbers have both signs and numbers, but for oxidation numbers the signs are written before the numbers. For uncombined atoms and monatomic ions, the oxidation number has the same value as the charge. Thus, the oxidation number of Na is 0, and the oxidation number of Na^+ is $+1$. In reaction (1), the oxidation number of sodium increased. We can now restate the definition of oxidation in terms of oxidation numbers: **oxidation** is an increase in the oxidation number of atoms or ions of an element.

In reaction (2), oxygen atoms changed to anions by gaining electrons. A reaction in which electrons are gained is called **reduction.** Thus, in this example, oxygen was reduced. Since oxygen atoms have no charge, the oxidation number of O is 0, but the O^{2-} anion has an oxidation number of -2. Thus, the oxidation number of oxygen decreased during the reaction. We can also define **reduction** as a decrease in oxidation number of atoms or ions of an element.

EXAMPLE 6.5 What is the oxidation number of each of the following?

a. Mg   **b.** S^{2-}

Solution   Since oxidation numbers of atoms and monatomic ions are the same as their charges (but written in reverse order), the oxidation number of Mg is 0 and that of S^{2-} is -2.

EXERCISE 6.5 What is the oxidation number of each of the following?

a. Cs^+ **b.** Al   **c.** I^- **d.** Se^{2-} ■

The loss of electrons by sodium and the gain of electrons by oxygen occurred at the same time; both processes were necessary for the formation of sodium oxide. Oxidation and reduction always go together, and we call the combination of the two an **oxidation-reduction reaction.** In an oxidation-reduction reaction, one substance is oxidized and another is reduced. The substance being oxidized loses electrons and undergoes an increase in oxidation number while the substance being reduced gains electrons and undergoes a decrease in oxidation number.

Other examples of oxidation-reduction reactions are given below, with oxidation numbers indicated below the formulas. You can see that oxygen is not a requirement for oxidation-reduction reactions.

$$K\cdot + :\ddot{\underset{\cdot}{C}l}: \longrightarrow K^+ :\ddot{\underset{\cdot\cdot}{C}l}:^-$$
$$\quad 0 \qquad 0 \qquad\qquad +1 \quad -1$$

$$:\overset{\cdot}{Al} + 3 :\ddot{\underset{\cdot}{F}}: \longrightarrow Al^{3+} (:\ddot{\underset{\cdot\cdot}{F}}:^-)_3$$
$$\quad 0 \qquad 0 \qquad\qquad +3 \quad -1$$

A more extensive discussion of oxidation and reduction is given in chapter 16.

6.6 COVALENT BONDS

H
$1s^1$

He
$1s^2$

FIGURE 6-6 Atoms of Hydrogen and Helium (The small p inside each nucleus refers to the protons in the nucleus.)

The second major category of chemical compounds is molecular compounds. These usually contain only nonmetals held together in molecules by covalent bonds. As an approach to understanding the nature of covalent bonds, let's compare the first two elements in the periodic table, hydrogen and helium. A hydrogen atom consists of a positively charged nucleus and one electron, while a helium atom has a positively charged nucleus and two electrons. Hence, the electronic structures of hydrogen and helium differ only by one electron (see Figure 6-6). Keep in mind, however, that helium, a noble gas, forms no compounds; yet hydrogen, so similar in electronic structure, forms countless compounds.

As the free element, hydrogen exists as two atoms bonded together in a diatomic molecule, H_2. Hydrogen molecules apparently form from pairs of hydrogen atoms because each nucleus has an attraction for the electron of the other atom (see Figure 6-7). The two hydrogen nuclei remain apart in the hydrogen molecule, but each still has an attraction for its own electron as well as for the electron of the other atom. Thus, in the hydrogen molecule, two nuclei attract two electrons, and vice versa. These mutual attractions between electrons and protons allow the hydrogen molecule to form. When it forms, the $1s$ orbitals of the two atoms overlap, and the two electrons then move about within the blended orbitals. This means that the two atoms of the molecule now share both electrons. A chemical bond composed of a pair of shared electrons is called a **covalent bond.** Compounds in which all of the bonds are covalent are composed of molecules. The sharing of the electron pair gives each atom in the hydrogen molecule the electron configuration of the noble gas helium, and the molecule is therefore more stable than individual hydrogen atoms.

FIGURE 6-7 The Formation of a Hydrogen Molecule from Two Hydrogen Atoms

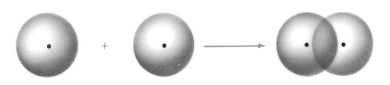

H

H

H_2

FIGURE 6-8 The Formation of a
Fluorine Molecule from Two Fluorine
Atoms (Only one 2*p* orbital is shown for
each fluorine atom.)

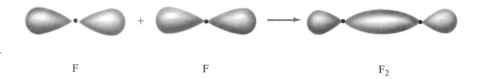

F   F   F$_2$

We can represent the hydrogen atom, the helium atom, and the hydrogen molecule by the electron dot formulas shown below:

H·   He: H:H   H—H

Hydrogen atom   Helium atom   Hydrogen molecule   Hydrogen molecule

The first electron dot formula of H$_2$ shows that a shared pair of electrons
bind the hydrogen molecule together. As a matter of convenience, we
usually designate a shared electron pair by a straight line drawn between
the two atoms, as shown in the second electron dot formula of H$_2$.

Since a hydrogen atom has only one electron, it is logical to expect that
hydrogen atoms form covalent bonds with atoms of other elements that
can contribute a single electron toward covalent bond formation. Furthermore, we can anticipate that many atoms are capable of forming
covalent bonds; their compounds will be stable if covalent bond formation gives each atom a noble gas configuration.

Now let's examine the nonmetal fluorine. In some ways, fluorine atoms
are similar to hydrogen atoms. A fluorine atom always bonds to only one
other atom, just as a hydrogen atom does. A second similarity is that a
fluorine atom can also achieve a noble gas configuration by sharing a pair
of electrons. The formation of a fluorine molecule from two fluorine
atoms is illustrated in Figure 6-8. Each atom contributes its single unpaired 2*p* electron toward covalent bond formation. The bond is formed
when a 2*p* orbital of one fluorine atom overlaps with a 2*p* orbital of the
other. The shared pair of electrons can then move about in the new orbital
that results from blending the two 2*p* orbitals. By sharing the pair of
electrons, each fluorine atom now has an octet of valence electrons, corresponding to the electron configuration of the noble gas neon.

The electron dot formulas of the fluorine atom and the fluorine molecule are shown below.

:F̤:   :F̤:F̤:   :F̤—F̤:

Fluorine atom   Fluorine molecule   Fluorine molecule

TABLE 6-2	Elements Consisting of Diatomic Molecules		
Element	Molecular Formula	Physical State (25° C)	Color
Hydrogen	H$_2$	Gas	Colorless
Nitrogen	N$_2$	Gas	Colorless
Oxygen	O$_2$	Gas	Colorless
Fluorine	F$_2$	Gas	Pale green
Chlorine	Cl$_2$	Gas	Pale greenish yellow
Bromine	Br$_2$	Liquid	Reddish brown
Iodine	I$_2$	Solid	Dark purple

Those valence electrons not involved in the covalent bond are called *nonbonding electrons, unshared pairs,* or *lone pairs.*

Seven elements exist in their uncombined states as diatomic molecules. These elements are listed in Table 6-2.

6.7 MOLECULAR COMPOUNDS

Now that you can write electron dot formulas for H_2 and F_2 molecules, let's proceed to a molecular *compound* composed of these two elements. Hydrogen fluoride, HF, is a gas formed when hydrogen and fluorine react. We can write an electron dot formula for the HF molecule by first writing electron dot formulas for the individual atoms. We see that hydrogen has one valence electron available for covalent bond formation, while fluorine has seven. By allowing the single electron of hydrogen to pair with the unpaired electron of fluorine, we can form the covalent bond on paper. We then draw the covalent bond between the two atoms and indicate the nonbonding electrons on the fluorine atom.

H·   :F̈:   H—F̈:

Hydrogen atom Fluorine atom Hydrogen fluoride molecule

Note once again that each atom in the molecule, by sharing the bonding electron pair, has a noble gas configuration.

Most molecular compounds are composed of nonmetals, because nonmetals have similar tendencies to gain or lose electrons, and thus, they share electrons with one another. Many molecular compounds are gases such as carbon dioxide (CO_2), the colorless, odorless gas we exhale, and nitrogen dioxide (NO_2), the reddish brown gas in smog. There are also many molecular compounds that are liquids and solids at room temperature. The liquids have relatively low boiling points (50–200° C), and the solids have relatively low melting points (100–300° C). Solid molecular compounds exist as crystals in which molecules occupy positions in the crystal lattice. Attractive forces among molecules are weak relative to ionic attractions, accounting for the low melting and boiling points of molecular compounds.

6.8 MULTIPLE COVALENT BONDS

Two atoms can share two or even three pairs of electrons, if they achieve an octet of valence electrons by doing so. Such bonds are called **multiple covalent bonds.**

When two pairs of electrons are shared by two atoms, the bond is referred to as a **double bond.** Carbon dioxide, CO_2, is an example of a compound with double bonds. The electron dot formula for carbon dioxide is given in two forms below.

Ö::C::Ö Ö=C=Ö

Two ways of writing the electron dot formula of the carbon dioxide molecule

In the second formula, which is preferred, the double bond is represented by a pair of straight lines between two atoms. There are two double bonds in the carbon dioxide molecule, each connecting the carbon atom to an oxygen atom. Notice that each atom has an octet of valence electrons.

If three pairs of electrons are shared between two atoms, the bond is called a **triple bond.** The nitrogen molecule, N_2, is held together by a triple bond,

$$:N{\equiv}N:$$
Nitrogen molecule

A triple bond is represented by three straight lines between two atoms; there is one triple bond in the nitrogen molecule. By sharing three pairs of electrons, each nitrogen atom attains an octet of valence electrons.

As you continue to study chemistry, you will find many compounds having double or triple bonds. The following are examples of common multiple bonds:

$\diagdown{C}{=}C\diagup$	carbon-carbon double bond
$-C{\equiv}C-$	carbon-carbon triple bond
$-\ddot{N}{=}\ddot{N}-$	nitrogen-nitrogen double bond
$:N{\equiv}N:$	nitrogen-nitrogen triple bond
$\diagdown{C}{=}\ddot{N}-$	carbon-nitrogen double bond
$-C{\equiv}N:$	carbon-nitrogen triple bond
$\diagdown{C}{=}\ddot{O}$	carbon-oxygen double bond

6.9 POLYATOMIC IONS AND COORDINATE COVALENT BONDS

Many ions are composed of two or more atoms. For example, the principal anion in bone tissue, PO_4^{3-}, contains one phosphorus atom and four oxygen atoms. In another example, the anion found in chalk, CO_3^{2-}, contains one carbon atom and three oxygen atoms. An ion containing two or more atoms is called a **polyatomic ion.** The atoms in polyatomic ions are covalently bonded together so that they behave as one unit. The charge on a polyatomic ion arises from an excess or deficiency of electrons relative to protons in the unit; the charge is not associated with any particular atom but belongs to the entire unit. Polyatomic ions may have positive or negative charges.

Poly-
Greek prefix meaning "many."

The most common polyatomic cation is the ammonium ion, NH_4^+. It is formed when a hydrogen cation adds to the ammonia molecule, NH_3.

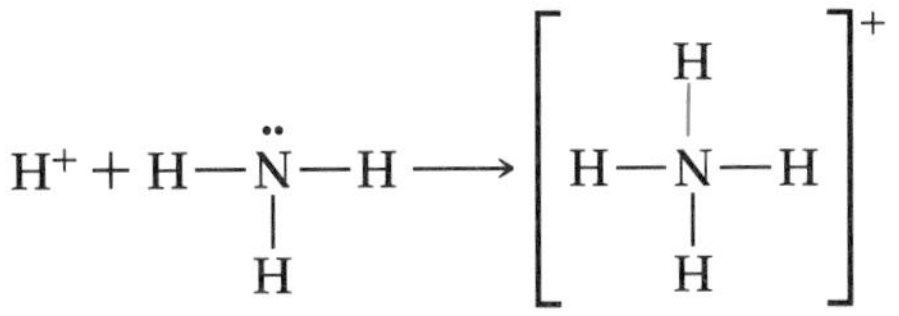

$$H^+ + H-\overset{..}{\underset{|}{N}}-H \longrightarrow \left[H-\overset{\overset{H}{|}}{\underset{\underset{H}{|}}{N}}-H \right]^+$$

The positive charge on the ammonium ion indicates that it contains one more proton than its number of electrons. In other words, the unit composed of one nitrogen atom and four hydrogen atoms has lost one valence electron. Brackets are used to emphasize that the charge applies to the whole unit.

The ammonium cation contains one covalent bond in which both electrons were contributed by nitrogen. When a covalent bond is composed of a pair of electrons provided by only one of the connected atoms, the bond is called a **coordinate covalent bond.** After it is formed, a coordinate covalent bond has all the properties of any other covalent bond and cannot be distinguished from other covalent bonds. The coordinate covalent bond in the ammonium cation above is indicated in color simply for purposes of illustration.

The total number of valence electrons in the ammonium cation is calculated as follows:

Valence electrons in a N atom	5
Valence electrons in 4 H atoms	4
Less 1 electron, as indicated by charge	-1
Total number of valence electrons	8

Polyatomic anions are much more common than polyatomic cations. Several of these are illustrated in Figure 6-9.

Some compounds containing polyatomic ions are found as natural deposits in the earth's crust; some examples are given in Table 6-3. The study of ionic compounds was once restricted to the field of **inorganic chemistry,** the chemistry of nonliving substances. However, modern research has demonstrated the importance of ionic compounds in living organisms, from single cells to complex organisms, and ionic compounds are now considered to be active participants in the chemistry of life.

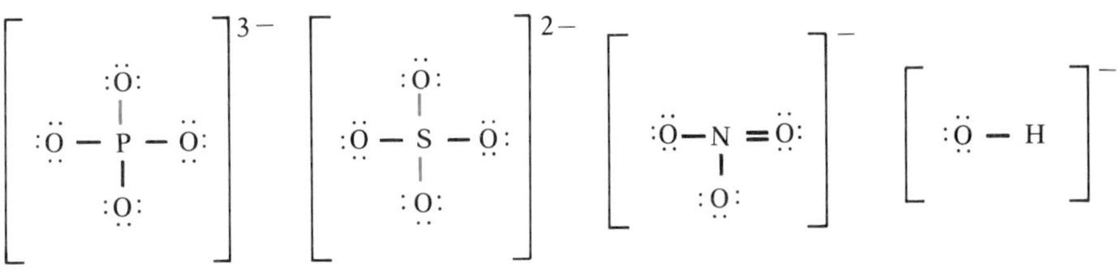

FIGURE 6-9 Examples of Polyatomic Anions (Coordinate covalent bonds are shown in color.)

PO_4^{3-}
Phosphate ion

SO_4^{2-}
Sulfate ion

NO_3^-
Nitrate ion

OH^-
Hydroxide ion

TABLE 6-3 Some Ionic Compounds in the Environment

Mineral Name	Chemical Name	Formula
Alumina	Aluminum oxide	Al_2O_3
Barite	Barium sulfate	$BaSO_4$
Calamine	Zinc carbonate	$ZnCO_3$
Chalk	Calcium carbonate	$CaCO_3$
Cinnebar	Mercury(II) sulfide	HgS
Hematite	Iron(III) oxide	Fe_2O_3
Lime	Calcium oxide	CaO
Magnesia	Magnesium oxide	MgO
Saltpeter	Potassium nitrate	KNO_3
Talc	Magnesium silicate	$Mg_3Si_4O_{10}(OH)_2$

6.10 ELECTRON DOT FORMULAS OF MOLECULAR COMPOUNDS AND POLYATOMIC IONS

A hydrogen atom can form only one covalent bond because it has only one electron to contribute toward a shared pair of bonding electrons. Thus, we could say that hydrogen has a combining power of one, since a hydrogen atom can combine with only one other atom. Atoms that combine with two other atoms therefore have a combining power of two, those that combine with three other atoms have a combining power of three, and so on. In our modern concept of bonding, we define combining power in terms of how many covalent bonds an atom can form, and we also give it another name—**valence.** Valence is the number of covalent bonds an atom can form.

Electron dot formulas for molecular compounds and polyatomic ions are based on the formulas of the molecules or ions and the common valences of the atoms. The following steps summarize how to arrive at the electron formula of a molecular compound or polyatomic ion.

1. Determine the atomic skeleton of the molecule or polyatomic ion.
   a. The first step in determining the atomic skeleton is to identify the central atom, that is, the atom that has the most other atoms bonded to it. For most binary molecules and polyatomic ions, the central atom appears only once in the formula. For example, in BH_3, B is the central atom, and in ClO_4^-, Cl is the central atom. For molecules containing carbon, the central atom is usually C; for example, C is the central atom in HCN. Hydrogen and fluorine can never be central atoms. (In some binary compounds, there is no central atom; in many of these, two or more identical nonmetal atoms are bonded together, as in H_2N-NH_2. Carbon atoms are frequently bonded together in covalent compounds that contain two or more carbon atoms per molecule such as propane, $H_3C-CH_2-CH_3$.)
   b. Once you have identified the central atom, the next step is to arrange as many of the other atoms as possible about the central atom. This often gives a symmetrical pattern about the central atom. Not all of

the other atoms will necessarily be bonded to the central atom; some may be bonded to each other. In arranging the atoms, use the following valences:

Carbon usually has a valence of 4.

Nitrogen usually has a valence of 3 but sometimes has a valence of 4.

Oxygen usually has a valence of 2.

Hydrogen usually has a valence of 1.

(Note that these are *common* valences; occasionally carbon, nitrogen, oxygen, and hydrogen show other valences.)

c. After arranging the atoms, connect them with single covalent bonds. For example, the molecule H_2O would be arranged as

$$H—O—H$$

2. Count the total number of valence electrons of all the atoms in the molecule or polyatomic ion. (You can use the group number of each element as a guide in this step.) Be sure to account for the charge on a polyatomic ion when counting its total number of valence electrons.

3. Subtract two valence electrons from the total for each single covalent bond written in step 1. Then distribute the remaining electrons as unshared pairs to give each atom an octet of valence electrons (except for hydrogen, which can have only two electrons).

4. If a single covalent bond does not give an atom an octet of electrons, change the single bond to a double or triple bond by shifting nonbonding electrons from nearby atoms.

Let's work out the electron dot formula of ammonia, NH_3, as an example. Since N must be the central atom, the only atomic skeleton that satisfies the usual valences of nitrogen (3) and hydrogen (1) is

$$\begin{array}{c} H—N—H \\ | \\ H \end{array}$$

Next, count the total number of valence electrons: 5 from the nitrogen atom and 1 from each of three hydrogen atoms gives a total of 8 valence electrons. Since we have already drawn three covalent bonds, we must subtract 6 electrons from the total of 8 (step 3), leaving two electrons, which we place on the nitrogen atom as nonbonding electrons.

$$\begin{array}{c} \overset{\displaystyle ..}{H—N—H} \\ | \\ H \end{array}$$

We check our electron dot formula by confirming that each atom has an octet of valence electrons (except for hydrogen, which can have only two valence electrons).

EXAMPLE 6.6   Write the electron dot formula of the water molecule, H_2O.

Solution   1. Draw the atomic skeleton to show O as the central atom, and join the atoms with single bonds, making sure that hydrogen has a valence of 1 and oxygen a valence of 2.

$$H—O—H$$

2. Count total valence electrons.

$$2\ H = 2(1) = 2$$
$$O = \underline{\quad 6}$$
$$8\ \text{total valence electrons}$$

3. Subtract two valence electrons for each single bond and distribute the remaining valence electrons as nonbonding electrons.

$8 - 4 = 4$ remaining valence electrons to be distributed as nonbonding electrons

This gives

$$H—\overset{\displaystyle ..}{\underset{\displaystyle ..}{O}}—H$$

4. Check to see that each atom has a noble gas valence shell configuration. They do (each hydrogen atom has two valence shell electrons and the oxygen atom has eight valence shell electrons), so our electron dot formula of H_2O is correct.

EXERCISE 6.6   Write the electron dot formula of each of the following molecules.

a. CH_4 b. NCl_3 c. H_2S ■

EXAMPLE 6.7   Write the electron dot formula of the SO_2 molecule.

Solution   1. Draw the atomic skeleton to show S as the central atom. In this case, oxygen apparently does not have a valence of 2, because this valence is not compatible with S as the central atom.

$$O—S—O$$

2. Count total valence electrons.

$$2\ O = 2(6) = 12$$
$$S = \underline{\quad 6}$$
$$18\ \text{total valence electrons}$$

3. Subtract two electrons for each single bond and distribute the remaining electrons as nonbonding electrons.

$18 - 4 = 14$ remaining electrons to be distributed as nonbonding electrons

This gives

$$:\overset{\displaystyle ..}{\underset{\displaystyle ..}{O}}—\overset{\displaystyle ..}{\underset{\displaystyle ..}{S}}—\overset{\displaystyle ..}{\underset{\displaystyle ..}{O}}:$$

4. If necessary, shift nonbonding electrons to create multiple bonds and give each atom a noble gas configuration.

$$:\overset{\displaystyle ..}{\underset{\displaystyle ..}{O}}—\overset{\displaystyle ..}{\underset{\displaystyle \overset{..}{\curvearrowright}}{S}}—\overset{\displaystyle ..}{\underset{\displaystyle ..}{O}} \longrightarrow :\overset{\displaystyle ..}{\underset{\displaystyle ..}{O}}—\overset{\displaystyle ..}{S}=\overset{\displaystyle ..}{\underset{\displaystyle ..}{O}}$$

(Note that one oxygen atom has a valence of 1 instead of the usual 2.)

EXERCISE 6.7 Write the electron dot formula of each of the following molecules.

a. CO   b. N_2O_4 c. CO_2 ◼

EXAMPLE 6.8 Write the electron dot formula of the nitrate ion, NO_3^-.

Solution
1. Draw the atomic skeleton to show N as the central atom. In this case, oxygen apparently does not have a valence of 2, because this valence is not compatible with N as the central atom.

$$\left[\begin{array}{c} O-N-O \\ | \\ O \end{array} \right]^{-}$$

2. Count total valence electrons.

$$\begin{array}{rl} 3\ O = 3(6) = & 18 \\ N = & 5 \\ charge = & \underline{1} \\ & 24 \ total\ valence\ electrons \end{array}$$

3. Find number of nonbonding electrons and distribute them.

$$24 - 6 = 18 \ nonbonding\ electrons$$

$$\left[\begin{array}{c} :\ddot{O}-\ddot{N}-\ddot{O} \\ | \\ :\ddot{O}: \end{array} \right]^{-}$$

4. Create multiple bonds if necessary.

$$\left[\begin{array}{c} :\ddot{O}-\ddot{N}{\overset{\cdot\cdot}{\curvearrowleft}}\ddot{O} \\ | \\ :\ddot{O}: \end{array} \right]^{-} \longrightarrow \left[\begin{array}{c} :\ddot{O}-N=\ddot{O} \\ | \\ :\ddot{O}: \end{array} \right]^{-}$$

(Note that two oxygen atoms have valences of 1 instead of the usual 2, and the nitrogen atom has a valence of 4 instead of its usual 3.)

EXERCISE 6.8 Write the electron dot formula of the carbonate anion, CO_3^{2-}.   ◼

6.11 MOLECULAR SHAPES

The physical properties of covalent compounds are greatly influenced by the shapes of their molecules; thus, it is important to understand the major factors that are responsible for molecular shape. You have learned to write electron dot formulas of molecules on paper, but these representations do not tell much about the three-dimensional shapes of molecules. To learn about molecular shapes, we must consider the angles at which atoms are bonded to one another.

Each covalent bond is a region of space made up of two overlapping atomic orbitals. A pair of electrons can occupy this region despite their like charges if the electrons have opposite spins. However, a set of paired electrons repels other nearby sets of paired electrons due to the high concentration of negative charge in each region. Thus, a covalent bond repels other nearby covalent bonds, and the valence shell electron pairs of an atom will stay as far away from each other as possible. This **valence shell electron pair repulsion theory,** or **VSEPR theory,** allows us to make predictions about bond angles and molecular shapes. For example, methane (CH_4) has the electron dot formula

$$
\begin{array}{c}
\text{H} \\
| \\
\text{H} - \text{C} - \text{H} \\
| \\
\text{H}
\end{array}
$$

Methane

How can the valence shell electron pairs of the carbon atom get as far away from each other as possible? The bonds to carbon must spread out into space to give the largest possible equal bond angles. If you were to build a model of the carbon atom and its bonds using a gumdrop and four toothpicks, you would find that the largest possible equal bond angles are 109.5°. Indeed, this is the value for all the bond angles in methane. If we consider that a hydrogen atom is attached to each bond, we find that the methane molecule has the shape of a regular tetrahedron (see Figure 6-10(b)). In this solid figure of four equivalent faces and equal corner angles, the carbon atom occupies the central position and the hydrogen atoms are at the four corners.

The VSEPR theory can also be used to deduce the shape of the ammonia molecule, NH_3, shown in Figure 6-11(b). In this case, we must consider the pair of nonbonding electrons in one of nitrogen's $2p$ orbitals. This pair of electrons occupies a region of space just as the bonding electron pairs do, and if we allow for repulsions among all of the pairs of valence shell electrons, we would expect the same bond angles as in methane. However, measurements have shown the bond angles in the ammonia molecule to be slightly less than the 109.5° predicted. The small deviation occurs because the unshared electron pair occupies a slightly larger region of space than a pair of bonding electrons. Since the nonbonding electrons are not shared by two atoms, they are more free to move around than bonding electrons. The larger region occupied by the nonbonding

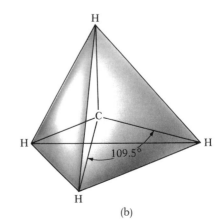

FIGURE 6-10 (a) The Electron Dot Formula of Methane (b) The Geometry of the Methane Molecule

(a)   (b)

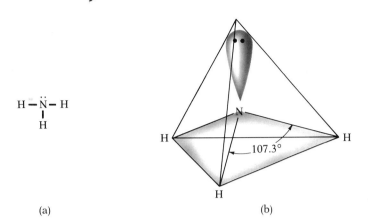

H — N̈ — H
|
H

(a)

FIGURE 6-11 (a) The Electron Dot Formula of Ammonia (b) The Geometry of the Ammonia Molecule

(b)

electron pair has the effect of squeezing the nitrogen-to-hydrogen bonds slightly closer together. Thus, the ammonia molecule has bond angles of 107.3°, and it is shaped like a pyramid (see Figure 6-11(b)). In this case, the three hydrogen atoms are located at corners of a slightly irregular tetrahedron, and the fourth corner is occupied by the pair of nonbonding electrons.

Let's examine the shape of the water molecule, H_2O, shown in Figure 6-12(b). The oxygen atom has two pairs of nonbonding electrons to consider, and each occupies its own region of space. Thus there are four individual pairs of mutually repulsive valence shell electrons on the oxygen atom, and we would predict tetrahedral geometry for them. However, the two pairs of unshared electrons compress the bond angles even more than the single pair did in the ammonia molecule. Thus, bond angles in the water molecule are 104.5°. This gives the water molecule the *bent* shape illustrated in Figure 6-12(b).

In all of the preceding examples, there have been eight valence electrons about a central atom; however, atoms of two representative elements, hydrogen and boron, are unable to achieve an octet of valence electrons when they form molecules. Hydrogen is limited to two valence electrons because it has only the $1s$ orbital in its ground state. Atoms of

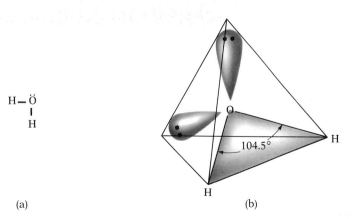

H — Ö
|
H

(a)

FIGURE 6-12 (a) The Electron Dot Formula of Water (b) The Geometry of the Water Molecule

(b)

FIGURE 6-13 (a) The Electron Dot Formula of Boron Trifluoride (b) The Geometry of the Boron Trifluoride Molecule

Trigonal

Shaped like a triangle.

boron, the nonmetallic metalloid in Group IIIA, have three valence electrons, and they form only covalent bonds. Thus when boron forms a molecule, there are only three covalent bonds to each boron atom and boron has only six valence electrons, but the molecule is stable. The three single covalent bonds to a boron atom mutually repel each other, as you would expect, and all of the bonds lie in the same plane separated by angles of 120°. Figure 6-13(b) illustrates the geometry of the boron trifluoride molecule, BF_3. Because its bonds point to the three corners of a triangle, the BF_3 molecule is said to be *trigonal*.

From reasoning based on VSEPR theory, we have deduced that when there are four individual pairs of valence electrons about an atom in a molecule, the bond angles to the atom will be either 109.5° or something very close to this value, and the pairs of valence electrons will be tetrahedrally distributed about the atom. The molecule will have a tetrahedral shape, like methane, if there are no nonbonding electron pairs on the atom. It will have a pyramidal shape, like ammonia, if there is one pair of nonbonding electrons on the atom, or it will have a bent shape, like the water molecule, if there are two pairs of nonbonding electrons on the atom. Similarly, we deduced that three individual pairs of bonding electrons about an atom produce bond angles of 120°, with the pairs of valence electrons distributed trigonally about the atom.

EXAMPLE 6.9

A molecule of NF_3 contains only single covalent bonds. Predict the shape of the NF_3 molecule.

Solution

First, derive the electron dot formula.

$$:\overset{..}{\underset{..}{F}}:$$
$$|$$
$$:\overset{..}{\underset{..}{F}}-N-\overset{..}{\underset{..}{F}}:$$

We now see that the nitrogen atom is bonded to each fluorine and that there is one nonbonding pair of electrons on nitrogen. Thus, there are four pairs of valence electrons about the nitrogen atom, and one of these pairs is a nonbonded pair. This means that the bond angles about nitrogen will be close to 109.5°, and the molecule is shaped like a pyramid.

$$\overset{..}{N}$$
$$\diagup \mid \diagdown$$
$$\overset{.}{.}F\overset{.}{.}\!:\!\overset{.}{.}F\!:\!\overset{.}{.}F\overset{.}{.}$$

EXERCISE 6.9 A molecule of OF_2 contains only single covalent bonds. Predict the shape of the OF_2 molecule. ∎

So far we have focused on molecules containing only single covalent bonds. What about the shapes of molecules that have double or triple bonds? In these cases, each multiple covalent bond behaves as if it were a single electron pair, and the geometry of the molecule is predicted in the same way as for a molecule containing only single covalent bonds. For example, the sulfur dioxide molecule, SO_2, is bent

because the sulfur atom is surrounded by one nonbonding pair of electrons, a pair of electrons in the single covalent bond, and two pairs of electrons in the double bond (but these electrons behave as if they were a single pair of bonding electrons).

In another example, the hydrogen cyanide molecule

$$H-C\equiv N\!:$$
180°

has a linear shape, with bond angles of 180° about the carbon atom. This is because the electrons of the triple bond behave as if they were a single bonding pair, and in order for the electrons of the triple bond to get as far away as possible from the electrons of the single bond, the two sets of electrons assume positions on opposite sides of the carbon atom. Thus, a molecule of HCN has *linear* geometry; that is, all atoms lie in a straight line. Many linear molecules contain at least one multiple covalent bond.

EXAMPLE 6.10 A molecule of CO_2 has at least one multiple covalent bond. Predict the shape of the CO_2 molecule.

Solution First, derive the electron dot formula.

$$\overset{\cdot\cdot}{O}=C=\overset{\cdot\cdot}{\underset{\cdot\cdot}{O}}$$

We now see that the carbon atom is bonded to each oxygen atom by a double bond. Since each double bond behaves as if it were a single covalent bond, the repulsion between the two sets of bonding electrons is at a minimum when they are 180° apart. Thus, we would predict the CO_2 molecule to have a linear shape.

EXERCISE 6.10 A molecule of SO_3 has at least one multiple covalent bond. Predict the shape of the SO_3 molecule. ∎

Since bond angles determine the overall shapes of molecules, we can now write more realistic formulas for simple molecules. Formulas that indicate the geometry of molecules are called **structural formulas.** Structural formulas of methane, ammonia, water, and boron trifluoride are given in Figure 6-14.

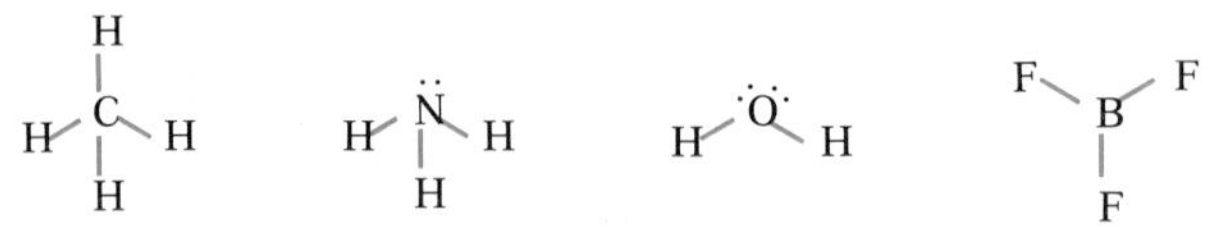

FIGURE 6-14 Structural Formulas of (a) Methane, (b) Ammonia, (c) Water, and (d) Boron Trifluoride

6.12 ELECTRONEGATIVITY

Until now we have classified chemical bonds as either covalent or ionic. In fact, these are two extremes, and there are many bonds intermediate between the two extremes. True covalent bonds result from equal sharing of an electron pair; this can happen only if the bond is between two atoms of the same element. Our picture of ionic bonds is also a bit oversimplified, in that transfer of an electron is never totally complete, even with atoms that have extremely different tendencies to gain and lose electrons. Thus, we must modify our ideas by considering **electronegativity,** the ability of an atom to attract bonding electrons toward itself. For example, fluorine atoms in molecules attract bonding electrons strongly; thus, fluorine has a high electronegativity. On the other hand, carbon atoms do not pull electrons in covalent bonds as strongly as fluorine atoms do, and carbon thus has a lower electronegativity than fluorine. These properties seem reasonable when we remember that carbon and fluorine atoms have the same number of principal energy levels but that a fluorine atom has more protons in its nucleus than does a carbon atom.

The American chemist Linus Pauling was the first scientist to assign numerical electronegativity values to elements. In the early 1930s, Pauling suggested that electronegativity could be determined for any element by measuring the energy required to break chemical bonds in several of its compounds. Thus, he derived the electronegativity values shown in Fig-

FIGURE 6-15 Pauling Electronegativity Values

ure 6-15. Pauling's method for calculating electronegativity values uses fluorine, the most electronegative element, as the standard, and all other elements are rated relative to fluorine. Figure 6-15 illustrates that electronegativity increases within a period as atomic size decreases. Within a chemical family, electronegativity decreases as atomic size increases. In general, metals have low electronegativities and nonmetals have high electronegativities.

The concept of electronegativity becomes very important in bonds between atoms of different elements. For example, the difference in electronegativities between lithium and fluorine is so large that, for all practical purposes, electron transfer is complete. On the other hand, there are many molecules in which covalent bonds represent unequal sharing of bonding electrons, as we will discuss in the following section.

6.13 POLAR COVALENT BONDS

A **polar covalent bond** is a covalent bond between atoms of different elements in which the shared pair of electrons is shifted toward the atom of higher electronegativity. Because of this shift, the more electronegative atom has a greater share of the bonding electrons than does the less electronegative atom. Hydrogen fluoride, HF, has a polar covalent bond between the hydrogen atom and the fluorine atom. The difference in electronegativity between hydrogen and fluorine ($4.0 - 2.1 = 1.9$) apparently is not great enough to produce ionic bonding, but it is large enough to cause the bonding electrons to be shifted toward fluorine. We indicate this shift as

$$\overset{\delta+}{H}\!\!-\!\!\overset{\delta-}{F}$$

where the small Greek letter *delta* (δ) indicates a partial charge, that is, a charge of less than one unit.

Differences in electronegativity between the two atoms in a bond can be used to predict the nature of the bond. This method is somewhat imprecise, but in general if the electronegativity difference is less than 0.5, the bond will be essentially nonpolar covalent; if the electronegativity difference is greater than 2.0, the bond is likely to be ionic. Electronegativity differences in the range of 0.5–2.0 generally lead to polar covalent bonds.

Because a polar covalent bond has two ends, or poles—a positive one and a negative one—the bond can also be called a **dipolar bond.** It is in fact an **electric dipole,** that is, a region of space where positive and negative electric charge are separated. Because the only bond in HF is a polar covalent bond, the molecule itself is an electric dipole. In this case, we say that the molecule is polar. Polar molecules align themselves as shown in Figure 6-16, with the partially positive end of one near the partially negative end of another, to form a network of interacting molecules.

We can predict that hydrogen atoms will form polar covalent bonds with elements such as the halogens, oxygen, and nitrogen. Water, H_2O, contains polar covalent bonds; in the water molecule, electrons in each

Di-
Greek prefix meaning "two."

Dipolar
Having two poles.

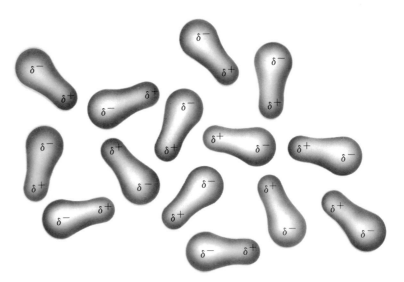

FIGURE 6-16　Interacting Polar
Molecules

covalent bond are shifted toward oxygen, as indicated in the structural
formula below. (The arrows indicate direction of electron shift.)

$$\overset{\delta-}{O}$$
$$\underset{\delta+}{H}\qquad\underset{\delta+}{H}$$

EXAMPLE 6.11　　Classify each bond as ionic, polar covalent, or nonpolar covalent. For any
polar covalent bonds, use delta notation ($\delta+$ and $\delta-$) to indicate direction
of polarity.

a. HBr　　**b.** Cl_2　　**c.** CaO

Solution　　**a.** First we must calculate the electronegativity difference in the HBr
bond.

Electronegativity difference = 2.8 − 2.1 = 0.7

The difference of 0.7 indicates that the HBr bond is probably polar
covalent. The direction of polarity is such that the more electronegative atom, Br, is surrounded by more negative charge than is the less
electronegative H atom. Thus, the bond polarity is indicated as follows:

$$\overset{\delta+}{H}-\overset{\delta-}{Br}$$

b. Since the two atoms of Cl_2 are identical, the difference in electronegativity is 0, and the bond is nonpolar covalent.
c. First we must calculate the electronegativity difference in the CaO
bond.

Electronegativity difference = 3.5 − 1.0 = 2.5

The value of 2.5 indicates that the CaO bond is probably ionic.

EXERCISE 6.11   Classify each bond as ionic, polar covalent, or nonpolar covalent. For any polar covalent bonds, use delta notation ($\delta+$ and $\delta-$) to indicate direction of polarity.

a. O_2 b. ClF   c. CO

6.14 POLARITY OF MOLECULES

Polar covalent bonds can cause an entire molecule to be polar. A **polar molecule** is a molecule with an unsymmetrical distribution of electric charge, which gives the molecule a + pole and a − pole. The simplest kind of polar molecule is one composed of two atoms joined by a polar covalent bond. Hydrogen fluoride,

$$\overset{\delta+}{H}\longrightarrow\overset{\delta-}{F}$$

as discussed in the preceding section, is an example of a polar molecule; the hydrogen atom is the positive pole and the fluorine atom is the negative pole. Other examples are HCl and HBr.

If a molecule contains three or more atoms, a combination of both molecular shape and bond polarities determines whether the molecule is polar. For example, the carbon dioxide molecule

$$\overset{\delta-}{O}=\overset{\delta+}{C}=\overset{\delta-}{O}$$
$$\longleftarrow \qquad \longrightarrow$$

contains two polar covalent bonds but is not a polar molecule. The molecule has a linear shape, and the oxygen atoms withdraw bonding electrons with equal force in opposite directions, resulting in a symmetrical distribution of charge through the CO_2 molecule. In contrast, consider the water molecule.

$$\overset{\delta-}{O}$$
$$\overset{H}{\underset{\delta+}{}} \qquad \overset{H}{\underset{\delta+}{}}$$

Like CO_2, it contains three atoms and two polar covalent bonds; however, unlike CO_2, the water molecule has a bent shape, and the electron withdrawing by oxygen in one direction is not offset by an equal withdrawing force in the opposite direction. Hence the electrical charge is distributed unsymmetrically through the water molecule, making it a polar molecule. The polarity of the water molecule can be illustrated as

$$\overset{-}{\underset{+}{O}}$$
$$H \qquad H$$

where the crossed arrow ($\longmapsto$) indicates the direction of the dipole. (The positive end is located between the two hydrogen atoms, and the negative end is on the opposite side of the oxygen atom.)

In still another example, the carbon tetrachloride molecule, CCl_4, con-

---

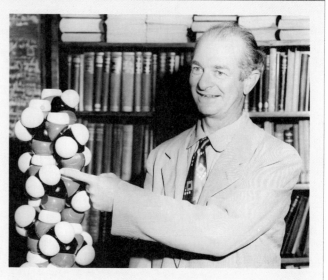
tains five atoms and four polar covalent bonds (see Figure 6-17). Even though each C—Cl bond is polar, the molecule is nonpolar because the four chlorine atoms are symmetrically positioned about the central carbon atom, and their electron-withdrawing effects are thus also symmetrical. These factors result in a symmetrical distribution of charge through the molecule, making carbon tetrachloride nonpolar. However, if one to three chlorine atoms are replaced by hydrogen atoms, the electrical

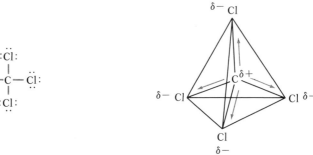

FIGURE 6-17 Carbon Tetrachloride, a Nonpolar Molecule Containing Polar Bonds

(a)   (b)

charge distribution becomes unsymmetrical, creating a polar molecule; for example, the chloroform molecule, $CHCl_3$, is polar.

Thus, for molecules containing three or more atoms and at least one polar covalent bond, symmetrical molecules will be nonpolar and unsymmetrical molecules will be polar.

EXAMPLE 6.12   The geometry of the boron trifluoride molecule is given in Figure 6-13(b). Would you expect the molecule to be polar? Why or why not?

Solution   To answer this question, we must decide whether the electrical charge is distributed symmetrically through the BF_3 molecule. We should first determine whether there are any polar covalent bonds in BF_3. Since all bonds are identical (three B—F bonds), we need only to find the electronegativity difference between B and F.

Electronegativity difference $= 4.0 - 2.0 = 2.0$

The value of 2.0 indicates that the B—F bond is probably polar covalent. Now, we must sketch the molecule, indicating electron-withdrawing effects, and determine whether they are symmetrical.

$$
\begin{array}{c}
\overset{\delta-}{F} \\
\Uparrow \\
B \\
\overset{\delta+}{\swarrow \quad \searrow} \\
\underset{\delta-}{F} \qquad \underset{\delta-}{F}
\end{array}
$$

The sketch suggests that the electron-withdrawing effect of the fluorine atoms is symmetrically distributed through the BF_3 molecule; thus, the BF_3 molecule is not expected to be polar.

EXERCISE 6.12   The geometry of the ammonia molecule, NH_3, is given in Figure 6-11(b). Would you expect the molecule to be polar? Why or why not?   ■

SUMMARY

The two major categories of chemical compounds are ionic compounds and molecular compounds. An **ionic compound** conducts electricity when dissolved in water and usually contains a metal and one or more nonmetals. The electrical attractions between the oppositely charged ions of ionic compounds are called **ionic bonds.** The valence electrons of atoms and monatomic ions can be represented by **electron dot formulas.** An ionic compound has the same amount of positive charge as negative charge and is electrically neutral. Ionic compounds exist in the solid state as crystals in which cations and anions are arranged in an orderly array called a **crystal lattice. Binary ionic compounds** are composed of ions of only two elements. In general, ionic compounds have high melting and boiling points.

Oxidation numbers are assigned to monatomic ions and uncombined atoms on the basis of charge; for these species, the oxidation number has the same value as the charge but is written in reverse order. **Oxidation** is a loss of electrons or an increase in oxidation number. **Reduction** is a gain of electrons or a decrease in oxidation number.

A **molecular compound** does not conduct electricity when dissolved in water and usually contains only nonmetals. Atoms in a molecule are held together by **covalent bonds,** chemical bonds composed of a pair of shared electrons. The valence electrons not used in covalent bonds of molecules are called nonbonding electrons, unshared pairs, or lone pairs. A covalent bond composed of a pair of electrons provided by only one of the connected atoms is called a **coordinate covalent bond.** Molecular compounds can be gases, liquids, or solids at room temperature. Solid molecular compounds exist as crystals in which molecules occupy positions in the crystal lattice. Molecular compounds have low melting and boiling points.

Two pairs of electrons shared by two atoms constitute a **double bond;** three pairs of electrons shared by two atoms constitute a **triple bond.** Double and triple bonds are referred to as **multiple covalent bonds.**

An ion containing two or more atoms is called a **polyatomic ion.** The atoms in polyatomic ions are covalently bonded and they behave as one unit; the charge on a polyatomic ion belongs to the entire unit.

Electron dot formulas of molecules and polyatomic ions are constructed on the basis of formulas and common valences of the atoms.

The concept that the covalent bonds of any one atom will stay as far away from each other as possible is called the **valence shell electron pair repulsion (VSEPR) theory.** This theory explains the tetrahedral shape of the methane molecule, the pyramidal shape of the ammonia molecule, the bent shape of the water molecule, and the trigonal shape of the boron trifluoride molecule.

Electronegativity is the ability of an atom to draw bonding electrons toward itself. The Pauling electronegativity values are based on fluorine, the most electronegative element, as the standard. A covalent bond between atoms of different elements in which the shared pair of electrons is shifted toward the atom of higher electronegativity is called a **polar covalent bond.** A molecule containing only one polar covalent bond will be polar, but a molecule containing two or more polar covalent bonds can be either polar or nonpolar, depending on the shape of the molecule.

STUDY QUESTIONS AND PROBLEMS

(More difficult questions and problems are marked with an asterisk.)

TYPES OF CHEMICAL BONDS

1. What are chemical bonds?
2. What are the two major types of chemical bonds?
3. What are the two major types of chemical compounds? How do they differ from each other?

4. Classify each of the following compounds as ionic or molecular.
   a. MnO_2 d. KI   g. Ag_2S j. $GaCl_3$
   b. SiH_4 e. $BaCl_2$ h. $AlPO_4$ k. $Mg(C_2H_3O_2)_2$
   c. $SOCl_2$ f. $PbBr_2$ i. HCl   l. $Fe_2(SO_4)_3$

IONIC BONDS

5. Why are cations smaller than their parent atoms?
6. Why are anions larger than their parent atoms?
7. What are ionic bonds?

ELECTRON DOT FORMULAS OF ATOMS AND MONATOMIC IONS

8. Give the electron dot formula of each of the following atoms and ions.
   a. Ba^{2+} d. P   g. N^{3-} j. Be
   b. I   e. Xe   h. Br^- k. Pb
   c. S^{2-} f. Al^{3+} i. Rb^+ l. O^{2-}
9. Give the electron dot formula for the ion most likely to be formed by each of the following atoms.
   a. Se   c. Br   e. S
   b. Cs   d. Sr   f. In
10. Illustrate with electron dot formulas the formation of an ionic compound from each of the following pairs of atoms.
    a. Ba and O   c. Tl and O   e. K and S
    b. Be and I   d. Mg and F   f. Ra and N
11. Give the formula of the ionic compound composed of each pair of elements.
    a. Cs and O   c. Ca and At   e. Na and N
    b. Al and S   d. Ba and Se   f. K and Se

IONIC COMPOUNDS

12. Define each of the following terms:
    a. Ionic compound   c. Binary ionic compound
    b. Crystal lattice
13. Why do ionic compounds usually have high melting points and boiling points?

AN INTRODUCTION TO OXIDATION AND REDUCTION

14. What is the oxidation number of each of the following?
    a. K   c. He   e. I^-
    b. Fe^{3+} d. Pt^{2+} f. P^{3-}
15. Define oxidation and reduction. Why do the two processes always go together?
16. For each of the following reactions, indicate the substance being oxidized and the substance being reduced.
    a. $Fe + 2\,H^+ \rightarrow Fe^{2+} + 2\,H$ d. $Cu + 2\,Ag^+ \rightarrow 2\,Ag + Cu^{2+}$
    b. $Ni + Cu^{2+} \rightarrow Ni^{2+} + Cu$ e. $Sn^{2+} + Pb \rightarrow Pb^{2+} + Sn$
    c. $Pb + Zn^{2+} \rightarrow Pb^{2+} + Zn$ f. $Zn + 2\,AgCl \rightarrow ZnCl_2 + 2\,Ag$

COVALENT BONDS

17. What is a covalent bond? How do covalent bonds differ from ionic bonds?
18. Explain what is meant by *nonbonding electrons.*
19. Using atomic orbital diagrams, illustrate the formation of a chlorine molecule, Cl_2, from two chlorine atoms.
20. Why is a hydrogen molecule, H_2, more stable than two individual hydrogen atoms?
21. Why don't the noble gases exist as molecules?

MOLECULAR COMPOUNDS

22. Why are most molecular compounds composed of nonmetals?
23. What particles occupy lattice positions in the crystals of molecular compounds?
24. Why do most molecular solids have low melting points?

MULTIPLE COVALENT BONDS

25. What is a double bond? A triple bond?
26. Why do some atoms form multiple bonds?
27. Write the electron dot formula of the nitrogen molecule, N_2.

POLYATOMIC IONS AND COORDINATE COVALENT BONDS

28. What is a polyatomic ion? What is the basis for the charge on a polyatomic ion?
29. What forces hold the atoms together in a polyatomic ion?
30. What is a coordinate covalent bond? How does it differ from a normal covalent bond?

ELECTRON DOT FORMULAS OF MOLECULAR COMPOUNDS AND POLYATOMIC IONS

31. What is valence? Why do you think carbon usually has a valence of 4? Nitrogen a valence of 3? Oxygen a valence of 2?
32. Write the electron dot formula of each of the following molecules.
    a. H_2S c. PCl_3 *e. P_2O_5
    *b. CH_3OH d. O_3 * f. H_2CO
33. Write the electron dot formula of each of the following ions.
    a. PO_4^{3-} c. HCO_3^- e. NH_4^+
    b. SO_4^{2-} *d. $C_2H_3O_2^-$ * f. $H_2PO_3^-$
*34. Write the electron dot formula of each of the following compounds.
    a. $CaSO_4$ c. K_3PO_4 e. $Mg(C_2H_3O_2)_2$
    b. $NaNO_3$ d. Na_2SO_4 f. $Sr(ClO_4)_2$

MOLECULAR SHAPES

35. Explain what is meant by the VSEPR theory.
36. Give the shape and bond angles of each of the following molecules.
    a. CH_4 b. NH_3 c. H_2O d. BF_3
37. Give the structural formula of each molecule in question 36.

38. Predict the shape and bond angles and give the structural formula of each of the following molecules.
    a. H_2S b. PCl_3 c. CF_4

ELECTRONEGATIVITY

39. Explain what is meant by electronegativity.
40. Describe the trend in electronegativity values within a period and within a group.
41. What is the most electronegative element? The least electronegative element?

POLAR COVALENT BONDS

42. What is a polar covalent bond?
43. Classify each bond in the following molecules as ionic, polar covalent, or nonpolar covalent. For any polar covalent bonds, use delta notation ($\delta+$ and $\delta-$) to indicate direction of polarity.
    a. H_2 d. NaI g. N_2 j. CCl_4
    b. MgO e. CF_4 h. AlF_3 k. SO_2
    c. H_2O f. HBr i. $CaCl_2$ l. NH_3
44. Arrange the following bonds in order of increasing polarity, and use delta notation ($\delta+$ and $\delta-$) to indicate direction of polarity.
    a. $S-O$ b. $H-O$ c. $N-Cl$ d. $P-F$

POLARITY OF MOLECULES

45. The HCN molecule has linear geometry. Would you expect the molecule to be polar? Why or why not?
46. Arrange the following molecules in order of increasing polarity, and explain how you arrived at your answer.
    a. CCl_4 b. CH_3Cl c. CH_2Cl_2 d. $CHCl_3$
47. Classify each of the following molecules as polar or nonpolar, and explain your answers.
    a. CBr_4 b. H_2S c. Br_2

GENERAL EXERCISES

48. In each of the following ionic compounds, determine the charge on the cation from the charge on the anion.
    a. $CoCl_2$ c. MnO_2 e. VCl_3
    b. FeO d. Cu_2O f. Ag_2S
49. Write the electron dot formulas of the bromine and iodine molecules, Br_2 and I_2. Why are these molecules more stable than their individual atoms?
50. Find the total number of valence electrons in each of the following.
    a. NO_3^- c. HSO_4^- e. $H_2PO_4^-$
    b. SO_4^{2-} d. PO_4^{3-} f. $C_2H_3O_2^-$
*51. Write the electron dot formula of each of the following compounds.
    a. C_2H_4 c. KCN e. $Mg(NO_2)_2$
    b. $Ca(BrO_3)_2$ d. Cl_2O f. PBr_3

*52. Each of the following molecules has at least one multiple bond. Write the electron dot formula and structural formula and predict the shape and bond angles of each.

   a. N_2 c. H_2CO e. SO_2

   b. HCN   d. C_2H_2 f. CS_2

53. Use delta notation ($\delta+$ and $\delta-$) to indicate the direction of polarity in each bond of the following molecules.

   a. HCl   c. ClBr   e. PCl_3

   b. NH_3 d. H_2S * f. CH_3OH

54. Classify each of the following molecules as polar or nonpolar, and explain your answers.

   a. PH_3 b. SO_3 c. CS_2

NAMING INORGANIC COMPOUNDS

OUTLINE

7.1 Oxidation Numbers
7.2 Binary Ionic Compounds
7.3 Ionic Compounds Containing
         Polyatomic Ions
7.4 Binary Molecular Compounds
7.5 Acids
         Perspective: The Origin of Scientific
         Nomenclature in Chemistry
7.6 Inorganic Hydrates
         Summary
         Study Questions and Problems

Is your name unique? Does it set you apart from all other people in the world? Chances are the answer is no. It is not unusual to find two or more people with the same name, even in a small town. If you know two people with the same name, you can keep them separate in your mind by remembering the individual characteristics of each person. But how would you like to contend with millions of chemical compounds if many of them had the same name? Even worse, what if their names gave little or no information about their compositions and hence their chemical characteristics?

The alchemists had quaint names for substances, names like *spirit of wine, sal ammoniac,* and *oil of vitriol.* As time passed, more substances became characterized and they were given names according to their uses, origins, or appearances. Such names are called **common names,** and we still use many of them today. Table 7-1 gives common names of some familiar substances.

By the early part of this century, chemists had grown dissatisfied with giving common names to substances. For one thing, they were running out of possible names for compounds, and for another, it was just not possible to remember all of the common names and the substances they identified. In 1921, an organization called the International Union of Pure

TABLE 7-1   Common Names of Familiar Substances

Common Name	Chemical Name	Formula
Alum	Potassium aluminum sulfate	$KAl(SO_4)_2 \cdot 12\ H_2O$
Baking soda	Sodium hydrogen carbonate	$NaHCO_3$
Brimstone	Sulfur	S
Cream of tartar	Potassium hydrogen tartrate	$KHC_4H_4O_6$
Epsom salts	Magnesium sulfate heptahydrate	$MgSO_4 \cdot 7\ H_2O$
Grain alcohol	Ethanol; ethyl alcohol	C_2H_5OH
Hypo	Sodium thiosulfate	$Na_2S_2O_3$
Laughing gas	Dinitrogen oxide; nitrous oxide	N_2O
Lime	Calcium oxide	CaO
Lye; caustic soda	Sodium hydroxide	NaOH
Milk of magnesia	Magnesium hydroxide	$Mg(OH)_2$
Muriatic acid	Hydrochloric acid	HCl
Plaster of paris	Calcium sulfate hemihydrate	$CaSO_4 \cdot \frac{1}{2}\ H_2O$
Quicksilver	Mercury	Hg
Saltpeter	Sodium nitrate	$NaNO_3$
Slaked lime	Calcium hydroxide	$Ca(OH)_2$
Table salt	Sodium chloride	NaCl
Table sugar	Sucrose	$C_{12}H_{22}O_{11}$
Washing soda	Sodium carbonate decahydrate	$Na_2CO_3 \cdot 10\ H_2O$
Wood alcohol	Methanol; methyl alcohol	CH_3OH

and Applied Chemistry (IUPAC) began to study systems of assigning individual, unambiguous names to chemical compounds, so that each compound had a unique name from which its formula could be written. Today we rely almost entirely on the **systematic nomenclature** devised by the IUPAC, and in this chapter, you will learn about its use in naming **inorganic compounds,** that is, all compounds except those containing carbon-hydrogen covalent bonds.

7.1 OXIDATION NUMBERS

In chapter 6, you were introduced to oxidation numbers and the fact that oxidation numbers of atoms and monatomic ions are derived from their charges. An **oxidation number** (also called **oxidation state**) is a number assigned to an element, either uncombined or in a compound, that allows us to keep track of bond formation and electron transfers between atoms. Oxidation numbers can be zeros or signed numbers. Since a knowledge of oxidation numbers is important in naming inorganic compounds, practice using the following rules for assigning oxidation numbers to elements:

1. Uncombined elements have an oxidation number of zero, regardless of whether they exist as individual atoms or as molecules. Thus, Na, Mg, H_2, O_2, and S_8 have oxidation numbers of zero.
2. The oxidation number of a monatomic ion is the same as its charge. The group number indicates the charge for the ions of most representative metals. Ions of indium (In), tin (Sn), thallium (Tl), lead (Pb), and transition elements have variable charges and therefore variable oxidation numbers, which must be deduced by rule 5 below.
3. Hydrogen in compounds has an oxidation number of $+1$, except when it is combined with a less electronegative element, usually a metal. In this case, hydrogen has an oxidation number of -1. Thus, in LiH, lithium has an oxidation number of $+1$ and hydrogen has an oxidation number of -1.
4. Oxygen in compounds has an oxidation number of -2, except in compounds called **peroxides,** where oxygen has an oxidation number of -1. Peroxides are compounds in which an oxygen atom is bonded to another oxygen atom, for example, hydrogen peroxide, $H-O-O-H$.
5. The sum of the oxidation numbers of all atoms or ions in a compound is zero, and the sum of the oxidation numbers of all atoms in a polyatomic ion is equal to the charge of the ion. Oxidation numbers of elements not included in rules 1–4 above can be deduced from formulas of their compounds.

EXAMPLE 7.1 Find the oxidation number of each element in magnesium chloride, $MgCl_2$.

Solution Magnesium chloride, composed of a metal and a nonmetal, is a binary ionic compound consisting of the monatomic ions Mg^{2+} and Cl^-. Therefore the oxidation numbers of their ions are the same as their charges (rule

2). Thus, the oxidation number of Mg^{2+} is $+2$ and that of Cl^- is -1. Notice that the sum of the oxidation numbers of the ions in a formula unit of $MgCl_2$ add up to zero:

$$MgCl_2$$
$$\underset{+2\,-1}{}$$

Ions	Individual Oxidation Numbers	Sum of Oxidation Numbers = 0
Mg^{2+}	$+2$	$+2$
$2\,Cl^-$	-1	$\underline{-2}$
		0

EXERCISE 7.1   Find the oxidation number of each element in Al_2O_3.   ■

EXAMPLE 7.2   Find the oxidation number of each element in the polyatomic ion $SO_4{}^{2-}$.

Solution   In polyatomic ions, the sum of the oxidation numbers must equal the charge on the ion (rule 5). Since we know the oxidation number of oxygen, -2, we can use this value and the charge on the polyatomic ion (2−) to deduce that the oxidation number of S is $+6$.

Atoms	Individual Oxidation Numbers	Sum of Oxidation Numbers = -2
$4\,O$	-2	-8
S	$+6$	$\underline{+6}$
		-2

Therefore, the oxidation numbers are assigned as follows.

$$SO_4{}^{2-}$$
$$\underset{+6\,-2}{}$$

EXERCISE 7.2   Find the oxidation number of each element in the polyatomic ion $NO_3{}^-$.   ■

EXAMPLE 7.3   Find the oxidation number of each element in carbon dioxide, CO_2.

Solution   Since the sum of oxidation numbers in a compound is equal to zero, we can deduce the oxidation number of carbon in CO_2 to be $+4$.

Atoms	Individual Oxidation Numbers	Sum of Oxidation Numbers = 0
$2\,O$	-2	-4
C	$+4$	$\underline{+4}$
		0

Therefore, the oxidation numbers are assigned as follows.

$$CO_2$$
$$\underset{+4\,-2}{}$$

EXERCISE 7.3 Find the oxidation number of each element in phosphoric acid, H_3PO_4.   ▪

7.2 BINARY IONIC COMPOUNDS

Most binary ionic compounds are composed of a metal and a nonmetal, and the metal exists as one or more cations while the nonmetal exists as one or more anions. The following steps summarize the systematic nomenclature of binary ionic compounds.

1. Name the cation by using the complete English name of the element. (In the formula, the symbol of the cation is written first and the symbol of the anion follows.)
2. Name the anion by using the English root of the element's name and then adding the ending -ide. Table 7-2 gives names of all monatomic anions with the English roots italicized.
3. Name the binary ionic compound by writing first the name of the cation and then, as a separate word, the name of the anion with its -ide ending, for example, sodium sulfide (Na_2S), calcium oxide (CaO), lithium iodide (LiI), strontium bromide ($SrBr_2$), and aluminum chloride ($AlCl_3$).
4. A cation of a transition metal may have several possible charges and therefore several possible oxidation numbers. The cations of the representative metals indium (In), tin (Sn), thallium (Tl), and lead (Pb) also may have more than one charge. In these cases the charge on the cation must be deduced from the formula of the compound and from the charge on the anion(s) combined with the cation. If a metal cation can have more than one charge, its oxidation number must be specified for each compound by Roman numerals in parentheses immediately following the name of the metal. There is no space between the metal name and the parentheses, for example, iron(III) oxide (Fe_2O_3) and copper(II) chloride ($CuCl_2$).

EXAMPLE 7.4 Give the systematic name of each compound.

a. KBr   **b.** CoN

Solution   **a.** Potassium bromide   **b.** Cobalt(III) nitride

TABLE 7-2 Names of Monatomic Anions (English roots are italicized.)

Element	Ion	Name of Anion	Element	Ion	Name of Anion
*Hydro*gen	H^-	Hydride	*Sulf*ur	S^{2-}	Sulfide
*Fluor*ine	F^-	Fluoride	*Tellur*ium	Te^{2-}	Telluride
*Chlor*ine	Cl^-	Chloride	*Selen*ium	Se^{2-}	Selenide
*Brom*ine	Br^-	Bromide	*Nitro*gen	N^{3-}	Nitride
*Iod*ine	I^-	Iodide	*Phosph*orus	P^{3-}	Phosphide
*Oxy*gen	O^{2-}	Oxide	*Carb*on	C^{4-}	Carbide

EXERCISE 7.4   Give the systematic name of each compound.

   a. CaSe   **b.** MnO_2 ■

EXAMPLE 7.5   Write the formula of each compound.

   a. Tin(II) chloride   **b.** Iron(III) sulfide

Solution   In writing formulas from names of ionic compounds, we must make certain that total positive charges are equal to total negative charges:

   a. Sn^{2+} ⤫ Cl^{1-} or $SnCl_2$

   b. Fe^{3+} ⤫ S^{2-} or Fe_2S_3

EXERCISE 7.5   Write the formula of each compound.

   a. Lead(II) sulfide   **b.** Aluminum oxide   ■

An older method also exists for naming binary ionic compounds of metals that have two oxidation states. This method is rapidly losing popularity but is still used occasionally. The common names in this method have characteristic endings attached to the root of the metal name. In this case, the root may be English or Latin, depending on the name from which the element symbol is derived. The ending -ous is used for the cation of lower oxidation state, and the ending -ic is used for the cation of higher oxidation state. Table 7-3 contrasts the systematic nomenclature of these cations with the older method of naming them. Note that the older method does not indicate the actual oxidation number of each pair of cations but only that two oxidation states exist.

EXAMPLE 7.6   Refer to Table 7-3 and give the common name of each compound.

   a. $PbCl_2$ **b.** Fe_2O_3

Solution   **a.** Plumbous chloride   **b.** Ferric oxide

TABLE 7-3	Names of Cations of Some Metals Having Two Oxidation States		
Element	Ion	Systematic Name	Common Name
Iron	Fe^{2+}	iron(II)	ferrous ion
	Fe^{3+}	iron(III)	ferric ion
Copper	Cu^+	copper(I)	cuprous ion
	Cu^{2+}	copper(II)	cupric ion
Lead	Pb^{2+}	lead(II)	plumbous ion
	Pb^{4+}	lead(IV)	plumbic ion
Mercury	Hg^+*	mercury(I)	mercurous ion
	Hg^{2+}	mercury(II)	mercuric ion
Tin	Sn^{2+}	tin(II)	stannous ion
	Sn^{4+}	tin(IV)	stannic ion

* Actually exists as Hg_2^{2+}

EXERCISE 7.6 Refer to Table 7-3 and give the common name of each compound.

a. HgO   b. FeCl$_2$ ■

EXAMPLE 7.7 Refer to Table 7-3 and write the formula of each compound.

a. Stannous fluoride   b. Ferrous oxide

Solution a. SnF$_2$ b. FeO

EXERCISE 7.7 Refer to Table 7-3 and write the formula of each compound.

a. Mercuric bromide   b. Stannic chloride   ■

7.3 IONIC COMPOUNDS CONTAINING POLYATOMIC IONS

Ionic compounds that contain polyatomic ions are named in much the same way as binary ionic compounds. The cation name is written first, followed by the anion name written as a separate word. Table 7-4 gives the names of common polyatomic ions. When two or more of the same polyatomic ion are present in a formula unit, the symbols of the polyatomic ion are enclosed in parentheses and a subscript is used to indicate the number of polyatomic ions present in the formula unit. An example is Al(C$_2$H$_3$O$_2$)$_3$.

Notice that most of the polyatomic ions in Table 7-4 contain oxygen and one other element. These are called **oxyanions.** Different oxyanions of the same element are named according to the oxidation state of the element in the particular anion. If there are only two oxyanions of the same element, the oxyanion in which the element has the lower oxidation state has the name ending -*ite*; that in which the element has the higher oxidation state has the name ending -*ate*. Examples are sulfite (SO$_3$$^{2-}$, the oxidation

TABLE 7-4 Polyatomic Ions

Ion	Name	Ion	Name
C$_2$H$_3$O$_2$$^-$	Acetate	OH$^-$	Hydroxide
NH$_4$$^+$	Ammonium	BrO$^-$	Hypobromite
BrO$_3$$^-$	Bromate	ClO$^-$	Hypochlorite
BrO$_2$$^-$	Bromite	IO$^-$	Hypoiodite
CO$_3$$^{2-}$	Carbonate	IO$_3$$^-$	Iodate
ClO$_3$$^-$	Chlorate	IO$_2$$^-$	Iodite
ClO$_2$$^-$	Chlorite	NO$_3$$^-$	Nitrate
CrO$_4$$^{2-}$	Chromate	NO$_2$$^-$	Nitrite
CN$^-$	Cyanide	C$_2$O$_4$$^{2-}$	Oxalate
Cr$_2$O$_7$$^{2-}$	Dichromate	BrO$_4$$^-$	Perbromate
H$_2$PO$_4$$^-$	Dihydrogen phosphate	ClO$_4$$^-$	Perchlorate
H$_2$PO$_3$$^-$	Dihydrogen phosphite	IO$_4$$^-$	Periodate
HCO$_3$$^-$	Hydrogen carbonate *or* bicarbonate	MnO$_4$$^-$	Permanganate
HPO$_4$$^{2-}$	Hydrogen phosphate	PO$_4$$^{3-}$	Phosphate
HPO$_3$$^{2-}$	Hydrogen phosphite	PO$_3$$^{2-}$	Phosphite
HSO$_4$$^-$	Hydrogen sulfate *or* bisulfate	SO$_4$$^{2-}$	Sulfate
HSO$_3$$^-$	Hydrogen sulfite *or* bisulfite	S$_2$O$_3$$^{2-}$	Thiosulfate

number of sulfur is $+4$) and sulfate (SO_4^{2-}, the oxidation number of sulfur is $+6$).

If there are four oxyanions of the same element, the oxyanion with the element in the lowest oxidation state has a name beginning with *hypo-* and ending with *-ite*. The oxyanion with the element in the highest oxidation state has a name beginning with *per-* and ending with *-ate*. The two intermediate oxyanions follow the system described in the preceding paragraph, with name endings of *-ite* and *-ate*. The oxyanions of chlorine shown below are examples.

ClO^-	*Hypochlorite*	(oxidation number of Cl is $+1$)
ClO_2^-	Chlor*ite*	(oxidation number of Cl is $+3$)
ClO_3^-	Chlor*ate*	(oxidation number of Cl is $+5$)
ClO_4^-	*Perchlorate*	(oxidation number of Cl is $+7$)

EXAMPLE 7.8   Refer to Table 7-4 and name each compound.

a. $NaHCO_3$ **b.** $Al_2(SO_4)_3$

Solution   **a.** Sodium hydrogen carbonate or sodium bicarbonate
b. Aluminum sulfate

EXERCISE 7.8   Refer to Table 7-4 and name each compound.

a. $NaCN$ **b.** $KMnO_4$ ■

EXAMPLE 7.9   Refer to Table 7-4 and write the formula of each compound.

a. Aluminum hydrogen phosphate
b. Iron(II) sulfate

Solution   **a.** In this example we must balance charges as before:
$$Al^{3+} \quad HPO_4^{2-} = Al_2(HPO_4)_3$$
b. In this example we must balance charges as before:
$$Fe^{2+} \quad SO_4^{2-} = FeSO_4$$

EXERCISE 7.9   Refer to Table 7-4 and write the formula of each compound.

a. Cesium phosphate   **b.** Aluminum hydroxide   ■

7.4 BINARY MOLECULAR COMPOUNDS

Molecular compounds are composed entirely of covalent bonds. In this section, we will be concerned with the systematic nomenclature of **binary molecular compounds,** that is, molecular compounds containing only two elements. The following rules apply:

1. Name the less electronegative (more metallic) element by using its complete English name and, if necessary, a Greek prefix (see Table 7-5)

TABLE 7-5 Greek Prefixes	
Prefix	Meaning
Mono-	One
Di-	Two
Tri-	Three
Tetra-	Four
Penta-	Five
Hexa-	Six
Hepta-	Seven
Octa-	Eight
Nona-	Nine
Deca-	Ten

to indicate the number of atoms of the element in the molecule. The prefix *mono-* is not normally used with the less electronegative element.

2. Name the more electronegative (more nonmetallic) element as the second word in the name of the compound; use its English root, add the ending *-ide* and, if necessary, use the appropriate Greek prefix to indicate the number of atoms of the element. The prefix *mono-* is not normally used for the more electronegative element, except in the name carbon monoxide. (Note the last *o* in *mono-* was dropped to make the pronunciation easier.) Examples of names of binary molecular compounds are sulfur trioxide (SO_3) and dinitrogen pentoxide (N_2O_5).

Some binary molecular compounds have names and formulas that originated before the rules of nomenclature were created. In many of these cases, we still use the common names and formulas, even though more proper ones exist. Some examples are water (H_2O), ammonia (NH_3), and methane (CH_4).

EXAMPLE 7.10 Name each compound.

a. S_2Cl_2 b. CCl_4

Solution a. Disulfur dichloride   b. Carbon tetrachloride

EXERCISE 7.10 Name each compound.

a. PCl_5 b. N_2O ∎

EXAMPLE 7.11 Write the formula of each compound.

a. Dinitrogen trioxide   b. Phosphorus trichloride

Solution a. N_2O_3 b. PCl_3

EXERCISE 7.11 Write the formula of each compound.

a. Hydrogen bromide   b. Carbon disulfide   ∎

7.5 ACIDS

Acids are molecular compounds that produce hydrogen ions (H^+) when dissolved in water. For example, the gas hydrogen chloride (HCl) dissolves in water, and the solution is an acid because it contains hydrogen ions. To distinguish the water solution of HCl from gaseous HCl, the solution is given the name hydrochloric acid. The same formula, HCl, is used for both the gaseous molecular compound and the acid.

Acids can usually be recognized by their formulas, which contain hydrogen as the element written first. (An exception is, of course, H_2O.) Thus, HCl, HBr, and HI are all acids when dissolved in water. These are examples of **binary acids,** acids composed of only two elements. Binary acids are named by adding the prefix *hydro-* and the suffix *-ic* to the root of the second element in the formula, followed by the word *acid*. Table 7-6 gives the names of the common molecular compounds that form binary acids in water.

Many acids contain three or more elements. One example is HCN, whose acid name is hydrocyanic acid; this acid is named as if it were a binary acid. Other complex acids contain an oxyanion combined with enough hydrogen ions to make a neutral molecule; these are known as **oxyacids.** An oxyacid is named on the basis of the oxyanion it contains. To name an oxyacid, change the *-ate* ending of the oxyanion name to *-ic* or the *-ite* ending of the oxyanion name to *-ous*, and add the word *acid*. (No prefixes are used in naming oxyacids.) Examples are sulfur*ic* acid (H_2SO_4), phosphor*ous* acid (H_3PO_3), and acet*ic* acid ($HC_2H_3O_2$).

EXAMPLE 7.12   Name each acid.

a. HI   **b.** H_2SO_3

Solution   **a.** Molecular compound name: hydrogen iodide
Acid name: hydroiodic acid
b. Oxyanion name: sulfite
Acid name: sulfurous acid

EXERCISE 7.12   Name each acid.

a. H_2CrO_4 **b.** H_2CO_3 ■

EXAMPLE 7.13   Write the formula of each acid.

a. Chloric acid   **b.** Nitric acid

TABLE 7-6 Common Molecular Compounds That Form Binary Acids		
Compound	Formula	Acid Name
Hydrogen sulfide	H_2S	Hydrosulfuric acid
Hydrogen fluoride	HF	Hydrofluoric acid
Hydrogen chloride	HCl	Hydrochloric acid
Hydrogen bromide	HBr	Hydrobromic acid
Hydrogen iodide	HI	Hydroiodic acid

PERSPECTIVE THE ORIGIN OF SCIENTIFIC NOMENCLATURE IN CHEMISTRY

Antoine Lavoisier (1743–1794), shown in the accompanying figure, is often referred to as the founder of modern chemistry. He is famous for his brilliant experiments that revolutionized chemistry in the eighteenth century. By 1783 he had found that burning sulfur, phosphorus, nitrogen, or carbon in moist air produced acids. Knowing that there was one principal reactant in the moist air, Lavoisier took the unprecedented step of changing the name of this substance from "dephlogisticated air" to "oxygen." He reasoned that since this gas generated acids when combined with certain nonmetals, the gas should have a name more indicative of its chemical properties; hence, he coined the name oxygen, derived from the Greek *oxy-*, meaning "acid," and *-gen*, meaning "production of." This name is thought to be the beginning of chemical nomenclature based on chemical principles. Lavoisier's action marked a turning point in chemistry, and he set about to revise chemical nomenclature. He was especially annoyed with names like *oil of vitriol, oil of tartar, butter of arsenic, flowers of zinc, liver of sulfur,* and *sugar of lead,* which were ridiculed because they suggested that chemists borrowed their language from the kitchen. Lavoisier proposed that chemical names should reflect composition and that they should be derived from Latin or Greek in order to be widely and easily understood. By 1787, Lavoisier and three other

Antoine Lavoisier and His Wife Marie-Anne in His Laboratory

scientists had published *Methode de Nomenclature Chimique* (Methods of Chemical Nomenclature), and soon they had completed a dictionary giving the old and new names of 700 substances.

Solution **a.** Since the name *chloric acid* has no *hydro-* prefix, the compound must be an oxyacid. Chlor*ic* acid must contain the chlor*ate* anion, so the formula must be $HClO_3$.

 b. Since the name *nitric acid* has no *hydro-* prefix, the compound must be an oxyacid. Nitr*ic* acid must contain the nitr*ate* anion, so the formula must be HNO_3.

EXERCISE 7.13 Write the formula of each acid.

a. Phosphorous acid **b.** Permanganic acid ∎

7.6 INORGANIC HYDRATES

Inorganic hydrates are ionic compounds that contain definite amounts of water in their crystals. They show no visible moisture, but each contains a fixed amount of water and has a definite composition. The water in hydrates, referred to as **water of hydration,** can usually be removed by heating. The compound is then said to be **anhydrous** (without water).

Formulas of hydrates are written to show the water of hydration. The formula of the anhydrous compound is written and a raised dot is added, followed by the number of water molecules in the simplest unit of the hydrate. Hydrates are named by writing the name of the anhydrous compound followed by the word *hydrate*, modified with the appropriate Greek prefix to indicate the number of water molecules in the formula unit. Table 7-7 gives examples of the names and formulas of inorganic hydrates.

EXAMPLE 7.14   Name each compound.

a. $CaCl_2 \cdot 2\ H_2O$ b. $NaC_2H_3O_2 \cdot 3\ H_2O$

Solution   a. Calcium chloride dihydrate   b. Sodium acetate trihydrate

EXERCISE 7.14   Name each compound.

a. $Ba(OH)_2 \cdot 8\ H_2O$ b. $(NH_4)_2C_2O_4 \cdot H_2O$

EXAMPLE 7.15   Write the formula of each compound.

a. Magnesium sulfate heptahydrate
b. Sodium carbonate decahydrate

Solution   a. $MgSO_4 \cdot 7\ H_2O$ b. $Na_2CO_3 \cdot 10\ H_2O$

EXERCISE 7.15   Write the formula of each compound.

a. Cobalt(II) chloride hexahydrate
b. Aluminum bromate nonahydrate

TABLE 7-7   Some Inorganic Hydrates

Formula	Name*
$CuSO_4 \cdot 5\ H_2O$	Copper sulfate pentahydrate (blue vitriol)
$Na_2SO_4 \cdot 10\ H_2O$	Sodium sulfate decahydrate (Glauber's salt)
$KAl(SO_4)_2 \cdot 12\ H_2O$	Potassium aluminum sulfate dodecahydrate (alum)
$Na_2B_4O_7 \cdot 10\ H_2O$	Sodium tetraborate decahydrate (borax)
$FeSO_4 \cdot 7\ H_2O$	Iron(II) sulfate heptahydrate (green vitriol)
$H_2SO_4 \cdot H_2O$	Sulfuric acid hydrate

* Common names are given in parentheses.

SUMMARY

An **oxidation number** (or **oxidation state**) is a number assigned to an element, either uncombined or in a compound, that allows us to keep track of bond formation and electron transfers between atoms. Oxidation numbers may be zero or signed numbers. Rules for assigning oxidation numbers are given on p. 159.

Most **binary ionic compounds** are composed of a metal and a nonmetal. The systematic nomenclature of binary ionic compounds is outlined on p. 161. An older method also exists for naming binary ionic compounds of metals that have two oxidation states; the ending *-ous* is used for the cation of lower oxidation state, and the ending *-ic* is used for the cation of higher oxidation state.

Ionic compounds that contain polyatomic ions are named in much the same way as binary ionic compounds. For elements having only two **oxyanions,** the oxyanion in which the element has the lower oxidation state has the name ending *-ite;* the oxyanion in which the element has the higher oxidation state has the name ending *-ate.* If there are four oxyanions of the same element, the oxyanion with the element in the lowest oxidation state has a name beginning with *hypo-* and ending with *-ite.* The oxyanion with the element in the highest oxidation state has a name beginning with *per-* and ending with *-ate.* Names for the two intermediate oxyanions follow the *-ite/-ate* system without use of prefixes.

Binary molecular compounds are covalent compounds containing only two elements. Rules for naming binary molecular compounds are given on pp. 164-165.

Acids are molecular compounds that produce hydrogen ions (H^+) when dissolved in water. **Binary acids,** those composed of only two elements, are named by adding the prefix *hydro-* and the suffix *-ic* to the root of the second element in the formula. **Oxyacids,** acids that contain oxyanions, are named by changing the *-ate* ending of the oxyanion name to *-ic* or the *-ite* ending of the oxyanion name to *-ous* and adding the word *acid.*

Inorganic hydrates are ionic compounds that contain fixed amounts of water in their crystals. They are named by writing the name of the anhydrous compound followed by the word *hydrate,* modified with the appropriate Greek prefix to indicate the number of water molecules in the formula unit.

STUDY QUESTIONS AND PROBLEMS

(More difficult questions and problems are marked with an asterisk.)

OXIDATION NUMBERS

1. Define each term:
   a. Oxidation number   b. Peroxide

2. Find the oxidation number of each element in the following compounds.
   a. $AlBr_3$ c. $NiCl_2$ e. Co_2S_3
   b. BaO_2 d. CaH_2

3. Find the oxidation number of each element in the following polyatomic ions.
   a. HCO_3^- c. $C_2O_4^{2-}$ e. NH_4^+
   b. $S_2O_3^{2-}$ d. OH^-

*4. Find the oxidation number of each element in the following compounds.
   a. $Rh_2(SO_4)_3$ c. $C_{12}H_{22}O_{11}$ e. Cl_2O_7
   b. $H_4P_2O_5$ d. $Po(IO_3)_4$

BINARY IONIC COMPOUNDS

5. Give the systematic name of each compound.
   a. GaI_3 c. Hg_2O
   b. FeS d. $PtCl_2$

6. Write the formula of each compound.
   a. Iron(II) iodide   d. Barium telluride
   b. Sodium hydride   e. Mercury(II) fluoride
   c. Zirconium(IV) oxide

7. Give the common name of each compound.
   a. FeF_2 c. $PbCl_4$ e. SnO_2
   b. Cu_2S d. Hg_2I_2

8. Write the formula of each compound.
   a. Ferric iodide   d. Mercurous selenide
   b. Cuprous oxide   e. Stannic bromide
   c. Plumbic sulfide

IONIC COMPOUNDS CONTAINING POLYATOMIC IONS

9. What is an oxyanion? How are oxyanions named?

10. Give the systematic name of each compound.
    a. $Al(HSO_4)_3$ b. $Ca(CN)_2$ c. $Fe_2(HPO_4)_3$ d. $Sr(IO_3)_2$

11. Write the formula of each compound.
    a. Barium hydroxide   d. Plumbous iodate
    b. Potassium hydrogen sulfite   e. Cupric cyanide
    c. Indium(III) perchlorate

BINARY MOLECULAR COMPOUNDS

12. Name each compound.
    a. B_2Br_4 b. BrF_3 c. $CaSe_2$ d. HF

13. Write the formula of each compound.
    a. Xenon tetroxide
    b. Tetrasulfur dinitride
    c. Hydrogen sulfide
    d. Disilicon hexaiodide
    e. Diphosphorus tetrachloride

ACIDS

14. Name each acid.
  a. HCN   **c.** $HC_2H_3O_2$ **e.** HF
  b. H_2SO_4 **d.** HNO_3

15. Write the formula of each compound.
  a. Sulfurous acid   **d.** Oxalic acid
  b. Hydrosulfuric acid   **e.** Chromic acid
  c. Periodic acid

INORGANIC HYDRATES

16. Define each term:
  a. Inorganic hydrate   **b.** Water of hydration

17. Give the systematic name of each compound.
  a. $Mg(ClO_3)_2 \cdot 6\,H_2O$ **c.** $Bi(NO_3)_3 \cdot 5\,H_2O$
  b. $FeSO_4 \cdot 4\,H_2O$ **d.** $(NH_4)_2SO_3 \cdot H_2O$

18. Write the formula of each compound.
  a. Cadmium(II) nitrate tetrahydrate
  b. Gold(III) cyanide trihydrate
  c. Lead(IV) perchlorate trihydrate
  d. Potassium carbonate dihydrate
  e. Sodium hydrogen phosphate heptahydrate

GENERAL EXERCISES

19. Find the oxidation number of each element in the following compounds.
  a. $CaSO_4$ **c.** K_2SO_3 **e.** $RbBrO_3$
  b. $Mg(OH)_2$ **d.** $Na_2S_2O_3$

20. Write the formula of each compound.
  a. Manganese(III)fluoride   **d.** Vanadium(IV) bromide
  b. Chromium(III) sulfate   **e.** Strontium bromate
  c. Potassium hypochlorite

*21. Name each compound.
  a. K_2Se **c.** N_2S_5 **e.** $Ga_2(C_2O_4)_3 \cdot 4\,H_2O$
  b. ZnC_2O_4 **d.** $HClO_3$

*22. Name each compound.
  a. Cl_2O_7 **c.** Hg_2CO_3 **e.** NH_4HSO_3
  b. $Cr(C_2H_3O_2)_3 \cdot H_2O$ **d.** H_3PO_4

*23. Write the formula of each compound.
  a. Platinum(II) bromide
  b. Rhenium(IV) oxide dihydrate
  c. Diphosphorous pentaselenide
  d. Acetic acid
  e. Potassium bicarbonate
  f. Sodium thiosulfate pentahydrate

CALCULATIONS BASED ON CHEMICAL FORMULAS

OUTLINE

8.1 The Chemical Mole
8.2 Molar Masses
8.3 Calculation of Molar Masses
8.4 Additional Calculations Based on the Mole Concept
8.5 Empirical and True Formulas
Perspective: A Simple Experiment to Determine Percentage Composition
8.6 Calculation of the Empirical Formula
Problem-Solving Skills: Manipulating Ratios
8.7 Calculation of the True Formula from the Empirical Formula
Summary
Study Questions and Problems

All of us work with chemical compounds each day, perhaps without realizing it. The sugar that you put on your cereal is sucrose, $C_{12}H_{22}O_{11}$, and the salt you season with is sodium chloride, NaCl. These are only two examples of the many chemical compounds common to daily life. We generally use both of these substances in quantities of grams or fractions of grams, although we may measure them with teaspoons. In fact, we seldom use quantities of chemical compounds smaller than fractions of a gram, even in the laboratory, because smaller quantities are usually impractical. Regardless of the quantity used, chemical compounds react as individual ions or molecules. For example, individual sucrose molecules trigger your taste buds to report the sensation of sweetness to your brain. But how do we know how many sucrose molecules are in a teaspoon of sugar (about 28 grams)? In this chapter you will learn how to make this calculation and several others that are based on chemical formulas. Before going on, you should review scientific notation, the unit-factor method, and the concept of mass in chapter 2, and formula weights and percentage composition in chapter 3. It would also be helpful to brush up on the mathematical skills in appendix A.

8.1   THE CHEMICAL MOLE

When chemical compounds react, they do so as individual ions or molecules, which are such small particles that we cannot count them individually. However, we can weigh a sample of a substance and then calculate the number of ions or molecules that are present. To do this, we rely on the concept of the chemical mole, abbreviated in the SI as mol.

A **mole** is defined as 6.02×10^{23} objects or particles. This very large number, known as **Avogadro's number**, is named for the nineteenth century Italian scientist Amedeo Avogadro. We can use the term *mole* to refer to any kind of chemical particles: ions, atoms, or molecules; it always stands for 6.02×10^{23} particles, just as *dozen* always stands for 12 units or objects. However, Avogadro's number is not an exact quantity; it has been measured experimentally and rounded off to three significant figures. By specifying the mole as a standard measure for a large number of small objects such as ions, atoms, and molecules, we have created a larger, more practical unit to work with.

8.2   MOLAR MASSES

The chemical mole is a unique and important unit in chemistry because *a mole of a substance has a mass that corresponds to the formula weight of the substance expressed in grams.* This mass, called the **molar mass** of the substance, has units of grams per mole, or g/mol. For example, the simplest unit of the element carbon is a carbon atom, and the formula for this simplest unit is C. The formula weight of carbon is 12.0 amu, but if we express this value in units of grams, 12.0 g, then this quantity contains 6.02×10^{23} carbon atoms, or a mole of carbon.

When the mass of a mole of a substance is expressed in grams, the mass can be technically referred to as a gram mole. The mass of a mole of a substance may be expressed in other mass or weight units, such as pounds or kilograms. In these cases, the mass could be referred to as a pound mole or a kilogram mole, respectively.

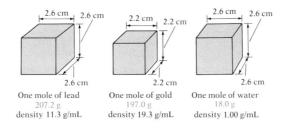

2.6 cm   2.6 cm

2.2 cm   2.2 cm

2.6 cm   2.6 cm

2.6 cm

2.2 cm

2.6 cm

One mole of lead
207.2 g
density 11.3 g/mL

One mole of gold
197.0 g
density 19.3 g/mL

One mole of water
18.0 g
density 1.00 g/mL

FIGURE 8-1 Mole Quantities of Lead, Gold, and Water

In general, when we speak of a mole of an element, we are referring to a mole of atoms of the element, even though many elements actually exist as molecules. Thus, a mole of any element contains the same number of atoms (6.02×10^{23}) as a mole of any other element; however, no two elements have the same molar mass. For example, a mole of hydrogen contains 6.02×10^{23} atoms and weighs 1.0 g; however, a mole of sodium weighs 23.0 g even though it also contains 6.02×10^{23} atoms. A useful analogy is the comparison of a dozen eggs to a dozen apples. There are the same number of each, but the dozen eggs will have a different mass than the dozen apples. Figure 8-1 compares mole quantities of lead, gold, and water in terms of mass, density, and volume. Note that while each is one mole, they do not necessarily have equal masses or volumes.

It is important to identify the particles being considered when referring to moles. For example, a mole of hydrogen molecules, H_2, contains 6.02×10^{23} molecules but it contains twice as many hydrogen atoms, or 12.0×10^{23} atoms, because each hydrogen molecule contains two atoms. For the molecular compound HCl, the formula weight expressed in grams is 36.5 g. (Formula weights of molecular compounds are also called **molecular weights**.) Thus 36.5 g is the weight of 6.02×10^{23} molecules of HCl, or one mole. Since each molecule of HCl contains one hydrogen atom and one chlorine atom, a mole of HCl contains one mole of hydrogen atoms and one mole of chlorine atoms. In the case of the ionic compound KCl, the formula weight expressed in grams, 74.6 g, is the weight of 6.02×10^{23} formula units of KCl, or one mole of KCl. Since each formula unit of KCl contains two ions, K^+ and Cl^-, a mole of KCl contains a mole of K^+ and a mole of Cl^-. Additional examples of mole relationships are given in Table 8-1.

TABLE 8-1 Examples of Mole Relationships

Particles	One Mole Contains	One Mole Weighs
C atoms	6.02×10^{23} atoms	12.0 g
H_2 molecules	6.02×10^{23} molecules	2.0 g
Na atoms	6.02×10^{23} atoms	23.0 g
NaCl units	6.02×10^{23} NaCl units	58.5 g
Na^+ ions	6.02×10^{23} Na^+ ions	23.0 g
Cl^- ions	6.02×10^{23} Cl^- ions	35.5 g

8.3   CALCULATION OF MOLAR MASSES

To calculate the molar mass of a substance, all we need is the formula and the atomic weights. From these we can calculate the formula weight, and the formula weight expressed in grams is the molar mass of the substance. The following example shows the calculation of the molar mass of fructose.

EXAMPLE 8.1   Fructose is a sugar found in many fruits. Its formula is $C_6H_{12}O_6$. What is the weight in grams of one mole of fructose?

Solution   From its formula, we can find the formula weight of fructose:

$$FW = (6 \text{ atoms C})\left(\frac{12.0 \text{ amu}}{\text{atom C}}\right) + (12 \text{ atoms H})\left(\frac{1.0 \text{ amu}}{\text{atom H}}\right)$$
$$+ (6 \text{ atoms O})\left(\frac{16.0 \text{ amu}}{\text{atom O}}\right)$$
$$= 72.0 \text{ amu} + 12.0 \text{ amu} + 96.0 \text{ amu}$$
$$= 180.0 \text{ amu}$$

The formula weight of fructose expressed in grams, 180.0 g, is the weight of one mole of fructose.

EXERCISE 8.1   Find the weight in grams of one mole of each substance.

a. H_2 c. $MgSO_4$
b. CO_2 d. K^+ ■

MATH TIP

When interconverting masses and number of particles, a two-step conversion is most convenient. This is apparent if you write a "map" following this format:

$$g \xleftrightarrow{\text{(molar mass)}} mol \xleftrightarrow{(6.02 \times 10^{23})} \text{particles}$$

Since 2.0 g Au is given in the problem, the map becomes

$$2.0 \text{ g Au} \longrightarrow ? \text{ mol Au} \longrightarrow ? \text{ Au atoms}$$

Refer to the map to determine the number of conversions necessary and the order in which to perform them.

8.4   ADDITIONAL CALCULATIONS BASED ON THE MOLE CONCEPT

By using the mole concept and Avogadro's number, we can interconvert masses and number of particles. For example, let's calculate the number of gold atoms in a pure gold nugget weighing 2.0 g. We solve this problem by using the unit-factor method. Here are our unit relationships:

$$1 \text{ mol Au} = 197.0 \text{ g Au}$$
$$1 \text{ mol Au} = 6.02 \times 10^{23} \text{ Au atoms}$$

Our mathematical expression is then

$$\text{Number of Au atoms} = (2.0 \text{ g Au})\left(\frac{1 \text{ mol Au}}{197.0 \text{ g Au}}\right)\left(\frac{6.02 \times 10^{23} \text{ Au atoms}}{1 \text{ mol Au}}\right)$$
$$= 6.1 \times 10^{21} \text{ Au atoms}$$

EXAMPLE 8.2   Sodium bicarbonate, $NaHCO_3$, is ordinary baking soda. How many sodium ions are in 17.0 g of sodium bicarbonate?

Solution First we must find the formula weight of sodium bicarbonate:

$$FW = (1 \text{ atom Na})\left(\frac{23.0 \text{ amu}}{\text{atom Na}}\right) + (1 \text{ atom H})\left(\frac{1.0 \text{ amu}}{\text{atom H}}\right)$$

$$(+ 1 \text{ atom C})\left(\frac{12.0 \text{ amu}}{\text{atom C}}\right) + (3 \text{ atoms O})\left(\frac{16.0 \text{ amu}}{\text{atom O}}\right)$$

$$= 23.0 \text{ amu} + 1.0 \text{ amu} + 12.0 \text{ amu} + 48.0 \text{ amu}$$

$$= 84.0 \text{ amu}$$

The formula weight of sodium bicarbonate expressed in grams (84.0 g) is the weight of one mole of sodium bicarbonate. We can use this relationship along with the others given below to find the number of sodium ions in 17.0 g of sodium bicarbonate.

$$1 \text{ mol NaHCO}_3 = 84.0 \text{ g NaHCO}_3$$

$$1 \text{ mol Na}^+ = 6.02 \times 10^{23} \text{ Na}^+$$

$$1 \text{ mol NaHCO}_3 = 1 \text{ mole Na}^+$$

We use these relationships to set up the solution to the problem:

$$\text{Number of Na}^+ = (17.0 \text{ g NaHCO}_3)\left(\frac{1 \text{ mol NaHCO}_3}{84.0 \text{ g NaHCO}_3}\right)\left(\frac{6.02 \times 10^{23} \text{ Na}^+}{\text{mol NaHCO}_3}\right)$$

$$= 1.22 \times 10^{23} \text{ Na}^+$$

MATH TIP

This problem requires an additional unit relationship because g $NaHCO_3$ are given and number of Na^+ ions are required. Examining the formula for the compound shows that 1 formula unit of $NaHCO_3$ contains 1 Na^+ ion; therefore 1 mol $NaHCO_3$ contains 1 mol Na^+ ions. The map requires an additional step.

17.0 g NaHCO₃ ⟶
        ? mol NaHCO₃ ⟶
              ? mol Na⁺ ions ⟶ ? Na⁺ ions

$17.0 \text{ g NaHCO}_3 \left(\dfrac{1 \text{ mol NaHCO}_3}{84.0 \text{ g NaHCO}_3}\right)$

$\left(\dfrac{1 \text{ mol Na}^+ \text{ ions}}{1 \text{ mol NaHCO}_3}\right)\left(\dfrac{6.02 \times 10^{23} \text{ Na}^+ \text{ ions}}{1 \text{ mol Na}^+}\right)$

Notice that the 1 mol Na^+ ions = 1 mol $NaHCO_3$ conversion can be condensed, making this a two-step conversion.

EXERCISE 8.2 How many molecules are in 26.3 g of carbon dioxide, CO_2? ■

EXAMPLE 8.3 What is the mass of 2.48×10^{32} nitrogen atoms?

Solution First, use the unit-factor method to find moles of nitrogen atoms.

$$1 \text{ mol N atoms} = 6.02 \times 10^{23} \text{ N atoms}$$

$$\text{moles N atoms} = (2.48 \times 10^{32} \text{ N atoms})\left(\frac{1 \text{ mol N atoms}}{6.02 \times 10^{23} \text{ N atoms}}\right)$$

$$= 4.12 \times 10^8 \text{ mol N atoms}$$

Then convert moles of nitrogen atoms to mass:

$$1 \text{ mol N atoms} = 14.0 \text{ g N}$$

$$\text{mass N} = (4.12 \times 10^8 \text{ mol N atoms})\left(\frac{14.0 \text{ g N}}{\text{mol N atoms}}\right) = 5.77 \times 10^9 \text{ g N}$$

Notice that we could have combined both of the preceding steps into one expression:

$$\text{mass N} = (2.48 \times 10^{32} \text{ N atoms})\left(\frac{1 \text{ mol N atoms}}{6.02 \times 10^{23} \text{ N atoms}}\right)\left(\frac{14.0 \text{ g N}}{\text{mol N atoms}}\right)$$

$$= 5.77 \times 10^9 \text{ g N}$$

MATH TIP

Once again a map can be written.

? g N ⟵ ? mol N ⟵ 2.48 × 10³² N atoms

The map shows that this problem will be worked in reverse order from the previous examples.

EXERCISE 8.3 What is the mass of 7.63×10^{20} molecules of water, H_2O? ■

We can also use the mole concept to interconvert mass and number of moles of a substance. The following examples are illustrations.

EXAMPLE 8.4

How many moles of sodium sulfate, Na_2SO_4, are in 10.0 g of the compound?

Solution

First, find the molar mass of Na_2SO_4:

$$FW = (2\ Na^+)\left(\frac{23.0\ amu}{Na^+}\right) + (1\ atom\ S)\left(\frac{32.1\ amu}{atom\ S}\right)$$
$$+ (4\ atoms\ O)\left(\frac{16.0\ amu}{atom\ O}\right)$$
$$= 46.0\ amu + 32.1\ amu + 64.0\ amu$$
$$= 142.1\ amu$$

Therefore, the molar mass of Na_2SO_4 is 142.1 g/mol.
Then use the unit-factor method to find moles of Na_2SO_4:

$$1\ mol\ Na_2SO_4 = 142.1\ g\ Na_2SO_4$$

$$moles\ Na_2SO_4 = (10.0\ g\ Na_2SO_4)\left(\frac{1\ mol\ Na_2SO_4}{142.1\ g\ Na_2SO_4}\right)$$
$$= 7.04 \times 10^{-2}\ mol\ Na_2SO_4$$

MATH TIP

The map

$$10.0\ g\ Na_2SO_4 \longrightarrow ?\ mol\ Na_2SO_4$$
$$\longrightarrow ?\ formula\ units\ Na_2SO_4$$

shows two conversions; however the required information in the problem is moles Na_2SO_4. The last conversion can be omitted in this problem.

EXERCISE 8.4

How many moles of acetic acid, $HC_2H_3O_2$, are in 17.9 g of the compound? ▪

EXAMPLE 8.5

What is the mass of 4.50 moles of dinitrogen pentoxide, N_2O_5?

Solution

First, find the molar mass of N_2O_5:

$$FW = (2\ atoms\ N)\left(\frac{14.0\ amu}{atom\ N}\right) + (5\ atoms\ O)\left(\frac{16.0\ amu}{atom\ O}\right)$$
$$= 28.0\ amu + 80.0\ amu$$
$$= 108.0\ amu$$

Therefore, the molar mass of N_2O_5 is 108.0 g/mol.
Now use the unit-factor method to find mass of N_2O_5:

$$1\ mol\ N_2O_5 = 108.0\ g\ N_2O_5$$

$$mass\ N_2O_5 = (4.50\ mol\ N_2O_5)\left(\frac{108.0\ g\ N_2O_5}{mol\ N_2O_5}\right) = 486\ g\ N_2O_5$$

EXERCISE 8.5

What is the mass of 2.83 moles of aluminum hydroxide, $Al(OH)_3$? ▪

8.5 EMPIRICAL AND TRUE FORMULAS

One way to characterize a compound is to decompose it and find the percentage of each constituent element. Finding percentage composition in this way is referred to as *performing an elemental analysis* of the com-

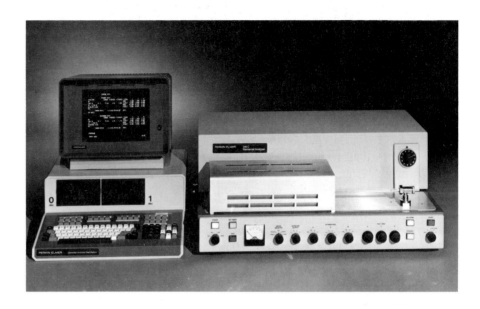

FIGURE 8-2 An Automated Elemental Analyzer (Courtesy of Perkin-Elmer Corporation)

pound. Figure 8-2 shows an automated elemental analyzer that can be used to measure percentages of carbon, hydrogen, nitrogen, oxygen, and sulfur in compounds.

The percentage composition of a compound can be used to calculate the simplest whole-number ratio of ions or atoms in the compound; this ratio is called the **empirical formula** of the compound. For example, the empirical formula of butane, the fuel of disposable cigarette lighters, is C_2H_5. This formula tells us that the simplest ratio of atoms in butane is two atoms of carbon for every five atoms of hydrogen, and the molar mass calculated from the empirical formula, 29.0 g/mol, is called the **apparent molar mass**. However, the molar mass of a compound can be directly measured by experimentation, and such measurements show butane to have a molar mass of 58.0 g/mol. This experimentally determined quantity is called the **actual molar mass**.

Why is the apparent molar mass of butane different from the actual molar mass? It is different because the empirical formula of butane is not the true formula of the compound. The **true formula** of a compound represents the total number of ions or atoms present in a formula unit of the compound. (For molecular compounds, the true formula is also referred to as the **molecular formula**.) For any compound, there could be more than one formula that agrees with the percentage composition, but the true formula must also indicate the correct molar mass. In the case of butane, the actual molar mass is twice the apparent molar mass (58.0 amu/29.0 amu = 2.00). Thus, in the true formula of butane, the atoms are present in a ratio that is a multiple of the simplest ratio; by comparing the actual molar mass to the apparent molar mass, we can see that the multiple must be 2. When we multiply the empirical formula by 2, we obtain the true formula of butane: $C_{(2\times2)}H_{(2\times5)}$ or C_4H_{10}. Table 8-2 shows several

If the percentage composition of a compound is known, calculating the empirical formula is straightforward. Although percentage composition is often determined by use of an instrument like the one shown in Figure 8-2, the percentage composition of some simple compounds may be determined by a laboratory experiment such as that illustrated in the accompanying figure. To analyze the metal oxide, we start with a weighed sample of the compound, perhaps 0.750 g. We then heat the sample in a stream of hydrogen. The oxygen released from the metal oxide is converted to water vapor, which leaves the system with the excess hydrogen. The solid residue left in the reaction tube at the end of the reaction is pure metal. By weighing the residue, we find that its mass is 0.603 g. The percentage of metal in the oxide must then be

$$\% \text{ metal} = \frac{0.603 \text{ g}}{0.750 \text{ g}} \times 100\% = 80.4\%$$

and the percentage of oxygen is calculated by subtraction:

$$\% \text{ oxygen} = 100.0\% - 80.4\% = 19.6\%$$

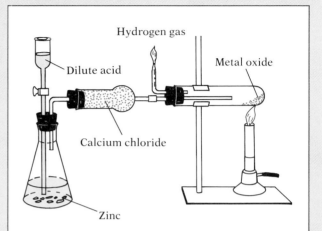

Determination of the Percentage Composition of a Metal Oxide. Hydrogen is produced in the flask by the reaction of acid, added by the dropping funnel, with zinc metal. The hydrogen then passes through the calcium chloride, which removes moisture, and on into the test tube containing the metal oxide. As the metal oxide is heated it releases oxygen, which reacts with the hydrogen to form water. The water produced by the reaction passes off with the excess hydrogen, which is burned as it leaves the reaction tube.

compounds having identical percentage compositions and empirical formulas but different true formulas.

As you can see, the true formula can never have fewer ions or atoms than the empirical formula, but the true formula may have more ions or atoms than the empirical formula. It is also possible for the true formula to be identical to the empirical formula. Table 8-3 lists empirical and true formulas of some common substances.

TABLE 8-2	Examples of Compounds Having Identical Percentage Compositions but Different True Formulas					
Name	%C	%H	Empirical Formula	Apparent Molar Mass	Actual Molar Mass	True Formula
Acetylene	92.3	7.7	CH	13.0	26.0	C_2H_2
Benzene	92.3	7.7	CH	13.0	78.0	C_6H_6
Styrene	92.3	7.7	CH	13.0	104.0	C_8H_8

TABLE 8-3	Empirical and True Formulas of Some Common Substances	
Substance	Empirical Formula	True Formula
Acetic acid	CH_2O	$C_2H_4O_2$
Formaldehyde	CH_2O	CH_2O
Glucose	CH_2O	$C_6H_{12}O_6$
Carbon dioxide	CO_2	CO_2
Hydrogen chloride	HCl	HCl
Hydrogen	H	H_2
Chlorine	Cl	Cl_2
Oxygen	O	O_2
Nitrogen	N	N_2

8.6 CALCULATION OF THE EMPIRICAL FORMULA

The percentage composition is the percentage of each element in a compound. Since the elements are combined in whole-number ratios of ions or atoms, the simplest ratio of elements and thus the empirical formula of a compound can be calculated from its percentage composition. Three steps are involved:

1. If the weight of the sample is given, convert the percentage of each element to actual weight. If the weight of the sample is not given, then assume a definite quantity of the compound (100 g is convenient) and convert the percentage of each element present to actual weight.
2. Convert the weight of each element to moles of the element.
3. Find the smallest whole-number ratio of the elements by dividing the moles of each element by the smallest number of moles calculated in step 2.

The following examples illustrate calculation of the empirical formula.

EXAMPLE 8.6 A compound consists of 74.2% sodium and 25.8% oxygen. What is the empirical formula of the compound?

Solution Step 1: Assume there is 100 g of the compound and calculate the actual weight of each element.

$$\text{wt Na} = (0.742)(100 \text{ g}) = 74.2 \text{ g}$$

$$\text{wt O} = (0.258)(100 \text{ g}) = 25.8 \text{ g}$$

Step 2: Calculate moles of each element.

$$\text{moles Na} = (74.2 \text{ g Na})\left(\frac{1 \text{ mol Na}}{23.0 \text{ g Na}}\right) = 3.23 \text{ mol Na}$$

$$\text{moles O} = (25.8 \text{ g O})\left(\frac{1 \text{ mol O}}{16.0 \text{ g O}}\right) = 1.61 \text{ mol O}$$

▽   **PROBLEM-SOLVING SKILLS**

MANIPULATING RATIOS

Dividing or multiplying a ratio by a quantity does not change the meaning of the ratio. To find the lowest whole-number ratio, divide every number of the ratio by the smallest number of the ratio. For example, consider the ratio $2.36 : 1.18$. Dividing by 1.18 gives

$$\frac{2.36}{1.18} : \frac{1.18}{1.18} = 2.00 : 1.00$$

In another example,

$$6.28 : 4.19 : 2.09 = \frac{6.28}{2.09} : \frac{4.19}{2.09} : \frac{2.09}{2.09}$$

$$= 3.00 : 2.00 : 1.00$$

In some cases, division by the smallest number of the ratio does not give a whole-number ratio. Then the numbers obtained by division must be multiplied by the smallest number that will convert all of them to whole numbers. For example, let's look at the ratio

$$0.278 : 0.418$$

Dividing by 0.278 does not give a whole-number ratio:

$$\frac{0.278}{0.278} : \frac{0.418}{0.278} = 1.00 : 1.50$$

The smallest number that will convert both numbers of the ratio to whole numbers is 2:

$$(2)(1.00) : (2)(1.50) = 2.00 : 3.00$$

Step 3:   Find the smallest whole-number ratio of elements.

$$\text{Na: } \frac{3.23}{1.61} = 2.01$$

$$\text{O: } \frac{1.61}{1.61} = 1.00$$

Therefore, the simplest whole-number ratio of Na : O is 2 : 1, and the empirical formula is Na_2O. (Notice that 2.01 was rounded off to 2 to give the ratio 2 : 1; this is acceptable when the ratio differs from a whole number *only* in the second decimal place.)

EXERCISE 8.6   A compound consists of 30.4% nitrogen and 69.6% oxygen. Find the empirical formula of the compound.   ▪

EXAMPLE 8.7   A compound consists of 1.59% hydrogen, 22.2% nitrogen, and 76.2% oxygen. Find the empirical formula of the compound.

Solution   Step 1:

$$wt\ H = (0.0159)(100\ g) = 1.59\ g$$
$$wt\ N = (0.222)(100\ g) = 22.2\ g$$
$$wt\ O = (0.762)(100\ g) = 76.2\ g$$

Step 2:

$$moles\ H = (1.59\ g\ H)\left(\frac{1\ mol\ H}{1.01\ g\ H}\right) = 1.57\ mol\ H$$

$$moles\ N = (22.2\ g\ N)\left(\frac{1\ mol\ N}{14.0\ g\ N}\right) = 1.59\ mol\ N$$

$$moles\ O = (76.2\ g\ O)\left(\frac{1\ mol\ O}{16.0\ g\ O}\right) = 4.76\ mol\ O$$

Step 3:

$$H: \frac{1.57}{1.57} = 1.00$$

$$N: \frac{1.59}{1.57} = 1.01$$

$$O: \frac{4.76}{1.57} = 3.03$$

Therefore, the simplest whole-number ratio of H : N : O is 1 : 1 : 3, and the empirical formula is HNO_3.

EXERCISE 8.7   A compound consists of 40.0% carbon, 6.67% hydrogen, and 53.3% oxygen. Find the empirical formula of the compound.   ▪

EXAMPLE 8.8   A sample of a compound consists of 7.51 g of aluminum and 6.69 g of oxygen. Find the empirical formula of the compound.

Solution   Step 1:   Since the weight of each element is given, we can skip step 1 and go on to step 2 to calculate moles of each element using the given weights.

Step 2:

$$moles\ Al = (7.51\ g\ Al)\left(\frac{1\ mol\ Al}{27.0\ g\ Al}\right) = 0.278\ mol\ Al$$

$$moles\ O = (6.69\ g\ O)\left(\frac{1\ mol\ O}{16.0\ g\ O}\right) = 0.418\ mol\ O$$

Step 3:

$$Al: \frac{0.278}{0.278} = 1.00$$

$$O: \frac{0.418}{0.278} = 1.50$$

In this case, we have not yet reached a whole-number ratio of elements, so we must multiply both values by the smallest number that will convert both to whole numbers. This number is 2:

$$\text{Al: } (1.00)(2) = 2.00$$
$$\text{O: } (1.50)(2) = 3.00$$

Thus, the smallest whole-number ratio of Al:O is 2:3, and the empirical formula is Al_2O_3.

EXERCISE 8.8   A sample of a compound consists of 15.5 g of oxygen and 12.3 g of fluorine. Find the empirical formula of the compound.

8.7 CALCULATION OF THE TRUE FORMULA FROM THE EMPIRICAL FORMULA

If the actual molar mass of a compound is known, the true formula can be calculated from the empirical formula by using the following relationship,

$$n = \frac{\text{actual molar mass}}{\text{apparent molar mass}}$$

where n represents a whole-number multiple (1, 2, 3, . . .) of the empirical formula. After finding n, simply multiply the empirical formula by n to find the true formula. This procedure is illustrated by the following examples.

EXAMPLE 8.9   A compound having a molar mass of 170 g/mol was found to have the empirical formula $HCrO_2$. What is the true formula of the compound?

Solution   First, find the apparent molar mass:

$$\text{Apparent FW} = (1\text{ atom H})\left(\frac{1.0 \text{ amu}}{\text{atom H}}\right) + (1\text{ atom Cr})\left(\frac{52.0 \text{ amu}}{\text{atom Cr}}\right)$$
$$+ (2\text{ atoms O})\left(\frac{16.0 \text{ amu}}{\text{atom O}}\right)$$
$$= 85.0 \text{ amu}$$

Thus, the apparent molar mass is 85.0 g/mol.
Next, find n:

$$n = \frac{\text{actual molar mass}}{\text{apparent molar mass}}$$
$$= \frac{170 \text{ g/mol}}{85.0 \text{ g/mol}}$$
$$= 2.0$$

Finally, multiply the empirical formula by n:

$$\text{True formula} = H_{(1\times2)}Cr_{(1\times2)}O_{(2\times2)} = H_2Cr_2O_4$$

EXERCISE 8.9   A compound having a molar mass of 34.0 g/mol was found to have the empirical formula HO. What is the true formula of the compound?   ■

EXAMPLE 8.10   A compound having a molar mass of 489.2 g/mol consists of 82.0% mercury, 4.91% carbon, and 13.1% oxygen. Find the true formula of the compound.

Solution   First, find the empirical formula:

$$\text{wt Hg} = (0.820)(100 \text{ g}) = 82.0 \text{ g}$$
$$\text{wt C} = (0.0491)(100 \text{ g}) = 4.91 \text{ g}$$
$$\text{wt O} = (0.131)(100 \text{ g}) = 13.1 \text{ g}$$

$$\text{moles Hg} = (82.0 \text{ g Hg})\left(\frac{1 \text{ mol Hg}}{200.6 \text{ g Hg}}\right) = 0.409 \text{ mol Hg}$$

$$\text{moles C} = (4.91 \text{ g C})\left(\frac{1 \text{ mol C}}{12.0 \text{ g C}}\right) = 0.409 \text{ mol C}$$

$$\text{moles O} = (13.1 \text{ g O})\left(\frac{1 \text{ mol O}}{16.0 \text{ g O}}\right) = 0.819 \text{ mol O}$$

$$\text{Hg: } \frac{0.409}{0.409} = 1.00$$

$$\text{C: } \frac{0.409}{0.409} = 1.00$$

$$\text{O: } \frac{0.819}{0.409} = 2.00$$

Therefore, the empirical formula is $HgCO_2$.
Next, find the apparent molar mass:

$$\text{Apparent FW} = (1 \text{ atom Hg})\left(\frac{200.6 \text{ amu}}{\text{atom Hg}}\right) + (1 \text{ atom C})\left(\frac{12.0 \text{ amu}}{\text{atom C}}\right)$$
$$+ (2 \text{ atoms O})\left(\frac{16.0 \text{ amu}}{\text{atom O}}\right)$$
$$= 244.6 \text{ amu}$$

Thus, the apparent molar mass is 244.6 g/mol. Then, find n:

$$n = \frac{\text{actual molar mass}}{\text{apparent molar mass}}$$
$$= \frac{489.2 \text{ g/mol}}{244.6 \text{ g/mol}}$$
$$= 2.000$$

Finally, multiply the empirical formula by n:

$$\text{True formula} = Hg_{(1\times2)}C_{(1\times2)}O_{(2\times2)} = Hg_2C_2O_4$$

EXERCISE 8.10   A compound having a molar mass of 180.0 g/mol consists of 40.0% carbon, 6.67% hydrogen, and 53.3% oxygen. Find the true formula of the compound.   ■

SUMMARY

A chemical **mole** is 6.02×10^{23} particles; a mole of an element refers to a mole of atoms of the element. A mole of a substance has a mass corresponding to the formula weight of the substance expressed in grams. This mass, called the **molar mass** of the substance, has units of g/mol. A molar mass can be calculated from a formula by first finding the formula weight and then expressing it in units of g/mol. The mole concept can be used to interconvert mass and number of particles or mass and number of moles of a substance.

The percentage composition of a compound can be used to find the **empirical formula**, the simplest whole-number ratio of ions or atoms in the compound. The molar mass calculated from the empirical formula is called the **apparent molar mass** of the compound. The actual molar mass of a compound can be measured experimentally. The **true formula** of a compound represents the total number of ions or atoms present in a formula unit of the compound. The true formula may be identical to the empirical formula or it may be a multiple of the empirical formula. The true formula of a compound can be calculated from the empirical formula if the actual molar mass is known.

STUDY QUESTIONS AND PROBLEMS

(More difficult questions and problems are marked with an asterisk.)

THE CHEMICAL MOLE

1. How many particles are in a mole? Who is this number named for?
2. Is the number of particles in a mole an exact number? Explain your answer.
3. Give the number of particles in each quantity.
   - a. 9.62 moles
   - b. 4.07×10^{-4} moles
   - c. 1.93 moles
   - d. 25 moles
4. Give the number of moles represented by each quantity.
   - a. 7.49×10^{11} particles
   - b. 9.21×10^{32} particles
   - c. 1.05×10^{6} particles
   - d. 1 particle

MOLAR MASSES

5. What is molar mass? What are its units?
6. Why does a mole of one element have a different mass than a mole of another element?
7. Complete the following table.

	Element	Weight (g)	Number of Moles	Number of Atoms
a.	Rb	14.9	_____	_____
b.	Kr	_____	_____	1.29×10^{20}
c.	N	_____	1.23	_____
d.	Cr	_____	_____	4.08×10^{39}
e.	S	40.5	_____	_____
f.	Mo	_____	7.86	_____

8. For each of the following compounds, give the number of molecules and the number of atoms of each element present in 2.00 moles of the compound.
   a. NO_2 b. PCl_3 c. $HC_2H_3O_2$
9. For each of the following compounds, give the number of cations and the number of anions present in 3.50 moles of the compound.
   a. $Na_2Cr_2O_7$ b. HgC_2O_4 c. K_2SO_4

CALCULATION OF MOLAR MASSES

10. Calculate the molar mass of each compound in question 8.
11. Calculate the molar mass of each compound in question 9.

ADDITIONAL CALCULATIONS BASED ON THE MOLE CONCEPT

12. The compound $MgSO_4 \cdot 7 H_2O$, commonly known as epsom salts, is used in dyeing fabrics. How many cations and how many anions are in 32.6 g of the compound? How many water molecules?
13. Lye, the common name for sodium hydroxide (NaOH), is a major ingredient in many drain cleaners. How many cations and how many anions are in 1.93×10^{-3} g of lye?
14. How many molecules are in 9.42 g of ozone, O_3?
15. Sodium hypochlorite, NaClO, is the active ingredient in household bleach. What is the mass of 6.57×10^{17} formula units of NaClO?
16. Complete the following table.

Compound	Weight (g)	Number of Moles	Number of Molecules or Formula Units
a. $SbBr_3$	_____	_____	1.49×10^{46}
b. $Cd(C_2H_3O_2)_2$	24.7	_____	_____
c. ISO_3F	_____	14.2	_____
d. $LiMnO_4$	2.66×10^{-5}	_____	_____
e. $MgCO_3 \cdot 5 H_2O$	_____	4.32×10^{-3}	_____
f. $Rb_2Cr_2O_7$	_____	_____	9.98×10^{16}

EMPIRICAL AND TRUE FORMULAS

17. Define the following terms:
   a. Elemental analysis   d. Actual molar mass
   b. Empirical formula   e. True formula
   c. Apparent molar mass
18. Why is it that several different compounds can have the same empirical formula?
19. Can the actual molar mass of a compound be the same as its apparent molar mass? Explain your answer.
20. Can the true formula of a compound have fewer ions or atoms than the empirical formula of the compound? Explain your answer.

CALCULATION OF THE EMPIRICAL FORMULA

21. Isooctane, the compound on which the rating scale of gasoline is based, consists of 84.2% carbon and 15.8% hydrogen. What is the empirical formula of isooctane?

22. The insecticide Lindane consists of 24.7% carbon, 2.06% hydrogen, and 73.2% chlorine. What is the empirical formula of Lindane?

23. An elemental analysis performed on a compound established the following composition: 2.81 g of bismuth, 0.417 g of phosphorus, and 0.863 g of oxygen. What is the empirical formula of the compound?

24. A red liquid compound consists of 14.9% nitrogen and 85.1% sulfur. What is the empirical formula of the compound?

25. A student burned a piece of magnesium in the laboratory and found that 3.04 g of magnesium combined with 2.00 g of oxygen to form a product. What was the empirical formula of the product?

CALCULATION OF THE TRUE FORMULA FROM THE EMPIRICAL FORMULA

26. Cyclopropane, a gas sometimes used as a general anesthetic, consists of 85.7% carbon and 14.3% hydrogen and has a molar mass of 42.0 g/mol. What is the true formula of cyclopropane?

27. Hydrazine, a liquid used as a rocket fuel, consists of 87.5% nitrogen and 12.5% hydrogen and has a molar mass of 32.0 g/mol. What is the true formula of hydrazine?

28. Butyric acid, one of the compounds responsible for the odor of goats, has a molar mass of 88.0 g/mol and consists of 54.5% carbon, 9.09% hydrogen, and 36.4% oxygen. What is the true formula of butyric acid?

GENERAL EXERCISES

29. For each of the following compounds, give the number of molecules and the number of atoms of each element present in 4.25 moles of the compound.
a. $XeOF_4$ b. C_2H_6 c. $PSBr_2Cl$

30. For each of the following compounds, give the number of cations and the number of anions present in 5.60 moles of the compound.
a. $AlCl_3$ b. $Mg(HSO_3)_2$ c. $Ba_3(PO_4)_2$

31. Dinitrogen oxide, N_2O, sometimes called laughing gas, is used as an anesthetic for dental patients. What is the mass of 4.92×10^{52} molecules of N_2O?

32. Freon is the trade name of a group of gases used in refrigeration systems. One type of freon consists of 8.73% carbon, 77.5% chlorine, and 13.8% fluorine. What is the empirical formula of this compound?

33. A compound consisting of carbon and hydrogen was burned. All of the carbon in the compound was converted to 13.2 g of CO_2 and all of the hydrogen was converted to 10.8 g of H_2O. What is the empirical formula of the compound?

*34. An oxide of phosphorus was prepared by burning 10.00 g of phosphorus. The weight of the oxide was 17.77 g and its molar mass was 220 g/mol. What is the true formula of the oxide?

CHEMICAL EQUATIONS

CHAPTER 9

OUTLINE

9.1 The Language of Chemical Equations
9.2 Conservation of Mass in Chemical
   Reactions
   Perspective: Antoine Lavoisier and
   an Early Demonstration of the
   Conservation of Mass in Chemical
   Reactions
9.3 Balancing Chemical Equations
9.4 Types of Chemical Reactions
   Decomposition
   Combination
   Single Replacement
   Double Replacement
9.5 Stoichiometry
9.6 Stoichiometric Calculations
   Mole-Mole Calculations
   Mole-Weight Calculations
   Weight-Mole Calculations
   Weight-Weight Calculations
   Problem-Solving Skills: Combining
      Steps in the Solutions to Problems
9.7 Percentage Yield
9.8 Limiting Reactant
   Summary
   Study Questions and Problems

We encounter chemical reactions everywhere we turn. When we burn wood to create warmth, the wood reacts with oxygen in the air to produce water vapor and carbon dioxide. Rust accumulates on iron because of a slow chemical reaction that converts iron and oxygen in the air to the reddish brown material iron(III) oxide. Other familiar examples of chemical reactions are the conversion of sugar to alcohol and carbon dioxide in the fermentation process, the tarnishing of silver and copper, the digestion of food, and the incorporation of fluoride into tooth enamel. In this chapter you will study chemical equations, which are symbolic descriptions of chemical reactions, and the calculations involving chemical equations. To prepare for this chapter, you will need to review dimensional analysis (section 2.6), ionic compounds (section 6.4), molecular compounds (section 6.7), polyatomic ions (section 6.9), naming inorganic compounds (chapter 7), and calculations based on chemical formulas (chapter 8).

9.1   THE LANGUAGE OF CHEMICAL EQUATIONS

Chemical equations are concise, accurate descriptions of chemical reactions. The formation of rust can be described by a chemical equation

$$4\,Fe + 3\,O_2 \longrightarrow 2\,Fe_2O_3,$$
$$\text{iron}\quad\text{oxygen}\qquad\text{iron(III) oxide}$$

and so can the process of fermentation

$$C_6H_{12}O_6 \longrightarrow 2\,C_2H_5OH + 2\,CO_2$$
$$\text{glucose}\qquad\text{ethanol}\qquad\text{carbon dioxide}$$

In chemical equations, formulas for the starting materials are shown on the left side of the arrow; these materials are called **reactants.** The substances produced by the chemical reaction are called **products;** their formulas are located to the right of the arrow. The arrow itself is read as "produce" or "yield." The numbers in front of the formulas are **coefficients** that indicate the smallest number of atoms or other chemical units that can participate in the chemical reaction. (If there is no coefficient before a formula, it is understood that the formula has a coefficient of 1.) The equation for the rusting of iron can be expressed in words: four atoms of iron react with three molecules of oxygen to produce two formula units of iron(III) oxide. Similarly, we can use words to paraphrase the equation for fermentation: one molecule of glucose yields two molecules of ethanol and two molecules of carbon dioxide.

Occasionally chemical equations may contain various other kinds of notation to indicate observations or conditions. Some examples are

(s) or ↓	solid
(l)	liquid
(g) or ↑	gas
Δ	heat
(aq)	dissolved in water (aqueous solution)

Coefficient

A number placed before a quantity to indicate multiplication.

As you read through the chemical equations in this chapter, practice putting them into words. By doing this, you will see that chemical equations actually represent complete statements of fact.

9.2 CONSERVATION OF MASS IN CHEMICAL REACTIONS

Combustion

A chemical reaction, in which oxygen is one reactant, characterized by rapid burning with the production of heat and light.

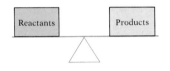

FIGURE 9-1 Conservation of Mass in Chemical Reactions

Let's consider the conservation of mass in the chemical reaction that occurs when charcoal is burned. Although charcoal contains a number of minor ingredients, for most purposes it can be considered to be composed of carbon. We write the equation for the combustion of charcoal as follows:

$$C(s) + O_2(g) \longrightarrow CO_2(g)$$

Here, carbon and oxygen are reactants, and carbon dioxide is the product. If we were to measure the total mass of reactants, we would find this mass to be equal to the mass of carbon dioxide produced. Thus, this chemical reaction, like all others, follows the law of conservation of mass. For all chemical reactions, the total mass of reactants is equal to the total mass of products (Figure 9-1).

EXAMPLE 9.1 The elements hydrogen and fluorine react to form hydrogen fluoride:

$$H_2(g) + F_2(g) \longrightarrow 2 HF(g)$$

If 2 grams of hydrogen and 38 grams of fluorine react, how many grams of hydrogen fluoride are formed?

Solution In this reaction, hydrogen and fluorine are the reactants, and hydrogen fluoride is the product. Since the total mass of reactants is 40 g, the mass of the product must also be 40 g:

$$\text{total mass of reactants} = \text{total mass of products}$$
$$2 \text{ g} + 38 \text{ g} = 40 \text{ g}$$

EXERCISE 9.1 When 5.00 g of carbon are burned, 13.3 g of oxygen are consumed. How many grams of carbon dioxide are produced?

$$C(s) + O_2(g) \longrightarrow CO_2(g)$$

9.3 BALANCING CHEMICAL EQUATIONS

In order for an equation to describe a reaction accurately, the equation must be balanced. For example, when carbon burns, carbon dioxide is formed:

$$\text{carbon} + \text{oxygen} \longrightarrow \text{carbon dioxide}$$

Experiments have demonstrated that for every atom of carbon used up, one molecule of oxygen is used and one molecule of carbon dioxide is formed. The chemical equation for the reaction is based on this knowl-

Because of his carefully controlled experiments and quantitative measurements, Antoine Lavoisier, a member of the French nobility and also a tax collector, is considered the father of modern chemistry. In 1772 he began experimenting with combustion. By weighing objects before and then again after combustion, he observed that burning objects gain weight. As he pursued these studies, he also found that when combustion is carried out in a closed container, there is no change in weight; the weight of substances in the container is the same after combustion as before, despite obvious changes of form by the substances. These results suggested to Lavoisier that when a sample is burned it gains something from the air, and that the weight gained by the burning sample is the same as the weight lost by the air. This is why no change in weight occurs when combustion is carried out in a closed container. On the basis of his experiments, Lavoisier proposed that a burning object removes oxygen from the air, and the oxygen combines with the burning object. Today we do indeed find that most common combustion reactions involve oxygen, as Lavoisier suggested some two hundred years ago.

Antoine Lavoisier, after a painting by Louis David (Courtesy of Burndy Library)

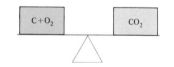

FIGURE 9-2   The Balanced Equation:
$C(s) + O_2(g) \rightarrow CO_2(g)$

edge; it shows one atom of C reacting with one molecule of O_2 to produce one molecule of CO_2:

$$C(s) + O_2(g) \longrightarrow CO_2(g)$$

Notice that our equation is balanced as it stands. That is, there is the same number of each kind of atom on each side of the arrow, and the equation satisfies the law of conservation of mass (see Figure 9-2).

Let's look at another example. Hydrogen and chlorine react to form hydrogen chloride. We can write an initial equation containing the formulas of reactants and product,

$$H_2(g) + Cl_2(g) \longrightarrow HCl(g)$$

but we see that this initial equation is not balanced. There are two hydrogen atoms on the left and only one on the right; the same is true for chlorine atoms. If we leave the equation this way, we are saying that one hydrogen atom and one chlorine atom cease to exist; this is a violation of the law of conservation of mass (see Figure 9-3) and thus impossible. We can correct our initial equation by inserting the number 2 as a coefficient for HCl. The equation now states the facts accurately: two molecules of

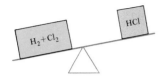

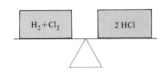

FIGURE 9-3 Balancing the Equation $H_2(g) + Cl_2(g) \rightarrow 2\,HCl(g)$

Remember that some elements exist as diatomic molecules. These are H_2, N_2, O_2, and the halogens.

hydrogen chloride are formed from one molecule of hydrogen and one molecule of chlorine.

$$H_2(g) + Cl_2(g) \longrightarrow 2\,HCl(g)$$

The equation is now balanced, illustrating that atoms are neither created nor destroyed in the chemical reaction. In fact, the number of atoms is the same before and after the reaction; only bonds have been rearranged to form the product:

$$H-H + Cl-Cl \longrightarrow 2\,H-Cl$$

Figure 9-3 shows the effect of balancing the equation.

As you can see, balancing equations is only a matter of accurate bookkeeping. You begin with the formulas for the reactants and products, and you can usually deduce these from your knowledge of the periodic table and bonding. You can then balance the equation by using the following rules:

1. Write the initial equation using correct formulas for reactants and products. Use + signs when there are two or more reactants or products, and separate reactants from products with an arrow.
2. Count the number of atoms of each element in the reactants and products.
3. Using the smallest possible whole numbers, insert coefficients for one element at a time to balance the number of atoms of each element in the reactants and products. Be sure not to change subscripts in the formulas; the coefficient of a formula multiplied by the subscripts determines the number of atoms of each element. It usually works best to balance metals first, nonmetals (except hydrogen and oxygen) second, hydrogen third, and oxygen last. Identical polyatomic ions that appear on both sides of the arrow can be balanced the same way as individual atoms.
4. Recount the number of atoms of each element in the reactants and products.
5. If you find that you have "unbalanced" one element while balancing another, change the coefficients until you achieve a balanced equation for all elements.

We can practice following these rules by balancing the equation for the formation of ammonia, a gas, from nitrogen and hydrogen.

Step 1:   Initial equation:

$$N_2(g) + H_2(g) \longrightarrow NH_3(g)$$

Step 2:   Count atoms:

Element	Reactants	Products
N	2	1
H	2	3

Step 3:   Insert coefficients for nitrogen:

$$N_2(g) + H_2(g) \longrightarrow 2\ NH_3(g)$$

Insert coefficients for hydrogen (note that 2 NH_3 contains 6 hydrogen atoms):

$$N_2(g) + 3\ H_2(g) \longrightarrow 2\ NH_3(g)$$

Step 4:   Recount atoms:

Element	Reactants	Products
N	2	2
H	6	6

Step 5:   Change coefficients if necessary:
This step is not necessary, since all elements are now balanced.

EXAMPLE 9.2   Write the balanced equation for the combustion of methane, $CH_4(g)$, producing water vapor and carbon dioxide gas.

Solution   Step 1:   Initial equation:

$$CH_4(g) + O_2(g) \longrightarrow H_2O(g) + CO_2(g)$$

Step 2:   Count atoms (note that oxygen is present in two products):

Element	Reactants	Products
C	1	1
O	2	3
H	4	2

Step 3:   Since carbon is already balanced, insert coefficients for hydrogen first and then for oxygen:

$$CH_4(g) + 2\ O_2(g) \longrightarrow 2\ H_2O(g) + CO_2(g)$$

Step 4:   Recount atoms:

Element	Reactants	Products
C	1	1
O	4	4
H	4	4

Step 5:   Change coefficients if necessary:
This step is not necessary.

EXERCISE 9.2   Write the balanced equation for each reaction.

a. Sodium metal reacts with chlorine gas to produce solid sodium chloride.
b. Magnesium metal reacts with oxygen gas to produce solid magnesium oxide. ■

EXAMPLE 9.3   Sodium metal reacts vigorously with liquid water to form sodium hydroxide, NaOH(s), and hydrogen gas. Write a balanced equation for this reaction.

Solution   Step 1:   Initial equation:

$$Na(s) + H_2O(l) \longrightarrow NaOH(s) + H_2(g)$$

Step 2:   Count atoms:

Element	Reactants	Products
Na	1	1
H	2	3
O	1	1

Step 3:   Insert coefficients:
Since sodium and oxygen are balanced, we need to insert coefficients only for hydrogen. Insertion of ½ before H_2 will balance hydrogen atoms.

$$Na(s) + H_2O(l) \longrightarrow NaOH(s) + ½ H_2(g)$$

However, we usually try to find the smallest whole-number coefficient, because ions, atoms, and formula units do not exist in fractional quantities. We can multiply the entire equation by 2 to balance it with whole-number coefficients.

$$2 Na(s) + 2 H_2O(l) \longrightarrow 2 NaOH(s) + H_2(g)$$

Step 4:   Recount atoms:

Element	Reactants	Products
Na	2	2
H	4	4
O	2	2

Step 5:   Change coefficients if necessary:
This step is not necessary.

EXERCISE 9.3   Write the balanced equation for each reaction.

a. Solid sulfur, S_8, burns to produce sulfur dioxide, a gas.
b. Ethane, $C_2H_6(g)$, burns to produce carbon dioxide and water vapor. ■

EXAMPLE 9.4

An aqueous solution of iron(III) chloride reacts with an aqueous solution of sodium carbonate to produce solid iron(III) carbonate and an aqueous solution of sodium chloride. Write the balanced equation for the reaction.

Solution Step 1: Initial equation:

$$FeCl_3(aq) + Na_2CO_3(aq) \longrightarrow Fe_2(CO_3)_3(s) + NaCl(aq)$$

Step 2: Count atoms:

Element	Reactants	Products
Fe	1	2
Cl	3	1
Na	2	1
C	1	3
O	3	9

Step 3: Insert coefficients:
   We first insert coefficients for the metals: 2 for Fe in the reactants and 2 for Na in the products.

$$2\ FeCl_3(aq) + Na_2CO_3(aq) \longrightarrow Fe_2(CO_3)_3(s) + 2\ NaCl(aq)$$

Next, we must balance the nonmetals: Cl, C, and O. For Cl, we must change the coefficient of NaCl from 2 to 6. (This "unbalances" Na, but we can adjust Na later if necessary.)

$$2\ FeCl_3(aq) + Na_2CO_3(aq) \longrightarrow Fe_2(CO_3)_3(s) + 6\ NaCl(aq)$$

Since the polyatomic ion CO_3^{2-} appears on both sides of the equation, we can treat it as a unit, rather than working with individual C and O atoms. Thus, we must insert a coefficient of 3 in front of Na_2CO_3 to balance CO_3^{2-} ions.

$$2\ FeCl_3(aq) + 3\ Na_2CO_3(aq) \longrightarrow Fe_2(CO_3)_3(s) + 6\ NaCl(aq)$$

Step 4: Recount atoms:

Element	Reactants	Products
Fe	2	2
Cl	6	6
Na	6	6
C	3	3
O	9	9

Step 5: Change coefficients if necessary:
   This step is not necessary, since balancing CO_3^{2-} ions also balanced Na.

EXERCISE 9.4 Write the balanced equation for each reaction.

a. $CaCl_2(aq) + Na_3PO_4(aq) \rightarrow Ca_3(PO_4)_2(s) + NaCl(aq)$

b. $H_2SO_4(aq) + NaOH(aq) \rightarrow Na_2SO_4(aq) + H_2O(l)$ ◼

9.4 TYPES OF CHEMICAL REACTIONS

Although there is a seemingly infinite number of chemical reactions, most of them can be classified according to four basic patterns. The following discussion summarizes the four fundamental reaction types.

Decomposition Decomposition reactions are easily recognized because a single reactant is broken down into two or more simpler products. The general pattern of a decomposition reaction is

$$AB \longrightarrow A + B$$

Some examples of decomposition reactions are:

$$CO_2(g) \xrightarrow{\Delta} C(s) + O_2(g)$$

$$2\ NH_3(g) \xrightarrow{\Delta} N_2(g) + 3\ H_2(g)$$

$$(NH_4)_2Cr_2O_7 \xrightarrow{\Delta} Cr_2O_3 + N_2 + 4\ H_2O$$

Combination Combination reactions form a compound by combining two or more atoms or simpler compounds. Thus, combination reactions are just reversals of decomposition reactions. The general pattern of combination reactions is

$$A + B \longrightarrow AB$$

The following reactions are examples of combination reactions.

$$4\ Al(s) + 3\ O_2(g) \longrightarrow 2\ Al_2O_3(s)$$
$$2\ H_2(g) + O_2(g) \longrightarrow 2\ H_2O(g)$$
$$N_2(g) + 3\ F_2(g) \longrightarrow 2\ NF_3(g)$$

Single replacement reactions involve oxidation-reduction.

Single Replacement In single replacement reactions, an ion, atom, or group of atoms replaces another ion, atom, or group of atoms in a compound. The general pattern is

$$A + BC \longrightarrow AC + B$$

One example is the replacement of silver in silver nitrate by copper:

$$Cu(s) + 2\ AgNO_3(aq) \longrightarrow Cu(NO_3)_2(aq) + 2\ Ag(s)$$

The replacement of *half* of the hydrogen in water by sodium is another example:

$$2\ Na(s) + 2\ HOH(l) \longrightarrow 2\ NaOH(s) + H_2(g)$$

(Note that H_2O is written as HOH to emphasize replacement of one H atom by Na.)

Double replacement reactions do not involve oxidation-reduction.

Double Replacement In double replacement reactions, two parts of two different reactants replace each other. The general pattern is

$$AB + CD \longrightarrow AD + CB$$

Ionic compounds often react in double replacement reactions. For example, barium chloride reacts with sodium sulfate

$$BaCl_2(aq) + Na_2SO_4(aq) \longrightarrow BaSO_4(s) + 2\ NaCl(aq)$$

and potassium carbonate reacts with lead(II) nitrate,

$$K_2CO_3(aq) + Pb(NO_3)_2(aq) \longrightarrow PbCO_3(s) + 2\ KNO_3(aq)$$

As you can see from the general pattern, double replacement reactions are a trading of partners between compounds. For ionic compounds, the reactants switch cation-anion partners.

EXAMPLE 9.5 Classify each of the following chemical reactions.

a. $Ca(s) + 2\ HOH(l) \rightarrow Ca(OH)_2(s) + H_2(g)$
b. $H_2CO_3(aq) \rightarrow CO_2(g) + H_2O(l)$
c. $HCl(g) + NaOH(s) \rightarrow HOH(l) + NaCl(s)$
d. $P_4(s) + 6\ H_2(g) \rightarrow 4\ PH_3(g)$

Solution a. This reaction follows the pattern $A + BC \rightarrow AC + B$, where A is Ca, BC is H—OH, AC is NaOH, and B is H_2. Thus, this is a single replacement reaction.

b. This reaction follows the pattern $AB \rightarrow A + B$, where A is H_2CO_3, B is CO_2, and C is H_2O. Thus, this is a decomposition reaction.

c. This reaction follows the pattern $AB + CD \rightarrow AD + CB$, where AB is HCl, CD is NaOH, AD is HOH, and CB is NaCl. Thus, this is a double replacement reaction.

d. This reaction follows the pattern $A + B \rightarrow AB$, where A is P_4, B is H_2, and AB is PH_3. Thus, this is a combination reaction.

EXERCISE 9.5 Classify each of the following reactions.

a. $P_4(s) + 5\ O_2(g) \rightarrow 2\ P_2O_5(s)$
b. $Fe_2O_3(s) + 3\ H_2(g) \rightarrow 2\ Fe(s) + 3\ H_2O(l)$
c. $2\ NaCl(l) \rightarrow 2\ Na(l) + Cl_2(g)$
d. $H_2SO_4(aq) + 2\ NaOH(aq) \rightarrow Na_2SO_4(aq) + 2\ H_2O(l)$ ▪

EXAMPLE 9.6 Predict the products of the following double replacement reactions of ionic compounds, and write a balanced equation for each reaction.

a. $AgNO_3 + NaCl \rightarrow$
b. $Na_2SO_4 + CuCl_2 \rightarrow$

Solution a. According to the general pattern for double replacement, $AgNO_3$ corresponds to AB and NaCl to CD:

$$AB\ +\ CD \longrightarrow AD + CB$$
$$AgNO_3 + NaCl \longrightarrow$$

Thus, $A = Ag^+$, $B = NO_3^-$, $C = Na^+$, and $D = Cl^-$. When the reactants trade ionic partners, the products are as shown below.

$$AB + CD \longrightarrow AD + CB$$
$$AgNO_3 + NaCl \longrightarrow AgCl + NaNO_3$$

The formulas of products are correct as written, and the equation is balanced as it stands.

b. According to the general pattern for double replacement, Na_2SO_4 corresponds to AB and $CuCl_2$ to CD.

$$AB + CD \longrightarrow AD + CB$$
$$Na_2SO_4 + CuCl_2 \longrightarrow$$

Thus, $A = Na^+$, $B = SO_4^{2-}$, $C = Cu^{2+}$, and $D = Cl^-$. When the reactants trade ionic partners, the products are

$$AD = Na^{\textcircled{1}+} \quad Cl^{\textcircled{1}-} = NaCl$$
$$CB = Cu^{\textcircled{2}+} \quad SO_4^{\textcircled{2}-} = CuSO_4$$

Therefore the preliminary equation is

$$AB + CD \longrightarrow AD + CB$$
$$Na_2SO_4 + CuCl_2 \longrightarrow NaCl + CuSO_4$$

To balance the equation, we must place a 2 in front of NaCl:

$$Na_2SO_4 + CuCl_2 \longrightarrow 2\ NaCl + CuSO_4$$

EXERCISE 9.6 Predict the products of the following double replacement reactions of ionic compounds, and write a balanced equation for each reaction.

a. $CaF_2 + H_2SO_4 \longrightarrow$
b. $NaNO_3 + H_2SO_4 \longrightarrow$ ■

9.5 STOICHIOMETRY

The numerical coefficients of a balanced equation tell us the number of ions, atoms, or formula units that react with each other and the number of ions, atoms, or formula units of product that forms. However, the numerical coefficients also tell us much more about the reaction. For example, let's look at the equation describing the formation of water from hydrogen and oxygen:

$$2\ H_2(g) + O_2(g) \longrightarrow 2\ H_2O(g)$$

It tells us that two molecules of hydrogen react with one molecule of oxygen to form two molecules of water. These are the smallest quantities of hydrogen and oxygen that can react. However, hydrogen and oxygen can react in larger quantities, as long as there are at least twice as many hydrogen molecules as oxygen molecules. Thus, two moles of H_2 ($2 \times 6.02 \times 10^{23}$ molecules) can react with one mole of O_2 ($1 \times 6.02 \times 10^{23}$

molecules), and they will produce two moles of H_2O ($2 \times 6.02 \times 10^{23}$ molecules). Since a mole is simply a fixed number of molecules, then the numerical coefficients of a balanced equation also tell us the number of moles of substances that can react and the number of moles of product that will be formed. In addition, the number of moles of reactants and products can be converted to weights by using molar masses. This is very practical information, since moles and weights are quantities we work with in the laboratory. Thus, balanced equations can give us useful relationships among quantities of reactants and products; we call these relationships **stoichiometry.**

Stoichiometry
(STOW-ih-key-AAH-meh-tree): derived from Greek words that mean ''to measure components.''

Since the numerical coefficients of a balanced equation indicate the number of moles of reactants and products, it is useful to think in terms of mole ratios. A **mole ratio** is the ratio of moles of any two substances — ions, atoms, or formula units — determined from the numerical coefficients of a balanced equation. Looking at the formation of water as our example,

$$2\,H_2(g) + O_2(g) \longrightarrow 2\,H_2O(g)$$

we can derive two mole ratios for each pair of substances:

$$\frac{2\text{ mol }H_2}{1\text{ mol }O_2}, \quad \frac{1\text{ mol }O_2}{2\text{ mol }H_2}$$

$$\frac{2\text{ mol }H_2}{2\text{ mol }H_2O}, \quad \frac{2\text{ mol }H_2O}{2\text{ mol }H_2}$$

$$\frac{1\text{ mol }O_2}{2\text{ mol }H_2O}, \quad \frac{2\text{ mol }H_2O}{1\text{ mol }O_2}$$

In fact, we can do the same with the reactants and products of any balanced equation. Then these mole ratios can be used as unit factors in solving problems. Notice that mole ratios contain exact numbers and thus can never limit the number of significant figures in our calculations.

EXAMPLE 9.7   For the following balanced equation,

$$2\,FeCl_3(aq) + 3\,H_2S(g) \longrightarrow Fe_2S_3(s) + 6\,HCl(g)$$

derive mole ratios for each pair of substances.

Solution   There will be two mole ratios for each different pair of substances. Since there are four substances, there are six different pairs of substances and a total of 12 mole ratios.

$$\frac{2\text{ mol }FeCl_3}{3\text{ mol }H_2S}, \quad \frac{3\text{ mol }H_2S}{2\text{ mol }FeCl_3}$$

$$\frac{2\text{ mol }FeCl_3}{1\text{ mol }Fe_2S_3}, \quad \frac{1\text{ mol }Fe_2S_3}{2\text{ mol }FeCl_3}$$

$$\frac{2\text{ mol }FeCl_3}{6\text{ mol }HCl}, \quad \frac{6\text{ mol }HCl}{2\text{ mol }FeCl_3}$$

$$\frac{3\text{ mol }H_2S}{1\text{ mol }Fe_2S_3}, \quad \frac{1\text{ mol }Fe_2S_3}{3\text{ mol }H_2S}$$

$$\frac{3 \text{ mol } H_2S}{6 \text{ mol } HCl}, \quad \frac{6 \text{ mol } HCl}{3 \text{ mol } H_2S}$$

$$\frac{1 \text{ mol } Fe_2S_3}{6 \text{ mol } HCl}, \quad \frac{6 \text{ mol } HCl}{1 \text{ mol } Fe_2S_3}$$

EXERCISE 9.7 For the following balanced equation,

$$3 \text{ NaOH(aq)} + H_3PO_4(aq) \longrightarrow Na_3PO_4(aq) + 3 \text{ H}_2O(l)$$

derive mole ratios for each pair of substances. ■

9.6 STOICHIOMETRIC CALCULATIONS

In doing stoichiometric calculations, it is essential to start with a balanced equation.

Calculations of quantities of reactants and products based on balanced equations are known as **stoichiometric calculations.** There are four categories of stoichiometric calculations: (1) calculating moles of one substance from moles of another substance; (2) calculating the weight of one substance from moles of another substance; (3) calculating moles of one substance from the weight of another substance; and (4) calculating the weight of one substance from the weight of another substance. These calculations are discussed in the following paragraphs.

MATH TIP

In stoichiometric calculations, it is very important to write both the unit and the label of a quantity: 2.25 mol Cl_2, rather than 2.25 mol. The unit and label must agree before two quantities can be cancelled.

Mole-Mole Calculations In calculating moles of one substance from moles of another substance, a method much like the unit-factor method is used. Two steps are needed.

1. Write down the number of moles given for the first substance.
2. Multiply the given number of moles by the mole ratio that will give moles of the substance you are seeking.

These steps are illustrated in the following examples.

EXAMPLE 9.8 Chlorine reacts with sodium to form sodium chloride:

$$Cl_2(g) + 2 \text{ Na(s)} \longrightarrow 2 \text{ NaCl(s)}$$

How many moles of NaCl will be produced from 2.25 moles of Cl_2?

Solution Step 1: Write down the given number of moles.

$$2.25 \text{ mol } Cl_2$$

Step 2: Multiply the given number of moles by the mole ratio that relates Cl_2 and NaCl and allows cancellation of Cl_2.

$$(2.25 \text{ mol } Cl_2) \left(\frac{2 \text{ mol NaCl}}{1 \text{ mol } Cl_2} \right) = 4.50 \text{ mol NaCl}$$

EXERCISE 9.8 Potassium reacts with oxygen to form potassium oxide:

$$4 \text{ K(s)} + O_2(g) \longrightarrow 2 \text{ K}_2O(s)$$

How many moles of K_2O will be formed from 0.283 moles of K? ■

EXAMPLE 9.9 Sodium hydroxide reacts with sulfuric acid to form sodium sulfate and water:

$$2\ NaOH(s) + H_2SO_4(aq) \longrightarrow Na_2SO_4(aq) + 2\ H_2O(l)$$

How many moles of H_2SO_4 are required to react with 3.58 moles of NaOH?

Solution Step 1: Write down the given number of moles.

$$3.58\ mol\ NaOH$$

Step 2: Multiply the given number of moles by the mole ratio that relates NaOH and H_2SO_4 and allows cancellation of NaOH.

$$(3.58\ \text{mol NaOH}) \left(\frac{1\ mol\ H_2SO_4}{2\ \text{mol NaOH}} \right) = 1.79\ mol\ H_2SO_4$$

EXERCISE 9.9 Potassium chlorate decomposes to potassium chloride and oxygen when heated:

$$2\ KClO_3(s) \xrightarrow{\Delta} 2\ KCl(s) + 3\ O_2(g)$$

How many moles of O_2 will be produced from 5.62 moles of $KClO_3$? ■

Mole-Weight Calculations In this type of calculation, the number of moles of one substance is used to calculate the weight of another substance. *The only direct relationship between two substances in a chemical reaction is their mole ratio;* therefore, we convert moles of the first substance to moles of the second substance before calculating the weight of the second substance. Three steps are involved:

1. Write down the given number of moles.
2. Convert moles of the first substance to moles of the second substance using the appropriate mole ratio.
3. Convert moles of the second substance to weight using the molar mass.

The following examples illustrate mole-weight calculations.

EXAMPLE 9.10 When carbon is burned in a limited supply of oxygen, carbon monoxide is produced:

$$2\ C(s) + O_2(g) \longrightarrow 2\ CO(g)$$

How many grams of CO are produced from 2.35 moles of C?

Solution Step 1: Write down the given number of moles.

$$2.35\ mol\ C$$

Step 2: Convert moles of C to moles of CO by use of the appropriate mole ratio.

$$(2.35\ \text{mol C}) \left(\frac{2\ mol\ CO}{2\ \text{mol C}} \right) = 2.35\ mol\ CO$$

Step 3:   Convert moles of CO to weight of CO by using the molar mass.

$$\text{FW of CO} = (1 \text{ atom C}) \left(\frac{12.0 \text{ amu}}{\text{atom C}} \right) + (1 \text{ atom O}) \left(\frac{16.0 \text{ amu}}{\text{atom O}} \right)$$

$$= 12.0 \text{ amu} + 16.0 \text{ amu}$$

$$= 28.0 \text{ amu}$$

Therefore the molar mass of CO is 28.0 g.

$$(2.35 \text{ mol CO}) \left(\frac{28.0 \text{ g CO}}{\text{mol CO}} \right) = 65.8 \text{ g CO}$$

EXERCISE 9.10   Ammonia reacts with oxygen to form nitrogen oxide and water vapor:

$$4 \text{ NH}_3(g) + 5 \text{ O}_2(g) \longrightarrow 4 \text{ NO}(g) + 6 \text{ H}_2\text{O}(g)$$

How many grams of NO are produced by 1.50 moles of NH_3?   ■

EXAMPLE 9.11   Aluminum reacts with sulfur to produce aluminum sulfide:

$$3 \text{ S}_8(s) + 16 \text{ Al}(s) \longrightarrow 8 \text{ Al}_2\text{S}_3(s)$$

How many grams of S_8 are required to react with 4.80 moles of Al?

Solution   Step 1:   Write down the given number of moles.

$$4.80 \text{ mol Al}$$

Step 2:   Convert moles of Al to moles of S_8 by use of the appropriate mole ratio.

$$(4.80 \text{ mol Al}) \left(\frac{3 \text{ mol S}_8}{16 \text{ mol Al}} \right) = 0.900 \text{ mol S}_8$$

Step 3:   Convert moles of S_8 to weight of S_8 using the molar mass.

$$\text{FW of S}_8 = (8 \text{ atoms S}) \left(\frac{32.1 \text{ amu}}{\text{atom S}} \right) = 257 \text{ amu}$$

Therefore the molar mass of S_8 is 257 g.

$$(0.900 \text{ mol S}_8) \left(\frac{257 \text{ g S}_8}{\text{mol S}_8} \right) = 231 \text{ g S}_8$$

EXERCISE 9.11   Calcium oxide reacts with water to form calcium hydroxide:

$$\text{CaO}(s) + \text{H}_2\text{O}(l) \longrightarrow \text{Ca(OH)}_2(s)$$

How many grams of H_2O would be required to react with 1.75 moles of CaO?   ■

Weight-Mole Calculations   In these calculations, the weight of one substance is used to calculate the moles of another substance. In this case, we must first calculate moles of the first substance and then convert this quantity to moles of the second substance using the appropriate mole ratio. Three steps are involved:

1. Write down the given weight.
2. Convert the given weight to moles by using the molar mass.
3. Convert moles of the first substance to moles of the substance you are seeking by using the appropriate mole ratio.

The following examples illustrate weight-mole calculations.

EXAMPLE 9.12   Burning methane produces carbon dioxide and steam (water vapor):

$$CH_4(g) + 2\ O_2(g) \longrightarrow CO_2(g) + 2\ H_2O(g)$$

How many moles of CO_2 are produced from 10.0 g of CH_4?

Solution   Step 1:   Write down the given weight.

$$10.0\ g\ CH_4$$

Step 2:   Convert weight of CH_4 to moles of CH_4 by using the molar mass.

$$(10.0\ g\ CH_4) \left(\frac{1\ mol\ CH_4}{16.0\ g\ CH_4} \right) = 0.625\ mol\ CH_4$$

Step 3:   Convert moles of CH_4 to moles of CO_2 by using the appropriate mole ratio.

$$(0.625\ mol\ CH_4) \left(\frac{1\ mol\ CO_2}{1\ mol\ CH_4} \right) = 0.625\ mol\ CO_2$$

EXERCISE 9.12   Nitrogen and hydrogen react to form ammonia:

$$N_2(g) + 3\ H_2(g) \xrightarrow{\Delta} 2\ NH_3(g)$$

How many moles of ammonia would be formed from 46.5 g of N_2?   ■

EXAMPLE 9.13   Phosphorus trichloride reacts with water to produce phosphorous acid and hydrogen chloride:

$$PCl_3(g) + 3\ H_2O(l) \longrightarrow H_3PO_3(s) + 3\ HCl(g)$$

How many moles of H_2O would be required to react with 23.8 g of PCl_3?

Solution   Step 1:   Write down the given weight.

$$23.8\ g\ PCl_3$$

Step 2:   Convert weight of PCl_3 to moles of PCl_3 by use of the molar mass.

$$(23.8\ g\ PCl_3) \left(\frac{1\ mol\ PCl_3}{137.5\ g\ PCl_3} \right) = 0.173\ mol\ PCl_3$$

Step 3:   Convert moles of PCl_3 to moles of H_2O by use of the appropriate mole ratio.

$$(0.173\ mol\ PCl_3) \left(\frac{3\ mol\ H_2O}{1\ mol\ PCl_3} \right) = 0.519\ mol\ H_2O$$

EXERCISE 9.13 Mercury(II) oxide decomposes to mercury and oxygen when heated:

$$2\ HgO(s) \xrightarrow{\Delta} 2\ Hg(l) + O_2(g)$$

If 17.2 g of Hg are produced, how many moles of O_2 are formed?   ■

Weight-Weight Calculations   In these calculations, the weight of one substance in a reaction is used to calculate the weight of another substance in the reaction. In this case, we first convert the given weight to moles and then relate the moles of the first substance to moles of the second substance using the appropriate mole ratio. Then the weight of the second substance can be calculated. Four steps are needed:

1. Write down the given weight.
2. Convert the given weight to moles using the molar mass.
3. Convert moles of the first substance to moles of the second substance using the appropriate mole ratio.
4. Convert moles of the second substance to weight by using the molar mass.

The fundamental relationships important to stoichiometric calculations are summarized in the "mole map" of Figure 9-4, and the following examples illustrate weight-weight calculations.

EXAMPLE 9.14 Phosphorus reacts with oxygen to produce diphosphorous pentoxide:

$$P_4(s) + 5\ O_2(g) \longrightarrow 2\ P_2O_5(s)$$

What weight of P_2O_5 will be formed from 20.0 g of P_4?

Solution   The mole map of Figure 9-4 indicates that we can make the following series of conversions:

$$g\ P_4 \longrightarrow moles\ P_4 \longrightarrow moles\ P_2O_5 \longrightarrow g\ P_2O_5$$

The four steps outlined below accomplish these conversions.

Step 1:   Write down the given weight

$$20.0\ g\ P_4$$

Step 2:   Convert the given weight to moles by use of the molar mass.

$$(20.0\ \cancel{g\ P_4}) \left(\frac{1\ mol\ P_4}{124.0\ \cancel{g\ P_4}} \right) = 0.161\ mol\ P_4$$

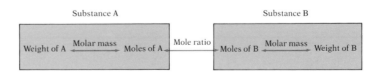

FIGURE 9-4 Fundamental Relationships Important to Stoichiometric Calculations. The weight and number of moles of the same substance can be interconverted by using the molar mass. The number of moles of two different substances can be interconverted by using the mole ratio from the balanced equation.

MATH TIP

Use the mole map to outline the problem:

$$g\ C \longrightarrow mol\ C \longrightarrow mol\ CO \longrightarrow g\ CO$$

Since the given information is mol C, you can begin at this point in the mole map.

MATH TIP

The mole map for this problem is

$$g\ CH_4 \longrightarrow mol\ CH_4 \longrightarrow$$
$$mol\ CO_2 \longrightarrow g\ CO_2$$

Since mol CO_2 are required, the last step in the map can be omitted.

MATH TIP

In multistep problems, enter into the calculator the number found on the left side of the problem. Then, perform multiplication and division operations as you proceed to the right. Press 20.0 ÷ 124.0 × 2 × 142 = . The display reads 45.80645161. Round off to 3 significant figures and report the answer as 45.8 g P_2O_5. Notice that the numbers 1 and 2 written with the mole quantities represent exact numbers. Also notice that the answer differs from the one determined using individual steps since that process required several intermediate answers that were rounded off each time.

▽ **PROBLEM-SOLVING SKILLS**

COMBINING STEPS IN THE SOLUTIONS TO PROBLEMS

When you first learn to solve stoichiometric problems, it is best to work in a series of individual steps, as outlined in examples 9.6 – 9.13. This will organize your thinking, and you will find that solving problems is largely a matter of developing a method. Do not try to memorize the method; instead, practice working the problems until the method becomes almost an automatic thought process. Studies of how people learn indicate that recalling methods and processes is much easier than recalling factual information; hence, rote memorization does little good in learning to solve problems.

As you become familiar with working stoichiometric problems in several individual steps, you will see that many of these steps can be combined to make one mathematical expression. For example, the solution given to example 9.10 involves three steps, which can be combined as

$$(2.35\ mol\ C)\left(\frac{2\ mol\ CO}{2\ mol\ C}\right)\left(\frac{28.0\ g\ CO}{mol\ CO}\right) = 65.8\ g\ CO$$

In another case, the steps in the solution to example 9.12 could be combined to give

$$(10.0\ g\ CH_4)\left(\frac{1\ mol\ CH_4}{16.0\ g\ CH_4}\right)\left(\frac{1\ mol\ CO_2}{1\ mol\ CH_4}\right) = 0.625\ mol\ CO_2$$

We could also combine the steps in the solution of example 9.14:

$$(20.0\ g\ P_4)\left(\frac{1\ mol\ P_4}{124.0\ g\ P_4}\right)\left(\frac{2\ mol\ P_2O_5}{1\ mol\ P_4}\right)\left(\frac{142\ g\ P_2O_5}{mol\ P_2O_5}\right) = 45.8\ g\ P_2O_5$$

In each of these examples, the combined expressions are very much like unit conversions in which several unit factors are used in series. Combining the steps allows you to perform all of the numerical calculations at one time and to round off only once instead of several times. After you feel comfortable with using several individual steps to work stoichiometric problems, try combining the steps as shown above. When you learn to do this, you will have developed considerable skill in solving problems.

Step 3:   Convert moles of P_4 to moles of P_2O_5 by using the appropriate mole ratio.

$$(0.161\ mol\ P_4)\left(\frac{2\ mol\ P_2O_5}{1\ mol\ P_4}\right) = 0.322\ mol\ P_2O_5$$

Step 4: Convert moles of P_2O_5 to weight of P_2O_5 by use of the molar mass.

$$(0.322 \text{ mol } P_2O_5) \left(\frac{142 \text{ g } P_2O_5}{\text{mol } P_2O_5} \right) = 45.7 \text{ g } P_2O_5$$

EXERCISE 9.14 Acetylene (C_2H_2) burns to produce carbon dioxide and water vapor:

$$2 \text{ C}_2\text{H}_2(g) + 5 \text{ O}_2(g) \longrightarrow 4 \text{ CO}_2(g) + 2 \text{ H}_2\text{O}(g)$$

What weight of CO_2 will be produced by 14.5 g of C_2H_2? ◼

EXAMPLE 9.15 Methane (CH_4) reacts with chlorine to form carbon tetrachloride and hydrogen chloride:

$$\text{CH}_4(g) + 4 \text{ Cl}_2(g) \xrightarrow{\Delta} \text{CCl}_4(l) + 4 \text{ HCl}(g)$$

What weight of Cl_2 would be required to react with 22.4 g of CH_4?

Solution The mole map of Figure 9-4 indicates that the following series of conversions is possible:

$$\text{g CH}_4 \longrightarrow \text{moles CH}_4 \longrightarrow \text{moles Cl}_2 \longrightarrow \text{g Cl}_2$$

The four steps outlined below accomplish these conversions.

Step 1: Write down the given weight.

$$22.4 \text{ g CH}_4$$

Step 2: Convert the given weight to moles using the molar mass.

$$(22.4 \text{ g CH}_4) \left(\frac{1 \text{ mol CH}_4}{16.0 \text{ g CH}_4} \right) = 1.40 \text{ mol CH}_4$$

Step 3: Convert moles of CH_4 to moles of Cl_2 using the appropriate mole fraction.

$$(1.40 \text{ mol CH}_4) \left(\frac{4 \text{ mol Cl}_2}{1 \text{ mol CH}_4} \right) = 5.60 \text{ mol Cl}_2$$

Step 4: Convert moles of Cl_2 to weight of Cl_2 using the molar mass.

$$(5.60 \text{ mol Cl}_2) \left(\frac{71.0 \text{ g Cl}_2}{\text{mol Cl}_2} \right) = 398 \text{ g Cl}_2$$

These four steps can be combined as follows:

$$(22.4 \text{ g CH}_4) \left(\frac{1 \text{ mol CH}_4}{16.0 \text{ g CH}_4} \right) \left(\frac{4 \text{ mol Cl}_2}{1 \text{ mol CH}_4} \right) \left(\frac{71.0 \text{ g Cl}_2}{\text{mol Cl}_2} \right) = 398 \text{ g Cl}_2$$

EXERCISE 9.15 Zinc reacts with hydrogen chloride to produce zinc(II) chloride and hydrogen:

$$\text{Zn}(s) + 2 \text{ HCl}(g) \longrightarrow \text{ZnCl}_2(s) + \text{H}_2(g)$$

What weight of HCl would be required to react with 14.5 g of Zn? ◼

9.7 PERCENTAGE YIELD

So far, our stoichiometric calculations have been based on theoretical considerations derived from balanced equations. However, each balanced equation describes only one reaction for each set of reactants. In reality, it is not unusual for a given set of reactants to undergo two or more different reactions simultaneously. For example, methane (the major component of natural gas) can burn in an abundant supply of oxygen to form carbon dioxide and water vapor:

$$CH_4(g) + 2\ O_2(g) \longrightarrow CO_2(g) + 2\ H_2O(g)$$

However, limited oxygen supplies result in incomplete combustion, producing carbon monoxide (a toxic gas) and water vapor:

$$2\ CH_4(g) + 3\ O_2(g) \longrightarrow 2\ CO(g) + 4\ H_2O(g)$$

(Notice that the mole ratio of O_2 to CH_4 is $2:1$ in the first reaction but only $3:2$ in the second reaction.) Under ordinary conditions, burning methane produces mostly carbon dioxide, but some carbon monoxide is formed, indicating that incomplete combustion occurs as a side reaction. (This is why good ventilation is required when burning natural gas indoors; otherwise, the carbon monoxide might build up to hazardous levels.) Thus methane and oxygen react to give a set of major products (CO_2 and H_2O) and a set of minor products (CO and H_2O). Because of the existence of the second reaction, the amount of CO_2 produced is seldom as much as the balanced equation predicts.

Because of the existence of side reactions, the actual yield of many chemical reactions is seldom as much as the balanced equations predict. In addition, many reactions do not go to completion, so some portions of the reactants remain unchanged. A third reason for not getting as much product as we would expect from the balanced equation is a very practical one. When we run reactions in the laboratory, we must isolate the products in order to weigh them. In most cases, small amounts of product are accidentally lost in handling and transferring from one container to another. *Thus, the balanced equation allows us to calculate the maximum amount of product to be expected* (the theoretical yield), but in reality, we seldom recover this maximum amount of product.

In reporting our results, we usually calculate the percentage yield of the reaction. **Percentage yield** is the actual amount of product obtained (the **actual yield**) divided by the yield calculated from the balanced equation (the **theoretical yield**) multiplied by 100%.

$$\% \text{ yield} = \frac{\text{actual yield}}{\text{theoretical yield}} \times 100\%$$

For example, if the actual yield of a product is 14.2 g and the calculated yield is 16.7 g, the percentage yield is

$$\% \text{ yield} = \frac{14.2\ \cancel{g}}{16.7\ \cancel{g}} \times 100\% = 85.0\%$$

Notice that both yields must have the same units, such as grams or moles. In calculating percentage yield, two major steps are used:

1. Calculate the theoretical yield from the balanced equation, using one of the categories of stoichiometric calculations described earlier.
2. Calculate the percentage yield using the actual yield and the theoretical yield.

EXAMPLE 9.16   Fluorine and ammonia react to form dinitrogen tetrafluoride and hydrogen fluoride:

$$5\ F_2(g) + 2\ NH_3(g) \longrightarrow N_2F_4(g) + 6\ HF(g)$$

If 10.0 g of F_2 are reacted and 4.95 g of N_2F_4 are formed, what is the percentage yield of N_2F_4?

Solution   Step 1:   Since the actual yield of N_2F_4 is given in units of grams, we must calculate theoretical yield in grams. Thus we must make a weight-weight stoichiometric calculation.

$$(10.0\ g\ F_2)\left(\frac{1\ mol\ F_2}{38.0\ g\ F_2}\right)\left(\frac{1\ mol\ N_2F_4}{5\ mol\ F_2}\right)\left(\frac{104\ g\ N_2F_4}{mol\ N_2F_4}\right) = 5.47\ g\ N_2F_4$$

Step 2:   Calculate the percentage yield using the actual yield and the theoretical yield.

$$\% \text{ yield} = \frac{\text{actual yield}}{\text{theoretical yield}} \times 100\%$$

$$= \frac{4.95\ g\ N_2F_4}{5.47\ g\ N_2F_4} \times 100\%$$

$$= 90.5\%$$

EXERCISE 9.16   Chromium(III) oxide reacts with carbon to produce chromium and carbon monoxide:

$$Cr_2O_3(s) + 3\ C(s) \xrightarrow{\Delta} 2\ Cr(s) + 3\ CO(g)$$

If 5.00 g of C are used and 12.0 g of Cr are produced, what is the percentage yield of Cr?   ∎

EXAMPLE 9.17   Sulfur dioxide reacts with oxygen and water to produce sulfuric acid:

$$2\ SO_2(g) + O_2(g) + 2\ H_2O(l) \longrightarrow 2\ H_2SO_4(l)$$

If 1.00 mole of O_2 is available for reaction and 1.75 moles of H_2SO_4 are formed, what is the percentage yield of H_2SO_4?

Solution   Step 1:   Since the actual yield of H_2SO_4 is given in units of moles, we must calculate the theoretical yield in moles. This amounts to a mole-mole stoichiometric calculation.

$$(1.00\ mol\ O_2)\left(\frac{2\ mol\ H_2SO_4}{1\ mol\ O_2}\right) = 2.00\ mol\ H_2SO_4$$

Step 2:   Calculate the percent yield:

$$\% \text{ yield} = \frac{\text{actual yield}}{\text{theoretical yield}} \times 100\%$$

$$= \frac{1.75 \text{ mol } H_2SO_4}{2.00 \text{ mol } H_2SO_4} \times 100\%$$

$$= 87.5\%$$

EXERCISE 9.17   Lead(II) nitrate and sodium chloride react to form lead(II) chloride and sodium nitrate:

$$Pb(NO_3)_2(aq) + 2\,NaCl(aq) \longrightarrow PbCl_2(s) + 2\,NaNO_3(aq)$$

If 0.0200 moles of $Pb(NO_3)_2$ are used and 4.34 g of $PbCl_2$ are formed, what is the percentage yield of $PbCl_2$?   ■

9.8   LIMITING REACTANT

As we have discussed, many reactions do not go to completion; some portions of the reactants remain unchanged. Some incomplete reactions occur because the reactants are not given enough time to react completely. (Different reactions proceed at different rates. Rates of reactions will be explored more fully in chapter 13.) Other reactions may be incomplete because they reverse themselves spontaneously and reform reactants from products. (Reversible reactions will also be discussed in chapter 13.) In either case, the actual yield of product can be improved if one reactant is used in excess, that is, in a greater quantity than indicated by the numerical coefficients in the balanced equation. In practice, we usually choose the less expensive or more easily obtained reactant as the one to use in excess. The reactant that is present in smaller quantities than indicated by the mole ratios of the balanced equation is called the **limiting reactant,** because it limits the amount of product that can form. The theoretical yield allowed by the limiting reactant is then used to calculate the percentage yield.

In many laboratory experiments, the limiting reactant is not immediately obvious, but we can find the limiting reactant by calculating the theoretical yield allowed by the actual amount of each reactant present. The reactant that gives the lowest theoretical yield is the limiting reactant. Thus two major steps are needed to find the limiting reactant:

1. Calculate the number of moles of each reactant.
2. For the same product, calculate the number of moles that would be produced by each reactant. The reactant that gives the least number of moles of the product is the limiting reactant.

Limiting reactant problems are actually stoichiometry problems, as the following examples illustrate.

EXAMPLE 9.18   In a reaction described by the following equation,

$$Si(s) + O_2(g) \longrightarrow SiO_2(s)$$

15.0 g of Si are allowed to react with 85.0 g of O_2. What is the limiting reactant?

Solution   Step 1: Calculate the number of moles of each reactant:

$$Si: (15.0 \text{ g Si}) \left(\frac{1 \text{ mol Si}}{28.1 \text{ g Si}} \right) = 0.534 \text{ mol Si}$$

$$O_2: (85.0 \text{ g } O_2) \left(\frac{1 \text{ mol } O_2}{32.0 \text{ g } O_2} \right) = 2.66 \text{ mol } O_2$$

Step 2: Calculate the number of moles of SiO_2 that would be produced by each reactant:

$$Si: (0.534 \text{ mol Si}) \left(\frac{1 \text{ mol } SiO_2}{1 \text{ mol Si}} \right) = 0.534 \text{ mol } SiO_2$$

$$O_2: (2.66 \text{ mol } O_2) \left(\frac{1 \text{ mol } SiO_2}{1 \text{ mol } O_2} \right) = 2.66 \text{ mol } SiO_2$$

Since Si produces the lower theoretical yield, Si is the limiting reactant.

EXERCISE 9.18   A mixture of 22.3 g of Al and 47.6 g of O_2 are allowed to react as described below:

$$4 \text{ Al(s)} + 3 \text{ } O_2(g) \longrightarrow 2 \text{ } Al_2O_3(s)$$

Find the limiting reactant.   ■

EXAMPLE 9.19   If 37.5 g of N_2 are allowed to react with an equal weight of H_2 to form NH_3,

$$N_2(g) + 3 \text{ } H_2(g) \xrightarrow{\Delta} 2 \text{ } NH_3(g)$$

what is the limiting reactant?

Solution   Step 1: Calculate the number of moles of each reactant:

$$N_2: (37.5 \text{ g } N_2) \left(\frac{1 \text{ mol } N_2}{28.0 \text{ g } N_2} \right) = 1.34 \text{ mol } N_2$$

$$H_2: (37.5 \text{ g } H_2) \left(\frac{1 \text{ mol } H_2}{2.0 \text{ g } H_2} \right) = 18.8 \text{ mol } H_2$$

Step 2: Calculate the number of moles of NH_3 that would be produced by each reactant:

$$N_2: (1.34 \text{ mol } N_2) \left(\frac{2 \text{ mol } NH_3}{1 \text{ mol } N_2} \right) = 2.68 \text{ mol } NH_3$$

$$H_2: (18.8 \text{ mol } H_2) \left(\frac{2 \text{ mol } NH_3}{3 \text{ mol } H_2} \right) = 12.5 \text{ mol } NH_3$$

Therefore N_2 is the limiting reactant.

EXERCISE 9.19   If 15.0 g of $AgNO_3$ is mixed with 12.0 g of $CaCl_2$, what is the limiting reactant in the following reaction?

$$2 \text{ } AgNO_3(aq) + CaCl_2(aq) \longrightarrow 2 \text{ } AgCl(s) + Ca(NO_3)_2(aq)$$ ■

The following three steps show how to use the amount of the limiting reactant to calculate the percentage yield of a reaction:

1. Calculate the number of moles of each reactant.
2. Find the limiting reactant based on its theoretical yield of product.
3. Calculate the percentage yield based on the theoretical yield of the limiting reactant.

These three steps are illustrated in the following example.

EXAMPLE 9.20 If 10.0 g of iron(III) sulfate and 12.0 g of barium chloride are allowed to react according to the following equation,

$$Fe_2(SO_4)_3(aq) + 3\ BaCl_2(aq) \longrightarrow 3\ BaSO_4(s) + 2\ FeCl_3(aq)$$

what is the limiting reactant? If 10.6 g of $BaSO_4$ is formed, what is the percentage yield?

Solution Step 1: Calculate the number of moles of each reactant:

$$Fe_2(SO_4)_3: (10.0\ g\ Fe_2(SO_4)_3) \left(\frac{1\ mol\ Fe_2(SO_4)_3}{400\ g\ Fe_2(SO_4)_3} \right)$$

$$= 0.0250\ mol\ Fe_2(SO_4)_3$$

$$BaCl_2: (12.0\ g\ BaCl_2) \left(\frac{1\ mol\ BaCl_2}{208\ g\ BaCl_2} \right) = 0.0577\ mol\ BaCl_2$$

Step 2: Find the limiting reactant on the basis of its theoretical yield of product. In this case, use the $BaSO_4$ product, since its actual yield is given.

$$Fe_2(SO_4)_3: (0.0250\ mol\ Fe_2(SO_4)_3) \left(\frac{3\ mol\ BaSO_4}{1\ mol\ Fe_2(SO_4)_3} \right)$$

$$= 0.0750\ mol\ BaSO_4$$

$$BaCl_2: (0.0577\ mol\ BaCl_2) \left(\frac{3\ mol\ BaSO_4}{3\ mol\ BaCl_2} \right) = 0.0577\ mol\ BaSO_4$$

Therefore $BaCl_2$ is the limiting reactant.

Step 3: Calculate the percentage yield based on the theoretical yield of $BaSO_4$ by $BaCl_2$:

$$Theoretical\ yield\ (in\ g) = (0.0577\ mol\ BaSO_4)$$

$$\left(\frac{233.4\ g\ BaSO_4}{mol\ BaSO_4} \right) = 13.5\ g\ BaSO_4$$

$$\%\ yield = \frac{10.6\ g\ BaSO_4}{13.5\ g\ BaSO_4} \times 100\% = 78.5\%$$

EXERCISE 9.20 If 15.0 g of H_3PO_4 and 20.0 g of $CaCO_3$ are allowed to react according to the following equation,

$$3\ CaCO_3(s) + 2\ H_3PO_4(aq) \longrightarrow Ca_3(PO_4)_2(s) + 3\ CO_2(g) + 3\ H_2O(l)$$

what is the limiting reactant? If 18.5 g of $Ca_3(PO_4)_2$ is formed, what is the percentage yield?

SUMMARY

Chemical equations are concise, accurate descriptions of chemical reactions. The numbers in front of the formulas in chemical equations are called **coefficients;** they indicate the smallest number of chemical units that can participate in the reaction. Chemical reactions follow the law of conservation of mass. In order for an equation to describe a reaction accurately, the equation must be balanced. Rules for balancing equations are given on p. 193. The four basic types of chemical reactions are **decomposition** reactions (AB → A + B), **combination** reactions (A + B → AB), **single replacement** reactions (A + BC → AC + B), and **double replacement** reactions (AB + CD → AD + CB).

Stoichiometry refers to the relationships among quantities of reactants and products that can be derived from balanced equations. The numerical coefficients of a balanced equation indicate the number of chemical units and the number of moles of reactants and products. A **mole ratio** is the ratio of moles of any two substances in a balanced equation; it is determined from the numerical coefficients of the equation. There are four types of **stoichiometric calculations: mole-mole** calculations, **mole-weight** calculations, **weight-mole** calculations, and **weight-weight** calculations.   These four types are discussed on pp. 210 – 207.

The actual yield of a chemical reaction is seldom as much as the balanced equation indicates; the yield is reduced by side reactions, incomplete reactions, and loss of product during handling and transfer. Percentage yield is the actual amount of product obtained **(actual yield)** divided by the yield calculated from the balanced equation **(theoretical yield)** multiplied by 100%.

A reactant that is present in smaller quantities than indicated by the mole ratios of the balanced equation is called the **limiting reactant.** When a limiting reactant is present, the percentage yield is calculated using the theoretical yield allowed by the limiting reactant.

STUDY QUESTIONS AND PROBLEMS

(More difficult questions and problems are marked with an asterisk.)

THE LANGUAGE OF CHEMICAL EQUATIONS

1. Define the following terms:
   a. Chemical equations   c. Product
   b. Reactant   d. Coefficient
2. Give the meaning of each notation.
   a. (s) or ↓   c. (g) or ↑   e. (aq)
   b. (l)   d. Δ
3. Express each of the following equations in a complete sentence.
   a. $Mg(s) + 2 HCl(g) \rightarrow MgCl_2(s) + H_2(g)$
   b. $HCl(g) + NH_3(g) \rightarrow NH_4Cl(s)$
   c. $2 KBr(aq) + Cl_2(g) \rightarrow 2 KCl(aq) + Br_2(l)$
   d. $H_2SO_4(aq) + 2 NaCN(s) \rightarrow 2 HCN(g) + Na_2SO_4(aq)$

CONSERVATION OF MASS IN CHEMICAL REACTIONS

4. What does the term *combustion* mean?
5. A sample of 8.50 g of sulfur is burned to produce 17.0 g of sulfur dioxide. How many grams of oxygen are used in the reaction?

$$S_8(s) + 8\ O_2(g) \longrightarrow 8\ SO_2(g)$$

6. A sample of 15.2 g of potassium nitrate produces 12.8 g of potassium nitrite when heated. How many grams of oxygen were produced?

$$2\ KNO_3(s) \xrightarrow{\Delta} 2\ KNO_2(s) + O_2(g)$$

7. A sample of magnesium sulfate heptahydrate was heated; 12.0 g of magnesium sulfate and 12.6 g of water vapor were obtained. What was the weight of the magnesium sulfate heptahydrate?

$$MgSO_4 \cdot 7\ H_2O(s) \longrightarrow MgSO_4(s) + 7\ H_2O(g)$$

BALANCING CHEMICAL EQUATIONS

8. Balance each equation.
   a. $CH_4O(l) + O_2(g) \rightarrow CO_2(g) + H_2O(g)$
   b. $Ag_2O(s) \xrightarrow{\Delta} Ag(s) + O_2(g)$
   c. $NH_4OH(aq) + H_2SO_4(aq) \rightarrow (NH_4)_2SO_4(aq) + H_2O(l)$
   d. $VO(s) + Fe_2O_3(s) \rightarrow FeO(s) + V_2O_5(s)$
   e. $C_4H_{10}(g) + O_2(g) \rightarrow CO_2(g) + H_2O(g)$
   f. $CaBr_2(aq) + H_2SO_4(aq) \rightarrow CaSO_4(s) + HBr(g)$
9. Balance each equation.
   a. $B_2O_3(s) + C(s) \xrightarrow{\Delta} B_4C(s) + CO(g)$
   b. $CaC_2(s) + H_2O(l) \rightarrow Ca(OH)_2(s) + C_2H_2(g)$
   c. $Ba(NO_3)_2(aq) + H_2SO_4(aq) \rightarrow BaSO_4(s) + HNO_3(aq)$
   d. $Bi(s) + O_2(g) \rightarrow Bi_2O_3(s)$
   e. $(NH_4)_2Cr_2O_7(s) \xrightarrow{\Delta} N_2(g) + H_2O(g) + Cr_2O_3(s)$

TYPES OF CHEMICAL REACTIONS

10. Classify the reactions in question 8, parts a–c and f.
11. Classify the reactions in question 9, parts b–e.
12. Predict the products of the following double replacement reactions involving ionic compounds, and write the balanced equation for each reaction.
    a. $Ni(NO_3)_2 + NaOH \rightarrow$　　　c. $Ba(OH)_2 + HCl \rightarrow$
    b. $KOH + H_3PO_4 \rightarrow$　　　d. $KCN + HCl \rightarrow$
*13. Fill in the blank in each reaction.
    a. $2\ PbO_2(s) \xrightarrow{\Delta} 2\ \underline{\hphantom{xxx}}\ (l) + 2\ O_2(g)$
    b. $Cl_2(g) + 2\ \underline{\hphantom{xxx}}\ (aq) \rightarrow I_2(s) + 2\ KCl(aq)$
    c. $\underline{\hphantom{xxx}}\ (s) + 2\ HNO_3(aq) \rightarrow Cu(NO_3)_2(aq) + H_2O(l)$
    d. $\underline{\hphantom{xxx}}\ (g) + H_2O(l) \rightarrow H_2SO_4(l)$
    e. $Fe(s) + \underline{\hphantom{xxx}}\ (aq) \rightarrow Cu(s) + FeSO_4(aq)$
    f. $2\ \underline{\hphantom{xxx}}\ (aq) + Zn(s) \rightarrow ZnCl_2(aq) + H_2(g)$

STOICHIOMETRY

14. For the following reaction,

$$CdSO_4(aq) + 2\ NaOH(aq) \longrightarrow Cd(OH)_2(s) + Na_2SO_4(aq)$$

give the mole ratio needed to calculate:
a. moles of Na_2SO_4 from moles of $CdSO_4$
b. moles of $CdSO_4$ from moles of $NaOH$
c. moles of $Cd(OH)_2$ from moles of $NaOH$
d. moles of Na_2SO_4 from moles of $NaOH$
e. moles of $NaOH$ from moles of $CdSO_4$
f. moles of $Cd(OH)_2$ from moles of $CdSO_4$

15. For the following reaction,

$$3\ Fe(s) + 4\ H_2O(l) \xrightarrow{\Delta} Fe_3O_4(s) + 4\ H_2(g)$$

give the mole ratio needed to calculate:
a. moles of Fe_3O_4 from moles of Fe
b. moles of H_2O from moles of Fe
c. moles of H_2 from moles of Fe_3O_4
d. moles of Fe_3O_4 from moles of H_2O
e. moles of Fe_3O_4 from moles of H_2
f. moles of H_2 from moles of Fe

STOICHIOMETRIC CALCULATIONS

16. For the following equation,

$$3\ KOH(aq) + H_3PO_4(aq) \longrightarrow K_3PO_4(aq) + 3\ H_2O(l)$$

a. how many moles of H_3PO_4 are required to react with 2.15 moles of KOH?
b. how many grams of K_3PO_4 will be produced by 17.8 g of H_3PO_4?
c. how many moles of KOH will be required to react with 24.6 g of H_3PO_4?
d. how many grams of H_2O will be produced when 34.7 g of K_3PO_4 are formed?

17. For the following reaction,

$$P_4(s) + 5\ O_2(g) \longrightarrow 2\ P_2O_5(s)$$

how many grams of oxygen are required by 3.10 g of phosphorus?

18. When a sample of potassium chlorate was heated, 0.96 g of oxygen were liberated. How many grams of potassium chlorate were present?

$$2\ KClO_3(s) \xrightarrow{\Delta} 2\ KCl(s) + 3\ O_2(g)$$

19. The following reaction is used for the commercial production of hydrogen chloride. How many grams of HCl can be obtained by heating 234 g of NaCl?

$$2\ NaCl(s) + H_2SO_4(aq) \xrightarrow{\Delta} Na_2SO_4(aq) + 2\ HCl(g)$$

20. The following reaction can be used to produce silver bromide, a substance used in photography.

$$AgNO_3(aq) + NaBr(aq) \longrightarrow AgBr(s) + NaNO_3(aq)$$

Calculate the number of grams of each reactant needed to produce 93.3 g of AgBr.

21. The leavening action of baking powder, a mixture of cream of tartar and baking soda, is described by the following equation.

$$KHC_4H_4O_6(aq) \; + \; NaHCO_3(aq) \longrightarrow$$

potassium hydrogen   sodium
      tartrate   bicarbonate
  "cream of tartar"   "baking soda"

$$KNaC_4H_4O_6(aq) + H_2O(l) + CO_2(g)$$

How many grams of baking soda would be required to make a sample of baking powder containing 50.0 g of cream of tartar?

22. When a piece of magnesium weighing 0.32 g was burned, the product weighed 0.53 g.
    a. Write the balanced equation for the reaction.
    b. Calculate the moles of oxygen required for the reaction.
    c. Calculate the number of grams of oxygen required for the reaction.

23. Sodium reacts vigorously with water to form sodium hydroxide and hydrogen. Write the balanced equation for this reaction and calculate the number of grams of hydrogen produced by 15.0 g of sodium.

24. Calculate the number of moles each of CO_2 and H_2O that are produced by burning 104 g of $C_2H_2(g)$ (acetylene).

25. How many grams of aluminum are needed to react with sulfur to form 600.0 g of aluminum sulfide, $Al_2S_3(s)$?

26. Heating sodium nitrate decomposes it to sodium nitrite and oxygen. How many grams of sodium nitrate are needed to produce 1.50 g of oxygen?

27. How many grams of oxygen would be needed to completely burn 500.0 g of gasoline, $C_8H_{18}(l)$? The combustion products are carbon dioxide and water vapor.

PERCENTAGE YIELD

28. In the manufacture of aspirin, the theoretical yield of aspirin in a reaction is 14.7 kg. If the actual yield is 12.1 kg, what is the percentage yield of aspirin?

29. Ethane, $C_2H_6(g)$, burns in excess oxygen to produce carbon dioxide and water vapor. If 72.0 g of ethane are burned and 105 g of carbon dioxide are formed, what is the percentage yield of carbon dioxide?

30. In the following reaction,

$$2 SO_2(g) + O_2(g) \longrightarrow 2 SO_3(g)$$

25.0 g of SO_2 produced 28.2 g of SO_3. What was the percentage yield of SO_3?

LIMITING REACTANT

31. If 25.0 g of SO_2 and 6.00 g of O_2 are allowed to react according to the following equation,

$$2\ SO_2(g) + O_2(g) \longrightarrow 2\ SO_3(g)$$

what is the limiting reactant?

32. If 5.23 g of $ZnCl_2$ and 2.12 g of $AgNO_3$ are allowed to react according to the following equation,

$$ZnCl_2(aq) + 2\ AgNO_3(aq) \longrightarrow Zn(NO_3)_2(aq) + 2\ AgCl(s)$$

what is the limiting reactant?

33. When 3.50 g of $NaNH_2$ and 3.50 g of $NaNO_3$ were allowed to react according to the following equation,

$$3\ NaNH_2(s) + NaNO_3(s) \longrightarrow NaN_3(s) + 3\ NaOH(s) + NH_3(g)$$

1.20 g of NaN_3 was formed. What was the percentage yield of NaN_3?

34. Milk of magnesia, a suspension of magnesium hydroxide, can be used to neutralize stomach acid, hydrochloric acid, by the following reaction.

$$Mg(OH)_2(aq) + 2\ HCl(aq) \longrightarrow MgCl_2(aq) + 2\ H_2O(l)$$

When 16.14 g of $Mg(OH)_2$ and 10.97 g of HCl were reacted, 12.23 g of $MgCl_2$ was formed. What was the percentage yield of $MgCl_2$?

35. DDT, $C_{14}H_9Cl_5$, can be prepared by the following reaction:

$$2\ C_6H_5Cl(l) + C_2HOCl_3(l) \longrightarrow C_{14}H_9Cl_5(s) + H_2O(l)$$

When 4.51 g of C_6H_5Cl was reacted with 9.00 g of C_2HOCl_3, 6.54 g of DDT was formed. What was the percentage yield of DDT?

GENERAL EXERCISES

*36. Write the balanced equation for each statement.
  a. Liquid octane, C_8H_{18}, burns to produce gaseous carbon dioxide and water vapor.
  b. Strontium metal reacts with chlorine gas to produce solid strontium chloride.
  c. Barium metal reacts with liquid water to produce solid barium hydroxide and hydrogen gas.
  d. Lead(II) acetate dissolved in water reacts with potassium sulfate dissolved in water to produce solid lead(II) sulfate and potassium acetate dissolved in water.
  e. Solid phosphorus triiodide reacts with liquid water to produce hydrogen iodide gas and solid phosphorous acid.

37. Calcium oxide, commonly called *lime*, reacts with water to produce calcium hydroxide, sometimes referred to as *slaked lime*. Write a balanced equation for this reaction and calculate the number of grams of slaked lime that would be produced by 100.0 g of lime.

38. Butane, $C_4H_{10}(g)$, is the fuel of disposable cigarette lighters. How many grams of carbon dioxide will be produced by burning 7.62 g of butane?

*39. Solid calcium oxide will absorb water vapor from the air to change completely to calcium hydroxide. If a sample of calcium oxide showed a weight gain of 0.289 g upon being exposed to moist air, how many moles of calcium hydroxide were formed?

40. The following reaction is used to remove calcium ions from hard water:

$$CaSO_4(aq) + Na_2CO_3(aq) \longrightarrow CaCO_3(s) + Na_2SO_4(aq)$$

If 10.75 g of $CaSO_4$ produced 7.02 g of $CaCO_3$, what was the percentage yield of $CaCO_3$?

*41. If 10.20 g of Al_2O_3 and 7.50 g of HNO_3 are allowed to react according to the following equation,

$$Al_2O_3(s) + 6\ HNO_3(aq) \longrightarrow 2\ Al(NO_3)_3(aq) + 3\ H_2O(l)$$

what is the limiting reactant? How many grams of the other reactant were left over?

*42. An impure sample of copper(II) sulfate was dissolved in water and allowed to react with excess zinc according to the following equation.

$$CuSO_4(aq) + Zn(s) \longrightarrow ZnSO_4(aq) + Cu(s)$$

If the impure sample weighed 5.52 g and 1.49 g of copper was produced, what was the percent of copper(II) sulfate in the sample?

*43. If 5.55 g of N_2 and 1.30 g of H_2 were allowed to react according to the following equation,

$$N_2(g) + 3\ H_2(g) \xrightarrow{\Delta} 2\ NH_3(g)$$

and the reaction gave a 56.3% yield, how many grams of ammonia were formed?

GASES

CHAPTER
10

OUTLINE

10.1 Properties of Gases
10.2 Volume and Pressure
      Relationships
10.3 Volume and Temperature
      Relationships
      Perspective: The Earth's
      Atmosphere
10.4 Pressure and Temperature
      Relationships
10.5 Combining the Gas Laws
10.6 Avogadro's Hypothesis
10.7 The Ideal Gas Equation
10.8 The Kinetic Molecular Theory
10.9 Gas Mixtures
10.10 Gas Density
10.11 Stoichiometric Calculations
      Involving Gas Volumes
      (optional)
      Mole-Volume Calculations
      Weight-Volume Calculations
      Volume-Volume Calculations
      Summary
      Study Questions and Problems

FIGURE 10-1 A Rainstorm over Farmland in Missouri

Each of the three states of matter has its own set of properties that are independent of the particular substance being studied. Consider water as an example. In the form of water vapor, it has certain properties because it is a gas. If water vapor is cooled so that it condenses into liquid water, it will have different properties as a liquid. If liquid water is frozen, it then takes on the properties of a solid.

In this chapter we will study the properties of gases. Several properties are typical of all gases. For example, a gas can be compressed; we compress air when we force it into a tire with a pump. When a gas is heated, it expands and rises, as we observe when hot air rises over a heater. If a gas is cooled sufficiently, it condenses to a liquid, as you have probably observed by breathing onto the cold surface of a window in winter. The earth's weather is largely the result of changes in the pressure and temperature of air. Weather phenomena such as clouds, storms (Figure 10-1), wind, and clear skies can all be explained by the properties of gases. To prepare for the study of gases, review the states of matter (chapter 1), density (chapter 1), and basic algebra (appendix A).

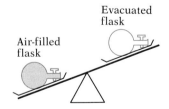

FIGURE 10-2 A Flask Filled with Air Weighs More Than an Identical Flask That Has Been Evacuated.

10.1 PROPERTIES OF GASES

Although our earth is surrounded by a blanket of gaseous atmosphere, scientists did not investigate the properties of gases until the seventeenth century. Theories of gaseous behavior were first put forth in the nineteenth century, and these theories are still evolving. In this section we will summarize the major properties of gases.

FIGURE 10-3 Von Guericke's Experimental Equipment. In the center is his air pump, and on the upper right are the famous Magdeburg hemispheres.

When weather reports give the barometer reading, this is the height in inches of a column of mercury supported by the atmosphere.

Torr

The pressure necessary to support a column of mercury 1 mm high at 0° C.

Mass and Volume Gases are a form of matter, and, as such, they have mass and occupy space. The fact that gases have volume permitted the invention of inflatable rubber tires to replace the solid rubber ones used on early automobiles. We can demonstrate that gases have mass by evacuating a rigid container and weighing it; the air-filled container weighs more than the evacuated container (Figure 10-2), indicating that air has mass.

Pressure Gases also exert pressure. The term *pressure* suggests a force, which is something that tends to move an object across a distance. In scientific terms, **pressure** is defined as a force that acts on a surface. The pound is one of the common units of force, and we sometimes express pressure (tire pressure, for example) in pounds per square inch (psi).

The most famous demonstration of the pressure exerted by a gas was the experiment performed in 1654 by Otto von Guericke with his Magdeburg hemispheres. Von Guericke placed two large metal cups (hemispheres) together to make a sphere; then he evacuated the sphere with an air pump (Figure 10-3). A team of horses attached to each hemisphere could not pull the two halves apart (Figure 10-4), yet the only force holding them together was the pressure of the atmosphere.

The pressure of the atmosphere is usually measured with a barometer. Our modern barometers are based on the original one invented by Evangelista Torricelli in 1643. Torricelli filled a 4-foot glass tube with mercury and inverted the tube into a dish (see Figure 10-5). He observed that some of the mercury did not flow out, and after much study, he concluded that the variation of the height of the mercury from day to day was caused by changes in atmospheric pressure. The mercury is prevented from running out of the tube by the pressure that the atmosphere exerts on the surface of the mercury in the dish.

Gas pressure is often expressed in units of millimeters of mercury (mm Hg), corresponding to the height of a column of mercury supported by the pressure of the gas on the surface of the mercury. One millimeter of mercury is called one **torr,** in honor of Torricelli. The normal pressure of the atmosphere at sea level supports a mercury column 760 mm high, so a pressure of 1 **atmosphere** (1 atm) is defined as 760 torr. The SI unit of

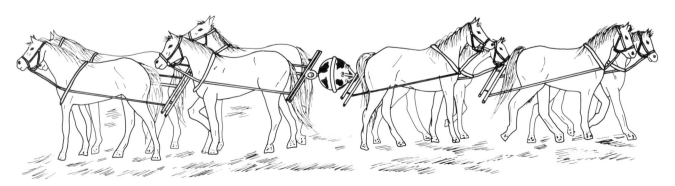

FIGURE 10-4 Horses Trying to Pull Von Guericke's Hemispheres Apart

Pascal

Named for Blaise Pascal (1623–1662), the French mathematician, physicist, and philosopher who was the founder of the modern theory of probability.

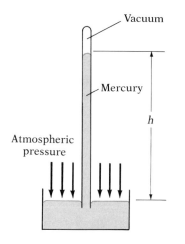

FIGURE 10-5 Torricelli's Barometer

pressure, the **pascal (Pa),** is not used as frequently as the torr and the atmosphere. One atmosphere is equivalent to 101,325 pascals.

Compressibility   Gases also have **compressibility,** that is, they can be *compressed.* When we fill a tire with air, we actually compress a large volume of air into the tire, and thus we create pressure on the tire walls from the air inside. Gases commonly used in hospitals and laboratories are stored by compressing large volumes into strong metal cylinders such as those shown in Figure 10-6.

Gaseous Expansion and Contraction   Temperature influences a fifth property of gases, the ability to expand and contract. When gases are heated, they expand and become less dense. When fire heats air (Figure 10-7), the less dense hot air rises and is replaced by more dense cool air from the surroundings.

Because gases expand when heated, gases confined in containers exert more pressure when warmed. Thus we should not incinerate aerosol spray cans because the intense heat of incineration increases the pressure inside the cans and they can easily explode. Another example is the increase in automobile tire pressure during high-speed driving. The friction created by the tire on the road surface generates heat, which in turn causes the air pressure to increase inside the tire. After the car has stopped, the tires cool down and the pressure returns to its original value.

Diffusion   The last property of gases that we will consider is **diffusion,** the ability of one gas to intermingle with another to form a homogeneous mixture. Two or more gases in any proportion form a homogeneous mix-

FIGURE 10-6 Compressed Gases Are Stored in Strong Metal Cylinders. These gases are used in chemical analysis; the gauges on the top of each cylinder register the pressure exerted by the gas inside.

FIGURE 10-7 Hot Air Rises over a Fire Because It Is Less Dense Than the Surrounding Air.

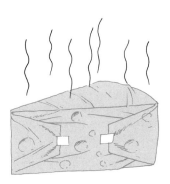

FIGURE 10-8 A Tight Plastic Wrapper Cannot Contain the Odor of Strong Cheese.

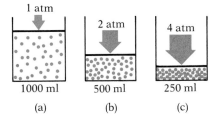

FIGURE 10-9 The Effect of Pressure on Gas Volume

ture, regardless of their identities. Anyone who has detected the odor of a skunk can confirm the ability of one gas to diffuse through another. Gases diffuse not only through each other but also through some solid substances. Thus, even a tight plastic wrapper cannot contain the odor of strong cheese (Figure 10-8).

10.2 VOLUME AND PRESSURE RELATIONSHIPS

You have seen that gases have mass, occupy volume, exert pressure, and are influenced by temperature. Thus, there are four variables to consider in working with gases: mass, volume, pressure, and temperature. In studying gases, two of the variables are held constant while the relationship between the other two is studied.

The English chemist Robert Boyle performed the first careful studies of gases. In fact, it was his work with gases in the seventeenth century that led him to the concept of the chemical elements. In 1662, Boyle investigated the relationship between the volume and pressure of a given mass of a gas at constant temperature. Boyle's experiments consisted of measuring the volume of a gas as he increased the external pressure on the gas. He found that *the pressure exerted on a fixed mass of a gas is inversely proportional to the volume of the gas.* As one increases, the other decreases by the same proportion. This relationship is illustrated in Figure 10-9, in which a cylinder of gas closed by a movable upper partition (a piston) has a volume of 1000 mL under 1 atm of pressure. When the pressure on the gas is doubled, the gas volume is reduced by one-half, provided that the mass and temperature remain constant. Thus, the same mass of gas has a volume of 500 mL at 2 atm of pressure and a volume of 250 mL at 4 atm of pressure. Increasing the applied pressure crowds the gas molecules closer together, causing the pressure exerted by the gas to increase. In each case, the pressure inside the cylinder is equal to the applied pressure when the piston is stationary.

The inverse relationship of pressure and volume, known as **Boyle's law,** is stated mathematically as

$$P \propto \frac{1}{V}$$

where the symbol $\propto$ means "proportional to."

We can use the data in Figure 10-9 to develop the mathematical form of Boyle's law. Tabulating the values of P and V from the figure we find

	Condition (a)	Condition (b)	Condition (c)
P	1 atm	2 atm	4 atm
V	1000 mL	500 mL	250 mL
$P \times V$	1000 atm mL	1000 atm mL	1000 atm mL

Notice that the result of multiplying P by V is always the same value. We can state this mathematically as

$$PV = \text{constant}$$

or

$$P_i V_i = P_f V_f$$

where P_i represents the initial pressure, V_i the initial volume, P_f the final pressure, and V_f the final volume. When using this relationship, P_i and P_f must have the same units of pressure and V_i and V_f must have the same units of volume.

Let's do a calculation based on Boyle's law. A gas occupies a volume of 2.0 L under a pressure of 1.0 atm. What will be the final pressure on the gas if its volume is allowed to increase to 4.0 L at constant temperature? We begin the solution of the problem by writing down the given information and what we are seeking:

$$\text{initial volume} = V_i = 2.0 \text{ L}$$
$$\text{initial pressure} = P_i = 1.0 \text{ atm}$$
$$\text{final volume} = V_f = 4.0 \text{ L}$$
$$\text{final pressure} = P_f = ? \text{ atm}$$

Our task is to calculate the final pressure after the gas has been allowed to expand from a volume of 2.0 liters to 4.0 L, at constant temperature and mass. Since the volume of the gas increases, we know (from Boyle's law) that the pressure on the gas must decrease. Thus, we must multiply the original pressure by a factor whose value is less than 1. Since the pressure change depends on the volume change, the factor is made up of the initial and final volumes, V_i and V_f; thus,

$$P_f = P_i \left(\frac{V_i}{V_f} \right)$$

and our factor is actually a ratio of volumes, 2.0 L/4.0 L. The calculation is as follows:

$$P_f = (1.0 \text{ atm}) \left(\frac{2.0 \text{ L}}{4.0 \text{ L}} \right) = 0.50 \text{ atm}$$

Note that the units of atm are in agreement with the quantity, P_f, we are seeking.

In problems involving pressure-volume relationships of gases, we need only remember that an increase in pressure causes a decrease in volume, and vice versa. Then we simply make a ratio of the initial and final values for one variable. The value of the ratio must be greater than 1 if the other variable should increase and less than 1 when the other variable should decrease. Thus, solving pressure-volume problems involves three steps:

1. State the problem by writing down the given information and the unknown variable.
2. Set up the solution by writing down the symbol of the unknown variable, followed by an equals sign, the corresponding known variable,

and the ratio that converts the known variable to the unknown variable. Then cancel units and do the calculations.

3. To evaluate your solution, examine your final answer to make sure that the unknown variable (P or V) changed in the correct direction.

EXAMPLE 10.1   A sample of a gas occupies a volume of 95.2 mL at a pressure of 710.0 torr. What will be its volume in mL at a pressure of 1.00 atm and constant temperature?

Solution   Step 1: State the problem:

$$V_i = 95.2 \text{ mL}$$
$$P_i = 710.0 \text{ torr}$$
$$P_f = 1.00 \text{ atm}$$
$$V_f = ? \text{ mL}$$

Step 2: Set up the solution and do the calculations. Since the pressure increases, the volume must decrease, and our pressure ratio must have a value of less than 1. Since one pressure is given in units of torr and the other in units of atm, we must make a conversion so that they will have the same units. It is easy to convert 1.00 atm to torr, since 1 atm is defined as 760 torr. Thus we can set up the solution

$$V_f = (95.2 \text{ mL}) \left(\frac{710.0 \text{ torr}}{760 \text{ torr}} \right) = 88.9 \text{ mL}$$

Step 3: As a check, compare the final volume to the initial volume. You can see that there was a volume decrease, as we had reasoned there should be.

EXERCISE 10.1   A sample of gas occupies a volume of 75.0 mL at a pressure of 720.0 torr. What will be its volume in mL if the pressure is decreased to 700.0 torr at constant temperature?   ■

EXAMPLE 10.2   A sample of gas occupies a volume of 100.0 mL at a pressure of 720.0 torr. What will be the pressure in torr if the gas volume is decreased to 65.0 mL at constant temperature?

Solution   Step 1: State the problem:

$$V_i = 100.0 \text{ mL}$$
$$P_i = 720.0 \text{ torr}$$
$$V_f = 65.0 \text{ mL}$$
$$P_f = ? \text{ torr}$$

Step 2: Set up the solution and do the calculations. Since the volume decreases, the pressure must increase, and our volume ratio must have a value of greater than 1.

$$P_f = (720.0 \text{ torr}) \left(\frac{100.0 \text{ mL}}{65.0 \text{ mL}} \right) = 1110 \text{ torr or } 1.11 \times 10^3 \text{ torr}$$

Step 3: When we compare final pressure to initial pressure, we see that there was a pressure increase, as we reasoned there should be.

EXERCISE 10.2 A sample of gas has a volume of 140.0 mL at a pressure of 1.00 atm. What will be the pressure in torr if the gas volume is allowed to increase to 200.0 mL at constant temperature?   ■

Boyle's law explains how breathing occurs. When we inhale, our lungs expand; this increase in volume causes a lower pressure inside the lungs. External air (at a higher pressure) then flows into the lungs until the internal pressure is equal to that of the external atmosphere. We exhale by contracting lung volume to increase the pressure of the air inside the lungs. Air then flows from the high-pressure environment within our lungs to the lower-pressure external atmosphere until the two pressures are again equal.

10.3 VOLUME AND TEMPERATURE RELATIONSHIPS

A second gas law was worked out in 1787 by the French physicist J. A. C. Charles. His studies were subsequently confirmed and published in 1802 by Joseph L. Gay-Lussac, a French physicist. **Charles' law** states that *the volume of a gas varies directly as its absolute temperature, if its pressure and mass are kept constant.*

$$V \propto T$$

The absolute temperature scale originated with the work of Lord Kelvin, an English physicist, who continued Charles' observations by measuring gas volume at very low temperatures. When Kelvin plotted gas volume as a function of Celsius temperature (see Figure 10-10), he found an apparent lower limit to the temperature of gases. He suggested that a gas at this limiting temperature, $-273.15°$ C, would have no volume and the energy of its molecules would become zero. (In actual fact, a gas will

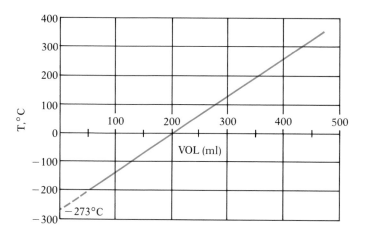

FIGURE 10-10 Plot of Temperature (°C) vs. Gas Volume

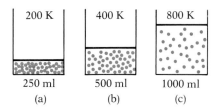

FIGURE 10-11 The Effect of Absolute Temperature on Gas Volume

condense to a liquid long before its temperature reaches $-273.15°$ C.) Since $-273.15°$ C signified the lowest possible temperature, it was given the name **absolute zero.** Kelvin's observations are the basis of the **Kelvin** (or **absolute**) **temperature scale,** whose units are named Kelvins (K). (The degree sign is not used for absolute temperature.) Celsius readings are converted to absolute temperature by the following relationship:

$$K = °C + 273$$

Because of the natural dependence of gas volume on absolute temperature, the Kelvin scale must be used in all gas law calculations involving temperature. Thus, Charles' law describes the relationship between gaseous volume and absolute temperature (Figure 10-11).

The mathematics of Charles' law can be established from the data in Figure 10-11. In tabular form, they are

	Condition (*a*)	Condition (*b*)	Condition (*c*)
V	250 mL	500 mL	1000 mL
T	200 K	400 K	800 K
$\dfrac{V}{T}$	1.25 mL/K	1.25 mL/K	1.25 mL/K

You can see that V/T gives the same quantity, regardless of the individual values of V and T. Thus,

$$\frac{V}{T} = \text{constant}$$

or

$$\frac{V_i}{T_i} = \frac{V_f}{T_f}$$

where V_i represents initial volume, T_i initial temperature, V_f the final volume, and T_f the final temperature. When using this relationship in calculations, V_i and V_f must have the same units, and T_i and T_f must be absolute temperature.

Volume-temperature problems can be solved much like pressure-volume problems. Let's suppose that 2.5 L of a gas at 25° C is heated to 300.0° C at constant pressure. What is the final volume of the gas? We can solve this problem in three steps:

1. State the problem by writing down the given information and the unknown variable.
2. Set up the solution by writing down the symbol for the unknown variable, followed by an equals sign, the corresponding known variable, and the ratio that converts the known variable to the unknown variable. Then cancel units and do the calculations.
3. To evaluate your work, examine your final answer to make sure that the unknown variable (V or T) changed in the correct direction.

The Kelvin scale is the *only* temperature scale used in gas law calculations.

Thus, our solution begins with stating the problem.

$$V_i = 2.5 \text{ L}$$
$$T_i = 25° \text{ C}$$
$$T_f = 300.0° \text{ C}$$
$$V_f = ? \text{ L}$$

However, before we can go on, we must convert the two temperatures to absolute temperature:

$$T_i = 25 + 273 = 298 \text{ K}$$
$$T_f = 300.0 + 273 = 573 \text{ K}$$

MATH TIP

The temperature must be expressed in Kelvins for this relationship to be valid. Both temperatures expressed in Celsius would allow for cancellation mathematically, but the answer would be incorrect.

Now we can derive the temperature ratio that will convert initial volume to final volume. Since the temperature increases, the volume must also increase, and our temperature ratio must be greater than 1. Hence, our complete solution is

$$V_f = (2.5 \text{ L})\left(\frac{573 \text{ K}}{298 \text{ K}}\right) = 4.8 \text{ L}$$

EXAMPLE 10.3   A gas has a volume of 4.50 L at 27° C. If the pressure is kept constant, at what temperature in °C will the volume be 6.00 L?

Solution   Step 1: State the problem:

$$V_i = 4.50 \text{ L}$$
$$T_i = 27° \text{ C}$$
$$V_f = 6.00 \text{ L}$$
$$T_f = ?° \text{ C}$$

Step 2: First we must convert T_i to absolute temperature:

$$T_i = 27 + 273 = 300 \text{ K}$$

Now we can set up the solution. Since the volume increases, the temperature must increase, and our volume ratio must have a value greater than 1.

MATH TIP

Subtract 273 from both sides of the equation:

$$K = °C + 273$$
$$K - 273 = °C + 273 - 273$$
$$K - 273 = °C$$
$$400 - 273 = °C$$
$$127 = °C$$

$$T_f = (3.00 \times 10^2 \text{ K})\left(\frac{6.00 \text{ L}}{4.50 \text{ L}}\right) = 4.00 \times 10^2 \text{ K}$$

Last, we must convert T_f to °C:

$$T_f = K - 273 = 400 - 273 = 127° \text{ C}$$

Step 3: A check of the final answer shows an increase in temperature, as we had reasoned there should be.

EXERCISE 10.3   A gas occupies a volume of 10.0 mL at 100.0° C. If the pressure is kept constant, at what temperature in °C will its volume be 8.00 mL?   ■

EXAMPLE 10.4   A gas has a volume of 100.0 mL at 50.0° C. If the pressure is kept constant, what will be its volume in mL at 30.0° C?

The earth's atmosphere is the gaseous environment in which we live. Its chemical composition at sea level is shown in the accompanying table. In many ways it is like the liquid sea of water that covers 75% of the earth's surface. Plants and animals living deep in the ocean experience a much different environment than those near the surface. We humans who live at the bottom of the atmospheric sea are seldom aware of the conditions higher up in the atmosphere, and we take for granted the fact that the atmosphere determines the environment we live in.

The pressure of the atmosphere decreases consistently with increasing altitude, as shown in the accompanying figure. Any area on the earth's surface experiences a pressure resulting from the entire column of atmosphere above it. This pressure, a consequence of the earth's gravitational attraction for the gases of the atmosphere, has an average value of 760 torr at sea level. At higher elevations, however, the pressure of the atmosphere is lower, because there is a smaller total mass of gas in the column above the ground. Thus, the normal atmospheric pressure at Denver, Colorado (elevation 1 mile), is usually about 700 torr. Also, at higher elevations, air is less dense and oxygen is less abundant. For this reason, aircraft flying above 10,000 feet are pressurized so that people in the aircraft will be able to breathe properly.

Average Composition of the Atmosphere (Dry Air at Sea Level)	
Gas	Percent by Volume
N_2	78.08
O_2	20.95
Ar	0.93
CO_2	0.033
Ne	0.0018
He	
CH_4	
Kr	trace of each
Xe	
H_2	
N_2O	

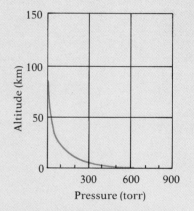

Variation of Atmospheric Pressure with Altitude

Solution   Step 1:   State the problem:

$$V_i = 100.0 \text{ mL}$$
$$T_i = 50.0° \text{ C}$$
$$T_f = 30.0° \text{ C}$$
$$V_f = ? \text{ mL}$$

Step 2:   First we must convert the Celsius temperatures to absolute temperatures:

$$T_i = 50.0 + 273 = 323 \text{ K}$$
$$T_f = 30.0 + 273 = 303 \text{ K}$$

Then, since temperature decreases, volume must also decrease, so our temperature ratio must have a value of less than 1.

$$V_f = (100.0 \text{ mL}) \left(\frac{303 \text{ K}}{323 \text{ K}} \right) = 93.8 \text{ mL}$$

Step 3:   We check our final answer to see that it represents a decrease in volume as we had reasoned there should be.

EXERCISE 10.4   A gas has a volume of 1.00 L at 25.0° C. If the pressure is kept constant, what will be its volume in L at 50.0° C?   ■

10.4   PRESSURE AND TEMPERATURE RELATIONSHIPS

The Kelvin scale is the *only* temperature scale used in gas law calculations.

The volume-temperature relationship of gases carries with it an implication about the effect of temperature on gas pressure. Consider what happens when we heat popcorn kernels. Steam forms in the seed cores, and the pressure created by the steam at high temperature causes the kernels to explode. This illustrates that when a confined gas is heated at constant volume, its pressure increases. *The pressure exerted by a gas at constant volume is directly proportional to absolute temperature:*

$$P \propto T$$

This is another form of Charles' law; it is illustrated by Figure 10-12. We can use the data in Figure 10-12 to establish the mathematical form of the pressure-temperature relationship.

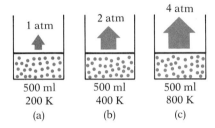

FIGURE 10-12   The Effect of Temperature on Gas Pressure

	Condition (*a*)	Condition (*b*)	Condition (*c*)
P	1 atm	2 atm	4 atm
T	200 K	400 K	800 K
$\dfrac{P}{T}$	5×10^{-3} atm/K	5×10^{-3} atm/K	5×10^{-3} atm/K

Since P/T gives a constant value,

$$\frac{P}{T} = \text{constant}$$

or

$$\frac{P_i}{T_i} = \frac{P_f}{T_f}$$

where P_i represents initial pressure, T_i initial temperature, P_f final pressure, and T_f final temperature. Pressure-temperature calculations are performed much like pressure-volume and volume-temperature calculations.

EXAMPLE 10.5   The temperature of a gas at 1.00 atm pressure is changed from 0.0° C to 200.0° C, while the gas volume is held constant. What is the final pressure of the gas in torr?

Solution Step 1: State the problem:

$$P_i = 1.00 \text{ atm}$$
$$T_i = 0.0° \text{ C}$$
$$T_f = 200.0° \text{ C}$$
$$P_f = ? \text{ torr}$$

Step 2: First we must convert Celsius temperature to absolute temperature:

$$T_i = 0.0 + 273 = 273 \text{ K}$$
$$T_f = 200.0 + 273 = 473 \text{ K}$$

Since temperature increases, the pressure must also increase, so our temperature ratio must have a value of greater than 1.

$$P_f = (1.00 \text{ atm}) \left(\frac{473 \text{ K}}{273 \text{ K}} \right) = 1.73 \text{ atm}$$

Finally, we must convert atmospheres to torr:

$$P_f = (1.73 \text{ atm}) \left(\frac{760 \text{ torr}}{1 \text{ atm}} \right) = 1310 \text{ torr or } 1.31 \times 10^3 \text{ torr}$$

Step 3: In conclusion, we check our final answer to see that it represents an increase in pressure as we had reasoned there should be.

EXERCISE 10.5 The temperature of a gas at 2.50 atm pressure is changed from 100.0° C to 50.0° C while the gas volume is held constant. What is the final pressure of the gas in torr? ▪

10.5 COMBINING THE GAS LAWS

You have learned to make calculations based on Boyle's law, in which mass and temperature are constant, and Charles' law, in which mass and either pressure or volume are constant. But what if pressure, volume, and temperature all change at the same time? For example, let's assume that we have 1.00 L of a gas at 1.50 atm pressure and 24.0° C, and we allow it to expand to a volume of 2.00 L at 10.0° C. What is the final pressure? We can solve this problem by breaking it into two parts. First, consider the temperature change to occur at constant pressure. Then,

$$V_i = 1.00 \text{ L}$$
$$T_i = 24.0° \text{ C} = 297 \text{ K}$$
$$T_f = 10.0° \text{ C} = 283 \text{ K}$$
$$V_f = ? \text{ L}$$

and

$$V_f = (1.00 \text{ L}) \left(\frac{283 \text{ K}}{297 \text{ K}} \right) = 0.953 \text{ L}$$

We have just calculated a new volume caused only by the change in temperature. Next, consider the effect of the calculated volume change on pressure. At this point, our given information is

$$V_i = 0.953 \text{ L}$$
$$P_i = 1.50 \text{ atm}$$
$$V_f = 2.00 \text{ L}$$
$$P_f = ? \text{ atm}$$

Our solution then becomes

$$P_f = (1.50 \text{ atm})\left(\frac{0.953 \cancel{\text{L}}}{2.00 \cancel{\text{L}}}\right) = 0.715 \text{ atm}$$

Thus, the combined effects of volume increase and temperature decrease caused a resulting decrease in gas pressure.

In this example we solved the problem with two sets of calculations. However, we could have achieved the same solution with one continuous calculation:

$$P_f = (1.50 \text{ atm})\left(\frac{283 \cancel{\text{K}}}{297 \cancel{\text{K}}}\right)\left(\frac{1.00 \cancel{\text{L}}}{2.00 \cancel{\text{L}}}\right) = 0.715 \text{ atm}$$

If we insert symbols (P_i, T_i, T_f, and so forth) for numerical values, then we can see that our calculation had the form

$$P_f = P_i\left(\frac{T_f}{T_i}\right)\left(\frac{V_i}{V_f}\right)$$

or

$$\frac{P_f}{P_i} = \frac{T_f}{T_i}\frac{V_i}{V_f}$$

Inversion gives

$$\frac{P_i}{P_f} = \frac{T_i}{T_f}\frac{V_f}{V_i}$$

and rearranging gives

$$\frac{P_i V_i}{T_i} = \frac{P_f V_f}{T_f} \tag{1}$$

This relationship is a combination of Boyle's and Charles' law. Boyle's law applies to constant temperature situations, in which $T_i = T_f = T$, and

$$\frac{P_i V_i}{\cancel{T}} = \frac{P_f V_f}{\cancel{T}}$$

or

$$P_i V_i = P_f V_f$$

Charles' law applies to constant pressure situations, in which $P_i = P_f = P$, and

$$\frac{\cancel{P} V_i}{T_i} = \frac{\cancel{P} V_f}{T_f}$$

or

$$\frac{V_i}{T_i} = \frac{V_f}{T_f}$$

Similarly, if volume remains constant, then $V_i = V_f = V$, and

$$\frac{P_i V}{T_i} = \frac{P_f V}{T_f}$$

or

$$\frac{P_i}{T_i} = \frac{P_f}{T_f}$$

Thus, equation (1) can be used for gas law problems involving any change in condition. However, it is essential that P_i and P_f have the same units and that V_i and V_f have the same units. Also, remember that only absolute temperature can be used in gas law calculations.

When we combine the gas laws, we are simply making a more compact mathematical expression to replace two steps in the solution of the problem. Calculations with the combined gas laws can be summarized as follows:

$$V_f = V_i \times P \text{ ratio} \times T \text{ ratio} \tag{2}$$
$$P_f = P_i \times V \text{ ratio} \times T \text{ ratio} \tag{3}$$
$$T_f = T_i \times V \text{ ratio} \times P \text{ ratio} \tag{4}$$

In equation (2), the final volume is equal to the initial volume multiplied by a pressure ratio and a temperature ratio. Pressure is inversely proportional to volume, so if the pressure increases, the pressure ratio must be less than 1 in order to decrease the initial volume. If the pressure decreases, the pressure ratio must be greater than 1 to increase the initial volume. Temperature is directly proportional to volume, so if the temperature decreases, the temperature ratio must be less than 1; if the temperature increases, the temperature ratio must be greater than 1. Similar reasoning can be applied to equations (3) and (4) to calculate final pressure and final temperature.

EXAMPLE 10.6

A gas has a volume of 400.0 mL at 760.0 torr and 0.0° C. What volume in mL will the gas have at 850.0 torr and 100.0° C?

Solution

$$V_i = 400.0 \text{ mL}$$
$$P_i = 760.0 \text{ torr}$$
$$T_i = 0.0° \text{ C} = 273 \text{ K}$$
$$P_f = 850.0 \text{ torr}$$
$$T_f = 100.0° \text{ C} = 373 \text{ K}$$
$$V_f = ? \text{ mL}$$

From equation (2),

$$V_f = V_i \times P \text{ ratio} \times T \text{ ratio}$$

Since the increase in pressure would cause a decrease in volume, our pressure ratio must have a value less than 1. The increase in temperature would cause an increase in volume, so our temperature ratio must have a value greater than 1.

$$V_f = (400.0 \text{ mL}) \left(\frac{760.0 \text{ torr}}{850.0 \text{ torr}} \right) \left(\frac{373 \text{ K}}{273 \text{ K}} \right)$$

$$= 489 \text{ mL}$$

EXERCISE 10.6   A gas has a volume of 1.50 L at 800.0 torr and 150.0° C. What will be its volume in L at 700.0 torr and 50.0° C?   ■

EXAMPLE 10.7   A gas has a volume of 20.0 L at 40.0°C and 780.0 torr. What will its pressure be in torr when it has a volume of 75.0 L at 0.0° C?

Solution

$$V_i = 20.0 \text{ L}$$
$$T_i = 40.0° \text{ C} = 313 \text{ K}$$
$$P_i = 780.0 \text{ torr}$$
$$V_f = 75.0 \text{ L}$$
$$T_f = 0.0° \text{ C} = 273 \text{ K}$$
$$P_f = ? \text{ torr}$$

From equation (3),

$$P_f = P_i \times V \text{ ratio} \times T \text{ ratio}$$

Since an increase in volume would cause a decrease in pressure, our volume ratio must have a value less than 1. The decrease in temperature would cause a decrease in pressure, so our temperature ratio must have a value less than 1.

$$P_f = (780.0 \text{ torr}) \left(\frac{20.0 \text{ L}}{75.0 \text{ L}} \right) \left(\frac{273 \text{ K}}{313 \text{ K}} \right) = 181 \text{ torr}$$

EXERCISE 10.7   A gas has a volume of 250.0 mL at 25.0° C and 760.0 torr. At what pressure in torr would this gas have a volume of 200.0 mL at 50.0° C?   ■

EXAMPLE 10.8   A gas has a volume of 10.0 L at 10.0° C and 600.0 torr. What will the temperature of the gas be in Celsius degrees when it has a volume of 5.00 L and a pressure of 700.0 torr?

Solution

$$V_i = 10.0 \text{ L}$$
$$T_i = 10.0° \text{ C} = 283 \text{ K}$$
$$P_i = 600.0 \text{ torr}$$
$$V_f = 5.00 \text{ L}$$
$$P_f = 700.0 \text{ torr}$$
$$T_f = ?° \text{ C}$$

From equation (4),

$$T_f = T_i \times V \text{ ratio} \times P \text{ ratio}$$

Since the decrease in volume would cause a decrease in temperature, our volume ratio must have a value less than 1. The increase in pressure would cause an increase in temperature, so our pressure ratio must have a value greater than 1.

$$T_f = (283 \text{ K}) \left(\frac{5.00 \text{ L}}{10.0 \text{ L}} \right) \left(\frac{700.0 \text{ torr}}{600.0 \text{ torr}} \right) = 165 \text{ K}$$

Since the problem asks for the final temperature in Celsius degrees, we must convert absolute temperature to Celsius:

$$T_f = \text{K} - 273 = 165 - 273 = -108° \text{ C}$$

EXERCISE 10.8 A gas has a volume of 7.85 L at 24° C and 545 torr. What will the temperature of the gas be in Celsius degrees when it has a volume of 9.60 L and a pressure of 435 torr? ■

10.6 AVOGADRO'S HYPOTHESIS

Boyle, Charles, and Gay-Lussac gave very specific descriptions of the behavior of gases influenced by changes in temperature, pressure, and volume. Their laws are derived from experimental observations. We will now look at an interpretation of the behavior of gases so that you can understand the physical basis of the gas laws.

Amedeo Avogadro, an Italian physicist, made the first attempt to explain the behavior of gases. Avogadro made two basic assumptions:

1. Gases are composed of molecules.
2. Each molecule of a gas occupies a certain volume of space at a given temperature and pressure.

From these assumptions came **Avogadro's hypothesis:** *at the same conditions of temperature and pressure, equal volumes of gases contain equal numbers of molecules.* Avogadro's hypothesis has been tested many times since it was first proposed, and we now accept it as one of the fundamental truths of chemistry.

Avogadro's hypothesis tells us that a mole of one gas will occupy the same volume as a mole of any other gas under the same conditions of temperature and pressure. Measurements of gas volumes are often converted to standard temperature and pressure (STP) to make it easy to compare measurements made under different sets of conditions. **Standard temperature and pressure (STP)** are defined as 1 atm and 273 K (0° C). Calculations show that *one mole of any ideal gas occupies 22.4 L at STP.* (An ideal gas is one whose particles do not interact with each other.) This volume of gas, called the **molar volume,** contains 6.02×10^{23} molecules (this is Avogadro's number, which was introduced in chapter 8). For noble gases, which are composed of individual atoms instead of molecules, 22.4 L contains 6.02×10^{23} atoms.

Note that 1 atm is an exact number.

10.7   THE IDEAL GAS EQUATION

The gas laws and Avogadro's hypothesis can all be expressed as proportionalities between volume and one other variable, with the remaining variables held constant,

Boyle's law:	$V \propto \dfrac{1}{P}$	T, n constant
Charles' law:	$V \propto T$	P, n constant
Avogadro's hypothesis:	$V \propto n$	P, T constant

where n stands for number of moles of the gas.

These three relationships can be combined into a single expression,

$$V \propto \frac{nT}{P}$$

To change the proportionality into an equation, a constant of proportionality, R, is inserted (this constant is called the molar gas constant).

$$V = \frac{nRT}{P}$$

Now we have arrived at a mathematical statement of Avogadro's hypothesis that a given number of moles of a gas will occupy the same volume as the same number of moles of any other gas under identical conditions of temperature and pressure.

$$PV = nRT \tag{5}$$

This expression says that the product of the pressure and volume of a gas is equal to the product of the number of moles of the gas, the molar gas constant, and the temperature of the gas. The **molar gas constant, *R*,** has the value of 0.0821 L atm/mol K. This expression (equation (5)) is called the **ideal gas equation.** It can be used to calculate P, V, n, or T, provided that the other three variables are known and that none of them changes. When using the ideal gas equation, P must be expressed in atmospheres, V in liters, and T in absolute temperature.

The ideal gas equation describes an ideal behavior for gases. The gases that we encounter in the world about us are real gases that do not behave ideally; instead, real gases deviate from ideal behavior when their pressures are above about 10 atm or when their temperatures are near their boiling points. However, these pressures and temperatures are not often encountered. At more normal pressures (1–10 atm) and temperatures, deviations from ideal behavior are small enough that the ideal gas equation can be used without serious error.

EXAMPLE 10.9   A gas sample has a volume of 7.0 L at 25° C and 1.0 atm. How many moles of gas are present?

Solution   This problem does not involve any changes in variables, so we can use the ideal gas equation to solve it. We must first convert Celsius temperature to absolute temperature:

$$T = 25° \text{ C} + 273 = 298 \text{ K}$$

Three of the variables (P, V, and T) are given, and we are asked to calculate the fourth, n. To do this, we must solve the ideal gas equation for n:

$$PV = nRT$$

$$n = \frac{PV}{RT}$$

Substituting numerical values for P, V, R, and T, we get

$$n = \frac{(1.0 \text{ atm})(7.0 \text{ L})}{(0.0821 \text{ L atm/mol K})(298 \text{ K})} = 0.29 \text{ mol}$$

MATH TIP

Separating the numbers from the units,

$$\frac{(1.0)(7.0)}{(0.0821)(298)} \times \frac{(\text{atm})(\text{L})}{\left(\frac{\text{L atm}}{\text{mol K}}\right)(\text{K})} = 0.29 \frac{1}{\frac{1}{\text{mol}}}$$

The unit can be further simplified by remembering that when you divide by a fraction, first invert the fraction, then multiply.

$$1 \div \frac{1}{\text{mol}} = 1 \times \frac{\text{mol}}{1} = \text{mol}$$

EXERCISE 10.9   A gas sample has a volume of 1.25 L at 30.0° C and 2.00 atm. How many moles of gas are present? ∎

A convenient way of determining the molar mass of a gas is to collect a sample of the gas and measure its mass, volume, temperature, and pressure. Then, its molar mass can be calculated using the ideal gas equation. This is illustrated in the following example.

EXAMPLE 10.10   A gas sample weighs 1.42 g and has a volume of 247 mL at 2.00 atm and 28° C. What is the molar mass of the gas?

Solution   Since there are no changes in variables, we can use the ideal gas equation to solve this problem. However, we must first convert the volume units of mL to L and the Celsius temperature to absolute temperature.

$$(247 \text{ mL}) \left(\frac{1 \text{ L}}{1000 \text{ mL}} \right) = 0.247 \text{ L}$$

$$T = 28° \text{ C} + 273 = 301 \text{ K}$$

Then we must solve the ideal gas equation for n, the quantity we are seeking, and substitute the numerical values for P, V, R, and T:

$$PV = nRT$$

$$n = \frac{PV}{RT}$$

$$= \frac{(2.00 \text{ atm})(0.247 \text{ L})}{(0.0821 \text{ L atm/mol K})(301 \text{ K})}$$

$$= 2.00 \times 10^{-2} \text{ mol}$$

At this point, we know that 1.42 g of the gas contains 2.00×10^{-2} mol. Now we can calculate the weight of one mole:

$$\text{Molar mass} = \frac{1.42 \text{ g}}{2.00 \times 10^{-2} \text{ mol}} = 71.0 \text{ g/mol}$$

EXERCISE 10.10   A gas sample weighs 220.0 g and has a volume of 80.1 L at 1.50 atm and 20.0° C. What is the molar mass of the gas?   ■

EXAMPLE 10.11   A gas sample containing 2.35 mol has a volume of 40.0 L at 1.00 atm. What is the temperature of the gas in °C?

Solution   Since there are no changes in variables, we can use the ideal gas equation to solve this problem. First, we must solve the ideal gas equation for T, the quantity we are seeking:

$$PV = nRT$$

$$T = \frac{PV}{nR}$$

MATH TIP
Separating the numbers from the units,

$$\frac{(1.0)(40.0)}{(2.35)(0.0821)} \frac{(atm)(L)}{(mol)\left(\dfrac{L\ atm}{mol\ K}\right)} = 207\ \frac{1}{\dfrac{1}{K}}$$

$$= 207\ K$$

Then, substituting numerical values,

$$T = \frac{(1.00\ atm)(40.0\ L)}{(2.35\ mol)(0.0821\ L\ atm/mol\ K)} = 207\ K$$

Converting absolute temperature to °C,

$$T = K - 273 = 207 - 273 = -64°\ C$$

EXERCISE 10.11   A gas sample containing 7.25 mol has a volume of 140.0 L at 1.20 atm. What is the temperature of the gas in °C?   ■

The ideal gas equation can be transformed to another expression when the number of moles (mass) of the gas remains constant but other variables change. For an initial set of variables,

$$P_i V_i = nRT_i$$

and

$$\frac{P_i V_i}{T_i} = nR$$

For a final set of values for P, V, and T,

$$\frac{P_f V_f}{T_f} = nR$$

Then,

$$\frac{P_i V_i}{T_i} = nR = \frac{P_f V_f}{T_f}$$

or

$$\frac{P_i V_i}{T_i} = \frac{P_f V_f}{T_f} \qquad (6)$$

which is the same as equation (1) in section 10.5.

Equation (6) is the last step in the development of the mathematical relationships among P, V, n, and T for gases. It demonstrates that there really is only one gas law

$$PV = nRT$$

and that all the other forms

$$\frac{P_iV_i}{T_i} = \frac{P_fV_f}{T_f}$$

$$P_iV_i = P_fV_f$$

$$\frac{P_i}{T_i} = \frac{P_f}{T_f}$$

$$\frac{V_i}{T_i} = \frac{V_f}{T_f}$$

and

$$\frac{V_i}{n_i} = \frac{V_i}{n_i}$$

are just special cases of the ideal gas equation.

EXAMPLE 10.12 A constant mass of a gas has a volume of 1.00 L at 1.30 atm and 25° C. What volume will it have at 2.00 atm and 30° C?

Solution Since three variables change in this problem, we must start with the relationship

$$\frac{P_iV_i}{T_i} = \frac{P_fV_f}{T_f}$$

and solve for V_f.

$$V_f = \frac{P_iV_iT_f}{T_iP_f}$$

$$= \frac{(1.30 \text{ atm})(1.00 \text{ L})(303 \text{ K})}{(298 \text{ K})(2.00 \text{ atm})}$$

$$= 0.661 \text{ L}$$

EXERCISE 10.12 A constant mass of gas occupies a volume of 1.35 L at 2.00 atm and 0.0° C. What will be its temperature in °C when it has a volume of 2.00 L at 1.50 atm?

10.8 THE KINETIC MOLECULAR THEORY

Avogadro's hypothesis is the basis for a general theory of gas behavior called the **kinetic molecular theory.** The main features of the theory are:

1. All gases consist of separate and distinct particles. These particles are molecules (hence the term kinetic *molecular* theory) for all gases except the noble gases, whose particles are atoms.

2. Individual gas molecules are so small that the space occupied by their nuclei and electrons is negligible. Thus, gases can be compressed into infinitely small volumes.
3. Molecules of a gas are very far apart (roughly 10 times the diameter of one molecule apart), and the volume occupied by a gas is mostly empty space.
4. Molecules of a gas are in random, straight-line motion, and thus they possess kinetic energy (hence the term *kinetic* molecular theory).
5. Molecules of a gas collide with each other and the walls of their container. These collisions are perfectly elastic; this means that the molecules neither attract nor repel each other or the walls of the container. The molecules behave like very hard billiard balls, and no energy is lost in a collision.
6. The average kinetic energy of the molecules of one gas is the same as for molecules of any other gas at a given temperature; this average kinetic energy is proportional to the absolute temperature. (The molecules of a gas do not all have the same kinetic energy; they have a wide range of kinetic energies because they have different velocities. However, the *average* kinetic energy is proportional to absolute temperature.)

We can now relate the kinetic molecular theory to the observed properties of gases.

1. Gases have mass because they are composed of particles of matter (molecules).
2. Gases occupy space because their molecules move randomly over a wide range of space.
3. Gases exert pressure because their molecules continuously collide with the walls of the container.
4. Gases are compressible because there are large empty spaces between their molecules.
5. Gases diffuse to fill any container because the individual molecules are in rapid, random motion and exert no attraction for each other.

The pressure of a gas is due to the collisions of its molecules with the walls of the container (see Figure 10-13). Crowding the molecules closer together by decreasing the volume of the container increases the number of collisions and thus the pressure. This is the basis of Boyle's law. Heating a gas gives each molecule a higher velocity. The higher velocities increase the frequency of collisions and give each collision greater impact, causing an increase in pressure (Charles' law).

Real gases do not behave ideally because of two facts ignored by the kinetic molecular theory: the nuclei and electrons of gas molecules occupy a certain amount of space, and gas molecules have short-range attractions for each other. These facts cause gases to depart from ideal behavior when their molecules are crowded closely together, as under conditions of high pressure and temperatures near the boiling points of the gases. Real gases have less compressibility than expected at high pressures and greater compressibility than expected at low temperatures. It is

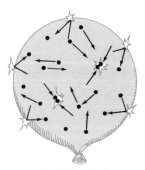

A Gas-Filled Balloon

FIGURE 10-13 Gas Molecules in Rapid, Random Motion. The pressure on the walls of the balloon is created by the forces of individual impacts.

this real behavior that allows many gases, propane and butane, for example, to be liquified by the combination of high pressures and low temperatures.

10.9 GAS MIXTURES

Because there are large spaces between gas molecules, you might expect a gas in a mixture of gases to act as if the others were not present. This expectation is usually true. For example, the 23% of our atmosphere that is oxygen behaves as if it were pure oxygen with a pressure of 0.23 atm (23% of 1.00 atm).

In 1801 John Dalton discovered that gases in a mixture behave independently of each other. Dalton's **law of partial pressures** states that *each component in a gas mixture exerts its own pressure (its* **partial pressure***) independent of other gases* and that *the total pressure of a gas mixture is the sum of the partial pressures of the individual gases.* Dalton's law of partial pressures is expressed mathematically as

$$P_{total} = P_1 + P_2 + P_3 + \ldots$$

where P_{total} is the total pressure of the gas mixture, and $P_1, P_2, P_3, \ldots$ are the partial pressures of the individual components of the mixture. The partial pressure of each component is directly related to the percentage of the component in the gas mixture.

If the partial pressures of the gases in a mixture are known, the total pressure is found by adding together all of the partial pressures. For example, the total pressure exerted by our atmosphere averages 760 torr at sea level; this value is the sum of the partial pressures of all the atmospheric components: nitrogen, oxygen, argon, carbon dioxide, water vapor, and traces of other gases.

EXAMPLE 10.13 The partial pressure of oxygen in dry air is 159 torr. Assume that the partial pressures of argon, carbon dioxide, and traces of other gases are negligible and calculate the partial pressure of nitrogen when the total pressure is 760 torr.

Solution Dry air is composed of nitrogen, oxygen, argon, carbon dioxide, and traces of other gases. Since only the partial pressures of nitrogen and oxygen need be considered, we can write

$$P_{total} = P_{nitrogen} + P_{oxygen}$$

Then,

$$P_{nitrogen} = P_{total} - P_{oxygen}$$
$$= 760 \text{ torr} - 159 \text{ torr}$$
$$= 601 \text{ torr}$$

EXERCISE 10.13 In a laboratory experiment, a mixture of hydrogen chloride gas and water vapor was collected in a sealed container. If the partial pressure of the water vapor was 24 torr and the total pressure was 750.0 torr, what was the partial pressure of the hydrogen chloride?

10.10 GAS DENSITY

Because molecules of gases are spaced much farther apart than molecules of liquids or solids, gases have very low densities relative to liquids and solids. For this reason, gas densities are usually expressed in units of g/L. Table 10-1 gives densities of common gases at 20° C and 1 atm. Since the volume of a gas depends on both temperature and pressure, the values of these variables must be stated along with the measured density of a gas.

The density of a gas at any temperature and pressure can be calculated using the ideal gas equation. As an example, consider nitrogen, N_2, at 1.00 atm and 25° C. First, we calculate the volume occupied by 1.00 mol of N_2:

$$PV = nRT$$

$$V = \frac{nRT}{P}$$

$$= \frac{(1.00 \text{ mol})(0.0821 \text{ L atm/mol K})(298 \text{ K})}{1.00 \text{ atm}}$$

$$= 24.4 \text{ L}$$

Since 1.00 mol of N_2 has a volume of 24.4 L at 1.00 atm and 25° C, the density is simply the mass of 1.00 mol divided by volume.

$$d = \frac{m}{v}$$

$$= \frac{28.0 \text{ g}}{24.4 \text{ L}} = 1.15 \text{ g/L}$$

Keep in mind, however, that such density calculations may vary slightly from actual measurements because real gases do not behave ideally.

MATH TIP

1.00 mol is the quantity selected here, since the mass of a mole of N_2 is known. This mass will be needed to calculate density below.

EXAMPLE 10.14 Calculate the density of helium at STP.

Solution First, calculate the volume of 1.00 mol of gas at the given conditions:

$$V = \frac{nRT}{P}$$

$$= \frac{(1.00 \text{ mol})(0.0821 \text{ L atm/mol K})(273 \text{ K})}{1.00 \text{ atm}}$$

$$= 22.4 \text{ L}$$

(Do you see the basis for the value of 22.4 L as the molar volume of a gas?) Next, calculate gas density:

$$d = \frac{m}{v}$$

$$= \frac{4.00 \text{ g}}{22.4 \text{ L}} = 0.179 \text{ g/L}$$

MATH TIP

Future calculations requiring the volume of one mole of any gas at STP would not be necessary, since the volume will always be 22.4 L.

TABLE 10-1 Densities of Common Gases at 20° C and 1 atm		
Gas	Formula	Density (g/L)
Acetylene	C_2H_2	1.12
Ammonia	NH_3	0.7188
Carbon dioxide	CO_2	1.975
Carbon monoxide	CO	1.250
Chlorine	Cl_2	2.98
Fluorine	F_2	1.580
Helium	He	0.17847
Hydrogen	H_2	0.08987
Hydrogen bromide	HBr	3.388
Hydrogen chloride	HCl	1.526
Hydrogen cyanide	HCN	0.901
Hydrogen fluoride	HF	0.922
Hydrogen iodide	HI	5.37
Methane	CH_4	0.667
Neon	Ne	0.8899
Nitrogen	N_2	1.165
Nitrogen dioxide	NO_2	1.447
Oxygen	O_2	1.331
Ozone	O_3	1.998
Sulfur dioxide	SO_2	2.716

Note that our result is close to the value given in Table 10-1, which is the density at 20° C.

EXERCISE 10.14   Calculate the density of ammonia at 2.00 atm and 30° C.   ■

10.11 STOICHIOMETRIC CALCULATIONS INVOLVING GAS VOLUMES *(optional)*

There are three types of stoichiometric calculations involving gaseous volume: mole-volume calculations, weight-volume calculations, and volume-volume calculations. These calculations are based on the principles we developed in chapter 9, and we are now adding use of the ideal gas equation.

When reactions involve gases, gas volumes are usually measured at conditions other than STP. However, the measured volumes can be used in stoichiometric calculations since the ideal gas equation relates volume to number of moles and is not restricted to STP.

Mole-Volume Calculations   Mole-volume calculations are very similar to mole-weight calculations (section 9.6). Three steps are involved:

1. Write down the given number of moles.
2. Convert moles of the first substance to moles of the second substance by use of the appropriate mole ratio.

3. Convert moles of the second substance to volume using the ideal gas equation.

EXAMPLE 10.15

Calculate the volume of oxygen at STP produced by heating 1.48 mol of $KClO_3$.

$$2 \, KClO_3(s) \longrightarrow 2 \, KCl(s) + 3 \, O_2(g)$$

Solution

Step 1: Write down the given number of moles:

$$1.48 \text{ mol } KClO_3$$

Step 2: Convert moles of $KClO_3$ to moles of O_2 using the mole ratio $3 \text{ mol } O_2/2 \text{ mol } KClO_3$:

$$(1.48 \cancel{\text{ mol } KClO_3}) \left(\frac{3 \text{ mol } O_2}{2 \cancel{\text{ mol } KClO_3}} \right) = 2.22 \text{ mol } O_2$$

Step 3: Convert moles of O_2 to volume of O_2 using the ideal gas equation:

$$PV = nRT$$

$$V = \frac{nRT}{P}$$

$$= \frac{(2.22 \cancel{\text{ mol}})(0.0821 \text{ L} \cdot \text{atm}/\cancel{\text{mol}} \cdot \cancel{K})(273 \cancel{K})}{1 \cancel{\text{ atm}}}$$

$$= 49.8 \text{ L}$$

(Remember that 1 atm is an exact number and thus does not limit the significant figures of the answer.) Note that the three steps outlined above could be combined as follows:

$$\left(\frac{3 \cancel{\text{ mol}}}{2 \cancel{\text{ mol}}} \right) \frac{(1.48 \cancel{\text{ mol}})(0.0821 \text{ L} \cdot \text{atm}/\cancel{\text{mol}} \cdot \cancel{K})(273 \cancel{K})}{(1 \cancel{\text{ atm}})} = 49.8 \text{ L}$$

MATH TIP

Notice the cancellation of mole units in this problem. Both unit and label must agree to cancel.

$$(1.48 \cancel{\text{ mol } KClO_3}) \left(\frac{3 \text{ mol } O_2}{2 \cancel{\text{ mol } KClO_3}} \right)$$

$$\frac{(0.0821 \text{ L atm}/(\text{mol})K)(273 \text{ K})}{(1 \text{ atm})}$$

The mol circled in the unit above refers to mol of O_2 since this is the gas in the problem. Rewriting the remaining units will show the cancellation of mol O_2.

$$\frac{\cancel{\text{mol } O_2} \frac{\text{L atm}}{\cancel{\text{mol}} K} K}{\text{atm}} = L$$

EXERCISE 10.15

What volume of HCl at STP would be produced by 2.35 mol NaCl and excess H_2SO_4?

$$2 \, NaCl(s) + H_2SO_4(aq) \longrightarrow Na_2SO_4(aq) + 2 \, HCl(g) \qquad ∎$$

Weight-Volume Calculations These calculations are quite similar to weight-mole calculations (section 9.6). Four steps are needed:

1. Write down the given weight.
2. Convert the given weight to moles using the molar mass.
3. Convert moles of the first substance to moles of the second substance using the appropriate mole ratio.
4. Convert moles of the second substance to volume using the ideal gas equation.

EXAMPLE 10.16

What volume of CO_2 at 308 K and 1.75 atm would be produced by the complete combustion of 14.2 g of CH_4?

$$CH_4(g) + 2 \, O_2(g) \longrightarrow CO_2(g) + 2 \, H_2O(g)$$

Solution Step 1: Write down the given weight:

$$14.2 \text{ g CH}_4$$

Step 2: Convert weight of CH_4 to moles of CH_4 using the molar mass:

$$(14.2 \text{ g CH}_4)\left(\frac{1 \text{ mol CH}_4}{16.0 \text{ g CH}_4}\right) = 0.888 \text{ mol CH}_4$$

Step 3: Convert moles of CH_4 to moles of CO_2 using the mole ratio 1 mol CO_2/1 mol CH_4:

$$(0.888 \text{ mol CH}_4)\left(\frac{1 \text{ mol CO}_2}{1 \text{ mol CH}_4}\right) = 0.888 \text{ mol CO}_2$$

Step 4: Convert moles of CO_2 to volume of CO_2 using the ideal gas equation:

$$V \text{ (of CO}_2) = \frac{nRT}{P}$$

$$= \frac{(0.888 \text{ mol})(0.0821 \text{ L atm/mol K})(308 \text{ K})}{1.75 \text{ atm}}$$

$$= 12.8 \text{ L}$$

Note that the four steps outlined above can be combined as follows:

$$(14.2 \text{ g CH}_4)\left(\frac{1 \text{ mol CH}_4}{16.0 \text{ g CH}_4}\right)\left(\frac{1 \text{ mol CO}_2}{1 \text{ mol CH}_4}\right)\frac{(0.0821 \text{ L atm/mol K})(308 \text{ K})}{(1.75 \text{ atm})}$$

$$= 12.8 \text{ L}$$

EXERCISE 10.16   What volume of H_2 at 325 K and 2.35 atm would be required to react completely with 10.5 g of N_2 in the following reaction?

$$N_2(g) + 3 \text{ H}_2(g) \longrightarrow 2 \text{ NH}_3(g) \qquad \blacksquare$$

Volume-Volume Calculations These calculations can be done much like weight-weight calculations (section 9.6), using the following four steps:

1. Write down the given volume.
2. Convert the given volume to moles using the ideal gas equation.
3. Convert moles of the first substance to moles of the second substance using the appropriate mole ratio.
4. Convert moles of the second substance to volume using the ideal gas equation.

Consider, as an example, the complete combustion of acetylene, C_2H_2.

$$2 \text{ C}_2\text{H}_2(g) + 5 \text{ O}_2(g) \longrightarrow 4 \text{ CO}_2(g) + 2 \text{ H}_2\text{O}(g)$$

What volume of O_2 at STP would be required by 37.9 L of C_2H_2 at STP? Using the four steps given above,

Step 1: Write down the given volume:

$$37.9 \text{ L C}_2\text{H}_2$$

Step 2: Convert volume of C_2H_2 to moles of C_2H_2 using the ideal gas equation:

$$PV = nRT$$

$$\text{moles of } C_2H_2 = n = \frac{PV}{RT}$$

$$= \frac{(1 \text{ atm})(37.9 \text{ L})}{(0.0821 \text{ L atm/mol K})(273 \text{ K})}$$

$$= 1.69 \text{ mol}$$

Step 3: Convert moles of C_2H_2 to moles of O_2 using the mole ratio 5 mol O_2/2 mol C_2H_2:

$$(1.69 \text{ mol } C_2H_2)\left(\frac{5 \text{ mol } O_2}{2 \text{ mol } C_2H_2}\right) = 4.22 \text{ mol } O_2$$

Step 4: Convert moles of O_2 to volume of O_2 using the ideal gas equation:

$$PV = nRT$$

$$V = \frac{nRT}{P}$$

$$= \frac{(4.22 \text{ mol})(0.0821 \text{ L atm/mol K})(273 \text{ K})}{1 \text{ atm}}$$

$$= 94.6 \text{ L}$$

Notice, however, that if we combine all of the steps,

$$\frac{(1 \text{ atm})(37.9 \text{ L})}{(0.0821 \text{ L atm/mol K})(273 \text{ K})}\left(\frac{5 \text{ mol } O_2}{2 \text{ mol } C_2H_2}\right)$$

$$\frac{0.0821 \text{ L atm/mol K})(273 \text{ K})}{(1 \text{ atm})} = 94.8 \text{ L } O_2$$

it is obvious everything cancels except

$$(37.9 \text{ L})\left(\frac{5 \text{ mol } O_2}{2 \text{ mol } C_2H_2}\right)$$

Indeed, since the molar volume at STP is the same for any gas, then moles and volume of any two gases are directly related. *Thus, at STP or any other constant conditions of temperature and pressure, the volume of the second gas can be calculated from the volume of the first gas simply by multiplying by the mole ratio.* This statement is a natural consequence of Avogadro's hypothesis. Knowing this, we can shorten the number of steps in volume-volume calculations to two:

1. Write down the given volume; include its units but do not label it.
2. Convert the given volume to volume of the second gas by multiplying by the appropriate mole ratio; do not use labels in the mole ratio.

Important relationships for stoichiometric calculations involving gases are summarized on the mole map of Figure 10-14.

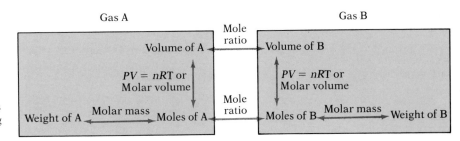

FIGURE 10-14 Important Relationships for Stoichiometric Calculations Involving Gases

EXAMPLE 10.17   What volume of O_2 would be required for the complete combustion of 37.9 L of C_2H_2 at the same conditions of temperature and pressure?

$$2 \ C_2H_2(g) + 5 \ O_2(g) \longrightarrow 4 \ CO_2(g) + 2 \ H_2O(g)$$

Solution   Step 1:   Write down the given volume with units but no label:

37.9 L

Step 2:   Convert the volume of C_2H_2 to volume of O_2 using the mole ratio 5 mol/2 mol:

$$(37.9 \ \text{L}) \left(\frac{5 \ \text{mol}}{2 \ \text{mol}} \right) = 94.8 \ \text{L}$$

EXERCISE 10.17   What volume of HCl would be produced by complete reaction of 44.3 L of CH_4 at the same conditions of temperature and pressure?

$$CH_4(g) + 4 \ Cl_2(g) \longrightarrow CCl_4(l) + 4 \ HCl(g)$$

SUMMARY

Gases are one of the three states of matter, and all gases exhibit certain properties. Gases have mass and occupy space. Gases also exert pressure, which is often measured with a barometer and expressed in units of **torr.** Gases can be compressed by applying external pressure. Lowering the temperature also causes gas volume to decrease, and heating gases causes them to expand. Gases have the ability to diffuse; that is, two or more gases can intermix completely in all proportions.

In 1662, Robert Boyle found that the pressure and volume of a given quantity of gas at constant temperature are inversely proportional **(Boyle's law).** A second gas law, **Charles' law,** states that the volume of a gas varies directly with its absolute temperature if its pressure and mass are constant. A variation of Charles' law states the relationship between gas pressure and temperature: at constant volume, the pressure of a gas is proportional to absolute temperature. In 1811 Avogadro hypothesized that gases are composed of particles and that each particle occupies a certain amount of space at a given temperature and pressure. **Avogadro's hypothesis** stated that equal volumes of gases contain equal numbers of molecules at the same temperature and pressure. A mole of gas occupies 22.4 L (the **molar volume**) at standard pressure and temperature **(STP).**

Avogadro's hypothesis can be combined with Boyle's law and Charles' law to give the **ideal gas equation,** $PV = nRT$. Avogadro's hypothesis is the basis for a general theory of gaseous matter, the **kinetic molecular theory,** which provides a conceptual foundation for understanding the observed properties of gases.

Each gas in a mixture behaves independently of the others present. Dalton's **law of partial pressures** states that each component of a gas mixture exerts its own **partial pressure** independent of the others and that the total pressure of the mixture is the sum of the partial pressures.

Gases have very low densities relative to liquids and solids. Gas densities are usually expressed in units of g/L. The density of a gas at any temperature and pressure can be calculated from the ideal gas equation. There are three types of stoichiometric calculations involving gaseous volume: **mole-volume, weight-volume,** and **volume-volume** calculations. Steps used in these calculations are summarized on pp. 243–246.

STUDY QUESTIONS AND PROBLEMS

(More difficult questions and problems are marked with an asterisk.)

PROPERTIES OF GASES

1. Define the following terms:
   a. Pressure   b. Torr   c. Compressibility   d. Diffusion
2. Describe the six major properties of gases.
3. Describe a way of demonstrating that gases have mass.
4. How does temperature affect gas density?
5. Explain why aerosol spray cans should not be heated.
6. A hot air balloon is simply an inverted sack with a heater located underneath the opening of the sack. Explain why a hot air balloon "flies."

VOLUME AND PRESSURE RELATIONSHIPS

7. Describe how the volume and pressure of a gas are related when the mass and temperature are constant.
8. Use Boyle's law to explain the movement of air during breathing.
9. Explain why air spontaneously flows out of a filled balloon unless the balloon is tightly closed.
10. A sample of gas has a volume of 360 mL at a pressure of 0.750 atm. If the temperature is constant, what volume in mL will the gas occupy at 1.00 atm?
11. The volume of a sample of gas is 4.00 L at 4.00 atm. If the temperature is constant, what will be the volume of the sample in L at each of the following pressures?
    a. 1.00 atm   b. 0.400 atm   c. 10.0 atm
12. The volume of a sample of gas is 200.0 mL at 1.00 atm. If the temperature is constant, what will be the pressure of the gas in atm at each of the following volumes?
    a. 250.0 mL   b. 100.0 mL   c. 1.00 L

13. The volume of a sample of gas is 25.0 L at 1.75 atm. If the temperature is constant, what will be the pressure of the gas in atm at each of the following volumes?
    a. 10.0 L   b. 35.0 L   c. 2.50 L
14. What pressure is required to compress 10.0 L of carbon dioxide at 1.00 atm to 1.00 L at constant temperature?

VOLUME AND TEMPERATURE RELATIONSHIPS

15. Describe how the volume and temperature of a gas are related when mass and pressure are constant.
16. Summarize Kelvin's observations of gaseous volume at low temperatures.
17. What is meant by absolute zero? Absolute temperature?
18. A sample of gas has a volume of 79.5 mL at 45° C. If the pressure is constant, what will be the volume of the sample in mL at 0° C?
19. The volume of a sample of gas is 100.0 mL at 90.0° C. If the pressure is constant, what will be the volume of the sample in mL at each of the following temperatures?
    a. 50.0° C   b. 145° C   c. −10.0° C
20. The average adult human lung capacity is 6.00×10^3 mL. What volume of air at a temperature of 68° F can be inhaled by an adult whose body temperature is 98.6° F?

PRESSURE AND TEMPERATURE RELATIONSHIPS

21. Describe how the pressure and temperature of a gas are related if mass and volume are constant.
22. A container is filled with gas at a pressure of 1.80 atm at 0° C. At what temperature in °C will the pressure of the gas be 2.50 atm?
23. A container is filled with a gas at a pressure of 3.00 atm at 25° C.
    a. What will be the gas pressure in atm at 100.0° C?
    b. At what temperature in °C will the gas pressure be 2.50 atm?
    c. At what temperature in °C will the gas pressure be 3.50 atm?
24. The pressure in an automobile tire is found to be 2280 torr when measured at 32° F in the winter. What will be the pressure of the same tire during the summer when the temperature is 110.0° F (assume there are no air leaks)?

COMBINING THE GAS LAWS

25. The volume of a sample of gas is 450.0 mL at 35° C and 1.10 atm. Calculate the volume of the gas in mL at STP.
26. A gas sample was heated from −5.0° C to 90.0° C, and the volume increased from 1.00 L to 3.00 L. If the initial pressure was 0.800 atm, what was the final pressure in atm?
27. The volume of a sample of gas is 155 mL at 30.0° C and 0.75 atm. Calculate the volume of the sample when its temperature is 45.0° C and its pressure is 0.50 atm.
28. The volume of a sample of gas is 800.0 mL at 80.0° C and 0.700 atm. At what temperature in °C will the sample have a volume of 1.00 L and a pressure of 1.00 atm?

29. State Avogadro's hypothesis and the two assumptions on which it is based.
30. Explain what *STP* means.
31. Explain what *molar volume* means.
32. What is the volume of 4.16×10^{20} molecules of carbon monoxide at STP?

THE IDEAL GAS EQUATION

33. State the ideal gas equation in words. Does it apply to real gases? Explain your answer.
34. At what pressure will 0.300 mol of N_2 have a volume of 10.0 L at 90.0° C?
35. How many moles of CO are present in 600.0 mL of the gas collected at 60.0° C and 1.25 atm?
36. A sample of N_2 has a volume of 47.3 L at 27° C and 1.50 atm. How many moles of N_2 are present? What is the mass of the nitrogen?
37. A neon sign contains 0.100 mol Ne at 2.00 torr and 25° C. What is the volume of Ne in the sign?
38. What volume will 10.0 g of CO_2 have at 25° C and 1.75 atm?

THE KINETIC MOLECULAR THEORY

39. Summarize the kinetic molecular theory.
40. Summarize the relationship between the kinetic molecular theory and the observed properties of gases.
41. Use the kinetic molecular theory to explain Boyle's law.
42. Use the kinetic molecular theory to explain Charles' law.
43. Account for deviations of real gases from ideal behavior.

GAS MIXTURES

44. Summarize Dalton's law of partial pressures.
45. Explain what is meant by the partial pressure of a gas.
46. In a cyclopropane-oxygen mixture used as a general anesthetic, the partial pressure of oxygen is 570 torr. If the total pressure of the mixture is 1.00 atm, what is the partial pressure of cyclopropane?
47. Gases inside the lungs of a typical adult have the following partial pressures:

$$O_2: 100.0 \text{ torr}$$
$$CO_2: 40.0 \text{ torr}$$
$$H_2O: 47.0 \text{ torr}$$
$$N_2: 573.0 \text{ torr}$$

What is the percentage of each gas inside the lungs?
48. If the breathing rate is slowed in a human, the partial pressure of CO_2 in the lungs is increased from 40.0 torr to 60.0 torr. Assuming that the partial pressures of other lung gas components (water vapor and N_2) do not change, what will be the effect on the partial pressure of oxygen

in the lungs if constant total pressure is maintained? The normal partial pressure of oxygen in the lungs is 100.0 torr.

GAS DENSITY

49. Why do gases have low densities compared to liquids and solids?
50. Calculate the density of each gas at STP.
   a. CO_2 **b.** Cl_2 **c.** SO_3
51. A sample of air has a volume of 5.00 L and has a density of 1.80 g/L. If the air is compressed to a volume of 1.00 L, what will be its density?

STOICHIOMETRIC CALCULATIONS INVOLVING GAS VOLUMES

52. In the following reaction,

$$2\ HCN(g) + NO_2(g) \longrightarrow C_2N_2(g) + NO(g) + H_2O(g)$$

how many mL each of HCN and NO_2 at STP are required to form 7.00 g of C_2N_2?
53. In the following reaction,

$$Al_4C_3(s) + 12\ H_2O(l) \longrightarrow 3\ CH_4(g) + 4\ Al(OH)_3(s)$$

what volume of CH_4 at STP would be obtained from 2.00 g of Al_4C_3?
54. In the complete combustion of octane (C_8H_{18}),

$$2\ C_8H_{18}(g) + 25\ O_2(g) \longrightarrow 16\ CO_2(g) + 18\ H_2O(g)$$

what volume of CO_2 is produced from 0.750 g of octane at 400.0° C and 10.0 atm?
55. The following reaction is a preliminary step in the commercial production of nitric acid:

$$4\ NH_3(g) + 5\ O_2(g) \longrightarrow 4\ NO(g) + 6\ H_2O(g)$$

What volume of NO would be produced by the complete reaction of 475 L of NH_3 if both gases are under the same conditions of pressure and temperature?

GENERAL EXERCISES

*56. How would you expect temperature to affect the rate of diffusion of a gas? Explain your answer.
*57. A typical weather balloon has a volume of 10.0 L. How many weather balloons at 27° C and 1.00 atm could be filled from a tank containing 12.0 L of helium at 150.0 atm and 27° C?
58. A sample of air had a volume of 2.00 L at −40.0° F. If the pressure was constant, what was the temperature of the air in °F when its volume was 2.50 L?
59. The temperature of 75.0 mL of a gas sample is 92° C. If the pressure is constant, what will be the temperature of the gas when it has each of the following volumes?
   a. 225 mL   **b.** 30.0 mL   **c.** 89.2 mL
60. Some aerosol spray cans will explode if their internal pressures exceed 3.0 atm. If an aerosol can has a pressure of 2.0 atm at 27° C,

would it be safe to leave it in an automobile on a hot summer day when the temperature of the automobile interior is 120° F?

61. A sample of gas has a volume of 1.00 L at 25° C and 1.50 atm. What will be the pressure in atm when the sample has a volume of 4.50 L and a temperature of 180° C?

62. Cyclopropane is a gas that is used as a general anesthetic. If 1.56 g of cyclopropane has a volume of 1.00 L at 0.984 atm and 50.0° C, what is the molar mass of cyclopropane?

63. A sample of a gas weighs 4.08 g and has a volume of 2.00 L at 0.850 atm and 32° C. What is the molar mass of the gas?

*64. A sample of neon collected over water at 28° C has a total pressure of 0.985 atm and a volume of 1.00 L. If all water vapor is then removed from the neon, and the neon is placed in a 2.00 L container at 45° C, what will be its pressure? (The partial pressure of water vapor at 28° C is 0.0373 atm.)

65. Carbon monoxide is converted to carbon dioxide by the following reaction:

$$2 \, CO(g) + O_2(g) \longrightarrow 2 \, CO_2(g)$$

What volume of O_2 is required for complete reaction of 10.0 mL of CO if both gases are under the same conditions of temperature and pressure?

LIQUIDS AND SOLIDS

OUTLINE

11.1 Intermolecular Forces
    Dipolar Attractions
    Hydrogen Bonds
    London Dispersion Forces
11.2 General Classes of Liquids
11.3 Densities of Liquids
11.4 Surface Tension
11.5 Viscosity
11.6 Vaporization
11.7 Boiling Point
11.8 Heat of Vaporization
11.9 The Solid State
    Crystalline Solids
    Amorphous Solids
11.10 The Melting Process
11.11 Heat of Fusion
11.12 Melting Point
    Perspective: Liquid Crystals
11.13 Calculations Based on Phase
        Changes (optional)
    Summary
    Study Questions and Problems

In addition to the gaseous state, matter exists in the liquid state and the solid state. A solid object has a definite shape and volume that cannot be easily changed, but liquids and gases are fluids, that is, they have the ability to flow and assume the shapes of their containers. When a liquid is poured from a container of one shape into a container of another shape, the shape of the liquid changes from that of the first container to that of the second container, but the liquid volume does not change. This behavior is in contrast to that of gases, which diffuse freely to fill a container of any size and thus to take on the shape of the container in all dimensions.

On the molecular level, properties of each of the three states of matter can be attributed to the space between particles and how much the particles move about, as illustrated in Figure 11-1. In gases, the particles (molecules or, in the case of the noble gases, atoms) are very far apart and have little attraction for each other. In contrast, the particles of liquids and solids are closer together and they interact with each other more than particles of gases do. Particles of solids are very close together and tightly packed, allowing for only minimal movement. In this chapter you will study the liquid and solid states of matter and explore the attractive forces that exist in these condensed states. To prepare for this study, you should review shapes of molecules (chapter 6), polar covalent bonds and polar molecules (chapter 6), density (chapter 2), and specific heat (chapter 2).

Intermolecular
Between molecules.

11.1 INTERMOLECULAR FORCES

The kinetic molecular theory applies not only to gases but also to liquids. Recall that gases are composed of widely separated particles (molecules) that have rapid, random motion. Temperature is a measure of the average kinetic energy of the particles. Thus, when a gas is cooled, the average kinetic energy of the particles is reduced, and they move more slowly. If cooling is continued, the molecules of a gas will coalesce into a liquid. This happens because gas molecules exert small, short-range attractions for each other, and if enough energy is lost, these attractions will cause the molecules to condense and form a liquid. The process in which a gas condenses into a liquid is called **liquefaction** or **condensation**. Liquefac-

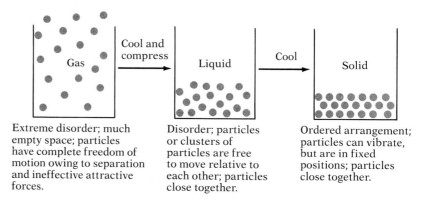

Gas — Cool and compress → Liquid — Cool → Solid

Extreme disorder; much empty space; particles have complete freedom of motion owing to separation and ineffective attractive forces.

Disorder; particles or clusters of particles are free to move relative to each other; particles close together.

Ordered arrangement; particles can vibrate, but are in fixed positions; particles close together.

FIGURE 11-1   The Three States of Matter

tion can also be brought about by applying extremely high pressures to a gas. The high pressures crowd the gas molecules closer to each other, reducing their movement and allowing the short-range attractions to draw the molecules closer together into the liquid state.

We can visualize a liquid as consisting of particles clustered closely together but moving freely about throughout the volume of the liquid. As in the case of gases, the average kinetic energy of the particles in a liquid is proportional to the temperature. Since virtually all liquids at ordinary temperatures are composed of molecules, we will study the kinds of forces that attract one molecule to another. These forces are called **intermolecular forces**.

Dipolar Attractions   The intermolecular forces between polar molecules are called **dipolar attractions**; they are simply attractions between oppositely charged ends of dipoles. One example of a polar molecule is the hydrogen chloride molecule, HCl. The positive end of one HCl molecule is attracted to the negative end of another because of their opposite partial charges. This attraction is illustrated below and in Figure 11-2.

$$\overset{\delta+}{H}-\overset{\delta-}{Cl} \qquad \overset{\delta+}{H}-\overset{\delta-}{Cl} \quad \overset{\delta+}{H}-\overset{\delta-}{Cl}$$

$$\underset{\delta-}{Cl}-\underset{\delta+}{H}$$

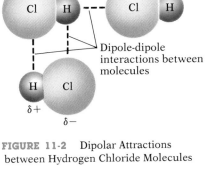

$\delta-$ $\delta-$
$\delta+$ $\delta+$

Cl   H ---- Cl   H

Dipole-dipole
interactions between
molecules

H   Cl

$\delta+$

$\delta-$

FIGURE 11-2   Dipolar Attractions between Hydrogen Chloride Molecules

Because there are only partial charges on atoms of polar molecules, dipolar attractions are not as strong as ionic attractions, and dipolar attractions operate over shorter distances than ionic attractions.

For molecules of approximately equal molar mass, the intermolecular forces increase with increasing polarity of the molecule. The strength of the intermolecular forces is reflected in the boiling points of liquids — the stronger the intermolecular attractions, the higher the boiling point. Table 11-1 illustrates this relationship.

Hydrogen Bonds   The polar attractions between molecules in which hydrogen is bonded to the highly electronegative atoms N, O, and F are much stronger than those in other molecules of similar formula weight and polarity. These unusually large intermolecular attractions are a special class of dipolar attractions. When hydrogen is bonded to a highly electronegative atom, the bond is very polar, and the bonding electrons

TABLE 11-1	The Boiling Points and Dipole Moments* of Selected Liquids			
Substance	Formula	Formula Weight	Boiling Point (°C)	Dipole Moment (D)
Methane	CH_4	16.0	-161.5	0
Ammonia	NH_3	17.0	-33.4	1.49
Bromine	Br_2	159.8	58.8	0
Iodine Chloride	ICl	162.4	97.8	0.65

* The dipole moment of a molecule is a measure of the polarity of the molecule.

TABLE 11-2	The Effect of Hydrogen Bonding on the Boiling Points of Liquids of Similar Formula Weights			
Compound	Formula	Formula Weight	Boiling Point (°C)	Dipole Moment (D)
Methane	CH_4	16.0	−161.5	0
Ammonia*	NH_3	17.0	−33.4	1.49
Hydrogen Fluoride*	HF	20.0	19.5	1.9
Water*	H_2O	18.0	100	1.84
Ethane	C_2H_6	30.0	−88.6	0
Fluoromethane	CH_3F	34.0	−78.4	1.81
Methyl amine*	CH_3NH_2	31.0	−6.7	1.23
Methanol*	CH_3OH	32.0	64.7	1.66

* Compounds capable of forming hydrogen bonds.

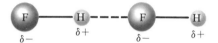

FIGURE 11-3 A Hydrogen Bond between Two Hydrogen Fluoride Molecules

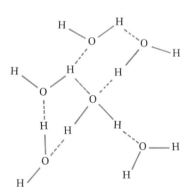

FIGURE 11-4 Hydrogen Bonds in Liquid Water

are drawn much closer to the electronegative atom. The hydrogen atom seems almost like a nucleus stripped of its electron, very much like a proton but without a full unit of positive charge. The hydrogen atom is thus strongly attracted to nonbonded electrons on electronegative atoms of nearby molecules. Because hydrogen is always involved in these attractions, they are called **hydrogen bonds** (H–bonds).

In HF, where hydrogen bonds are especially strong (remember that fluorine is the most electronegative element), the hydrogen behaves almost as if it were bonded to two fluorine atoms (see Figure 11-3). A hydrogen bond is represented by a dashed straight line between the hydrogen atom and an electronegative atom in another molecule.

Hydrogen bonds are also an important feature of liquid water, as illustrated in Figure 11-4. Each water molecule in the liquid is surrounded by other water molecules with their hydrogen and oxygen atoms oriented so that a maximum number of hydrogen bonds is formed. For this reason, water has an unusually high boiling point for a molecule of its size and mass.

Table 11-2 illustrates the effect of hydrogen bonds on boiling points; it compares compounds having similar formula weights, some of which form hydrogen bonds.

London Dispersion Forces   Even nonpolar molecules (which have no dipolar attractions) group together to form liquids. The intermolecular attractions between nonpolar molecules can be explained by the idea that the distribution of electrons in any molecule is always in a state of fluctuation (see Figure 11-5). The electron density cloud of a molecule continuously changes over time, and at any instant the electron density cloud is

FIGURE 11-5 Formation of an Instantaneous Dipole in a Diatomic Molecule. The small spheres represent atomic nuclei, and the shaded areas represent electron density clouds.

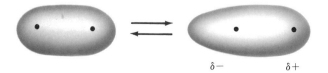

FIGURE 11-6 Synchronized
Instantaneous Dipoles of Two Diatomic
Molecules. The small spheres represent
atomic nuclei, and the shaded areas
represent electron density clouds.

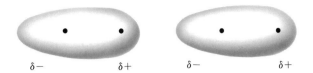

$\delta-$ $\delta+$ $\delta-$ $\delta+$

likely to be unevenly distributed about the molecule, resulting in an **instantaneous dipole**. The formation of an instantaneous dipole is facilitated by other nearby molecules. The electrons of one molecule repel electrons of a neighboring molecule, causing shifts in electron density of both molecules. Because these instantaneous dipoles are continuously changing, they cause the molecule to have only a temporary polarity that cannot be measured. Nonpolar molecules in a liquid synchronize the fluctuations of their instantaneous dipoles so that there is a net attraction among the molecules (see Figure 11-6). The attractions between instantaneous dipoles are called **London dispersion forces**, in honor of Fritz London, a German scientist who developed the theory for this attraction in 1930.

The strength of London dispersion forces depends on how easily the electron distribution in molecules can fluctuate. Since there are more electrons in larger molecules and they are farther away from the nuclei than in smaller molecules, the electron distribution in a larger molecule fluctuates to greater extremes than that in a smaller molecule. Therefore, the intermolecular attractions caused by London dispersion forces increase with molecular mass and size, as illustrated in Table 11-3, where the boiling points of some nonpolar compounds can be seen to increase with increasing formula weight.

London dispersion forces are, however, the weakest of all intermolecular forces, and, although they exist in all molecules, their effects are overshadowed if other attractions, such as dipolar forces, exist.

EXAMPLE 11.1

Classify each molecule by the type(s) of intermolecular attraction it would have for other molecules of the same kind. (There may be more than one kind of intermolecular force for some molecules.)

a. CH_4 b. HBr c. NH_3 d. CO_2

TABLE 11-3 Boiling Points and Formula Weights for Molecules of Approximately Equal Polarity

Compound	Formula	Formula Weight	Boiling Point (°C)
Methane	CH_4	16.0	−161.5
Ethane	C_2H_6	30.0	−88.6
Propane	C_3H_8	44.0	−42.1
Butane	C_4H_{10}	58.0	−0.5
Pentane	C_5H_{12}	72.0	36.1

Solution   To arrive at our answer, we must remember that there are three kinds of intermolecular forces: dipolar attractions, hydrogen bonds, and London dispersion forces. We must then determine the type(s) of intermolecular attraction exerted by a molecule on the basis of its structure. If the molecule is not polar, it can exert only London dispersion forces; molecules (*a*) (CH_4) and (*d*) (CO_2) fit this category. If the molecule is polar but does not have hydrogen attached to nitrogen, oxygen, or fluorine, the molecule will exert both London dispersion forces and dipolar attractions. Molecule (*b*) (HBr) is such a molecule. Finally, if the molecule is polar, with polar bonds between hydrogen and nitrogen, oxygen, or fluorine, the molecule will exert London dispersion forces and will form hydrogen bonds. Molecule (*c*) (NH_3) fits this category.

EXERCISE 11.1   Classify each molecule by the type(s) of intermolecular attraction it would exert for other molecules of the same kind. (There may be more than one kind of intermolecular force for some molecules.)

   a. CO   **b.** H_2S (bent shape)   **c.** H_2 **d.** HF   ■

11.2 GENERAL CLASSES OF LIQUIDS

Liquids can be divided into two general classes—polar and nonpolar—on the basis of their molecular structures. The molecules of nonpolar liquids are attracted to each other by only London dispersion forces; these liquids usually have low boiling points relative to their formula weights. Molecules of polar liquids are attracted to each other by dipolar forces as well as London dispersion forces, and because polar attractions are stronger than London dispersion forces, polar liquids tend to have higher boiling points than predicted by their formula weights. When the boiling points of polar and nonpolar liquids are compared for molecules of similar size, the polar liquids will have the higher boiling points. The most extreme kind of polar liquid is one in which there are hydrogen bonds. When comparing compounds of similar formula weights, hydrogen-bonded liquids show even higher boiling points than polar liquids that are incapable of forming hydrogen bonds.

11.3 DENSITIES OF LIQUIDS

Although a great deal is known about the properties of liquids, much less is known about the organization of molecules in liquids. It appears, however, that liquids contain small regions in which there is a fairly ordered arrangement of molecules and other regions in which the molecules are completely disordered. The ordered regions change as the molecules move about through the liquid. Because the molecules cannot pack very close together in the disordered regions, liquids almost always have lower densities than solids.

Water, however, is an exception. Liquid water has an unusually high density because each water molecule in the liquid state forms an average of three hydrogen bonds with neighboring water molecules. At best, molecules of most other liquids can form only one hydrogen bond each, and many cannot form any hydrogen bonds at all. Hence, the molecules of liquid water are packed together much more closely than molecules of most other liquids, giving liquid water an unusually high density.

Because of their lower densities, liquids that are insoluble in water float on the surface of liquid water. For example, salad oil floats on the water-based vinegar solution of some salad dressings. Differences in density also account for the formation of oil slicks after oil spills in the ocean.

11.4 SURFACE TENSION

Surface tension

The force acting on the surface of a liquid that tends to minimize the surface area.

The intermolecular forces among molecules of a liquid are responsible for **surface tension**, the property of a liquid that causes its surface to act as if it were a stretched elastic membrane. Surface tension causes water to form beads on a freshly waxed automobile. It also allows a glass of water to be slightly overfilled without spilling (Figure 11-7) and allows insects to "walk on water."

Surface tension is caused by an imbalance of forces at the surface of a liquid, as shown in Figure 11-8. The molecules in the interior of the liquid are attracted to other molecules in all directions. But those molecules on the surface of the liquid can experience attractions only by molecules on the sides and underneath. There are no attractions from above the surface of the liquid because there are no molecules of the liquid there. Because of this imbalance of forces, there is a net inward pull on the surface that contracts it almost as if it were a skinlike coating. Thus, a small amount of water on a waxy surface will form beads to minimize its surface area, and drops of a liquid are spherical for the same reason.

FIGURE 11-7 Because of Surface Tension, This Slightly Overfilled Glass of Water Has a Bulge on Top.

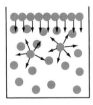

FIGURE 11-8 Surface Tension Is Caused by the Downward Attraction of Surface Molecules by Molecules in the Interior of the Liquid.

11.5 VISCOSITY

The resistance to flow exerted by a fluid is called **viscosity**. The higher the viscosity of a fluid, the more slowly it flows. Viscosity is related to the ease with which individual molecules can move past each other in fluids, and large, unsymmetrical molecules are likely to be more strongly attracted to each other than are small, compact molecules. Liquids with small, compact molecules have low viscosities. Water, for example, has a rather low viscosity. In contrast, liquids having large, unsymmetrical molecules flow slowly. Molasses and cooking oil are examples of highly viscous liquids. Motor oils are classified by viscosity (20 wt., 30 wt., 40 wt., and so forth). Viscosity increases as temperature decreases, since the ability to flow depends on the kinetic energy of the molecules of the liquid. For this reason, people who live in cold climates use low-viscosity motor oils so that the oil will flow even at the low temperatures. Table 11-4 gives viscosities of common liquids.

TABLE 11-4	Viscosities of Common Liquids at 20°C
Substance	Viscosity (kg/m-sec)
Benzene	0.65×10^{-3}
Castor Oil	1027.2×10^{-3}
Ethanol	1.20×10^{-3}
Ether	0.23×10^{-3}
Glycerol (Glycerin)	1490×10^{-3}
Mercury	1.55×10^{-3}
Olive Oil	100.8×10^{-3}
Water	1.00×10^{-3}

11.6 VAPORIZATION

Kinetic energy is related to velocity by the equation

$$\text{kinetic energy} = 1/2 \ mv^2$$

where m represents the mass of an object and v represents its velocity. The molecules of a liquid are continuously moving about at various speeds and thus possess a variety of kinetic energies.

At any given time, a few of the molecules have enough kinetic energy to break away from the surface of the liquid into the atmosphere above the liquid (see Figure 11-9). The process in which molecules pass from the liquid to the gaseous state is called **vaporization**. If vaporization significantly diminishes the volume of the liquid, the process is referred to as **evaporation**. If the liquid is in an open container, vaporization will continue until all of the liquid has been converted to vapor. Consider a sample of liquid water in an open container. As time passes, the volume of liquid water grows smaller and smaller, until the liquid water disappears; vaporization has then converted all of the liquid water to water vapor, and the vapor molecules are dispersed throughout the atmosphere. Thus the water has completely evaporated.

Although the motion of particles in the solid state is greatly restricted in comparison to the liquid state, molecules of some solids have enough energy to break away from the solid and become dispersed through the atmosphere. The process in which solid matter is converted directly to gaseous matter without passing through the liquid state is called **sublimation**. Although there are not many substances that sublime at standard conditions, two familiar examples are mothballs (*para*-dichlorobenzene) and "dry ice" (solid CO_2). These solid substances will evaporate completely if left in open containers.

To prevent evaporation of a liquid, we can store it in a tightly closed container. Then the molecules that escape from the surface of the liquid cannot disperse through the atmosphere; since they are moving randomly, they bombard the walls of the container and the surface of the liquid, and some of them will return to the liquid, aided by intermolecular attractions. Thus, in the closed container, molecules are always moving

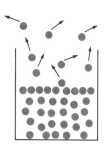

FIGURE 11-9 Vaporization of a Liquid

back and forth between the vapor phase and the liquid phase. When the liquid is first placed in the container, the number of molecules escaping from the liquid phase is greater than the number returning to the liquid phase. But as time goes on, the number of molecules in the vapor phase increases. As their number builds up, more and more molecules in the vapor phase are available to return to the liquid phase. Eventually, the number of molecules escaping from the liquid phase each second will be the same as the number of vapor-phase molecules recaptured by the liquid phase. From this time on, the liquid phase and vapor phase are in a state of equilibrium, as illustrated in Figure 11-10. Unless the container is very large compared to the initial volume of the liquid, no noticeable loss of liquid volume occurs.

The liquid-vapor equilibrium is a **dynamic equilibrium**—molecules are constantly moving between liquid and vapor, but the overall number of molecules in each phase remains constant. This dynamic equilibrium occurs whenever a liquid is stored in a closed container. The molecules that have escaped into the vapor phase behave as any gaseous substance and exert pressure if restricted to a constant volume. The pressure exerted by a vapor in equilibrium with its liquid is called the **vapor pressure** of the liquid.

The vapor pressure of a liquid depends on the strength of the intermolecular forces in the liquid. For example, liquid water has a rather low vapor pressure because of its extensive network of hydrogen bonds. Ethyl alcohol, which also contains hydrogen bonds,

$$CH_3OH \qquad CH_3OCH_3$$
ethyl alcohol   ethyl ether

has a higher vapor pressure than liquid water, because a molecule of ethyl alcohol cannot form as many hydrogen bonds as a molecule of liquid water can. Ethyl ether, sometimes used as a general anesthetic, is composed of bent polar molecules incapable of forming hydrogen bonds. Thus, the vapor pressure of ethyl ether is higher than that of ethyl alcohol. Table 11-5 gives the vapor pressures of these three liquids at various temperatures.

Substances with high vapor pressures tend to have low boiling points and evaporate easily. Such substances are said to be **volatile**. Ethyl ether is

Equilibrium
A state of balance due to equal action of opposing forces.

Dynamic equilibrium
A state of equilibrium in which there is continuous movement of particles in opposite directions in equal numbers per unit time.

 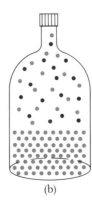

FIGURE 11-10 Vaporization of a Liquid in a Closed Container (a) Before Equilibrium and (b) After Equilibrium. The black spheres represent air molecules, and the colored spheres represent liquid and its vapor molecules.

(a)   (b)

TABLE 11-5   The Vapor Pressures of Liquid Water, Ethyl Alcohol, and Ethyl Ether at Various Temperatures

	Vapor Pressure (Torr)		
Temperature (°C)	Water[1]	Ethyl alcohol[2]	Ethyl ether[3]
0	4.6	12.2	185.3
10	9.2	23.6	291.7
20	17.5	43.9	442.2
30	31.8	78.8	647.3
35	42.2	102.9	760.0
40	55.3	135.3	921.3
50	92.5	222.2	1276.8
60	152.9	352.7	1729.0
70	233.7	542.5	2296.0
78	327.3	760.0	2654.7
80	355.1	812.6	2993.6
90	525.8	1187.1	3841.0
100	760.0	1693.3	4859.4

[1] Normal boiling point 100° C
[2] Normal boiling point 78° C
[3] Normal boiling point 35° C

quite volatile and is also highly flammable; thus, it should always be stored in a tightly closed container.

Solid substances, except for those that sublime, have extremely low vapor pressures.

11.7 BOILING POINT

As illustrated in Table 11-5 and Figure 11-11, the vapor pressure of a substance increases with temperature. As temperature increases, more and more molecules in the liquid phase acquire enough energy to overcome intermolecular forces and escape to the vapor phase. If a liquid is heated in an open container, vapor formation is opposed by the pressure exerted by the atmosphere. In order for the liquid to boil, the liquid's vapor pressure must overcome the effects of the atmospheric pressure. In other words, the **boiling point** of the liquid is the temperature at which the vapor pressure of the liquid is the same as the pressure of the atmosphere.

Notice the values for the vapor pressures of liquid water, ethyl alcohol, and ethyl ether at their boiling points in Table 11-5. For example, boiling water has a vapor pressure of 760 torr at sea level. But because atmospheric pressure decreases with altitude, the boiling point of water (or any other liquid) will also decrease with altitude. For this reason, water boils at 95° C in Denver, Colorado, where the elevation is 1 mile above sea level. Table 11-6 shows the decrease in the boiling point of water at elevations above sea level.

When a liquid is boiled in a sealed container, the pressure inside the container increases as the liquid vaporizes because of the increasing num-

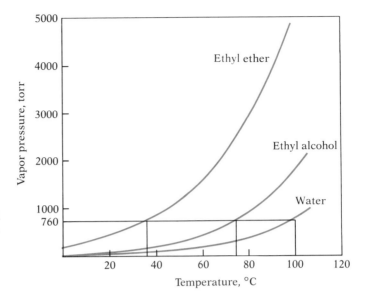

FIGURE 11-11 The Increase in Vapor Pressure with Temperature. The normal boiling points of water, ethyl alcohol, and ethyl ether are the temperatures at which their vapor pressures are equal to atmospheric pressure: 100° C for water, 78° C for ethyl alcohol, and 35° C for ethyl ether.

FIGURE 11-12 A Pressure Cooker

ber of vapor molecules inside the container. The resulting increased pressure raises the boiling point of the liquid. For example, water boils at 120° C when the pressure is 2 atm. This increase in boiling point with increasing pressure is the operating principle of a pressure cooker (Figure 11-12).

11.8 HEAT OF VAPORIZATION

If we heat water at its boiling point in an open container, boiling continues until all of the liquid water is converted to water vapor. However, if we measure the temperature of the boiling water at any time during the process, it is always the same, even though we are continually adding heat. Where has the heat gone, and why does the temperature remain constant? The answer is that the heat is used to overcome the intermolecular forces among the liquid molecules, allowing them to become vapor molecules. Heat used in this way does not increase the average kinetic energy of the molecules, and the temperature of the liquid does not change. The amount of heat required to vaporize one mole of a boiling liquid under one atmosphere of pressure is called the **molar heat of vaporization** (ΔH_{vap}). (The symbol Δ is the capital Greek letter *delta*; it means "change," in this

TABLE 11-6 Variation of the Boiling Point of Water with Pressure		
Altitude Above Sea Level (ft)	Atmospheric Pressure	Boiling Point of Water (°C)
0	760 torr	100
5,280 (Denver, Colo.)	630 torr	95
14,500 (Mt. Whitney, Calif.)	450 torr	86
29,000 (Mt. Everest)	253 torr	71

TABLE 11-7   Molar Heats of Vaporization

Substance	Formula	Formula Weight	Boiling Point (°C)	ΔH_{vap} kJ/mol	ΔH_{vap} kcal/mol
Chloroform	$CHCl_3$	119.5	61.7	31.9	7.62
Ethyl alcohol	C_2H_5OH	46.0	78.5	40.5	9.67
Glycerol (glycerin)	$C_3H_8O_3$	92.0	290	76.2	18.2
Mercury	Hg	200.6	356	56.9	13.6
Methane	CH_4	16.0	−161.5	8.16	1.95
Water	H_2O	18.0	100	40.6	9.71

case "change in heat.") In the past, molar heats of vaporization were usually expressed in units of kcal/mol, but the SI units of kJ/mol (kilojoules per mole) are now more popular. (1 J = 4.184 cal; 1 kJ = 4.184 kcal).

The molar heat of vaporization is a characteristic of a given liquid. Its magnitude is related to the type of intermolecular forces in the liquid. For example, water, with its extensive network of hydrogen bonds, has a molar heat of vaporization of 40.6 kJ/mol. On the other hand, methane molecules (CH_4) exert only London dispersion forces, and thus methane has a molar heat of vaporization of only 8.16 kJ/mol. Table 11-7 lists molar heats of vaporization for various liquids.

Our bodies make practical use of water's high molar heat of vaporization when we perspire. Perspiration, being mostly water, evaporates from our skin and absorbs large quantities of heat from our bodies; this heat loss cools our bodies. This same evaporative cooling effect accounts for the lower air temperature found near lakes and oceans.

Although molar heat of vaporization is defined as the heat *required* to vaporize one mole of a liquid, ΔH_{vap} is also the amount of heat *lost* by one mole of vapor when it condenses to the liquid state. When enough heat energy is removed from a vapor by cooling, the molecules coalesce to form liquid. For a given mass of vapor, the amount of heat energy lost in this process is identical to the amount of heat energy gained by the same mass of liquid when it vaporizes.

EXAMPLE 11.2   Calculate the amount of heat in kilojoules required to vaporize 1.00 g of liquid water at 100.0° C.

Solution   From Table 11-7, we see that the molar heat of vaporization for liquid water is 40.6 kJ/mol. In order to find the amount of heat needed to vaporize 1.00 g of liquid water, we must first find the number of moles in 1.00 g of H_2O:

$$(1.00 \text{ g } H_2O)\left(\frac{1 \text{ mol } H_2O}{18.0 \text{ g } H_2O}\right) = 0.0556 \text{ mol } H_2O$$

Then we can calculate the heat (number of kilojoules) needed to vaporize 0.0556 mol H_2O at its boiling point (100.0°C):

$$\text{Heat required} = (0.0556 \text{ mol } H_2O)\left(\frac{40.6 \text{ kJ}}{\text{mol } H_2O}\right) = 2.26 \text{ kJ}$$

Note that we could have combined the two steps as

$$\text{Heat required} = (1.00 \text{ g H}_2\text{O})\left(\frac{1 \text{ mol H}_2\text{O}}{18.0 \text{ g H}_2\text{O}}\right)\left(\frac{40.6 \text{ kJ}}{\text{mol H}_2\text{O}}\right) = 2.26 \text{ kJ}$$

EXERCISE 11.2   Calculate the amount of heat in kilojoules needed to vaporize 2.50 g of mercury at its boiling point.   ■

11.9 THE SOLID STATE

If a liquid is cooled, enough energy can be removed to cause the liquid to freeze into a solid. In the solid state of matter, particles occupy fixed positions in space, with their only movement being slow, gentle vibration about these fixed positions. Because of this limited movement of particles, solid matter maintains a rigid shape and constant volume. There are two major types of solid matter, crystalline solids and amorphous solids.

Crystalline Solids   Solids that exist in the form of crystals are called **crystalline solids**. Crystals are composed of ions, molecules, or atoms located at definite positions in the crystal lattice; they are held together by forces that we refer to as **crystal lattice forces**.

Although we tend to think of crystals as perfectly ordered arrays of particles, real crystals contain imperfections and defects that occur during crystal formation, such as the defect shown in Figure 11-13. If a crystal is allowed to form relatively slowly, the number of defects will be minimal and the crystal will have a large size. On the other hand, if a crystal forms relatively quickly, it is more likely to have imperfections and a small size. The situation is similar to a crowd of people finding their reserved seats in a theatre. If the people are admitted all at once just before the performance, many will be misplaced initially; but if they are allowed to take their seats over a longer period of time, confusion will be minimized.

The fact that small, imperfect crystals are formed by rapid cooling is used to great advantage by the frozen food industry. Food is usually quick-frozen so that the ice crystals formed will be very small. Slow freezing would cause the growth of large ice crystals, which would rupture the food cells, spilling their contents and reducing the quality of the flavor and the

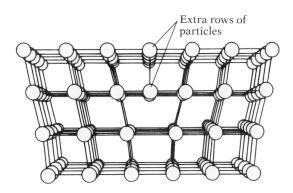

FIGURE 11-13   One Type of Crystal Defect

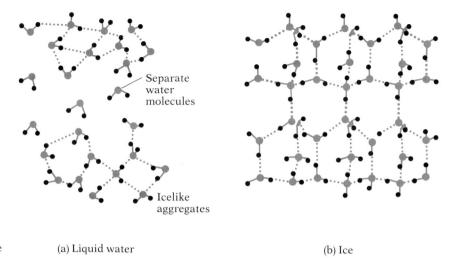

Separate water molecules

Icelike aggregates

(a) Liquid water

(b) Ice

FIGURE 11-14 The Arrangement of Molecules in (a) Liquid Water and (b) Ice

appearance of the product. Quick-freezing avoids this problem and produces frozen food of better taste and appearance.

The most common types of crystals are composed of ions or molecules, but some crystals have single atoms located at lattice positions. There are five major kinds of crystals.

1. **Ionic crystals** are composed of ions held at lattice points by electrical forces of attraction between cations and anions. Since ionic bonds are strong, ionic crystals have high melting points (see Table 6-1), usually higher than 300° C.

2. **Molecular crystals** are composed of molecules held at lattice points by intermolecular forces (London dispersion forces, dipolar attractions, hydrogen bonds). Most molecular elements and compounds form molecular crystals when they solidify. Figure 11-14 compares the arrangement of molecules in liquid water and in ice crystals. The intermolecular forces that hold molecular crystals together are weaker than ionic forces, so melting points of molecular crystals are low (usually lower than 300° C) compared to those of ionic crystals.

3. A **covalent crystal** is a network of atoms held at lattice points by covalent bonds. Structurally each covalent crystal is a giant molecule composed of billions of atoms. A familiar example of a covalent crystal is graphite, a form of carbon (Figure 11-15). Because of its layered planes of carbon atoms, graphite is used as a solid lubricant. It has a slippery feeling because it flakes easily as the planes of atoms slip across one another. Graphite is also used for pencil "lead" because its black color rubs onto paper easily. Because of their extensive networks of covalent bonds, covalent crystals are often extremely hard and have very high melting points. For example, diamond, another form of carbon and an example of a covalent crystal, melts above 3500° C, and graphite does not melt but passes directly into the vapor state at 4000° C.

4. **Metallic crystals** are composed of metal atoms arranged in a crystal lattice. The forces that hold the metal atoms in place are unlike any

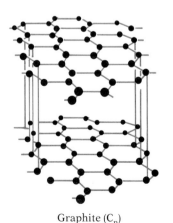

Graphite (C_n)

FIGURE 11-15 The Crystal Structure of Graphite

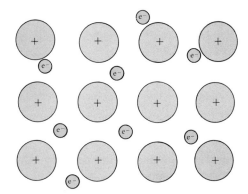

FIGURE 11-16 Metallic Crystals: Cations Embedded in a "Sea of Electrons"

discussed up to this point. Because there are not enough valence electrons in a metal atom to allow a network of covalent bonds, the valence electrons of metal atoms do not form definite bonds between any two atoms in a metallic crystal. Instead, the valence electrons bind each atom to many of its neighbors without actually forming bonds. Metallic crystals are considered to consist of cations embedded in a "sea of electrons," as illustrated in Figure 11-16. Metals conduct electricity because an electron in a metallic crystal can easily move from its original position to a nearby site. Under the influence of an electric current, electrons in metallic crystals move from one atom to another, carrying the electric current through the crystals.

5. **Atomic crystals** are composed of single nonmetallic atoms located at lattice positions. Only the noble gases can form atomic crystals, and they do so when cooled sufficiently. Atomic crystals are held together by London dispersion forces.

Amorphous

(ah-MOR-fus): lacking definite shape or form.

Amorphous Solids Extremely fast cooling of a liquid sometimes results in the formation of amorphous solids. **Amorphous solids** do not have ordered arrays of particles and thus do not exist as crystals. One familiar example of an amorphous solid is glass. Glass is composed largely of silicon dioxide (SiO_2) obtained from sand. Glass is manufactured by heating purified sand to very high temperatures until it melts. It is then cooled quickly into sheets or other shapes. When glass breaks, it does not form any particular geometric pattern as a crystalline material would. Because of its disordered arrangement of SiO_2 molecules (Figure 11-17), glass will not hold its shape completely over a long period of time — windowpanes

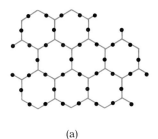
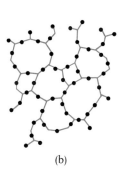

FIGURE 11-17 A Comparison of the Arrangement of Silicon Dioxide Molecules in (a) Silicon Dioxide Crystals and (b) Glass

(a)   (b)

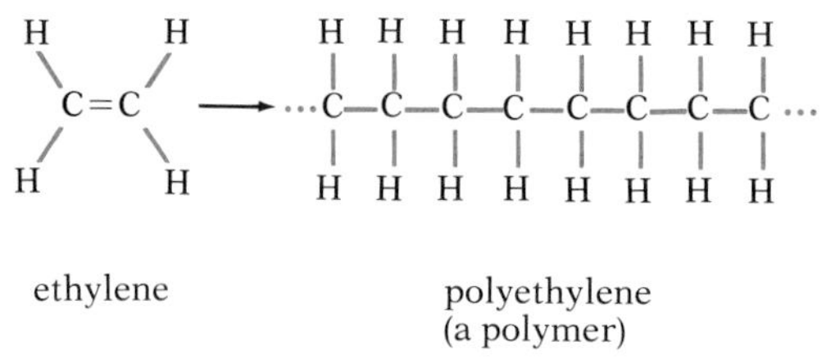

FIGURE 11-18 The Formation of Polyethylene from Ethylene

ethylene   polyethylene
                              (a polymer)

made in the eighteenth century now show evidence of glass flowing from top to bottom.

The synthetic plastics are also amorphous solids. These materials are formed when small molecules react to produce enormous molecules called polymers. A **polymer** is a large molecule composed of many repeating units. Polyethylene, for example, is a long, chainlike molecule formed from hundreds or even thousands of ethylene units (see Figure 11-18). These large molecules easily become tangled and are unable to pack into an orderly arrangement in the solid state. The disorder of molecules in solid plastics like polyethylene causes the plastics to be deformable, allowing them to be molded easily.

11.10   THE MELTING PROCESS

If heat is added to a crystalline solid, the particles within the solid acquire more and more energy until they are able to disrupt the crystal lattice forces. The process in which the crystal lattice is disrupted so that the particles move about freely and the solid becomes a liquid is called **melting**. After melting, the particles have more movement than they had in the solid state but less than they would have in the vapor state.

When solid and liquid phases of a substance are maintained at the melting point, the particles are in dynamic equilibrium, in a situation similar to liquid and vapor phases at the boiling point. Particles are moving from the liquid onto the crystal surfaces at the same rate that particles are passing from the crystal into the liquid.

11.11   HEAT OF FUSION

Fusion

The process of particles blending together at elevated temperatures.

As a crystalline solid is heated, its temperature increases until the melting point is reached. The **melting point** is the temperature at which liquid and solid are in equilibrium. At that point, the temperature remains constant until all of the solid is converted to liquid, even though heat is being added continuously. This situation is similar to the conversion of a liquid to a vapor. The amount of heat needed to melt one mole of the crystalline solid at its melting point is called the **molar heat of fusion** (ΔH_{fus}).

The molar heat of fusion is usually much smaller than the molar heat of vaporization of a substance. The heat of fusion is the amount of energy needed to disrupt the crystal lattice but still leave the particles in contact

TABLE 11-8   Molar Heats of Fusion

Substance	Formula	Formula Weight	Melting Point (°C)	ΔH_{fus} kJ/mol	ΔH_{fus} kcal/mol
Benzene	C_6H_6	78.0	5.5	9.83	2.35
Decane	$C_{10}H_{22}$	142.0	−29.7	28	6.8
Iodine	I_2	253.8	113.6	17	4.0
Naphthalene	$C_{10}H_8$	128.0	80.2	19	4.5
Water	H_2O	18.0	0	5.98	1.43

with each other, but a much larger amount of energy is required to vaporize the liquid because the attraction between particles must be completely overcome. Like heats of vaporization, heats of fusion are also applied to the reverse process; that is, cooling a liquid until it solidifies (freezes) involves removal of an amount of energy equal to the amount of energy needed to melt the substance. Thus, when a liquid is cooled to its freezing point (which is the same temperature as the melting point of the solid), an additional amount of energy must be lost by the liquid before it will solidify. Molar heats of fusion for a variety of substances are given in Table 11-8.

EXAMPLE 11.3   Calculate the amount of heat in kilojoules necessary to melt 62.0 g of ice at 0° C.

Solution   The molar heat of fusion of ice is given in Table 11-8 as 5.98 kJ/mol H_2O. Thus, we solve the problem as follows:

$$\text{Heat required} = (62.0 \text{ g } H_2O)\left(\frac{1 \text{ mol } H_2O}{18.0 \text{ g } H_2O}\right)\left(\frac{5.98 \text{ kJ}}{\text{mol } H_2O}\right) = 20.6 \text{ kJ}$$

EXERCISE 11.3   Calculate the amount of heat in kilojoules needed to melt 15.0 g of benzene at its melting point. ▪

11.12   MELTING POINT

Conversion of pure crystalline solids to the liquid state occurs over a very narrow temperature range, usually one or two Celsius degrees; this narrow temperature range is called the **melting point.** The melting point is sharp because the individual forces holding a crystal together are approximately equal, and all of the particles achieve the amount of energy necessary to disrupt these forces at very nearly the same temperature. Melting points are related to the strength of crystal lattice forces; thus, molecular crystals have much lower melting points than ionic crystals.

The story is somewhat different for amorphous solids. Because the forces that hold amorphous solids together are of many different types and strengths, the particles require a large range of energy to disrupt these forces. Thus, amorphous solids do not have sharp melting points. Instead, they slowly soften as they are heated until eventually a free-flowing liquid

Certain molecular crystals do not melt directly when heated but instead turn from a crystalline solid into what is called a liquid crystal. The liquid crystal state is a temporary one, and with further heating the liquid crystal completely melts to the liquid state. The liquid crystal state has some characteristics of a crystalline solid and some of a liquid. For example, liquid crystals reflect light like crystals, but they flow like liquids. This combination of crystalline and liquid properties gives liquid crystals unique characteristics.

One of the more interesting features of certain liquid crystals is that they react to an electric field by turning from clear to opaque; these liquid crystals have been used in the optical displays of clocks, wristwatches, and calculators. Other liquid crystals react to small changes in temperature by changing color; these liquid crystals are used in color-indicating thermometers to measure air temperature. They are also components of a temperature-indicating paste used to locate blood vessels. When the paste is applied to the skin over a blood vessel, the image of the blood vessel appears in the paste because the blood vessel is warmer than the tissue surrounding it. This facilitates location of a vein by doctors or nurses who need to give an injection or withdraw a blood sample.

is produced. For some amorphous solids such as ordinary glass, the temperature range over which softening begins and melting is complete may be several hundred Celsius degrees.

11.13 CALCULATIONS BASED ON PHASE CHANGES (optional)

When a pure substance is heated, its temperature increases according to its specific heat unless it is undergoing a phase change such as changing from solid to liquid. Its temperature remains constant during the time it is changing phases. Figure 11-19 shows the temperature of a sample of pure water as a function of time as heat is added at a constant rate. This diagram is called a **heating curve**. Beginning at $-20°$ C, as the ice is heated, its temperature rises according to its specific heat until the melting point, $0°$ C, is reached. At this point, the temperature levels off and remains constant until all the ice is melted. Then, as heating continues, the temperature of the liquid water rises according to its specific heat until the boiling point, $100°$ C, is reached. At this point, there is another temperature plateau, corresponding to vaporization of the liquid water. When all of the liquid water is converted to steam, the temperature rises again.

We can calculate the heat energy associated with heating curves by using specific heat and molar heats of fusion and vaporization. This is illustrated in the following example.

EXAMPLE 11.4

Calculate the amount of heat in kilojoules required to convert 1.00 mole of ice at $-10.0°$ C to steam at $100.0°$ C.

Solution

In this problem, we are asked to determine the total heat that must be added to 1.00 mole of ice at $-10.0°$ C to

1. heat the ice from $-10.0°$ C to its melting point, $0°$ C,
2. convert the ice to water at $0°$ C,

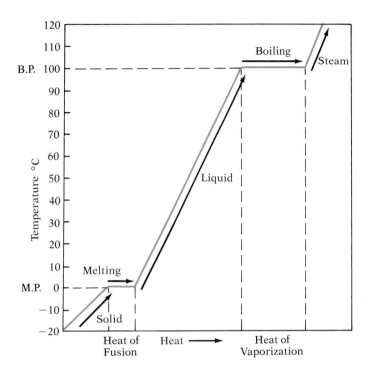

FIGURE 11-19 The Heating Curve for Water

3. heat the water from 0° C to its boiling point, 100.0° C, and
4. convert the water to steam at 100.0° C.

Step 1:   First, we must consult Table 2-5 for the specific heat of ice; then we can calculate the heat required to heat the ice from −10° C to 0° C:

$$\text{Heat required} = \left(\frac{0.492 \text{ cal}}{\text{g H}_2\text{O °C}}\right)(18.0 \text{ g H}_2\text{O})(10.0° \text{ C}) = 88.6 \text{ cal}$$

Then, we must convert cal to kJ:

$$(88.6 \text{ cal})\left(\frac{4.184 \text{ J}}{\text{cal}}\right)\left(\frac{1 \text{ kJ}}{1000 \text{ J}}\right) = 0.371 \text{ kJ}$$

Step 2:   Next, we calculate the amount of heat required to melt the ice at 0° C by using the molar heat of fusion (Table 11-8):

$$\text{Heat required} = (1.00 \text{ mol H}_2\text{O})\left(\frac{5.98 \text{ kJ}}{\text{mol H}_2\text{O}}\right) = 5.98 \text{ kJ}$$

Step 3:   Then, we calculate the amount of heat required to heat the water from 0° C to 100.0° C by using the specific heat of liquid water (Table 2-5):

$$\text{Heat required} = \left(\frac{1.00 \text{ cal}}{\text{g H}_2\text{O °C}}\right)(18.0 \text{ g H}_2\text{O})(100.0° \text{ C})$$

$$= 1800 \text{ cal or } 1.8 \times 10^3 \text{ cal}$$

and we must also convert cal to kJ:

$$(1.80 \times 10^3 \text{ cal})\left(\frac{4.184 \text{ J}}{1 \text{ cal}}\right)\left(\frac{1 \text{ kJ}}{1000 \text{ J}}\right) = 7.53 \text{ kJ}$$

MATH TIP
The answer has only one decimal place since step 4 has only one decimal place, 40.6 kJ. When adding, the answer must contain the same number of decimal places as that term with the least number of decimal places. Calculation yields 54.481 kJ so the answer is rounded off to 54.5 kJ.

Step 4:   Then, we calculate the amount of heat required to vaporize the water at 100.0° C by using the molar heat of vaporization (Table 11-7):

$$\text{Heat required} = (1.00 \text{ mol } H_2O)\left(\frac{40.6 \text{ kJ}}{\text{mol } H_2O}\right) = 40.6 \text{ kJ}$$

Step 5:   Finally, we add together all of the heat requirements:

Step 1:	0.371 kJ
Step 2:	5.98 kJ
Step 3:	7.53 kJ
Step 4:	40.6 kJ
Total	54.5 kJ

Thus, the entire process requires 54.5 kJ of heat.

EXERCISE 11.4   Calculate the amount of heat in kilojoules required to convert 10.0 g of ice at −5.0° C to steam at 100.0° C.   ■

The underlying principles of phase changes have applications in the fields of medicine and food preservation. One example is the technique of freeze-drying. In this process, a water-containing substance is frozen, and then its container is evacuated to very low pressures, in the range of 0.1 torr. This extremely low pressure allows the ice to sublime, removing the water from the substance and avoiding the heat damage that would occur if the liquid water were evaporated at higher temperatures. Freeze-drying is used extensively to prepare vaccines and antibiotics and to store skin and blood plasma. It has also gained wide acceptance as a technique for producing instant coffee and nonperishable dried foods and has been used to preserve food used in the NASA space program.

SUMMARY

Molecules are attracted to each other by various **intermolecular forces. Dipolar attractions** occur between molecules possessing permanent polarity. **London dispersion forces** are very weak attractions resulting from alignment of instantaneous dipoles that exist in all molecules. **Hydrogen bonds,** the strongest of all intermolecular forces, involve molecules that have hydrogen atoms bonded to nitrogen, oxygen, or fluorine atoms.

There are two general classes of liquids: polar and nonpolar. Liquids seem to contain small regions of orderly molecular arrangement and other regions in which the molecules are disordered; these regions change as the molecules move about in the liquid. Because of the relatively loose packing of molecules in the disordered regions, liquids are usually less dense than solids. **Surface tension** and **viscosity** are properties that depend on the structure of liquid molecules.

Vaporization occurs because some liquid molecules have enough energy to break away from the liquid surface and become vapor molecules. If a liquid is stored in a closed container, dynamic equilibrium between the vapor and liquid phases will occur. The pressure of vapor molecules in equilibrium with liquid is called the **vapor pressure** of the liquid. When the temperature of a liquid reaches the **boiling point,** additional **heat of vaporization** must be added to convert the liquid to vapor. A boiling liquid has a vapor pressure equal to atmospheric pressure.

Solid matter consists of particles packed very closely together and occupying fixed positions in space. When these particles occupy positions in a highly ordered arrangement, the solid is a **crystalline solid.** If the particles are in disarray, the solid is said to be an **amorphous solid.** The forces that hold solid matter together may be ionic attractions, dipolar attractions, London dispersion forces, hydrogen bonds, covalent bonds, or the electrical forces that exist in metallic crystals. Addition of enough heat to disrupt the attractive forces in solid matter causes **melting.** When the solid and liquid phases of a substance are maintained at the melting point, dynamic equilibrium occurs. When the temperature of a solid reaches the melting point, additional **heat of fusion** must be added to melt the solid.

When a substance is heated or cooled, its temperature changes until a phase change occurs. At the melting and boiling points, the temperature remains constant until the phase change is complete, even though heat is being added to or taken away from the substance continuously.

STUDY QUESTIONS AND PROBLEMS

(More difficult questions and problems are marked with an asterisk.)

INTERMOLECULAR FORCES

1. Define the following terms:
   a. Liquefaction
   b. Condensation
   c. Dipolar attractions
   d. London dispersion forces
   e. Hydrogen bonds
2. Describe the three kinds of intermolecular forces.
3. Predict the type(s) of intermolecular forces that would be exerted by each of the following molecules for others of the same kind. (There may be more than one kind of intermolecular force for some molecules.)
   a. CCl_4
   *b. CH_3-OH
   c. Ne (a one-atom molecule)
   d. CH_3-CH_3
   e. N_2
   *f. CH_3-NH_2

GENERAL CLASSES OF LIQUIDS

4. What are the two general classes of liquids? What is the basis of the classification?
5. Why do nonpolar liquids have lower boiling points than polar liquids?
6. Why do liquids with hydrogen bonds have higher boiling points than polar liquids without hydrogen bonds?

DENSITIES OF LIQUIDS

7. Why are liquids usually less dense than solids?
8. Why is water unusually dense in comparison to other liquids?
9. Why does oil float on water?

SURFACE TENSION

10. Explain what *surface tension* means.
11. Why is the surface tension of a liquid affected by the molecular structure of the liquid?
*12. Why does water coat a clean glass surface instead of forming beads on the glass? (Hint: Consider the relative forces of attraction among water molecules and those between water molecules and glass molecules.)

VISCOSITY

13. Explain what *viscosity* means.
14. How is the viscosity of a liquid affected by the molecular structure of the liquid?
15. Describe the effect of temperature on viscosity.

VAPORIZATION

16. Define the following terms:
    a. Vaporization   d. Dynamic equilibrium
    b. Evaporation   e. Vapor pressure of a liquid
    c. Sublimation   f. Volatile
17. Describe the dynamic equilibrium that exists in a closed container between a liquid and its vapor.
18. Why is vapor pressure affected by intermolecular forces?
19. Why does water have a lower vapor pressure than ethyl alcohol or ethyl ether at a given temperature?
20. Why is it especially necessary to store ethyl ether in a tightly closed container?

BOILING POINT

21. How is the vapor pressure of a liquid affected by temperature?
22. Explain what is meant by the *boiling point* of a liquid.
23. How does the boiling point of a liquid vary with atmospheric pressure? What is the reason for the variation?
24. Explain why the boiling point of water is lower than 100° C on a mountain where the atmospheric pressure is 610 torr.
25. Water has a higher boiling point than most other liquids of similar formula weight. Why?

HEAT OF VAPORIZATION

26. Why does the temperature of a boiling liquid remain constant, even though heat is being added?
27. Explain what is meant by *molar heat of vaporization* of a liquid.
28. How is the molar heat of vaporization of a liquid related to intermolecular forces in the liquid?
29. Water has a high molar heat of vaporization in comparison to other liquids. Why?
30. Explain how perspiration cools the human body.

31. Calculate the amount of heat in kilojoules needed to vaporize 3.75 g of H_2O at 100.0° C.
32. Calculate the amount of heat in kilojoules needed to vaporize 3.00×10^{23} molecules of liquid water at 100.0° C.

THE SOLID STATE

33. Why does solid matter maintain a rigid shape and constant volume?
34. Distinguish between the two major types of solid matter.
35. Why do slowly formed crystals contain fewer defects than quickly formed crystals?
36. Name the five types of crystals and describe the structure of each.
37. Why do molecular crystals usually have lower melting points than ionic crystals?
38. Why is graphite slippery?
39. Why do covalent crystals have such high melting points?
40. Name five types of crystal lattice forces and give an example of each.
41. Give two causes for the disordered internal structures of amorphous solids.
42. Explain why glass is classified as an amorphous solid.
43. What is a polymer? Why are some polymers classified as amorphous solids?

THE MELTING PROCESS

44. Describe what happens to the particles in a solid when it melts.
45. Describe the dynamic equilibrium that exists between a liquid and the corresponding solid at the melting point.

HEAT OF FUSION

46. Explain what is meant by the *melting point* of a solid.
47. Why does the temperature of a melting solid remain constant, even though heat is being added?
48. Explain what is meant by the *molar heat of fusion* of a solid.
49. Why is the molar heat of fusion of a substance smaller than the molar heat of vaporization of the substance?
50. How is the molar heat of fusion of a solid related to crystal lattice forces in the solid?
51. Calculate the amount of heat in kilojoules needed to melt 50.0 g of ice at 0.0° C.
52. Melting 15.0 g of benzene, C_6H_6, at 5.5° C requires 1.892 kJ of heat. Calculate the molar heat of fusion of benzene.

MELTING POINT

53. Why do pure crystalline solids have sharp melting points?
54. How are melting points related to crystal lattice forces?
55. Why do amorphous solids melt over broad temperature ranges?

CALCULATIONS BASED ON PHASE CHANGES

56. Sketch a heating curve to describe heating one mole of water from −25° C to 125° C.

57. Calculate the amount of heat in calories required to convert 25.0 g of ice at $-15°$ C to steam at $100.0°$ C.

58. Calculate the amount of heat in calories needed to convert 4.60×10^{23} molecules of ice at $-8.0°$ C to steam at $100.0°$ C.

59. Use the following data for oxygen to calculate the heat (kJ) liberated when 100.0 g of O_2 are cooled from $25.0°$ C to $-200.0°$ C.

melting point	$-219°$ C
boiling point	$-183°$ C
ΔH_{vap}	6.82 kJ/mol
ΔH_{fus}	0.444 kJ/mol
specific heat of O_2 (gas)	0.228 cal/g °C
specific heat of O_2 (liquid)	0.35 cal/g °C

60. Explain why freeze-drying is important in the preservation of blood plasma.

GENERAL EXERCISES

61. Compare liquids and solids in each of the following categories:
   a. Volume
   b. Shape
   c. Attractive forces among particles
   d. Relative distances between particles
   e. Relative mobilities of particles
   f. Organization of particles
   g. Types of particles

62. Why must food be cooked longer at high elevations than at low elevations?

63. Why must "dry ice" be stored in a closed container?

64. Stainless steel has a much greater density than water, yet a stainless steel razor blade can float on water. Why?

65. Vaporization of 10.0 g of liquid ammonia, NH_3, at the boiling point requires 13.7 kJ of heat. Calculate the molar heat of vaporization of ammonia.

66. Explain how quick-freezing helps to preserve the texture and taste of foods.

*67. Ice is less dense than water. Suggest a reason for this. (*Hint:* Consider the closeness of packing of molecules in liquid water and in ice.)

68. Calculate the amount of heat in kilojoules needed to melt 1.00×10^{23} molecules of solid I_2 at $114°$ C.

69. Many of us have learned from experience that steam burns are more severe than burns from boiling water, even though both heat sources are at $100.0°$ C. The following calculation will illustrate this point. Calculate the amount of heat in kilojoules released when 1.00 mol of steam at $100.0°$ C condenses and cools to $37.0°$ C (body temperature). Compare this value to the amount of heat in kilojoules released when 1.00 mol of liquid water at $100.0°$ C cools to $37.0°$ C.

WATER AND AQUEOUS SOLUTIONS

CHAPTER 12

OUTLINE

12.1 Water: Structure and Properties
12.2 Properties of Solutions
12.3 Solution Formation
12.4 Solubility
12.5 Factors Affecting Solubility
    Pressure
    Temperature
    Chemical Structure
12.6 Concentrations of Solutions:
    Molarity
    Molarity
    Perspective: Preparing Solutions
      of Specified Molarity
12.7 Concentrations of Solutions:
    Percent Concentration
12.8 Dilutions
12.9 Stoichiometric Calculations
    Involving Solutions
12.10 Colligative Properties of Solutions
    (optional)
    Vapor-Pressure Lowering
    Boiling-Point Elevation
    Freezing-Point Lowering
    Osmotic Pressure
12.11 Colloids and Suspensions
    (optional)
    Summary
    Study Questions and Problems

The unique and complex substance called water is probably the most familiar of all chemical compounds. Its importance was recognized early in history, and by the fifth century B.C., it was considered the principal "element." It was not until much later that Greek philosophers added fire, earth, and air as fundamental substances of the universe.

Water has the extraordinary ability to mix with countless substances to form solutions, dispersions, and suspensions. These mixtures span the range from homogeneity to heterogeneity. At one extreme is the true solution, in which the mixture is uniform throughout. At the other extreme is the suspension—a heterogeneous mixture in which the suspended particles are large enough to settle out by gravity. Between these extremes are colloidal dispersions, which are homogeneous mixtures containing particles larger than those in solutions but smaller than those in suspensions. In this chapter, we will explore the behavior of solutions, dispersions, and suspensions. To prepare for this chapter, you should review intermolecular forces (chapter 11), liquids and solids (chapter 11), homogeneous and heterogeneous matter (chapter 1), and stoichiometric calculations (chapter 9).

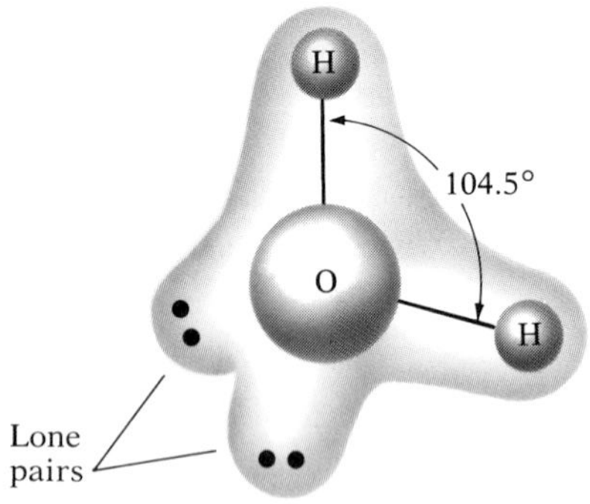

12.1 WATER: STRUCTURE AND PROPERTIES

The water molecule (Figure 12-1) has a bent shape, with an angle of 104.5° between the two hydrogen-oxygen bonds. It is thought that liquid water consists of aggregates of hydrogen-bonded molecules with individual molecules interspersed among these icelike aggregates and in dynamic equilibrium with them (see Figure 12-2). Thus, the aggregates are continuously forming, disintegrating, and reforming, and the molecular arrangement in liquid water is always in a state of change. On the average,

104.5°

Lone pairs

FIGURE 12-1   The Water Molecule

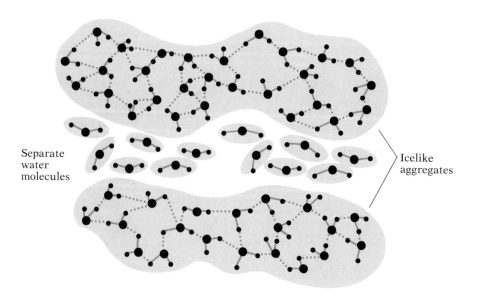

Separate water molecules

Icelike aggregates

FIGURE 12-2   The Arrangement of Molecules in Liquid Water

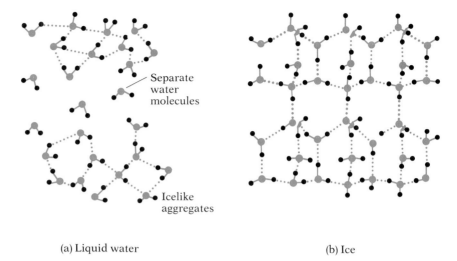

Separate
water
molecules

Icelike
aggregates

(a) Liquid water

(b) Ice

FIGURE 12-3 The Molecular
Arrangement in (a) Liquid Water and
(b) Ice

each molecule in liquid water forms three hydrogen bonds to other molecules.

The molecular arrangement in ice is more rigid and ordered than in
liquid water. In ice (see Figure 12-3), each molecule forms four hydrogen
bonds, compared to the average of three formed by each molecule in
liquid water. Usually the solid state of a substance is more dense than the
liquid, but for water the reverse is true. The highly ordered molecular
arrangement in ice causes ice to have larger spaces between its molecules
than liquid water has. Thus, ice is less dense than liquid water and floats on
the surface of liquid water. In cold climates, aquatic life survives in lakes
and oceans whose surfaces are covered with floating ice because the ice
insulates the water below and maintains a temperature suitable for life in
the depths of the water.

The extensive networks of hydrogen bonds in liquid and solid water
result in relatively high values for the freezing point, boiling point, heat of
vaporization, and heat of fusion. In addition, the geometry of water molecules and the effect of hydrogen bonding cause molecules in liquid water
to pack together more tightly than molecules of most liquids. Thus, water
has a higher density than would be expected on the basis of its formula
weight. Water occupies its smallest volume and thus has its highest density
at 3.98° C; at this temperature, the density of water is 1.00 g/mL. Table
12-1 summarizes the physical properties of water.

TABLE 12-1 Physical Properties of Water	
Appearance	Colorless liquid
Melting Point	0.00°C (1 atm)
Boiling Point	100.00°C (1 atm)
Density of liquid	0.999 g/mL (0°C)
Density of solid	0.917 g/mL (0°C)
Molar heat of fusion	1.44 kcal/mole
Molar heat of vaporization	9.72 kcal/mole
Dipole moment	1.85 D

The physical properties of water are ideal for its role in nature. Its high boiling point and heat of vaporization help to ensure that enough water is present in the liquid state to support life. These two properties and water's high specific heat allow water to cool our bodies as well as the air around us. Because of its ability to dissolve so many substances, water serves as an internal transportation system for vital compounds in plants and animals.

12.2 PROPERTIES OF SOLUTIONS

A **solution** is a mixture of two or more components uniformly distributed through each other. Thus, solutions are homogeneous mixtures. The major component is called the **solvent,** and the minor component or components are called the **solutes.** Solutions can be composed of gases, liquids, or solids. Examples of the types of solutions are given in Table 12-2.

The earth's atmosphere is an example of a gaseous solution. Dental amalgam, consisting of mercury, silver, tin, copper, and zinc, is a solid solution. A solution usually has the same physical state as its solvent. Liquid solutions, particularly those having water as the solvent, are the solutions most familiar to us. A solution in which water is the solvent is called an **aqueous solution.** The liquid solution existing in greatest quantity on earth is seawater. It is an aqueous solution, and sodium chloride is the most abundant of the many solutes. Table 12-3 gives the average composition of seawater (excluding dissolved gases).

Solutions are homogeneous throughout—this means that all samples taken from a solution will have the same composition and the same physical appearance. Solutions can be colorless, like a solution of NaCl in water, or they can be colored. For example, copper(II) sulfate pentahydrate, $CuSO_4 \cdot 5 H_2O$, is a brilliant blue crystalline solid; when it dissolves in water, the solution takes on its blue color. The solution is clear, and it retains its clarity even when light is shining through it. This clarity is a property of all solutions, regardless of the presence or absence of color.

The dissolved particles in a solution can be ions, atoms, or molecules. These particles usually have diameters of less than 1 nanometer (nm).

TABLE 12-2 Types of Solutions	
Type	Examples
gas in a gas	air, some anesthetic gases
gas in a liquid	carbonated beverages (CO_2 in H_2O)
gas in a solid	—
liquid in a gas	humid air
liquid in a liquid	rubbing alcohol, antifreeze
liquid in a solid	dental amalgam (tooth fillings)
solid in a gas	—
solid in a liquid	salt water
solid in a solid	brass, alloys, jewelry gold and silver

TABLE 12-3 The Average Composition of Seawater (excluding dissolved gases)

Constituent	g/kg of Seawater	Proportion of Total Salt Content (%)
Chloride (Cl^-)	18.980	55.044
Sulfate (SO_4^{2-})	2.649	7.682
Bicarbonate (HCO_3^-)	0.140	0.406
Bromide (Br^-)	0.065	0.189
Fluoride (F^-)	0.001	0.003
Boric acid (H_3BO_3)	0.026	0.075
Sodium (Na^+)	10.556	30.613
Magnesium (Mg^{2+})	1.272	3.689
Calcium (Ca^{2+})	0.400	1.160
Potassium (K^+)	0.380	1.102
Strontium (Sr^{2+})	0.013	0.038
Total	34.482	100.000
Water (with traces of other substances)	965.518	
Total	1,000.000	

(1 nm $= 10^{-9}$m.) Such small particles readily pass through filters, so methods other than filtration must be used to separate the components of a solution. Distillation can be used to separate liquid components of a solution according to their boiling points; a simple laboratory distillation is shown in Figure 12-4. The simplest way to separate solid solutes from liquid solvent is by evaporation of the solvent. When the liquid solvent has been completely evaporated, any dissolved solids are left as residue because they are nonvolatile. This method is used to obtain salt from sea water.

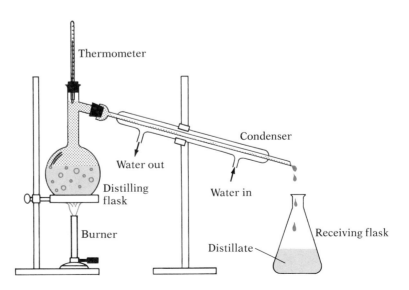

FIGURE 12-4 A Simple Laboratory Distillation. Cool water circulating through the jacket of the condenser causes the vapor formed by the boiling liquid to condense into the receiving flask.

Thermometer

Condenser

Water out

Water in

Distilling flask

Burner

Distillate

Receiving flask

12.3   SOLUTION FORMATION

Solutions are formed when one or more solutes dissolves in a solvent. Consider what happens when a liquid or solid solute dissolves in a liquid solvent. The solute particles must separate from each other and become dispersed uniformly through the solvent. At the same time, solvent molecules must separate so that solute particles can become dispersed among them. Thus there must be very close interaction between solute particles and solvent molecules. If the attractive forces between solute particles and solvent molecules are stronger than those among solute particles and among solvent molecules, this interaction will be favorable and a solution will form.

One of the most remarkable features of liquid water is its versatility as a solvent. More substances will dissolve in water than in any other existing liquid. Polar covalent and ionic substances are most likely to dissolve in water because their dipoles and ions are attracted to the polar water molecules. These interactions allow water molecules to surround dissolved particles, as shown in Figure 12-5 for the ions of dissolved NaCl. The resulting solutions are thus homogeneous mixtures.

Figure 12-6 shows how an ionic substance dissolves in water. Sodium chloride ions are attracted so strongly to the polar water molecules that the crystal lattice forces are disrupted. When a crystal of NaCl is added to water, the water molecules begin to bombard the surface of the crystal. Eventually a colliding water molecule will be oriented so that its partially negative oxygen atom is next to a positively charged sodium ion. The attraction between the two is strong enough to draw the sodium ion out of the crystal and into the water. Similarly, the negatively charged chloride ions are attracted to the partially positive hydrogen atoms of the water molecules. In this way, the crystal is broken apart and its ions are dispersed through the water. Each ion becomes surrounded by water mole-

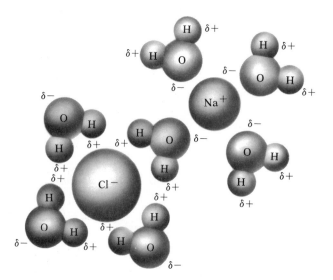

FIGURE 12-5   Water Molecules Surrounding the Ions of Dissolved Sodium Chloride, NaCl

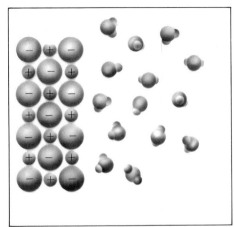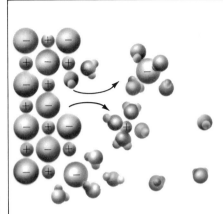

FIGURE 12-6 Sodium Chloride Dissolving in Water

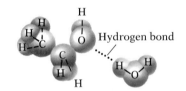

FIGURE 12-7 Hydrogen Bonding between an Ethyl Alcohol Molecule and a Water Molecule

cules; the exact number of water molecules varies with temperature. The process in which water molecules surround solute particles in a solution is called **hydration.** (For liquid solutions having solvents other than water, the more general term **solvation** is used.) Hydration of sodium chloride can be described by the following equation:

$$NaCl(s) \xrightarrow{H_2O} Na^+(aq) + Cl^-(aq)$$

Many ionic compounds dissolve in water, as do many polar covalent compounds. Polar covalent compounds interact favorably with water molecules through dipolar attractions or, in some cases, hydrogen bonds. Ethyl alcohol, CH_3CH_2OH, forms hydrogen bonds with water molecules when it dissolves in water, as illustrated in Figure 12-7.

Saying that a solute dissolves in a solvent gives no indication of how long the dissolving process might take. The *rate* at which a solute dissolves in a liquid solvent depends on three factors:

1. Surface area of the undissolved solute. The solvent comes into contact with the surface of the undissolved solute. For a given mass of solute, the greater the surface area, the more exposure the solute has to the solvent and the faster the solute will dissolve. Crushing or grinding a solid solute into tiny bits increases its surface area and thus increases its rate of dissolving. For liquid solutes, shaking breaks the liquid into small droplets, thereby increasing its surface area and causing faster dissolving.

2. Agitation. Agitation by stirring or shaking disperses the solute particles through the solution more quickly and thus increases the rate of dissolving.

3. Temperature. An increase in temperature causes solvent molecules to move faster. This increases the frequency of bombardment of undissolved solute by solvent molecules, and dissolving occurs faster than it would at lower temperatures.

12.4 SOLUBILITY

The maximum amount of a solute that will dissolve in a given quantity of a solvent at a particular temperature is called the **solubility** of the solute. Table 12-4 lists solubilities of some common substances in water; as you can see, there is quite a range of solubilities. The terms **soluble** and **insoluble** are commonly used to describe the extent to which a solute dissolves in a solvent, even though the meanings of these terms are not very precise. For our purposes, we will consider substances having water solubilities of 1 g/100 mL or greater to be soluble, and those having solubilities of less than 1 g/100 mL to be insoluble.

Many liquids are soluble in other liquids, for example, ethyl alcohol in water. The word **miscible** is used to refer to mutually soluble liquids.

If more solute than can be dissolved at a particular temperature is added to a solvent, a dynamic equilibrium will be established between the dissolved particles and the undissolved solute. For example, if we add 36 g of NaCl to 100 mL of water at 20° C, all of the salt will dissolve. But if we continue adding salt, we will not be able to exceed the solubility limit of 36 g/100 mL at 20° C, and the additional salt will remain undissolved. However, a dynamic equilibrium will be established between the dissolved salt and the undissolved salt, and ions will be exchanged between the crystals and the solution at equal rates (see Figure 12-8). When a solution is in dynamic equilibrium with undissolved solute, the solution is said to be **saturated.** The solubility of a solute can be determined in the laboratory by making an obviously saturated solution (one in which there is a second phase of undissolved solute) and then measuring the amount of solute in solution.

For a series of different solutes dissolving in the same solvent, it is often true that the more heat given off during dissolving, the greater the solubility of the solute at a given temperature. The heat change that occurs when one mole of a solute dissolves in a solvent is called the **molar heat of solution** ($\Delta H_{solution}$) of the substance. The heat change can be a release of heat or an absorption of heat, depending on the particular solute. If heat is released during dissolving, the solution becomes warm, the process is **exothermic,** and $\Delta H_{solution}$ has a negative sign. If heat is absorbed during dissolving, the solution becomes cool, the process is **endothermic,** and

FIGURE 12-8 The Dynamic Equilibrium of a Saturated Solution

TABLE 12-4	Solubilities of Common Substances in Water at 20° C
Substance	Solubility (g/100 mL)
Barium hydroxide, $Ba(OH)_2$	3.89
Iodine, I_2	0.029
Silver chloride, AgCl	0.00015
Sodium chloride, NaCl	36.0
Sodium hydroxide, NaOH	109
Sugar (sucrose, $C_{12}H_{22}O_{11}$)	203.9

TABLE 12-5 Molar Heats of Solution (Water as Solvent) and Aqueous Solubilities at 25° C

Substance	$\Delta H_{solution}$ (kJ/mol)	Solubility (mol/L)
Sodium chloride (NaCl)	3.9	6.18
Sodium iodide (NaI)	−7.5	12.3
Sodium hydroxide (NaOH)	−44.4	38.5
Silver iodide (AgI)	111.8	0.00000016
Silver chloride (AgCl)	66.16	0.000014
Silver fluoride (AgF)	−14	14.1
Potassium chloride (KCl)	17.2	4.76
Ammonium chloride (NH$_4$Cl)	14.8	7.35

$\Delta H_{solution}$ has a positive sign. Molar heats of solution for a variety of solutes in water are given in Table 12-5.

As illustrated in Table 12-5, formation of aqueous solutions can be exothermic or endothermic, depending on the particular solute. Ammonium nitrate, NH_4NO_3, has a large endothermic heat of solution in water (26 kJ/mol at 15° C) and can be used to make instant cold packs. The solid NH_4NO_3 is placed in a small plastic bag, which in turn is put into a larger plastic bag of water and sealed. Kneading the large bag ruptures the small bag and allows the NH_4NO_3 to dissolve in the water. As the solution forms, the pack gets very cold and can be used in place of a regular ice pack. Calcium chloride, $CaCl_2$, has a large exothermic heat of solution in water (−73 kJ/mol at 18° C) and is used for instant hot packs.

12.5 FACTORS AFFECTING SOLUBILITY

The most important factors that affect how much of a substance will dissolve in a solvent are pressure, temperature, and chemical structure.

Pressure External pressure affects solubility when solution formation causes a volume change. There is very little volume change when liquid and solid solutes dissolve in liquid and solid solvents, and thus pressure has only a negligible effect on these solubilities. Since a gaseous solution can change its volume to match the size of any container, there are not necessarily any volume changes when gases mix to form solutions. Hence, pressure has no effect on the solubilities of gases in one another. However, when a gas dissolves in a solid or in a liquid, as when CO_2 dissolves in water, there is a large volume decrease in going from the separate components to the solution. An increase in the pressure exerted on a gas will therefore cause more of the gas to dissolve in a liquid or a solid. The increased pressure simply forces more gas molecules to move into the solution (Figure 12-9) and thus increases the solubility of the gas. This principle is applied to the packaging of carbonated beverages under high pressure to force a greater amount of CO_2 to dissolve; when the beverage

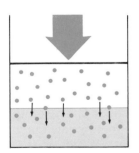

FIGURE 12-9 Increasing the Applied Pressure Forces More Gas Molecules to Dissolve in the Solvent.

container is opened, the high pressure is released, allowing some of the CO_2 to bubble out of solution.

Temperature   The effect of temperature on solubility depends on whether the heat of solution for a particular solute is positive (endothermic) or negative (exothermic). For solutes with positive heats of solution, dissolving requires that heat be put into the solution, and thus solubility increases with increasing temperature. Most solids and liquids have positive heats of solution in water, and their water solubilities increase at elevated temperatures. On the other hand, if a solute has a negative heat of solution, increasing the temperature will decrease the solubility of the solute. Most gases have exothermic heats of solution in water, so increased temperatures usually lower gas solubilities in water. Thus, when soda pop warms up after it has been opened, most of the CO_2 escapes and the beverage "goes flat."

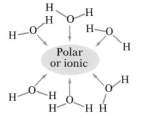

FIGURE 12-10 Hydrophilic Substances Are Readily Hydrated.

Chemical Structure   Substances having strong attractions for water molecules are said to be **hydrophilic** (water-loving). Such substances are readily hydrated (see Figure 12-10) and are usually quite soluble in water. Most hydrophilic substances are either ionic or polar covalent. Substances that can form hydrogen bonds with water molecules are especially hydrophilic. In contrast, there are some substances whose molecules have only London dispersion forces of attraction for water molecules. These attractions are not strong enough to overcome the attractions of water molecules for each other, and the substances are not soluble in water. Such substances are said to be **hydrophobic** (water-hating). They are usually nonpolar. If a hydrophobic substance is added to water, it will not dissolve. Gasoline and oil are examples of hydrophobic substances—they are not miscible with water to any extent. Both are composed of mixtures of nonpolar compounds derived from petroleum.

It is possible to make fairly reliable guesses about solubility by examining the chemical structures of the solute and solvent. If a solvent is composed of polar molecules, it is likely to dissolve ionic and polar covalent substances. If a solvent is capable of forming hydrogen bonds, it will show a pronounced preference for solutes that are also capable of forming hydrogen bonds. In contrast, if the solvent is nonpolar, it will probably dissolve only nonpolar solutes. These correlations are often summarized by the statement "like dissolves like."

12.6 CONCENTRATIONS OF SOLUTIONS: MOLARITY

Most of us prepare solutions during our daily activities. Cooking, for example, requires mixing the basic ingredients of sauces. Other solutions that we make and take for granted are chlorine solutions for swimming pools and liquid fertilizer for lawns and gardens. In most instances, the components of solutions are measured to achieve the correct solution concentration. The **concentration** of a solution is the amount of solute in a given quantity of solution. There are several ways to express solution concentrations, and each has its own particular advantages. We will start with the method called molarity.

Molarity Molarity (M) is solution concentration expressed as the number of moles of solute per liter of solution; molarity has units of mol/L. Thus,

$$M = \frac{\text{mol solute}}{\text{volume of solution}}$$

or

$$M = \frac{n}{V}$$

where n represents number of moles of solute and V represents volume expressed in liters. Thus, a 1.00 M (said as "one point zero zero molar") solution of NaCl contains 58.5 g (1.00 mol) of NaCl per liter of solution. Similarly, a 0.100 M solution of NaCl contains 5.85 g (0.100 mol) of NaCl per liter, and a 0.0500 M solution of NaCl contains 2.92 g (0.0500 mol) of NaCl per liter. This method of expressing concentration is especially useful for solutions containing solutes to be used in chemical reactions, since chemical species react in mole ratios.

Molarity is simply the ratio of moles of solute to volume of solution (mol/L). If we know the molarity of a solution, this ratio can be used much like a conversion factor to calculate the number of moles of solute, and hence the mass of solute, in any volume of the solution.

EXAMPLE 12.1 How many grams of $NaNO_3$ are needed to make 1.00 L of a 0.75 M solution?

Solution First, calculate the number of moles of $NaNO_3$ needed by using molarity as a conversion factor:

$$(1.00\,L)\left(\frac{0.75 \text{ mol } NaNO_3}{L}\right) = 0.75 \text{ mol } NaNO_3$$

Then, calculate the weight of $NaNO_3$ needed:

$$(0.75 \text{ mol } NaNO_3)\left(\frac{85.0 \text{ g } NaNO_3}{\text{mol } NaNO_3}\right) = 64 \text{ g } NaNO_3$$

Or, we could combine the steps:

$$(1.00\,L)\left(\frac{0.75 \text{ mol } NaNO_3}{L}\right)\left(\frac{85.0 \text{ g } NaNO_3}{\text{mol } NaNO_3}\right) = 64 \text{ g } NaNO_3$$

EXERCISE 12.1 How many grams of glucose, $C_6H_{12}O_6$, are needed to make 1.00 L of a 0.50 M solution?

EXAMPLE 12.2 How many grams of $MgCl_2$ are needed to make 500.0 mL of a 3.00 M solution?

Solution   First, convert 500.0 mL to L:

$$(500.0 \text{ mL})\left(\frac{1 \text{ L}}{1000 \text{ mL}}\right) = 0.5000 \text{ L}$$

Then, calculate weight of $MgCl_2$ needed by using molarity as a conversion factor:

$$(0.5000 \text{ L})\left(\frac{3.00 \text{ mol } MgCl_2}{L}\right) = 1.50 \text{ mol } MgCl_2$$

$$(1.50 \text{ mol } MgCl_2)\left(\frac{95.3 \text{ g } MgCl_2}{\text{mol } MgCl_2}\right) = 143 \text{ g } MgCl_2$$

Or, combining steps,

$$(0.5000 \text{ L})\left(\frac{3.00 \text{ mol } MgCl_2}{L}\right)\left(\frac{95.3 \text{ g } MgCl_2}{\text{mol } MgCl_2}\right) = 143 \text{ g } MgCl_2$$

EXERCISE 12.2   How many grams of $Mg_3(PO_4)_2$ are needed to make 250.0 mL of a 1.50 M solution? ▪

The definition of molarity contains three variables: molarity, moles of solute, and volume of solution. We can use the definition to calculate any one of the three variables if the other two are known. The following example illustrates how to calculate volume of solution from molarity and moles of solute.

EXAMPLE 12.3   What volume of 1.75 M $KClO_3$ solution contains 15.0 g of $KClO_3$?

Solution   First, calculate moles of $KClO_3$ in 15.0 g of $KClO_3$:

$$(15.0 \text{ g } KClO_3)\left(\frac{1 \text{ mol } KClO_3}{122.5 \text{ g } KClO_3}\right) = 0.122 \text{ mol } KClO_3$$

Then, solve the definition of molarity for volume and calculate volume:

$$M = \frac{\text{mol solute}}{\text{volume of solution}}$$

$$\text{volume of solution} = (\text{mol solute})\left(\frac{1}{M}\right)$$

$$= (0.122 \text{ mol } KClO_3)\left(\frac{L}{1.75 \text{ mol } KClO_3}\right)$$

$$= 0.0697 \text{ L}$$

Or,

$$\text{volume of solution} = (15.0 \text{ g } KClO_3)\left(\frac{1 \text{ mol } KClO_3}{122.5 \text{ g } KClO_3}\right)$$

$$\left(\frac{L}{1.75 \text{ mol } KClO_3}\right) = 0.0697 \text{ L}$$

MATH TIP

The problem requires a two-step conversion:

$$\text{g } KClO_3 \xrightarrow{\text{molar mass}} \text{moles } KClO_3 \xrightarrow{\text{molarity}}$$
$$\text{L } KClO_3 \text{ solution}$$

$$(15.0 \text{ g } KClO_3)\left(\frac{1 \text{ mol } KClO_3}{122.5 \text{ g } KClO_3}\right)$$

$$\left(\frac{? \text{ L solution}}{? \text{ mol } kClO_3}\right)$$

The arrangement of the units shown above will result in the required unit, L solution. Since the molarity is 1.75 M then $\frac{1.75 \text{ moles } KClO_3}{1 \text{ L solution}}$ or

$\frac{1 \text{ L solution}}{1.75 \text{ mol } KClO_3}$ can be used to make

conversions. For this problem the second arrangement is necessary for the units to cancel.

EXERCISE 12.3   What volume of 3.70 M $LiNO_3$ solution contains 26.2 g of $LiNO_3$? ▪

Preparing a solution of specified molarity would seem to be a simple task. For example, you might think that making a 1.00 *M* aqueous solution of NaCl would require only the addition of 1.00 mol (58.5 g) of NaCl to 1.00 L of pure water. However, this approach will not produce a 1.00 *M* solution, because adding solute to a measured volume of liquid solvent usually causes an increase in total volume of liquid. The extent of the volume increase is not easily predicted or easily measured, and thus the final volume and concentration are not accurately known.

Because of the volume increase, special pieces of glassware called volumetric flasks (see the accompanying figure) are used in preparing solutions of specified molarity. These flasks are constructed to contain a specific volume of liquid (10.00 mL,

100.0 mL, 500.0 mL, 1.00 L, and so forth) at 20° C; the neck of each volumetric flask is marked with a line that indicates the level to which it should be filled with liquid.

The general procedure for making an aqueous solution of specified molarity (in this case, 1.00 *M*) is illustrated in the accompanying figure. First, the solute is weighed. Then it is dissolved in a small quantity of pure water in a clean beaker. This solution is poured through a funnel into the volumetric flask. The beaker and funnel are then rinsed several times with pure water, and the rinsings are added to the volumetric flask. In this way, all traces of solute are transferred to the volumetric flask. Finally, the volumetric flask is filled to the mark with pure water. This procedure results in a solution of accurately known volume containing a precisely measured weight of solute.

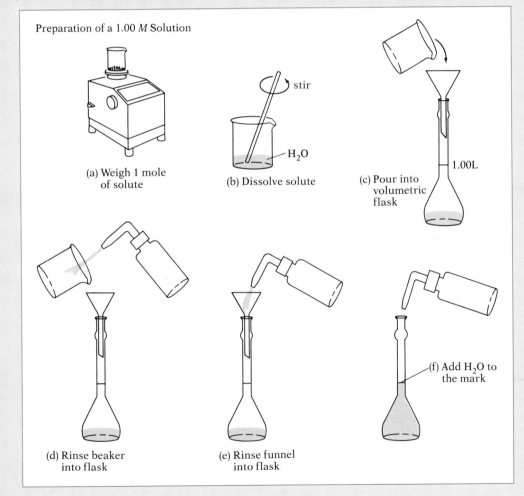

Preparation of a 1.00 *M* Solution

(a) Weigh 1 mole of solute

(b) Dissolve solute

(c) Pour into volumetric flask

1.00L

(d) Rinse beaker into flask

(e) Rinse funnel into flask

(f) Add H$_2$O to the mark

12.7 CONCENTRATIONS OF SOLUTIONS: PERCENT CONCENTRATION

In some instances, it is not necessary to specify the number of moles of solute in a given quantity of solution. Concentration is sometimes expressed as the mass of solute in a certain volume of solution. One way to express this kind of concentration is by percent. For example, an aqueous solution of NaCl contains 10.0 g of NaCl in 100.0 mL of solution. This concentration can be expressed as the ratio of g of NaCl to mL of solution (g/mL). If we multiply the ratio by 100%, we arrive at the **weight-to-volume percent concentration (%(w/v)).** This concentration is defined as

$$\%(w/v) = \frac{g \text{ solute}}{mL \text{ solution}} \times 100\%$$

Our example, the NaCl solution, thus has a concentration of 10.0%(w/v):

$$\frac{10.0 \text{ g solute}}{100.0 \text{ mL solution}} \times 100\% = 10.0\%(w/v)$$

It is important to note that weight-to-volume percent concentration is not a true percent because it is derived from a ratio of two different units, g and mL; however, the name weight-to-volume percent concentration is used for convenience.

EXAMPLE 12.4   What is the %(w/v) concentration of 16.5 g of glucose dissolved in 150.0 mL of solution?

Solution
$$\frac{16.5 \text{ g solute}}{150.0 \text{ mL solution}} \times 100\% = 11.0\%(w/v)$$

EXERCISE 12.4   What is the %(w/v) concentration of 25.0 g of KCl dissolved in 200.0 mL of solution?   ■

When we make solutions containing liquid solutes, it is more convenient to measure the solute's volume than its weight. Then the percent concentration is a volume-to-volume ratio. **Volume-to-volume percent concentration (%(v/v))** is defined as

$$\%(v/v) = \frac{mL \text{ solute}}{mL \text{ solution}} \times 100\%$$

Volume-to-volume percent concentration is frequently used to express concentrations of alcoholic beverages.

EXAMPLE 12.5   What is the %(v/v) concentration of 11.5 mL of ethyl alcohol dissolved in 80.0 mL of aqueous solution?

Solution
$$\frac{11.5 \text{ mL solute}}{80.0 \text{ mL solution}} \times 100\% = 14.4\%(v/v)$$

EXERCISE 12.5   What is the %(v/v) concentration of 4.50 mL of rubbing alcohol (isopropyl alcohol) dissolved in 55.0 mL of aqueous solution?   ■

The third method of expressing percent concentration is **weight-to-weight percent concentration (%(w/w)).** This method is valuable for solid solutions such as alloys. It is defined as

$$\%(w/w) = \frac{g\ solute}{g\ solution} \times 100\%$$

EXAMPLE 12.6   What is the %(w/w) concentration of 85.0 mg of silver dissolved in 8.46 g of an alloy (a solid solution of metals)?

Solution   We must first convert 85.0 mg to g:

$$(85.0\ mg)\left(\frac{1\ g}{1000\ mg}\right) = 0.0850\ g$$

Then,

$$\left(\frac{0.0850\ g\ solute}{8.46\ g\ solution}\right) \times 100\% = 1.00\%(w/w)$$

EXERCISE 12.6   Brass is an alloy composed of copper and zinc. What is the %(w/w) of 17.6 g of zinc dissolved in 50.0 g of brass?   ■

The term *percent* actually means parts per hundred, where the word *parts* refers to any unit of measure (g, mg, mL, and so forth); thus, percent concentration is used mostly for solutions that contain at least one part of solute per hundred parts of solution. In very dilute solutions, however, the solute may not be present to this extent. For example, pollutants in air and water are often present in exceedingly small amounts, yet these small amounts are significant for health and agriculture. The concentrations of such very small amounts are usually measured in parts per million (ppm) or parts per billion (ppb). A 1-ppm solution contains 1 part of solute for every million parts of solution, and a 1-ppb solution contains 1 part of solute for every billion parts of solution. Current usage of ppm and ppb requires that they be used for like units, for example 1 μg in 1 g (1 ppm) or 1 μL in 1 L (1 ppm).

12.8 DILUTIONS

Reagent
(ree-AY-gent): A *reactive chemical agent.*

In laboratory work, it is convenient to purchase or prepare concentrated stock solutions of reagents for later use. (A **reagent** is a reactive substance used in the laboratory for chemical analysis.) The stock reagents are diluted as needed to achieve lower concentrations for individual experiments. **Dilutions** are accomplished by adding solvent to a portion of the concentrated stock solution, thus lowering the concentration of the stock solution to a more desirable concentration.

For example, suppose you have 100.0 mL of an aqueous stock solution of 1.00 *M* glucose, and you wish to dilute the stock solution to prepare an aqueous solution of 0.100 *M* glucose. What should the final volume of the dilute solution be? We can solve this problem by reconsidering the mathematical definition of molarity.

$$M = \frac{n}{V}$$

For the concentrated stock solution, we can write

$$M_c = \frac{n_c}{V_c}$$

and

$$n_c = M_c V_c$$

where the subscript c indicates *concentrated*. For the dilute solution,

$$M_d = \frac{n_d}{V_d}$$

and

$$n_d = M_d V_d$$

where the subscript d indicates *dilute*. When water is added to the concentrated solution, the total amount of dissolved solute remains the same; it is just distributed through a larger volume of solution. Thus

$$n_c = n_d$$

and

$$M_c V_c = M_d V_d \qquad\qquad (1)$$

This relationship can now be used to solve our problem. Since we are seeking the final volume of the dilute solution, we must solve equation (1) for V_d:

$$M_c V_c = M_d V_d$$
$$\frac{M_c V_c}{M_d} = V_d$$

or

$$V_d = \frac{M_c V_c}{M_d}$$

Substituting into this expression and doing the calculation gives

$$V_d = \frac{(1.00 \ \text{mol/L})(100.0 \ \text{mL})}{0.100 \ \text{mol/L}}$$
$$= 1000 \ \text{mL or } 1.00 \times 10^3 \ \text{mL}$$

Thus we find that enough water must be added to the 100.0 mL of stock solution *to bring its final volume to 1.00 × 10³ mL*. (Note carefully the wording of this statement: The calculated value of 1.00×10^3 mL is final volume, not the volume of water to be added.)

Equation (1) says that the product of initial concentration and initial volume is equal to the product of final concentration and final volume. Thus, when volume goes up, concentration must go down, and vice versa. In fact, this principle holds true even when concentration is expressed in units other than molarity:

$$C_c V_c = C_d V_d \qquad (2)$$

where C_c and C_d represent concentrations of the concentrated and dilute solutions, respectively. In using this general expression, C_c and C_d may be expressed as molarity or as percent, as long as the units of the two are the same. Similarly, V_c and V_d may be expressed in any units of volume, as long as the units are the same.

EXAMPLE 12.7 What will be the concentration of a solution prepared by diluting 100.0 mL of a 2.0 M solution to 500.0 mL?

Solution We begin the solution of this problem by solving equation (2) for C_d:

$$C_c V_c = C_d V_d$$

$$C_d = C_c \left(\frac{V_c}{V_d} \right)$$

Then, substituting numerical values into the expression

$$C_d = 2.0\ M \left(\frac{100.0\ \text{mL}}{500.0\ \text{mL}} \right)$$

$$= 0.40\ M$$

EXERCISE 12.7 What will be the concentration of a solution prepared by diluting 250.0 mL of a 2.50 M solution to 800.0 mL? ◾

EXAMPLE 12.8 To what volume would 50.0 mL of a 15.0%(w/v) solution be diluted to prepare a 10.0%(w/v) solution?

Solution Again, we use equation (2) to solve the problem:

$$C_c V_c = C_d V_d$$

$$V_d = V_c \left(\frac{C_c}{C_d} \right)$$

$$= 50.0\ \text{mL} \left(\frac{15.0\%(\text{w/w})}{10.0\%(\text{w/w})} \right)$$

$$= 75.0\ \text{mL}$$

EXERCISE 12.8 To what volume would 100.0 mL of a 2.00 M solution be diluted to prepare a 0.100 M solution? ◾

12.9 STOICHIOMETRIC CALCULATIONS INVOLVING SOLUTIONS

Stoichiometric calculations involving solutions typically require inter-conversion of solution volume and moles of a solute; these two quantities are related by molarity (M = mol solute/volume solution). Thus, when molarity is known, there are two kinds of calculations: finding volume of solution from moles of solute, and finding moles of solute from volume of solution. The following examples demonstrate these calculations, and the "mole map" of Figure 12-11 summarizes relationships in the complete spectrum of stoichiometric problems.

EXAMPLE 12.9

What volume of 0.100 M $AgNO_3$ is needed to react completely with 4.62 g of NaCl?

$$NaCl(s) + AgNO_3(aq) \longrightarrow AgCl(s) + NaNO_3(aq)$$

Solution

First, convert weight of NaCl to moles of NaCl by using the molar mass:

$$(4.62 \text{ g NaCl}) \left(\frac{1 \text{ mol NaCl}}{58.5 \text{ g NaCl}} \right) = 0.0790 \text{ mol NaCl}$$

Next, convert moles of NaCl to moles of $AgNO_3$ by using the appropriate mole ratio from the balanced equation:

$$(0.0790 \text{ mol NaCl}) \left(\frac{1 \text{ mol AgNO}_3}{1 \text{ mol NaCl}} \right) = 0.0790 \text{ mol AgNO}_3$$

Last, convert moles of $AgNO_3$ to volume of solution, using the concentration as a conversion factor:

$$(0.0790 \text{ mol AgNO}_3) \left(\frac{L}{0.100 \text{ mol AgNO}_3} \right) = 0.790 \text{ L}$$

Or,

$$(4.62 \text{ g NaCl}) \left(\frac{1 \text{ mol NaCl}}{58.5 \text{ g NaCl}} \right) \left(\frac{1 \text{ mol AgNO}_3}{1 \text{ mol NaCl}} \right)$$
$$\left(\frac{L}{0.100 \text{ mol AgNO}_3} \right) = 0.790 \text{ L}$$

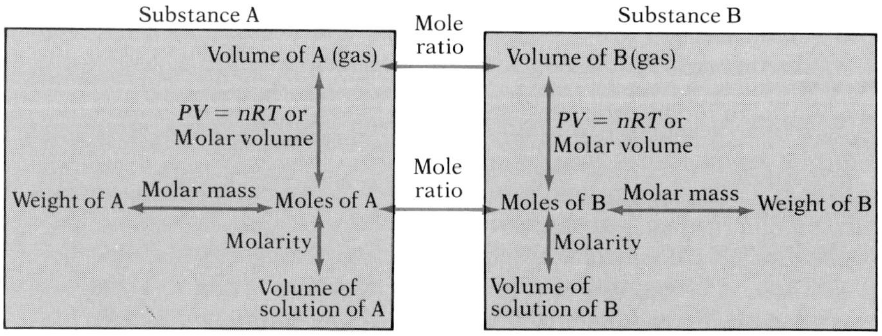

MATH TIP

Use the mole map in Figure 12-11 to outline the problem:

g NaCl $\longrightarrow$ mol NaCl $\longrightarrow$

           mol $AgNO_3$ $\longrightarrow$ g $AgNO_3$

                  L $AgNO_3$ solution

Arrange units so that the required units can be found:

$$(4.62 \text{ g NaCl}) \left(\frac{? \text{ mol NaCl}}{? \text{ g NaCl}} \right) \left(\frac{? \text{ mol AgNO}_3}{? \text{ mol NaCl}} \right)$$
$$\left(\frac{1 \text{ L AgNO}_3 \text{ solution}}{? \text{ mol AgNO}_3} \right)$$

Once the units are arranged, the quantities can be written using the molar masses: mol ratios and molarity.

FIGURE 12-11   Important Relationships for Solving Stoichiometric Problems

EXERCISE 12.9 What volume of 0.4600 *M* NaOH is needed to react completely with 1.000 g of $HC_2H_3O_2$?

$$HC_2H_3O_2(l) + NaOH(aq) \longrightarrow NaC_2H_3O_2(aq) + H_2O(l)$$ ■

EXAMPLE 12.10 What is the molarity of a solution of $Ba(OH)_2$ if 20.00 mL of this solution require 17.98 mL of 0.1000 *M* HCl for complete reaction?

$$Ba(OH)_2(aq) + 2\ HCl(aq) \longrightarrow BaCl_2(aq) + 2\ H_2O(l)$$

Solution First, convert volume of HCl solution to moles of HCl, using the concentration as a conversion factor:

$$(0.01798\ \cancel{L})\left(\frac{0.1000\ \text{mol HCl}}{\cancel{L}}\right) = 0.001798\ \text{mol HCl}$$

Next, convert moles of HCl to moles of $Ba(OH)_2$:

$$(0.001798\ \cancel{\text{mol HCl}})\left(\frac{1\ \text{mol } Ba(OH)_2}{2\ \cancel{\text{mol HCl}}}\right) = 0.0008990\ \text{mol } Ba(OH)_2$$

Last, calculate the molarity of the $Ba(OH)_2$ solution:

$$M = \frac{0.0008990\ \text{mol } Ba(OH)_2}{0.02000\ \text{L}} = 0.04495\ \text{mol } Ba(OH)_2/\text{L}$$

Or,

$$M = (0.01798\ \cancel{L})\left(\frac{0.1000\ \cancel{\text{mol HCl}}}{\cancel{L}}\right)\left(\frac{1\ \text{mol } Ba(OH)_2}{2\ \cancel{\text{mol HCl}}}\right)\left(\frac{1}{0.02000\ \text{L}}\right)$$

$$= 0.04495\ \text{mol } Ba(OH)_2/\text{L}$$

EXERCISE 12.10 What is the molarity of a solution of oxalic acid, $H_2C_2O_4$, if 25.00 mL of this solution require 27.05 mL of 0.1800 *M* NaOH for complete reaction?

$$H_2C_2O_4(aq) + 2\ NaOH(aq) \longrightarrow Na_2C_2O_4(aq) + 2\ H_2O(l)$$ ■

12.10 COLLIGATIVE PROPERTIES OF SOLUTIONS *(optional)*

There are some properties of solutions that depend only on the number of solute particles and are thus completely independent of the identity of the solute. These properties are called **colligative properties.** Colligative properties include vapor-pressure lowering, boiling-point elevation, freezing-point lowering, and osmotic pressure.

Colligative
(co-LIH-gah-tive)

Vapor-Pressure Lowering If a nonvolatile solute such as glucose, $C_6H_{12}O_6$, is dissolved in water, the vapor pressure of the solution is lower than the vapor pressure of pure water. This vapor-pressure lowering occurs in any liquid solution containing a nonvolatile solute. Thus solvent evaporates more slowly from such solutions than from the pure liquid. The amount by which the vapor pressure is lowered depends only on the number of nonvolatile solute particles present. In the case of the glucose

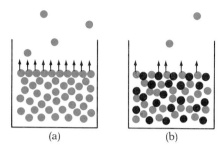

FIGURE 12-12 Vapor-Pressure Lowering by a Nonvolatile Solute. (a) Solvent molecules (colored spheres) occupy the entire surface in a pure liquid. (b) Solute particles (black spheres) occupy some of the surface positions in a solution, lowering the number of solvent molecules on the liquid surface and causing a lower vapor pressure than that of pure solvent.

solution, the number of glucose molecules determines the amount of vapor-pressure lowering. Vapor-pressure lowering can be explained by considering that solute particles are evenly distributed through the solution. Thus the surface of the solution contains both solute particles and solvent molecules (see Figure 12-12). Since solvent molecules are present in smaller numbers on the surface of the solution than they would be on the surface of pure solvent, fewer solvent molecules are available to escape from the surface of the solution than from the surface of pure solvent. Thus, the vapor pressure is decreased.

Boiling-Point Elevation   The vapor pressure of a liquid increases with temperature, and when the vapor pressure reaches atmospheric pressure, the liquid boils. When a nonvolatile solute is present in a liquid solution, the diminished vapor pressure requires a higher temperature to reach atmospheric pressure. Thus, the boiling point of the solution is elevated from that of the pure liquid. From this knowledge, we can expect that seawater has a higher boiling point than pure water.

Freezing-Point Lowering   At the freezing point, the vapor pressure of a solid and liquid are equal. However, a solution containing a nonvolatile solute has a lower vapor pressure than the pure liquid, and this causes a lower freezing point for the solution than for the pure solvent. The extent of freezing-point lowering depends only on the concentration of solute particles in the solution. One application of freezing-point lowering is the use of salt to melt ice on streets. When salt is poured on ice, some of the salt dissolves and causes the ice to melt, since the salt solution will still be a liquid at temperatures that would freeze pure water. The freezing-point lowering caused by nonvolatile solutes also explains why antifreeze in the water in automobile radiators prevents freezing in winter.

The freezing-point lowering caused by nonvolatile solutes has a very practical application in the laboratory, because it can be used to determine the molar mass of a nonvolatile solute. To find the molar mass of a solute, we must first learn to express the solute concentration in **molality** (m), defined as moles of solute per kilogram of solvent:

$$m = \frac{\text{mol solute}}{\text{kg solvent}}$$

(Note that molality has a lowercase m for its symbol, while molarity is symbolized by an uppercase M.)

EXAMPLE 12.11

What is the molality of a solution prepared by dissolving 5.00 g of glucose, $C_6H_{12}O_6$, in 50.0 g of H_2O?

Solution

First, find moles of glucose in the solution:

$$(5.00 \text{ g } C_6H_{12}O_6)\left(\frac{1 \text{ mol } C_6H_{12}O_6}{180.0 \text{ g } C_6H_{12}O_6}\right) = 0.0278 \text{ mol } C_6H_{12}O_6$$

Now, convert grams of water to kilograms and calculate the molality of the glucose solution:

$$(50.0 \text{ g})\left(\frac{1 \text{ kg}}{1000 \text{ g}}\right) = 0.0500 \text{ kg}$$

$$\frac{0.0278 \text{ mol solute}}{0.0500 \text{ kg solvent}} = 0.556 \; m$$

EXERCISE 12.11

Ethylene glycol, a nonvolatile molecular compound, is the principal ingredient of antifreeze. What is the molality of a solution prepared by dissolving 9.55 g of ethylene glycol, $C_2H_6O_2$, in 100.0 g of water? ■

For a molecular solute, the amount of freezing-point lowering is directly proportional to the molality of the solute:

$$\Delta t_f = k_f m, \tag{3}$$

where Δt_f is the amount by which the freezing point is lowered (in Celsius degrees), k_f is the freezing-point depression constant for the solvent, and m is the molality of the solute. Each solvent is affected to a different extent by nonvolatile solutes, but a solvent has a characteristic freezing-point depression constant. We can substitute the definition of molality and the following relationship

$$\text{mol solute} = \frac{\text{g solute}}{\text{molar mass of solute}}$$

into equation (3):

$$\Delta t_f = k_f m$$

$$= k_f(\text{mol solute})\left(\frac{1}{\text{kg solvent}}\right)$$

$$= k_f\left(\frac{\text{g solute}}{\text{molar mass of solute}}\right)\left(\frac{1}{\text{kg solvent}}\right) \tag{4}$$

We can now rearrange equation (4) to solve for molar mass of solute:

$$\text{molar mass of solute} = k_f\left(\frac{\text{g solute}}{\Delta t_f}\right)\left(\frac{1}{\text{kg solvent}}\right) \tag{5}$$

MATH TIP

The freezing-point depression constant, k_f, has units of $(°C)(\text{kg solvent})(\text{mol solute})^{-1}$. These units may also be written as

$$\frac{(°C)(\text{kg solvent})}{\text{mol solute}}$$

MATH TIP

To rearrange, multiply both sides of the equation by molar mass of solute.

$$(\Delta T_f)(\text{molar mass of solute})$$

$$= k_f\left(\frac{\text{g solute}}{\text{molar mass solute}}\right)\left(\frac{1}{\text{kg solvent}}\right)$$

$$(\text{molar mass solute})$$

Divide both sides by ΔT_f

$$\frac{(\Delta T_f)(\text{molar mass of solute})}{(\Delta T_f)}$$

$$= k_f\left(\frac{\text{g solute}}{\Delta T_f}\right)\left(\frac{1}{\text{kg solvent}}\right)$$

TABLE 12-6 Freezing-Point Depression Constants

Solvent	Freezing Point (°C)	k_f $(°C)(\text{kg solvent})(\text{mol solute})^{-1}$
Acetic acid	16.6	3.90
Benzene	5.5	5.12
Camphor	179.5	39.7
Carbon tetrachloride	−22.8	29.8
Chloroform	−63.5	4.68
Ethyl alcohol	−114.6	1.99
Naphthalene	80.2	6.80
Water	0.0	1.86

Thus, if k_f, grams of solute, Δt_f, and kilograms of solvent are known, we can calculate the molar mass of the solute. The laboratory determination of molar mass is done by measuring the freezing-point depression of a solution prepared from measured masses of solvent and nonvolatile solute.

Freezing-point depression constants for various solvents can be found in reference tables; Table 12-6 lists the freezing-point depression constants of a number of solvents.

EXAMPLE 12.12

A solution of 2.40 g of biphenyl (a nonvolatile molecular substance) dissolved in 75.0 g of benzene has a freezing point of 4.4° C. What is the molar mass of biphenyl?

Solution

First, find the normal freezing point of benzene in Table 12-6 and calculate Δt_f:

$$\Delta t_f = \text{normal freezing point} - \text{freezing point of solution}$$
$$= 5.5° \text{ C} - 4.4° \text{ C} = 1.1° \text{ C}$$

Next, convert grams of solvent to kilograms of solvent:

$$(75.0 \text{ g})\left(\frac{1 \text{ kg}}{1000 \text{ g}}\right) = 0.0750 \text{ g}$$

Last, find k_f for benzene in Table 12-6 and calculate molar mass from equation (5):

$$\text{molar mass} = k_f\left(\frac{\text{g solute}}{\Delta t_f}\right)\left(\frac{1}{\text{kg solvent}}\right)$$
$$= 5.12(°C)(\text{kg solvent})(\text{mol solute})^{-1}$$
$$\left(\frac{2.40 \text{ g solute}}{1.1° \text{ C}}\right)\left(\frac{1}{0.075 \text{ kg solvent}}\right)$$
$$= 150 \text{ g solute/mol solute or } 1.5 \times 10^2 \text{ g solute/mol solute}$$

EXERCISE 12.12

A solution prepared by dissolving 50.0 g of a nonvolatile molecular solute in 150.0 g of water has a freezing point of $-10.0°$ C. What is the molar mass of the solute? ∎

Osmotic Pressure   The concentration of a solution is uniform throughout, but consider what would happen if we were to add more solute to a small region of a solution. At first, the concentration of solute will be higher in the immediate vicinity of addition than elsewhere in the solution. But eventually, the solute particles and solvent molecules will diffuse through the bulk of the solution until the concentration is once again uniform throughout. In this diffusion process, solute particles move spontaneously from a region of higher concentration to one of lower concentration. Such movement also occurs with solvent molecules.

Most natural membranes and some synthetic membranes are structured so that only water molecules and certain solute particles can diffuse through the membranes. Such membranes are called **semipermeable**

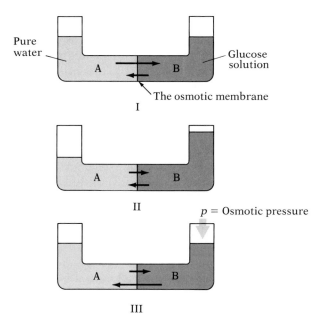

FIGURE 12-13 Osmosis. In I, a glucose solution is separated from pure water by an osmotic membrane. In II, water passes through the osmotic membrane to increase the volume of the glucose solution. In III, pressure is applied to prevent the passage of water across the osmotic membrane.

membranes. Osmotic membranes are a special class of semipermeable membranes that allow only water molecules to pass through. As you might expect, water will move across an osmotic membrane from a region of higher water concentration to one of lower water concentration. This movement is called **osmosis.**

In Figure 12-13 an osmotic membrane separates compartment A from compartment B. Initially, compartment A contains pure water and compartment B contains a glucose solution. As time passes, water will pass through the membrane from compartment A to compartment B in response to its lower concentration in compartment B. Hence, the level of solution in compartment B will rise. It is possible to counteract this volume increase by applying pressure to the surface of the glucose solution in compartment B. The amount of external pressure (measured in torr) needed to prevent the passage of water across the osmotic membrane (thus preventing an increase in volume of the glucose solution) is called the **osmotic pressure** of the glucose solution. The osmotic pressure of a solution is directly proportional to the number of nonvolatile solute particles, as are other colligative properties.

12.11 COLLOIDS AND SUSPENSIONS *(optional)*

If we collect a sample of muddy water and allow it to stand undisturbed, a lot of the mud will settle to the bottom of the container in a short while, but there will be smaller particles that remain suspended for days and perhaps indefinitely. These particles give the water a murky appearance, but they cannot be seen individually, even under a microscope, and ordinary filters do not remove them. This cloudy mixture is an example of a colloidal

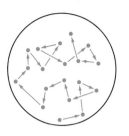

FIGURE 12-14 Brownian Motion

dispersion, or a colloid. **Colloids** are homogeneous mixtures of dispersed particles larger than most molecules; the particles can be dispersed in a gas, liquid, or solid. The diameters of the colloidal particles range from about 1 nm to 100 nm; they can be very large individual particles or groups of smaller particles. Colloidal dispersions appear cloudy or murky and are not true solutions. Some everyday examples of colloids are homogenized milk, whipped cream, jellies, and fog.

Colloidal particles move rapidly and randomly through the fluids in which they are dispersed. This motion is called **Brownian motion** after the Scottish botanist Robert Brown, who discovered it in 1827. Brownian motion is the constant but irregular motion of small particles, as illustrated in Figure 12-14. It is caused by bombardment of the dispersed particles by rapidly moving molecules of the fluid. Brownian motion prevents colloidal particles from settling.

When bright light passes through a colloidal dispersion, as shown in Figure 12-15, the light beam is clearly visible from the side; this is called the **Tyndall effect,** and it is sometimes used to distinguish colloidal dispersions from true solutions. True solute particles are not large enough to reflect light, so a light beam passing through a solution is not visible from the side. A familiar example of the Tyndall effect is the shaft of light between a movie projector and the screen; the shaft of light is visible from the side because light is reflected in all directions by dust dispersed as colloidal particles in the air.

Instead of the terms solute and solvent, which are usually reserved for solutions, we speak of the **dispersed substance** and the **dispersing medium** in describing colloids. All three states of matter can function in either role. Table 12-7 gives examples of the various types of colloidal dispersions.

Particles with larger dimensions than those of colloidal particles are visible to the eye and will settle out by gravity when dispersed in a fluid. When these larger particles are scattered through a fluid, the mixture is called a **suspension.** Suspensions are heterogeneous mixtures and can be separated by filtration. Blood contains dissolved substances (ions and molecules), dispersed colloids (large protein molecules and aggregates of fat molecules), and suspended solid matter (blood cells and platelets); the heart's pumping action keeps the solid matter suspended. Other familiar examples of suspensions are milk of magnesia ($Mg(OH)_2$ suspended in water) and orange juice containing suspended bits of orange pulp. A suspension of barium sulfate, $BaSO_4$, is used for making gastrointestinal X rays.

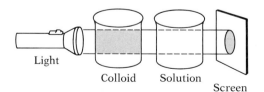

Light

Colloid   Solution

Screen

FIGURE 12-15 The Tyndall Effect

TABLE 12-7   Types of Colloidal Dispersions

Type	Examples
gas in a gas	none (all are solutions)
gas in a liquid	whipped cream
gas in a solid	marshmallow, soap foam
liquid in a gas	fog
liquid in a liquid	milk, mayonnaise
liquid in a solid	butter
solid in a gas	smoke
solid in a liquid	paint, gelatin
solid in a solid	ruby glass (dispersion of gold in a glass)

SUMMARY

Liquid water consists of aggregates of hydrogen-bonded molecules, with individual molecules interspersed among the aggregates and in equilibrium with them. The more rigid and orderly molecular arrangement in ice causes ice to be less dense than liquid water. Because of its extensive network of hydrogen bonds, water has relatively high values for its freezing point, boiling point, heat of vaporization, heat of fusion, specific heat, and surface tension. Liquid water dissolves more substances than any other existing liquid.

A **solution** is a homogeneous mixture consisting of a major component, the **solvent,** and one or more minor components, the **solutes.** Dissolved solute particles can be ions, atoms, or molecules. A solution will form if the attractive forces between solute particles and solvent molecules are stronger than those among solute particles and among solvent molecules. When a solute dissolves in liquid water, water molecules surround the solute particles in a process called **hydration.** (The term **solvation** is used for solutions having solvents other than water.) The rate at which a solute dissolves in a liquid solvent depends on the surface area of the undissolved solute, the amount of agitation, and the temperature. The heat change that occurs when one mole of solute dissolves in a solvent is called the **molar heat of solution** ($\Delta H_{\text{solution}}$) of the solute.

The maximum amount of a solute that dissolves in a given quantity of a solvent at a particular temperature is called the **solubility** of the solute. The word **miscible** is used to refer to mutually soluble liquids. When a solute is added to a solvent in amounts in excess of the solute's solubility, dynamic equilibrium is established between the solution and the undissolved solute, and the solution is said to be **saturated.** The factors that affect solubility are pressure, temperature, and chemical structure. An increase in pressure exerted on a gaseous solute increases the solubility of the gas in liquid and solid solvents. Pressure has a negligible effect on the solubilities of liquid and solid solutes. The effect of temperature on solubility depends on whether the solute's heat of solution is positive or negative. **Hydrophilic substances** are soluble in water, whereas **hydrophobic**

substances are not. In general, polar solvents are likely to dissolve polar and ionic solutes; nonpolar solvents dissolve only nonpolar solutes.

The **concentration** of a solution is the amount of solute in a given quantity of solution. **Molarity** (*M*) is concentration expressed as moles of solute per liter of solution. **Percent concentration** can be expressed as **weight-to-volume percent (%(w/v)), volume-to-volume percent (%(v/v)),** or **weight-to-weight percent (%(w/w)).** A **dilution** is accomplished by adding solvent to concentrated stock solution, thus lowering the concentration of the stock solution. Stoichiometric calculations involving solutions require interconversion of solution volume and moles of a solute.

The properties of solutions that depend only on the number of solute particles and are independent of solute identity are called **colligative properties.** The colligative properties include vapor-pressure lowering, boiling-point elevation, freezing-point lowering, and osmotic pressure. The extent of freezing-point lowering can be used to determine the molar mass of a nonvolatile solute. For this calculation, solute concentration must be expressed as **molality** (*m*), the number of moles of solute per kilogram of solvent. **Osmotic pressure** is the minimum amount of pressure (applied to the surface of a solution) needed to prevent the passage of water across an osmotic membrane into a solution.

Colloids are homogeneous mixtures of particles 1–100 nm in diameter dispersed in a gas, liquid, or solid. Colloidal dispersions in fluids exhibit **Brownian motion** and can be detected by the **Tyndall effect.** When particles having diameters greater than 100 nm are scattered through a fluid, the resulting heterogeneous mixture is called a **suspension.** These particles will settle by gravity and can be separated from the mixture by filtration.

STUDY QUESTIONS AND PROBLEMS

(More difficult questions and problems are marked with an asterisk.)

WATER: STRUCTURE AND PROPERTIES

1. Compare the molecular arrangement of liquid water to that of ice.
2. Why does a water molecule form more hydrogen bonds in ice than in liquid water?
3. Why is ice less dense than liquid water?
4. What is the basis for the high freezing point, boiling point, heat of vaporization, and heat of fusion of water?

PROPERTIES OF SOLUTIONS

5. Define the following terms:
   a. Solution   c. Solvent
   b. Aqueous solution   d. Solute
6. Give one example each of a gaseous solution, a liquid solution, and a solid solution.
7. Distinguish between the color of a solution and the clarity of a solution.

8. What are the sizes of the dissolved particles in a solution? How can the liquid and solid components of liquid solutions be separated?
9. Describe how salt can be obtained from seawater.

SOLUTION FORMATION

10. Describe the process by which NaCl dissolves in liquid water.
11. Explain what is meant by the terms *hydration* and *solvation*.
12. Write an equation to describe the hydration of each of the following ionic compounds:
    a. $MgCl_2$ c. $Al_2(SO_4)_3$
    b. NaOH   d. $LiNO_3$
13. Explain how each of the following factors affects the rate of dissolving:
    a. Surface area of the undissolved solute
    b. Agitation
    c. Temperature

SOLUBILITY

14. Define the following terms:
    a. Solubility   c. Saturated solution
    b. Miscible   d. Molar heat of solution
15. Distinguish between soluble and insoluble substances.
16. Describe the dynamic equilibrium that exists in a saturated solution.
17. Explain the significance of the sign of the molar heat of solution of a substance.
18. Predict the relative solubilities of the following compounds in water at 25° C, the temperature for which their molar heats of solution are reported.

NaCl $\Delta H_{solution} = 3.89$ kJ/mol

NaI $\Delta H_{solution} = -7.5$ kJ/mol

NaOH $\Delta H_{solution} = -44.3$ kJ/mol

*19. Which of the compounds in question 18 would make the best instant hot pack? Why?

FACTORS AFFECTING SOLUBILITY

20. Explain how each of the following factors affects solubility.
    a. Pressure   c. Chemical structure
    b. Temperature
21. Explain why soda pop bubbles vigorously when first opened.
22. Distinguish between hydrophilic and hydrophobic substances.
23. Explain what is meant by the statement "like dissolves like."

CONCENTRATIONS OF SOLUTIONS: MOLARITY

24. Explain what is meant by the terms *concentration* and *molarity*.
25. Calculate the molarity of each of the following solutions.
    a. 10.0 g of NaCl in 1.00 L of solution
    b. 20.0 g of KCl in 100.0 mL of solution
    c. 15.0 g of LiBr in 500.0 mL of solution

d. 25.0 g of $Mg(NO_3)_2$ in 650.0 mL of solution

e. 12.5 g of Na_3PO_4 in 300.0 mL of solution

26. Calculate the number of grams of solute needed to prepare each of the following solutions.

a. 1.00 L of 0.500 M glucose, $C_6H_{12}O_6$

b. 800.0 mL of 0.100 M NaCl

c. 420.0 mL of 0.250 M Na_3PO_4

d. 150.0 mL of 1.50 M $Al_2(HPO_4)_3$

e. 2.50 L of 0.0200 M $MgCl_2$

f. 750.0 mL of 2.00 M KCl

27. Calculate the number of moles of solute in each of the following solutions.

a. 150.0 mL of 0.100 M NaCl

b. 200.0 mL of 0.250 M KNO_3

c. 600.0 mL of 1.20 M $MgCl_2$

d. 400.0 mL of 0.750 M LiBr

e. 750.0 mL of 3.70 M KOH

f. 450.0 mL of 0.100 M NaOH

CONCENTRATIONS OF SOLUTIONS: PERCENT CONCENTRATION

28. Calculate the %(w/v) concentration of each of the following solutions.

a. 17.0 g of solute in 150.0 mL of solution

b. 24.3 g of solute in 1.00 L of solution

c. 35.7 g of solute in 900.0 mL of solution

d. 14.6 g of solute in 500.0 mL of solution

e. 30.6 g of solute in 745 mL of solution

f. 9.75 g of solute in 200.0 mL of solution

29. Hydrogen peroxide solution for bleaching hair is usually at a concentration of 5%(w/v) in water. How many grams of hydrogen peroxide, H_2O_2, are present in 80.0 mL of this solution?

30. Human blood contains NaCl at a concentration of 0.15 M. What is the %(w/v) concentration of NaCl in human blood?

31. Calculate the %(v/v) concentration of each of the following solutions.

a. 17.0 mL of solute in 150.0 mL of solution

b. 24.7 mL of solute in 250.0 mL of solution

c. 7.80 mL of solute in 125 mL of solution

d. 37.4 mL of solute in 750.0 mL of solution

e. 340.0 mL of solute in 1.50 L of solution

f. 17.2 mL of solute in 3.00 L of solution

32. Calculate the %(w/w) concentration of each of the following solutions.

a. 17.0 g of solute in 150.0 g of solution

b. 24.8 g of solute in 900.0 g of solution

c. 12.3 g of solute in 49.0 g of solution

d. 43.6 g of solute in 275 g of solution

e. 2.35 g of solute in 120.0 g of solution

f. 458 g of solute in 3.25 kg of solution

DILUTIONS

33. Define the following terms:
    a. Reagent   b. Dilution   c. Dilution factor
34. Calculate the final volumes for diluting each of the solutions in question 27 to 0.0500 M.
35. Calculate the final volumes for diluting each of the following solutions to 2.00%(w/v).
    a. 10.0 mL of a 7.50%(w/v) solution
    b. 25.0 mL of a 14.3%(w/v) solution
    c. 20.0 mL of a 10.0%(w/v) solution
    d. 5.00 mL of a 17.5%(w/v) solution
    e. 1.00 mL of a 25.0%(w/v) solution
36. Calculate the volume of each solution in question 27 needed to prepare 1.00 L of 0.0500 M solution.

STOICHIOMETRIC CALCULATIONS INVOLVING SOLUTIONS

37. What volume of 0.150 M NaOH is required to react completely with 2.75 g of H_2SO_4?

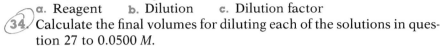

$$2 \text{ NaOH(aq)} + H_2SO_4(l) \longrightarrow Na_2SO_4(aq) + 2 H_2O(l)$$

38. What volume of 0.250 M KCl is required to react completely with 165 mL of 0.300 M $Pb(NO_3)_2$ solution?

$$2 \text{ KCl(aq)} + Pb(NO_3)_2(aq) \longrightarrow PbCl_2(s) + 2 KNO_3(aq)$$

39. Calculate the molarity of $PbNO_3$ if 3.75 g of Na_2SO_4 are required for complete reaction with 500.0 mL of $Pb(NO_3)_2$ solution.

$$Pb(NO_3)_2(aq) + Na_2SO_4(s) \longrightarrow PbSO_4(s) + 2 NaNO_3(aq)$$

40. Calculate the molarity of $CuSO_4$ if 0.195 g of CuS are formed by complete reaction of 750.0 mL of $CuSO_4$ solution.

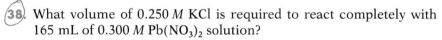

$$Na_2S(aq) + CuSO_4(s) \longrightarrow Na_2SO_4(aq) + CuS(s)$$

COLLIGATIVE PROPERTIES OF SOLUTIONS

41. Define the following terms:
    a. Colligative properties   d. Osmotic membrane
    b. Molality   e. Osmosis
    c. Semipermeable membrane   f. Osmotic pressure
42. Explain why the vapor pressure of a liquid solution containing nonvolatile solute is lower than that of the pure solvent.
43. Explain why the boiling point of a solution containing a nonvolatile solute is higher than that of the pure solvent.
44. How does the freezing point of a liquid solution containing a nonvolatile solute compare with that of the pure solvent?
45. Calculate the molality of each of the following solutions.
    a. 2.30 g of C_2H_5OH dissolved in 400.0 g of water
    b. 49.0 g of H_2SO_4 dissolved in 200.0 g of water

      c. 29.2 g of NaCl dissolved in 2.00 kg of water

      d. 10.0 g of table sugar, $C_{12}H_{22}O_{11}$, dissolved in 300.0 g of water

46. A solution of 12.0 g of sulfur dissolved in 250.0 g of naphthalene has a freezing point of 78.9° C. What is the molar mass of sulfur?

*47. Why does salty soil kill plants? (*Clue:* Consider the effects of osmosis.)

COLLOIDS AND SUSPENSIONS

48. Define the following terms:
      a. Colloid   c. Tyndall effect
      b. Brownian motion   d. Suspension

49. Explain why solutions are perfectly clear but colloidal dispersions are cloudy.

50. Explain how you would distinguish between members of the following pairs.
      a. A solution and a suspension   c. A colloid and a suspension
      b. A solution and a colloid

51. Why don't colloidal particles settle out of colloidal dispersions?

GENERAL EXERCISES

52. What types of substances are most likely to dissolve in water? Why?

53. Predict the relative solubilities of the following compounds in water at 18° C, the temperature for which their molar heats of solution are reported.

Na_2SO_4	$\Delta H_{solution} = 1.9 \text{ kJ/mol}$
$MgSO_4$	$\Delta H_{solution} = 84.87 \text{ kJ/mol}$
$Al_2(SO_4)_3 \cdot 6\ H_2O$	$\Delta H_{solution} = 234.0 \text{ kJ/mol}$

*54. Which of the compounds in question 53 would make the best instant cold pack? Why?

55. For each solution, calculate the volume that contains 10.0 g of solute.
      a. 0.200 M $Al_2(HPO_4)_3$ c. 2.00 M NaCl   e. 0.300 M $C_6H_{12}O_6$
      b. 0.150 M Li_2CO_3 d. 0.150 M K_2SO_4 f. 2.50 M $Ca(NO_3)_2$

*56. Calculate the number of grams of solute needed to prepare each of the following solutions.
      a. 175 mL of 30.0%(w/v) NaOH
      b. 900.0 mL of 10.0%(w/v) NaCl
      c. 775 mL of 20.0%(w/v) $Mg(NO_3)_2$
      d. 300.0 mL of 15.0%(w/v) $CaCl_2$
      e. 650.0 mL of 12.5%(w/v) KBr
      f. 450.0 mL of 17.0%(w/v) K_2SO_4

57. Whiskeys are often described by their *proof*, defined as twice the %(v/v) concentration of ethyl alcohol in solution at 60° F. How many mL of ethyl alcohol, C_2H_5OH, are in 1.00 L of a 90-proof whiskey?

58. In an analysis of NaCl in waste water, 10.00 mL of the waste water required 29.35 mL of 0.1000 M $AgNO_3$ for complete reaction.

$$NaCl(aq) + AgNO_3(aq) \longrightarrow AgCl(s) + NaNO_3(aq)$$

What was the molarity of NaCl in the waste water?

59. A solution of 14.5 g of fructose (a sugar found in fruits and honey) dissolved in 150.0 g of water has a freezing point of $-1.0°$ C. What is the molar mass of fructose?

*60. In making pickles, cucumbers are soaked in a 16%(w/v) solution of NaCl. Why are the resulting pickles shrunken and shriveled? (*Clue:* Consider the effects of osmosis.)

61. Why is homogenized whole milk more opaque than nonfat milk?

CHEMICAL EQUILIBRIUM AND REACTION RATES

CHAPTER 13

OUTLINE

13.1 Reversible Reactions
13.2 Rates of Chemical Reactions
13.3 Activation Energy
13.4 Factors Affecting Reaction Rates
    Temperature
    Concentration
    The Nature of the Reactants
    Catalysis
13.5 Equilibrium Constants
13.6 Application of Equilibrium Constants
13.7 LeChatelier's Principle
    Change in Concentration
    Change in Temperature
13.8 Energy and Chemical Change
    Perspective: Energy Use in the
        United States—Past, Present,
        and Future
13.9 The Relationship between
    Chemical Equilibrium and
    Reaction Rates
    Summary
    Study Questions and Problems

In chapter 9 we mentioned that not all reactions go to completion. In fact most reactions do not use up all of the reactants; instead, they reach a state of dynamic equilibrium in which the reactants and products are continuously interchanging but their concentrations remain constant. These reactions are called *reversible reactions*. Reversible reactions are similar in concept to other reversible processes, such as vaporization of a liquid by heating and subsequent liquefaction of the vapor by cooling. In this chapter we will focus on reversible reactions with gaseous components.

Chemical reactions occur at various rates. Some, like the rusting of iron and the formation of petroleum, are notoriously slow. Others, such as the explosion of a firecracker, happen in the blink of an eye. In this chapter, you will learn about factors that affect rates of reactions and how those rates are related to chemical equilibrium.

13.1 REVERSIBLE REACTIONS

Reversible reactions are reactions that proceed in both the forward and the reverse direction at the same time. If left undisturbed, reversible reactions progress toward a state of dynamic equilibrium in which the reactants and products are continuously interchanging but the concentrations of reactants and products remain constant. All reversible reactions will achieve this equilibrium if left undisturbed for a long enough period of time. Consider a hypothetical reaction in which reactant A is reversibly converted to product B. Let's suppose that we start the reaction with only substance A present. Then, at the beginning of the reaction, A produces B:

$$A \longrightarrow B$$

If this were the only reaction that occurred, then all of substance A would be converted to substance B, and the reaction would go to completion. Since our hypothetical reaction is reversible, there must also be a reaction in the reverse direction; that is, some of substance B must revert to substance A. Thus, as soon as some of substance B is formed, the reverse reaction begins:

$$B \longrightarrow A$$

In other words, formation of substance B allowed the reverse reaction to occur. If we were to analyze the reaction in the laboratory, we would find that the amount of substance A decreases initially, and the amount of substance B increases initially. But as time passes, the amounts of these two substances change less and less until they become constant (see Figure 13-1). From that time on, the reaction is in a state of dynamic equilibrium. Formula units of A and B are continuously interchanging, but the rate of the forward reaction is equal to the rate of the reverse reaction at equilibrium.

We can describe our hypothetical reaction with the following equation,

$$A \Longrightarrow B$$

Hypothetical

Assumed by hypothesis.

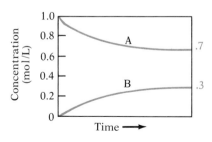

FIGURE 13-1 Changes in Concentrations of A and B as the Equilibrium A ⇌ B Is Approached. At the beginning there is 1.0 mol/L of A and no B.

where the two opposing arrows ($\rightleftharpoons$) indicate that the reaction is reversible.

13.2 RATES OF CHEMICAL REACTIONS

The rusting of iron proceeds at an imperceptibly slow rate, as does the tarnishing of silver. Why is it that some spontaneous events are slow, while others, such as combustion, are fast? To answer this question, we must consider rates of chemical reactions and the factors that influence those rates. The **rate of a chemical reaction** is the number of formula units of a product formed in a given period of time. The study of chemical reaction rates is referred to as **chemical kinetics**.

13.3 ACTIVATION ENERGY

Energy changes in chemical reactions are the result of energy differences between the reactants and products. However, in order to understand why some reactions are slow and others are fast, we must examine the energy changes that occur at various stages of the reaction between reactants and products. The hypothetical route by which the reactants in a chemical reaction are converted to products is called the **reaction pathway**.

We can compare a reaction pathway to a bicycle route. Suppose that you want to ride to a destination across town. One route, path A, has a high hill between your point of origin and your destination (see Figure 13-2). Another route, path B, is the same length, but it runs beside the hill so there are no major changes in elevation along the way. You know that if you take path A, you will have to pedal up the hill, and the rate of travel will be slow. On the other hand, if you take path B, the distance will be the same, but the ride will take less time because there is not such a steep uphill portion. You would probably choose path B, because you would not have to expend as much energy and would arrive at your destination more quickly than if you took path A. Path A has a high energy barrier that must be crossed to reach your destination, while path B contains only a negligible energy barrier.

Chemical reactions proceed across similar energy barriers; some of these are high and some are low. Figure 13-3 shows reaction pathways for

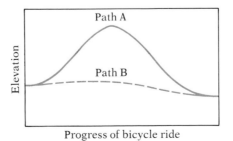

FIGURE 13.2 A Bicycle Ride across Town. Path A has a high hill, but path B has no major changes in elevation.

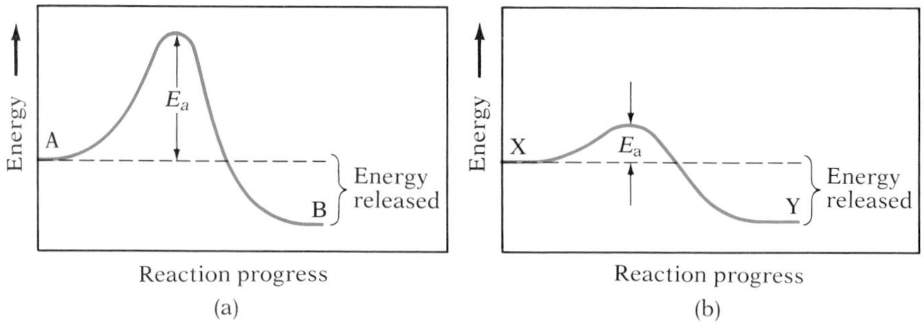

FIGURE 13-3 Reaction Pathways. In (a), reaction A → B has a relatively high activation energy. In (b), reaction X → Y has a relatively low activation energy.

two hypothetical reactions, A → B and X → Y. In each case, energy is released by the reaction, but reaction X → Y will occur at a faster rate because the energy barrier is lower. In order for reactants to be converted to products in any chemical reaction, the reactant formula units must acquire enough energy to cross the energy barrier in the reaction pathway. The amount of energy needed to cross the barrier is called the **activation energy** (E_a). Reactions with relatively low activation energies will proceed at faster rates than those with relatively high activation energies.

In order to be transformed into product formula units, reactant formula units must acquire enough energy (the activation energy) to overcome forces that tend to keep them as they are. We might imagine that when the reactant formula units have acquired their activation energy, their structures are in a transition form midway between the structures of unreacted starting formula units and the structures of product formula units. The particular structure (whatever it might be) that is midway between the structures of reactants and products and possesses the energy of activation is called the **activated complex** or the **transition state**.

13.4 FACTORS AFFECTING REACTION RATES

Chemical reactions show a wide range of rates; some take weeks or months to proceed to a measurable extent, while others are over in a fraction of a second. The most important factors that influence chemical reaction rates are discussed in this section.

Temperature An increase in the temperature of a substance is an indication that the substance has gained heat energy. The atoms within molecules vibrate about their bonds, and when heat energy is absorbed by the molecules, they move faster and their internal vibrations become faster and stronger. For a decomposition reaction such as

$$CO_2(g) \longrightarrow C(s) + O_2(g)$$

an increase in temperature causes the atoms within the reactant molecules to vibrate more strongly, so the molecules are more easily broken apart. Thus, a temperature increase speeds up the reaction by providing energy to the reactant and allowing more of its molecules to acquire the activation energy necessary for decomposition.

In a chemical system in which two molecules form a product, such as

$$H_2(g) + I_2(g) \longrightarrow 2\ HI(g)$$

the reactant molecules must collide with a certain amount of energy in order to react. Thus, when temperature is increased, the reactant molecules move faster, causing them to collide more frequently and with greater impact. As in the previous case, the higher temperature increases the reaction rate. In fact, this is true for all chemical reactions, and the converse is also true: cooling decreases rates of chemical reactions.

Concentration For reactions involving two or more reacting compounds, increasing the concentration of one or more reactants brings about more frequent collisions and thus faster reaction rates. In reactions where there is only one reacting compound, increasing the concentration of the reactant provides more molecules to react, and thus there are more molecules reacting during a given time period. Thus, increased concentrations of reactants increase rates of chemical reactions. For a reaction occurring in the gaseous state, decreasing the container volume confines the reacting molecules to a smaller space, increasing the concentration of reactants and thus the reaction rate.

Rates of chemical reactions are measured experimentally; they are usually expressed in units of concentration change per unit time, such as moles/liter per second ($mol\ L^{-1}\ sec^{-1}$) or moles/liter per minute ($mol\ L^{-1}\ min^{-1}$). It is possible to determine experimentally how much the rate of a reaction depends on concentration.

A reaction whose rate is dependent on only one reactant concentration is called a **unimolecular reaction.** Unimolecular reactions are described by the general equation

$$A \longrightarrow B$$

For unimolecular reactions, the concentration dependence of the reaction rate is described by the following expression

$$rate = k[A]$$

where the brackets denote concentration expressed in mol/L. Thus, the rate of the reaction varies directly with the concentration of reactant (at constant temperature). The mathematical relationship between the rate of a chemical reaction and concentration is called a **rate equation** or **rate law.** The constant of proportionality, k, is called the **rate constant.** Rate constants, like rates of reactions, are experimentally measured quantities.

A reaction whose rate is determined by the collision of two reactant formula units is called a **bimolecular reaction.** Rates of bimolecular reactions vary directly with the concentrations of both reactants:

$$A + B \longrightarrow C$$
$$rate = k[A][B]$$

In both unimolecular and bimolecular reactions, the rate is related to concentration by a rate constant. Every reaction has its own characteristic

MATH TIP
Remember that n^{-1} is the same as $1/n$; therefore,

$$mol\ L^{-1}\ sec^{-1} = \frac{mol}{L\ sec}$$

and

$$mol\ L^{-1}\ min^{-1} = \frac{mol}{L\ min}$$

rate constant, and only a change in temperature will alter the value of a rate constant.

EXAMPLE 13.1   For the following hypothetical reaction,

$$2\,A \longrightarrow B$$

a. write the rate equation.
b. describe the effect on the rate caused by doubling the concentration of A.

Solution   a. Since this is a bimolecular reaction, the rate will depend on the square of the concentration of A:

$$rate = k[A][A] = k[A]^2$$

b. Assume that the original concentration of A was 1 mol/L; then the original rate was

$$rate = k(1\ mol/L)^2 = 1\ k\ mol^2/L^2$$

If the concentration of A is doubled, then the new concentration is 2 mol/L, and the new rate is

$$rate = k(2\ mol/L)^2 = 4\ k\ mol^2/L^2$$

Thus, doubling the concentration of A gives a new rate that is four times as high as the first rate.

MATH TIP
Be sure to square all quantities in the parentheses.

$$rate = k(2\ mol/L)^2$$
$$= k(2)^2(mol/L)^2$$
$$= (k)(4)(mol^2/L^2)$$

EXERCISE 13.1   For the following reaction,

$$CH_3CHO(g) \longrightarrow CH_4(g) + CO(g)$$

a. write the rate equation.
b. describe the effect on the rate caused by decreasing the concentration of CH_3CHO to one-half its original value.   ■

The Nature of the Reactants   If temperature and concentration were the only factors affecting reaction rates, then all reactions at the same temperature and concentration would have the same rate. However, experimental evidence demonstrates a variety of reaction rates. Reactions differ in their rates because each individual reaction has its own characteristic activation energy and its own rate constant. These quantities are unique to each reaction because of the nature of the reactants, i.e., their individual structures and how susceptible their structures are to change. In general, reactions between ions have low activation energies and hence fast rates. In contrast, reactions of molecules are relatively slow because the activation energies for breaking covalent bonds and forming new ones are high; large amounts of energy are needed to overcome the original bonding forces. Thus, reactions between molecules are usually slower than reactions between ions.

Catalysis   A **catalyst** is a substance that does not undergo permanent change but affects the rate of a chemical reaction by its presence in the reaction mixture. Thus, a catalyst can be recovered in its original form

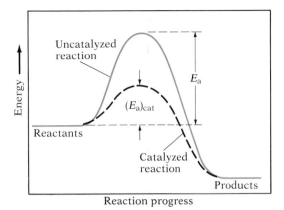

FIGURE 13-4 Comparison of the Reaction Pathways of a Catalyzed Reaction and the Corresponding Uncatalyzed Reaction

after the reaction is over. The action of a catalyst is referred to as **catalysis**. Most catalysts increase the rates of specific chemical reactions.

The question of how catalysts work is the subject of much current research. There may be many possible answers, but it appears that a catalyst speeds up a reaction by providing a new reaction pathway in which the activation energy is lower than in the absence of the catalyst, as illustrated in Figure 13-4. For reversible reactions, catalysts affect the forward and reverse rates to the same extent.

The commercial production of ammonia by the Haber process

$$N_2(g) + 3\ H_2(g) \rightleftharpoons 2\ NH_3(g)$$

uses a catalyst consisting of iron (Fe), iron(III) oxide (Fe_2O_3), potassium oxide (K_2O), and aluminum oxide (Al_2O_3). The catalyst accelerates attainment of equilibrium, so that formation of the maximum yield of ammonia occurs fast enough to make commercial production economically feasible.

Some of the most interesting catalysts are the naturally occurring ones called enzymes. Enzymes are protein molecules in living organisms. They speed up the rates of life-sustaining chemical reactions, giving living organisms the ability to respond very quickly to changes in their environments. Enzymes catalyze virtually all chemical reactions in living cells, including reactions responsible for digestion, energy storage and release, and vision.

Enzyme
(EN-zime)

13.5 EQUILIBRIUM CONSTANTS

In our hypothetical reversible reaction of section 13.1,

$$A \rightleftharpoons B$$

the concentration ratio of A to B at equilibrium was 7 : 3. For real reactions, the concentration ratios of reactant(s) to product(s) can be determined in the laboratory, and they are found to be independent of the initial amounts of reactants and products. In a reversible reaction such as A $\rightleftharpoons$ B, we would find exactly the same equilibrium ratio of concentrations regardless of whether we started with one component or any mixture of the two components. The ratio of component concentrations in an equilibrium mixture of a reversible reaction varies only with the temperature of the mixture. This constant ratio of component concentrations in the equilibrium mixture formed by a reversible reaction at a specific temperature is called the **equilibrium constant** for the reaction. The mathematical expression for an equilibrium constant is always written in the following form,

$$K_{eq} = \frac{[\text{product}]}{[\text{reactant}]}$$

where the brackets denote concentrations expressed in mol/L. Thus, the equilibrium constant of our hypothetical reversible reaction is

$$K_{eq} = \frac{[\text{B}]}{[\text{A}]}$$

This is the standard form of the equilibrium constant of a reversible reaction having one reactant and one product. If A and B are gases with equilibrium concentrations of 0.7 mol/L and 0.3 mol/L, respectively, we can insert the equilibrium concentrations into the equilibrium expression and calculate the numerical value of the equilibrium constant:

$$K_{eq} = \frac{[\text{B}]}{[\text{A}]}$$
$$= \frac{0.3 \ \cancel{\text{mol/L}}}{0.7 \ \cancel{\text{mol/L}}}$$
$$= 0.43$$

(In this case, K_{eq} has no units.)

The equation for our hypothetical reaction can also be written as

$$\text{B} \rightleftharpoons \text{A},$$

since both A and B are present in the equilibrium mixture. When the equation is written in this direction, the expression for the equilibrium constant becomes

$$K_{eq} = \frac{[\text{A}]}{[\text{B}]}$$

System

An isolated collection of substances.

Our hypothetical reversible reaction is a very simple **system**, because it contains only one reactant and only one product. Let's consider a slightly more complex reaction. Hydrogen reacts reversibly with iodine vapor according to the following equation:

$$\text{H}_2(g) + \text{I}_2(g) \rightleftharpoons 2\ \text{HI}(g)$$

By analysis of the reaction mixture at equilibrium, it has been found that the mathematical expression for the equilibrium constant is

$$K_{eq} = \frac{[\text{HI}]^2}{[\text{H}_2][\text{I}_2]}$$

Notice that the concentration of the product has an exponent of 2, and the concentrations of the two reactants are multiplied together. Analysis of mixtures of reactants and products of countless reversible reactions has disclosed a general pattern for the equilibrium constant expression. Using the following as a generalized reversible reaction,

$$a\text{A} + b\text{B} + c\text{C} + \cdots \rightleftharpoons d\text{D} + e\text{E} + f\text{F} + \cdots$$

any equilibrium constant can be expressed as

$$K_{eq} = \frac{[\text{D}]^d[\text{E}]^e[\text{F}]^f \cdots}{[\text{A}]^a[\text{B}]^b[\text{C}]^c \cdots} \tag{1}$$

The uppercase letters (A, B, C, D, E, F) represent chemical substances, and the lowercase letters (a, b, c, d, e, f) are their coefficients in the balanced equation. This general expression for the equilibrium constant can be used for any number of reactants and products provided they form a homogeneous mixture. The units of the calculated equilibrium constant will vary, depending on the exponents in the equilibrium constant expression. The following examples illustrate formulation of the equilibrium constant expression and calculation of the equilibrium constant.

EXAMPLE 13.2   Write the expression for the equilibrium constant of the reversible formation of nitrogen oxide:

$$\text{N}_2(g) + \text{O}_2(g) \rightleftharpoons 2\,\text{NO}(g)$$

Solution   Referring to the general expression for an equilibrium constant and substituting actual reactants and products,

$$K_{eq} = \frac{[\text{NO}]^2}{[\text{N}_2][\text{O}_2]}$$

EXERCISE 13.2   About 14 million tons of ammonia are used in this country each year as fertilizer. The large-scale production of ammonia is carried out at high temperature and pressure by the Haber process:

$$\text{N}_2(g) + 3\,\text{H}_2(g) \rightleftharpoons 2\,\text{NH}_3(g)$$

Write the expression for the equilibrium constant of the reversible formation of ammonia by the Haber process.   ■

EXAMPLE 13.3   Find the value of the equilibrium constant for the following reaction,

$$\text{N}_2(g) + \text{O}_2(g) \rightleftharpoons 2\,\text{NO}(g)$$

if the equilibrium concentrations are as follows:

$$[\text{N}_2] = 0.95 \text{ mol/L}$$
$$[\text{O}_2] = 0.95 \text{ mol/L}$$
$$[\text{NO}] = 0.10 \text{ mol/L}$$

Solution

In example 13.2 we found that the equilibrium constant expression for this reaction is

$$K_{eq} = \frac{[NO]^2}{[N_2][O_2]}$$

By substituting actual concentration values into this expression, we can calculate the equilibrium constant:

$$K_{eq} = \frac{(0.10 \text{ mol/L})^2}{(0.95 \text{ mol/L})(0.95 \text{ mol/L})}$$

$$= \frac{(0.10 \text{ mol/L})(0.10 \text{ mol/L})}{(0.95 \text{ mol/L})(0.95 \text{ mol/L})}$$

$$= 1.1 \times 10^{-2}$$

MATH TIP

This problem can be solved on the calculator two ways: Press 0.10 $\boxed{\div}$ 0.95 $\boxed{\times}$ 0.10 $\boxed{\div}$ 0.95 $\boxed{=}$. The display reads 0.011080332. The answer can be written as 1.1×10^{-2}. The other method involves using the square key $\boxed{x^2}$. Press 0.10 $\boxed{x^2}$ $\boxed{\div}$ 0.95 $\boxed{x^2}$ $\boxed{=}$. The display is the same as the other method. Either method is acceptable; however use of the square key reduces the number of calculator entries.

EXERCISE 13.3

Nitrogen and hydrogen were mixed at 1000 K and allowed to establish equilibrium with ammonia:

$$N_2(g) + 3 H_2(g) \rightleftharpoons 2 NH_3(g)$$

Calculate the equilibrium constant at 1000 K if the equilibrium concentrations were as follows:

$$[N_2] = 0.9783 \text{ mol/L}$$
$$[H_2] = 0.935 \text{ mol/L}$$
$$[NH_3] = 0.0434 \text{ mol/L}$$

13.6 APPLICATION OF EQUILIBRIUM CONSTANTS

The magnitude of an equilibrium constant can be used to estimate the extent of product formation in the corresponding reaction. Since the equilibrium constant is a ratio of molar concentrations of product(s) and reactant(s), its value provides information about relative amounts of product(s) and reactant(s) that exist at equilibrium. In example 13.3, we calculated an equilibrium constant for the reaction of nitrogen and oxygen to form nitrogen oxide:

$$N_2(g) + O_2(g) \rightleftharpoons 2 NO(g)$$
$$K_{eq} = 1.1 \times 10^{-2}$$

The small value of the equilibrium constant indicates that the equilibrium concentration of product is much less than the product of the equilibrium concentrations of reactants when all concentrations are raised to their appropriate powers. If we use the term **position of equilibrium** to designate the relative concentrations of products and reactants at equilibrium, we can say that the position of equilibrium for the formation of NO lies to the left, favoring formation of N_2 and O_2 from NO. Arrows of different lengths are sometimes used to indicate the position of equilibrium:

$$N_2(g) + O_2(g) \rightleftharpoons 2 NO(g)$$

Now consider another reversible reaction and its equilibrium constant at 1073 K:

$$CO(g) + H_2O(g) \rightleftharpoons CO_2(g) + H_2(g)$$
$$K_{eq} = 1.25$$

In this case, the value of the equilibrium constant is greater than 1. This means that the product of the equilibrium concentrations of products are greater than that of reactants, when all are raised to their appropriate powers. In this instance, the position of equilibrium lies to the right, favoring product formation. Arrows of different lengths can be used in this equation also:

$$CO(g) + H_2O(g) \rightleftharpoons CO_2(g) + H_2(g)$$

In general, when the value of an equilibrium constant is less than 1, the equilibrium favors reactants, and the position of equilibrium lies on the left. When the value of an equilibrium constant is greater than 1, product formation is favored, and the position of equilibrium lies on the right. When values of equilibrium constants are very large, for example, 10^6, the reaction is considered to go to completion. On the other hand, when the value of an equilibrium constant is very small, for example, 10^{-6}, the reaction is not considered to occur.

13.7 LECHATELIER'S PRINCIPLE

LeChatelier
(leh-shah-tel-YAY)

In 1888, Henri Louis LeChatelier, a French chemist, formulated what he called the "law of stability of chemical equilibrium." Today we refer to his law as **LeChatelier's principle.** It is summarized as follows: *if a chemical system in equilibrium is disturbed, the system will readjust in such a way as to offset the disturbance partially and restore equilibrium.* To understand LeChatelier's principle, we must remember that a chemical equilibrium is a dynamic state. The forward and reverse reactions occur at equal rates, and the system is in balance. However, if conditions are altered, causing an imbalance of forward and reverse rates, the rates of those processes change in order to achieve a new state of balance.

The two major factors that can disturb a chemical equilibrium are a change in reactant or product concentration and a change in temperature. These factors are discussed in the following paragraphs.

Change in Concentration   Consider the formation of ammonia described in exercise 13.2:

$$N_2(g) + 3 H_2(g) \rightleftharpoons 2 NH_3(g)$$

If we add H_2 to the equilibrium mixture, the concentrations of N_2 and NH_3 will readjust themselves spontaneously so that the concentration of H_2 is reduced back toward its equilibrium concentration. In the readjustment, N_2 and H_2 are used to form additional NH_3. Thus, increasing the concentration of H_2 above the equilibrium concentration causes formation of more NH_3, and the equilibrium is said to be shifted to the right,

$$N_2(g) + 3\ H_2(g) \rightleftharpoons 2\ NH_3(g)$$
$$\text{increased } [H_2] \longrightarrow$$

forming more product. The equilibrium constant does not change, but the system establishes a new equilibrium in which the concentrations of components are different from those of the original equilibrium. Since H_2 is used to form the additional NH_3, the concentration of H_2 decreases toward its original equilibrium value. N_2 is also used to form the additional NH_3, and thus its equilibrium concentration decreases below the original equilibrium value.

Let's look at the equilibrium constant in this situation. Remember that the value of the equilibrium constant for the reaction is always the same at any given temperature.

$$K_{eq} = \frac{[NH_3]^2}{[N_2][H_2]^3}$$

If $[H_2]$ is increased beyond its first equilibrium value, the value of the denominator of the equilibrium constant expression is increased. In order for K_{eq} to remain constant, $[NH_3]$ must increase and $[N_2]$ must decrease. Hence, more NH_3 is formed at the expense of reactants.

Following the reasoning used in this example, we can conclude that for a reaction at equilibrium, increasing the concentration of any or all of the reactants will cause formation of additional product(s) and shift the equilibrium to the right. The reverse situation is also true: for a reaction at equilibrium, increasing the concentration of product(s) will cause an increase in the amount of reactant(s) and shift the equilibrium to the left. As long as the temperature is constant, the equilibrium constant will not change, but the equilibrium will shift in the direction indicated below in response to changes in concentrations.

$$\text{reactants} \rightleftharpoons \text{products}$$
$$\text{increased [reactants]} \longrightarrow$$
$$\longleftarrow \text{increased [products]}$$

EXAMPLE 13.4   Predict the concentration changes that would occur and the direction in which each of the following equilibria would shift in response to the indicated change.

a. $H_2(g) + I_2(g) \rightleftharpoons 2\ HI(g)$
   Increase in $[H_2]$
b. $H_2(g) + CO_2(g) \rightleftharpoons H_2O(g) + CO(g)$
   Increase in $[H_2O]$
c. $N_2(g) + O_2(g) \rightleftharpoons 2\ NO(g)$
   Decrease in $[O_2]$

Solution   a.

$$K_{eq} = \frac{[HI]^2}{[H_2][I_2]}$$

An increase in $[H_2]$ would cause an increase in $[HI]$, a decrease in $[I_2]$, and a shift of the equilibrium to the right.

b.
$$K_{eq} = \frac{[H_2O][CO]}{[H_2][CO_2]}$$

An increase in $[H_2O]$ would cause an increase in $[H_2]$ and $[CO_2]$, a decrease in $[CO]$, and a shift of the equilibrium to the left.

c.
$$K_{eq} = \frac{[NO]^2}{[N_2][O_2]}$$

A decrease in $[O_2]$ would cause a decrease in $[NO]$, an increase in $[N_2]$, and a shift of the equilibrium to the left.

EXERCISE 13.4 Predict the concentration changes that would occur and the direction in which each of the following equilibria would shift in response to the indicated change.

a. $N_2O_4(g) \rightleftharpoons 2\ NO_2(g)$
Increase in $[NO_2]$
b. $2\ CO_2(g) \rightleftharpoons 2\ CO(g) + O_2(g)$
Increase in $[CO_2]$
c. $2\ H_2(g) + O_2(g) \rightleftharpoons 2\ H_2O(g)$
Decrease in $[H_2O]$ ■

In chemical equilibria containing one or more gases, a change in volume is equivalent to a change in concentration of the gas(es). For example, decreasing the volume of a gas confines the gaseous molecules to a smaller volume and thus increases their concentration. Conversely, allowing a gas to occupy a larger volume decreases the concentration of the gas(es). In a chemical equilibrium composed entirely of gases, a decrease in container volume shifts the equilibrium in whichever direction requires a smaller total gaseous volume. Since the gaseous volume is directly related to the number of moles of gases at constant temperature, the equilibrium will shift to the side of the balanced equation that contains fewer moles of gas. In the Haber synthesis of ammonia,

$$N_2(g) + 3\ H_2(g) \rightleftharpoons 2\ NH_3(g)$$

decreasing the volume of the container will shift the equilibrium to the right, because the two moles of NH_3 require less volume than the four moles of reactant gases. Increasing the volume of the container has just the opposite effect on gaseous equilibria. However, if the number of moles of gaseous reactant and product are equal in the balanced equation, a change in applied pressure (and thus a change in volume) will have no effect on the equilibrium.

EXAMPLE 13.5 How will each of the following equilibria respond to a decrease in container volume?

a. $2\ NO_2Cl(g) \rightleftharpoons 2\ NO_2(g) + Cl_2(g)$
b. $2\ IBr(g) \rightleftharpoons I_2(g) + Br_2(g)$
c. $2\ SO_2(g) + O_2(g) \rightleftharpoons 2\ SO_3(g)$

Solution a. The equilibrium will shift to the left, because two moles of NO_2Cl requires less volume than the three moles of products.

b. The equilibrium will remain unchanged, because in the balanced equation the number of moles of reactant is the same as the number of moles of products.

c. The equilibrium will shift to the right, because two moles of SO_3 requires less volume than the three moles of reactants.

EXERCISE 13.5 How will each of the following equilibria respond if the container volume is increased?

a. $CH_4(g) + 2\ H_2S(g) \rightleftharpoons CS_2(g) + 4\ H_2(g)$
b. $PCl_5(g) \rightleftharpoons PCl_3(g) + Cl_2(g)$
c. $N_2(g) + O_2(g) \rightleftharpoons 2\ NO_2(g)$ ■

Change in Temperature Almost every equilibrium constant changes when the temperature of the system is changed, because some chemical reactions release heat and others absorb heat. For example, heat release accompanies formation of ammonia,

$$N_2(g) + 3\ H_2(g) \rightleftharpoons 2\ NH_3(g) + heat$$

while heat absorption is necessary for decomposition of carbon dioxide,

$$2\ CO_2(g) + heat \rightleftharpoons 2\ CO(g) + O_2(g)$$

As illustrated in the equations, heat energy can be considered a reactant or a product in chemical reactions. For exothermic reactions, such as the formation of ammonia, increasing the temperature of the system by adding heat can be thought of as increasing the amount of a product (except that the heat does not appear in the expression for the equilibrium constant). In response to the stress of the higher temperature, the equilibrium changes by forming larger amounts of reactants and smaller amounts of products. The concentration ratio of product to reactant decreases, and the actual position of equilibrium is shifted to the left; thus, K_{eq} is decreased. The opposite effect occurs with endothermic reactions, such as the decomposition of carbon dioxide. In these cases, an increase in the temperature of the system causes an increase in the amount of product(s), a decrease in the amount of reactant(s), and an increase in K_{eq}. In both exothermic and endothermic reactions, K_{eq} changes with temperature. The effect of temperature on the position of equilibrium is another manifestation of LeChatelier's principle.

13.8 ENERGY AND CHEMICAL CHANGE

Most chemical reactions occur in the presence of constant pressure from the atmosphere. In these cases, any heat absorbed or released by the chemical reaction is called the **enthalpy change (ΔH).** For example, burning natural gas (methane) produces heat energy and is thus classified as an exothermic reaction.

$$CH_4(g) + 2\ O_2(g) \longrightarrow CO_2(g) + 2\ H_2O(l) + heat$$
$$\Delta H = -882.2\ kJ/mol$$

On the average, each person in the United States uses about 1.3 million kilojoules of energy each day. This amounts to about 100 times the average food-energy requirement per person. The use of energy increases each year, and in the United States it is expected to reach nearly twice its current value by the year 2000.

The sources of our energy have changed over time. Until 1850, wood supplied about 90 percent of energy used in the United States. Coal then became important as an energy source, and its use reached a maximum at about 1910, when coal supplied 75% of our energy. At that time other sources came into large-scale use, and at present oil supplies 46% of our energy, as compared to 31% by natural gas, 19% by coal, 2% by hydroelectric power, and 2% by nuclear power.

On the basis of past energy use, it is predicted that the United States could run out of oil and natural gas by the year 2000 unless other energy sources are found and energy use remains constant. For this reason, much effort is going into developing alternate energy sources. Among these are nuclear energy, solar energy, geothermal energy (heat from inside the earth), and energy from the winds and tides. Although all of these are important alternate energy sources, perhaps the one to show greatest promise is coal energy. Coal is our most plentiful fossil fuel (fuel formed from decayed plant and animal material). One reason that coal is not used to a larger extent now is that it produces more air pollution than do other fuels. Another is that most of the coal left in the United States is in the west, and shipping it to the East Coast, where energy use is heaviest, is very expensive.

One way of solving both these problems is to convert coal into a gaseous form, called synthetic natural gas. The conversion, called coal gasification, removes sulfur, the principal pollutant, and the resulting gas could be easily transported by pipelines. Coal gasification involves the addition of hydrogen and superheated steam to pulverized coal, producing a mixture of carbon monoxide (CO), unreacted hydrogen (H_2), and methane (CH_4). All of these are gases that could serve as fuel to provide energy much the same way as natural gas does currently.

The heat production per mole of methane is the change in enthalpy (ΔH) for the reaction. We designate release of heat by placing a negative sign before the value of the enthalpy change; thus negative ΔH values mean that the reaction is exothermic.

We interpret heat release in a chemical reaction to mean that the reactants (the initial state) contain more energy than the products (the final state). The energy difference between reactants and products is therefore equal to the heat release (see Figure 13-5).

When the reverse reaction occurs, 882.2 kJ of heat would have to be supplied to the reaction for each mole of CO_2 used up:

$$CO_2(g) + 2\ H_2O(l) + heat \longrightarrow CH_4(g) + 2\ O_2(g)$$
$$\Delta H = 882.2\ kJ/mol$$

The heat production of the first reaction is the same as the heat required by the reverse reaction because energy is neither created nor destroyed in a chemical reaction.

Enthalpy changes are measured in a **calorimeter,** illustrated in Figure 13-6. In this device, a vessel containing reactants is immersed in a measured quantity of water. The heat absorbed or released by the reaction is reflected by a change in water temperature, and the specific heat of water is used to calculate calories of heat responsible for the change in water temperature.

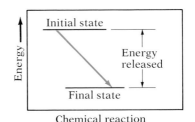

FIGURE 13-5 The Energy Change in a Chemical Reaction

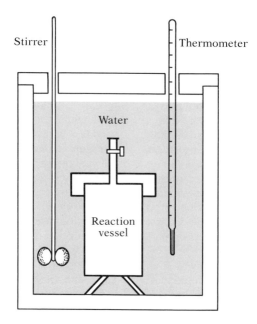

FIGURE 13-6   Diagram of a Calorimeter

13.9   THE RELATIONSHIP BETWEEN CHEMICAL EQUILIBRIUM AND REACTION RATES

The equilibrium constant for a chemical reaction is usually determined by measuring the concentrations of all substances present in the equilibrium mixture. But let's consider the relationship of the equilibrium constant to reaction rates. The reaction between hydrogen and iodine is a good example:

$$H_2(g) + I_2(g) \rightleftharpoons 2\ HI(g)$$

When the reaction is at equilibrium, the rates of the two opposing reactions are equal:

rate of forward reaction = rate of reverse reaction

If we substitute rate expressions into this relationship, we get

$$k_f[H_2][I_2] = k_r[HI][HI] = k_r[HI]^2$$

where k_f is the rate constant for the forward reaction and k_r is the rate constant for the reverse reaction. Rearrangement of terms gives

$$\frac{[HI]^2}{[H_2][I_2]} = \frac{k_f}{k_r}$$

We recognize that

$$\frac{[HI]^2}{[H_2][I_2]} = K_{eq}$$

We can make one final substitution and arrive at an expression that relates the equilibrium constant to the forward and reverse rate constants:

$$K_{eq} = \frac{k_f}{k_r}$$

We have just proved that the equilibrium constant for the reaction is the ratio of rate constants for the forward and reverse reactions. *All equilibrium constants are ratios of rate constants.* Thus there is another way to determine equilibrium constants. If it is possible to measure individual rates for the forward and reverse reactions and thus calculate rate constants for the two reactions, then those values can be used to calculate the equilibrium constant.

The relationship between rates of reaction and chemical equilibrium provides a deeper understanding of LeChatelier's principle. If the equilibrium of a chemical reaction is temporarily disrupted by a change in concentration of one or more components, the system is able to readjust itself because the forward and reverse rates change in response to changes in concentrations. (Remember that reaction rates are directly proportional to concentration.) Thus, if concentration changes occur, reaction rates will change in response, but the rate constants and the equilibrium constants remain the same. Rate constants change only when the temperature of the system changes; thus equilibrium constants vary only with temperature.

SUMMARY

Reversible reactions, reactions that proceed in both forward and reverse directions simultaneously, will achieve a state of dynamic equilibrium if left undisturbed.

Each chemical reaction passes through a **reaction pathway** containing an energy barrier called the **activation energy.** Reactions with relatively low activation energies proceed at faster rates than those with relatively high activation energies. An increase in temperature speeds up a reaction, as does an increase in the concentration of one or more reactants. Rates of chemical reactions are directly related to reactant concentrations by a constant of proportionality called the **rate constant.** The chemical nature of the reactants is an important influence on the rate of a chemical reaction. Catalysts also affect reaction rates. Most catalysts increase reaction rates and do so by lowering the activation energy of the reaction. Enzymes are protein molecules that catalyze chemical reactions in life processes.

The **equilibrium constant** for a reaction is the ratio of concentrations of components in an equilibrium mixture. When the value of an equilibrium constant is less than 1 ($K_{eq} < 1$), formation of reactants is favored. When the value of an equilibrium constant is greater than 1 ($K_{eq} > 1$), formation of products is favored. **LeChatelier's principle** states that if a chemical equilibrium is disturbed, the system will readjust in such a way as to offset the disturbance partially and return to equilibrium. A chemical equilibrium will respond to concentration changes by changing the forward and reverse reaction rates but maintaining K_{eq}; it will respond to a temperature change with an alteration in the forward and reverse rate constants and in K_{eq}.

Heat absorbed or released by a chemical reaction is called the **enthalpy change (ΔH)**. Exothermic reactions have negative enthalpy changes, and endothermic reactions have positive enthalpy changes.

STUDY QUESTIONS AND PROBLEMS

(More difficult questions and problems are marked with an asterisk.)

REVERSIBLE REACTIONS

1. What is meant by the term *reversible reaction?*
2. Describe the dynamic equilibrium of a reversible reaction.
3. The equilibrium mixture of the following hypothetical reaction

$$X \rightleftharpoons Y$$

contains the same concentration of X and Y. Sketch a graph showing how concentrations of X and Y change over time
   a. if the reaction is started with 1.00 mol/L of X.
   b. if the reaction is started with 1.00 mol/L of Y.
   c. if the reaction is started with a mixture of 0.25 mol/L of X and 0.75 mol/L of Y.

RATES OF CHEMICAL REACTIONS

4. Do spontaneous processes always occur rapidly? Explain your answer.
5. What do the terms *rate of reaction* and *chemical kinetics* mean?

ACTIVATION ENERGY

6. Define the following terms:
   a. Reaction pathway
   b. Activation energy
   c. Activated complex (or transition state)
7. How are rates of chemical reactions related to activation energies?
8. Reaction 1 has a rate of 1.75×10^4 mol L^{-1} sec^{-1}, while reaction 2 has a rate of 2.98×10^2 mol L^{-1} sec^{-1}. Both rates were measured at the same temperature. Which reaction has the higher activation energy? Explain your reasoning.

FACTORS AFFECTING REACTION RATES

9. Define the following terms:
   a. Unimolecular reaction   d. Rate constant
   b. Bimolecular reaction   e. Catalyst
   c. Rate equation (or rate law)   f. Catalysis
10. How does each of the following factors affect rates of chemical reactions?
    a. Temperature   c. Nature of reactants
    b. Concentration   d. Catalysis

11. For the general reaction

$$A + B \longrightarrow C + D$$

what is the effect on the rate of
a. doubling the concentration of A?
b. doubling the concentration of B?
c. doubling the concentration of both A and B?

12. For the following reaction:

$$H_2(g) + Cl_2(g) \longrightarrow 2 HCl(g)$$

a. Write the rate equation.
b. Suggest possible units for the rate constant.

13. Answer the following questions about a catalyzed reaction, and give reasons for your answers.
a. If the catalyst increases the rate of the forward reaction by a factor of 1000, what is the effect on the rate of the reverse reaction?
b. If the catalyst increases the rate of the forward reaction by a factor of 500, what is the effect on the equilibrium constant?

EQUILIBRIUM CONSTANTS

14. Explain what an equilibrium constant is.

15. Write the expression for the equilibrium constant for each reaction.
a. $4 NH_3(g) + 5 O_2(g) \rightleftharpoons 4 NO(g) + 6 H_2O(g)$
b. $2 NO_2(g) \rightleftharpoons N_2O_4(g)$
c. $2 SO_3(g) \rightleftharpoons 2 SO_2(g) + O_2(g)$
d. $PCl_5(g) \rightleftharpoons PCl_3(g) + Cl_2(g)$
e. $H_2(g) + Cl_2 \rightleftharpoons 2 HCl(g)$

16. An equilibrium mixture for the reaction

$$2 H_2S(g) \rightleftharpoons 2 H_2(g) + S_2(g)$$

had the following concentrations:

$$[H_2S] = 1.0 \text{ mol/L}$$
$$[H_2] = 0.20 \text{ mol/L}$$
$$[S_2] = 0.80 \text{ mol/L}$$

Calculate the equilibrium constant.

17. Calculate the equilibrium constant for the reaction

$$H_2(g) + I_2(g) \rightleftharpoons 2 HI(g)$$

on the basis of the following equilibrium concentrations:

$$[H_2] = 0.90 \text{ mol/L}$$

$$[I_2] = 0.40 \text{ mol/L}$$

$$[HI] = 0.60 \text{ mol/L}$$

18. An equilibrium mixture for the reaction

$$N_2O_4(g) \rightleftharpoons 2 NO_2(g)$$

had the following concentrations:

$$[N_2O_4] = 1.40 \times 10^{-3} \text{ mol/L}$$
$$[NO_2] = 1.72 \times 10^{-2} \text{ mol/L}$$

Calculate the equilibrium constant.

APPLICATION OF EQUILIBRIUM CONSTANTS

19. What does the term *position of equilibrium* mean?

20. How is the magnitude of an equilibrium constant used to predict the position of equilibrium of a reversible reaction?

21. Where is the position of equilibrium for the reactions characterized by the following equilibrium constants?
   a. 1.79×10^2 mol/L   d. 7.86×10^{10} mol²/L²
   b. 4.33×10^{-8} e. 0.321
   c. 1.03 L/mol

22. Classify the reactions characterized by the equilibrium constants in question 21 as complete, incomplete, or nonoccurring.

LECHATELIER'S PRINCIPLE

23. Summarize LeChatelier's principle.

24. Describe the response of a chemical equilibrium to each of the following:
   a. An increase in the concentration of a reactant
   b. An increase in the concentration of a product
   c. A decrease in the concentration of a reactant
   d. A decrease in the concentration of a product

25. Predict the concentration changes that would occur and the direction in which each of the following equilibria would shift in response to the indicated changes:
   a. $N_2(g) + O_2(g) \rightleftharpoons 2\ NO(g)$
      increase in $[O_2]$
   b. $N_2(g) + 3\ H_2(g) \rightleftharpoons 2\ NH_3(g)$
      increase in $[NH_3]$
   c. $CO(g) + H_2O(g) \rightleftharpoons CO_2(g) + H_2(g)$
      decrease in $[H_2]$
   d. $PCl_5(g) \rightleftharpoons PCl_3(g) + Cl_2(g)$
      decrease in $[PCl_5]$

26. How will each of the equilibria in question 25 respond to a decrease in container volume?

27. How does increased temperature affect K_{eq} for an exothermic reaction? For an endothermic reaction?

28. A mixture of $H_2(g)$, $O_2(g)$, and $H_2O(g)$ is in equilibrium in a closed container. Assuming that the formation of water

$$2\ H_2(g) + O_2(g) \rightleftharpoons 2\ H_2O(g)$$

is exothermic, predict the effect of each of the following changes on the concentration of $H_2(g)$:
   a. addition of H_2O to the system

b. removal of O_2 from the system
c. increasing the temperature of the system

ENERGY AND CHEMICAL CHANGE

29. Define the following terms:
    a. Enthalpy change   b. Calorimeter
30. Describe how enthalpy changes for chemical reactions are measured in a calorimeter.

THE RELATIONSHIP BETWEEN CHEMICAL EQUILIBRIUM AND REACTION RATES

31. How can an equilibrium constant be calculated from rate constants?
32. In terms of reaction rates, explain how a chemical equilibrium responds to a change in concentration of one or more components.
33. In terms of rate constants, explain why a change in temperature causes a change in the K_{eq} of a reversible reaction?

GENERAL EXERCISES

34. A rule of thumb is that increasing the temperature 10° C will double the rate of any chemical reaction. How much will a reaction rate increase if the temperature is raised 100° C?
35. For the general reaction

$$A + B \longrightarrow C$$

   a. Write the rate equation.
   b. If the rate of formation of C is 2.5×10^{-5} mol L^{-1} sec^{-1}, what is the value of the rate constant under the following conditions?

$$[A] = 0.20 \text{ mol/L}$$
$$[B] = 0.30 \text{ mol/L}$$

36. The rate of a certain biochemical reaction at body temperature was measured in a laboratory experiment in the absence of enzyme. The rate for the same reaction is 10^5 faster in the human body, where it is catalyzed by an enzyme. Answer the following questions about the reaction, and give reasons for your answers.
   a. How does the enzyme affect the activation energy of the reaction?
   b. How does the enzyme affect the equilibrium constant of the reaction?
37. New automobiles are required by law to have catalytic converters. These devices convert carbon monoxide and unburned gasoline in automobile exhaust to carbon dioxide, thereby minimizing air pollution due to the exhaust. Why do you think it is necessary for a catalyst to be present in the converter?
38. An equilibrium mixture for the reaction

$$PCl_5(g) \rightleftharpoons PCl_3(g) + Cl_2(g)$$

had the following concentrations:

$$[PCl_5] = 0.158 \text{ mol/L}$$
$$[PCl_3] = 0.081 \text{ mol/L}$$
$$[Cl_2] = 0.081 \text{ mol/L}$$

Calculate the equilibrium constant.

*39. At 25° C, K_{eq} for the following reaction is 4.0 L²/mol².

$$3 \text{ C}_2\text{H}_2(g) \rightleftharpoons \text{C}_6\text{H}_6(g)$$

If the equilibrium concentration of C_2H_2 is 0.70 mol/L, what is the concentration of C_6H_6?

 The following reaction is endothermic:

$$PCl_5(g) \rightleftharpoons PCl_3(g) + Cl_2(g)$$

Predict the effect on the equilibrium of each of the following changes:
a. addition of Cl_2 to the system
b. increasing the temperature of the system
c. removal of PCl_3 from the system

ACIDS, BASES, AND SALTS

CHAPTER 14

OUTLINE

14.1   Concepts of Acids and Bases
14.2   Properties of Acids and Bases
14.3   Common Acids and Bases
14.4   Acid Strength
14.5   Polyprotic Acids
14.6   Base Strength
14.7   Acid and Basic Anhydrides
      Perspective: Acid Rain
14.8   Acid-Base Titrations
14.9   Normality *(optional)*
14.10 Salts
14.11 Solubilities of Salts and Other
       Ionic Compounds
14.12 Solubility Product Constants
       (optional)
       Summary
       Study Questions and Problems

The origin of the idea that some substances are acids and others are bases is buried deeply in history. Acids were recognized first because of their sour taste. Bases were noticed at a later date as bitter-tasting compounds that react with acids to form salty-tasting substances. In the modern world, acids and bases are important in industry and in daily life. Sulfuric acid and sodium hydroxide (a base) are among the most commonly manufactured chemical products, and many chemical processes involve acids or bases as reactants or catalysts. Many reactions that occur in solution are closely tied to acid-base levels, and the acidity or alkalinity (level of base) of soil and water are of great importance for plants and animals. In this chapter, you will learn about the fundamental properties of acids and bases and how they neutralize each other to form salts. Before beginning, you should review oxidation numbers (chapter 7), stoichiometric calculations involving solutions (chapter 12), balancing equations (chapter 9), solubility (chapter 12), and calculations involving exponents (appendix A).

14.1 CONCEPTS OF ACIDS AND BASES

Arrhenius
(are-REH-nius)

Acids and bases were first defined in 1887 by a young Swedish graduate student named Svante Arrhenius, and his ideas are referred to as the Arrhenius theory of acids and bases. Arrhenius worked mostly with aqueous solutions, and he came to know acids and bases from their behavior in water. He defined **acids** as substances that produce hydrogen ions (H^+) in water. An aqueous solution of hydrogen chloride is an example of an acid. When hydrogen chloride, a gas at room temperature, dissolves in water, hydrogen ions are produced:

$$HCl(g) \xrightarrow{H_2O} H^+(aq) + Cl^-(aq)$$

Thus, aqueous solutions of hydrogen chloride came to be known as hydrochloric acid.

Arrhenius defined **bases** as substances that dissolve in water to produce hydroxide ions (OH^-). Sodium hydroxide is an example:

$$NaOH(s) \xrightarrow{H_2O} Na^+(aq) + OH^-(aq)$$

The Arrhenius concept of an acid is illustrated by the following general equation,

$$HA \xrightarrow{H_2O} H^+(aq) + A^-(aq)$$

where HA represents any acid and A^- represents the corresponding anion. Similarly, the Arrhenius concept of a base is illustrated by the general equation,

$$MOH \xrightarrow{H_2O} M^+(aq) + OH^-(aq)$$

where M represents any metallic element and M^+ represents the corresponding cation.

In 1923, two chemists, Johannes Bronsted from Denmark and Thomas Lowry from England, independently arrived at a more general theory of acids and bases. According to Bronsted and Lowry, *an acid is any substance that can donate a proton to another substance,* and *a base is any substance that can accept a proton.* (Remember that a proton is simply a hydrogen ion.) Thus, the reaction between an acid and a base can be illustrated as follows, where hydrogen chloride is the acid and ammonia is the base:

$$HCl(g) + NH_3(g) \longrightarrow NH_4Cl(s)$$

The Bronsted-Lowry theory does not restrict acids and bases to aqueous solutions, and it does not require that a base produce hydroxide ions; a base simply has to be capable of accepting a proton from a donor. Thus, the Bronsted-Lowry theory applies to a greater variety of situations than does the Arrhenius theory.

A general equation can also be written to illustrate Bronsted-Lowry theory:

$$\text{H-acid} + \text{base} \rightleftharpoons \text{H-base}^+ + \text{acid}^-$$

where H-base$^+$ is the protonated base and acid$^-$ is the deprotonated acid. Thus, an acid-base reaction can be pictured as a reversible proton transfer; the proton donor is the acid, and the proton acceptor is the base. Proton transfer reactions have very fast rates and thus achieve equilibrium almost instantaneously. As you learn more about acids and bases, you will see that the Bronsted-Lowry theory is perfectly compatible with the Arrhenius theory.

In chapter 12, you learned that very polar molecules, like HCl, dissoci-

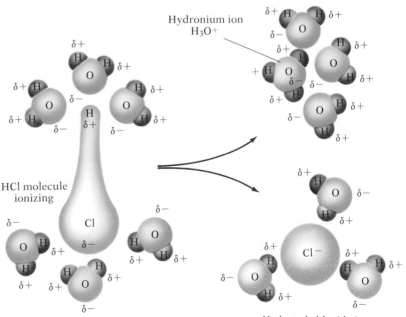

FIGURE 14-1 Ionization of an HCl Molecule in Water

ate into ions when dissolved in water. As HCl dissolves in water, the attractions between water molecules and the polar ends of the HCl molecule are strong enough to cause the HCl molecules to break apart into ions (see Figure 14-1). One way to show this process in an equation is to write

$$HCl(g) \xrightarrow{H_2O} H^+(aq) + Cl^-(aq)$$

but the following equation is a more accurate description of what happens:

$$HCl(g) + H_2O(l) \rightleftharpoons H_3O^+(aq) + Cl^-(aq)$$

Both equations mean that HCl molecules dissociate into protons and chloride ions, which become hydrated in the aqueous solution. But the second equation is a better description because when HCl dissolves, the protons released by the acid become bonded to water molecules to form **hydronium ions, H_3O^+:**

$$H^+ + \quad \overset{..}{\underset{H \qquad H}{O}} \quad \longrightarrow \quad \left[\overset{\overset{H}{|}}{\underset{H \qquad H}{O}} \right]^+$$

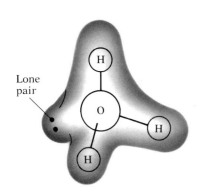

Lone pair

FIGURE 14-2 The Hydronium Ion, H_3O^+

The shape of the hydronium ion is shown in Figure 14-2.

Hydronium ions are formed because free protons are attracted to the nonbonded electrons of the oxygen atoms of water molecules. This attraction results in formation of a coordinate covalent bond between a proton and the oxygen atom of a water molecule. Thus, the covalent bond formed between a proton and a water molecule is composed of a pair of electrons donated by the oxygen atom. When HCl dissolves in water, hydronium ions are formed by transfer of protons from the acid, HCl, to the base, H_2O. Thus, for aqueous solutions, the Bronsted-Lowry theory is in total agreement with the Arrhenius theory. Both the hydronium ions and the chloride ions are hydrated by an indeterminate number of water molecules.

Acids and bases are defined by their behavior. In order for an acid to be recognized as such, it must donate a proton to a base; similarly, a base must accept a proton from an acid to be recognized as a base. Thus, acids and bases act in pairs, as illustrated by the example of dissolving HCl in water:

$$\underset{\text{acid}_1}{HCl(g)} + \underset{\text{base}_1}{H_2O(l)} \rightleftharpoons \underset{\text{acid}_2}{H_3O^+(aq)} + \underset{\text{base}_2}{Cl^-(aq)}$$

But after the acid, HCl, has transferred its proton to the base, H_2O, the products are also an acid and a base. That is, after H_2O accepts a proton, H_3O^+ then becomes an acid, because it is capable of donating a proton, and Cl^-, having lost a proton, is capable of accepting a proton. Thus, we can write a general equation to indicate this reciprocal relationship:

$$\text{acid}_1 + \text{base}_1 \rightleftharpoons \text{acid}_2 + \text{base}_2$$

The acid-base pairs connected by brackets are said to be conjugates of

Conjugate

Joined together, especially in pairs.

each other. Thus, Cl^- is the **conjugate base** of the acid HCl, and H_3O^+ is the **conjugate acid** of H_2O. We say that each pair of bracketed substances, HCl/Cl^- and H_2O/H_3O^+, is an **acid-base conjugate pair.** An acid forms its conjugate base by losing a proton, and a base forms its conjugate acid by gaining a proton. Thus, the only structural difference between an acid and its conjugate base or between a base and its conjugate acid is a proton. Most common acids are either molecules or cations, whereas bases are usually molecules or anions.

EXAMPLE 14.1 For each of the following acids and bases, write the formula of the conjugate base or acid.

ACIDS	BASES
a. HBr	d. H_2O
b. H_2O	e. OH^-
c. NH_4^+	f. ClO^-

Solution The conjugate base of each acid is the ion or molecule formed when the acid loses a proton. Thus, the conjugate bases are:

a. $\underset{\text{acid}}{HBr} \longrightarrow H^+ + \underset{\substack{\text{conjugate} \\ \text{base}}}{Br^-}$

b. $\underset{\text{acid}}{H_2O} \longrightarrow H^+ + \underset{\substack{\text{conjugate} \\ \text{base}}}{OH^-}$

c. $\underset{\text{acid}}{NH_4^+} \longrightarrow H^+ + \underset{\substack{\text{conjugate} \\ \text{base}}}{NH_3}$

The conjugate acid of each base is the ion or molecule formed when the base gains a proton. Thus, the conjugate acids are

d. $\underset{\text{base}}{H_2O} + H^+ \longrightarrow \underset{\substack{\text{conjugate} \\ \text{acid}}}{H_3O^+}$

e. $\underset{\text{base}}{OH^-} + H^+ \longrightarrow \underset{\substack{\text{conjugate} \\ \text{acid}}}{H_2O}$

f. $\underset{\text{base}}{ClO^-} + H^+ \longrightarrow \underset{\substack{\text{conjugate} \\ \text{acid}}}{HClO}$

Amphoteric

(am-fo-TEH-rick)

From parts b and d of example 14.1, you can see that H_2O can act as both an acid and a base. Such substances are said to be **amphoteric.** Another example of an amphoteric substance is the bicarbonate ion, HCO_3^-.

EXERCISE 14.1   For each of the following acids and bases, write the formula of the conjugate acid or base.

ACID	BASE
a. HCO_3^-	d. ClO_4^-
b. H_2SO_4	e. HCO_3^-
c. HF	f. NH_3

14.2 PROPERTIES OF ACIDS AND BASES

Most common acids exist as aqueous solutions, and thus these solutions contain hydronium ions. Hydronium ions give water a sour taste, like the taste of vinegar and lemon juice, which are both aqueous solutions of acids. We identify acids in the laboratory by their ability to change the color of certain dyes called acid-base indicators. An **acid-base indicator** is a substance that has one color in acid solution and another color in basic solution. One such indicator is litmus, a compound obtained from plants. Small strips of paper impregnated with litmus are called litmus paper; if a drop of aqueous solution changes the color of litmus paper from blue to red, the solution is acidic. This test actually corresponds to an acid-base reaction in which blue litmus is the base:

$$H_3O^+ + litmus \rightleftharpoons H\text{-}litmus^+ + H_2O$$

$$\text{acid}_1 \quad \text{base}_1 \qquad \text{acid}_2 \quad \text{base}_2$$

$$\text{(blue)} \qquad \text{(red)}$$

The reaction with litmus represents still another property of acids—they react with bases in **neutralization reactions.** When an acid reacts with a base, the characteristic properties of the acid and base disappear; thus, the acid and base are neutralized.

Another property of acids is that they react with many metals by dissolving the metal and producing hydrogen. An example is the reaction of hydrochloric acid and zinc:

$$2\,HCl(aq) + Zn(s) \longrightarrow ZnCl_2(aq) + H_2(g)$$

Thus acids should not be stored in metal containers or allowed to contact metal objects unless a chemical reaction is desired.

Like acids, the more familiar bases are usually dissolved in water. These solutions contain hydroxide ions (OH^-), which give the aqueous solutions a bitter taste and a slippery feeling (like soapy water). Bases are identified in the laboratory by their ability to change the color of litmus paper from red to blue, a reversal of the acid reaction with litmus:

$$OH^- + H\text{-}litmus^+ \rightleftharpoons litmus + H_2O$$

$$\text{base}_1 \quad \text{acid}_1 \qquad \text{base}_2 \quad \text{acid}_2$$

$$\text{(red)} \qquad \text{(blue)}$$

TABLE 14-1 Properties of Acids and Bases	
Acids	Bases
Taste sour	Taste bitter and feel slippery
Turn blue litmus red	Turn red litmus blue
Neutralize bases	Neutralize acids
Dissolve many metals	

As mentioned earlier, this is an example of an acid-base neutralization reaction. The properties of acids and bases are summarized in Table 14-1.

Concentrated solutions of acids and bases are harmful to all living tissue, and they should be handled cautiously in order to prevent physical contact. If ingested, they can cause severe damage to the gastrointestinal tract.

14.3   COMMON ACIDS AND BASES

Although there are many acids and bases, a few are widely known for their uses in medicine, industry, agriculture, and commercial products. For example, vinegar is a dilute aqueous solution of acetic acid, $HC_2H_3O_2$, and soft drinks are often flavored with small amounts of phosphoric acid, H_3PO_4. All carbonated beverages contain carbonic acid, H_2CO_3, which forms when CO_2 dissolves in water:

$$CO_2(g) + H_2O(l) \rightleftharpoons H_2CO_3(aq)$$

Carbonic acid is unstable and does not exist in pure form. When CO_2 dissolves in water, about 99% exists as hydrated CO_2 molecules, and the other 1% reacts to form carbonic acid. Carbonic acid solutions are so dilute that they are considered relatively safe to handle and ingest.

The most common bases are sodium hydroxide, $NaOH$, and ammonia, NH_3. Sodium hydroxide is also called caustic soda or lye; it is a white solid that is quite harmful to skin and can cause blindness if it contacts the eyes. Sodium hydroxide is the active ingredient in most oven and drain cleaners. Ammonia dissolves in water to form ammonium hydroxide, NH_4OH:

$$NH_3(g) + H_2O(l) \rightleftharpoons NH_4OH(aq)$$

As with carbon dioxide, most of the ammonia in an aqueous solution exists as hydrated molecules, but a small percentage undergoes reaction to form NH_4OH. Aqueous solutions of ammonia are called ammonium hydroxide, but the compound NH_4OH has never been isolated. Another name for aqueous solutions of ammonia is "aqua ammonia." Ordinary household ammonia is dilute ammonium hydroxide.

Many common acids and bases can be purchased from chemical supply companies. Concentrations of such commercially available acids and bases are given in Table 14-2.

TABLE 14-2	Concentrations of Commercially Available Acids and Bases
Substance	Molarity
Acetic acid (99.5%)	17.4
Hydrochloric acid (38%)	12.0
Perchloric acid (72%)	11.6
Hydrofluoric acid (45%)	25.7
Nitric acid (69%)	15.4
Phosphoric acid (85%)	14.7
Sulfuric acid (94%)	17.6
Ammonium hydroxide (27%)	14.3

14.4   ACID STRENGTH

Acids are designated as strong or weak depending on the extent to which they dissociate into ions when dissolved in water. Consider the ionization of a general acid, HA, in water:

$$HA(aq) + H_2O(l) \rightleftharpoons H_3O^+(aq) + A^-(aq)$$

(HA can be a molecule, like HF, or an anion, like HS^-.) The expression for the equilibrium constant of this ionization is

$$K_{eq} = \frac{[H_3O^+][A^-]}{[HA][H_2O]}$$

where the brackets stand for concentration in mol/L; for liquid solutions, mol/L is usually expressed as molarity, M. The molarity of water in most aqueous solutions is very nearly the same as for pure water, 55.5 M, so we can consider this value to be a constant. Thus, if we combine the concentration of water with K_{eq},

$$K_{eq}[H_2O] = \frac{[H_3O^+][A^-]}{[HA]}$$

we can derive an expression for a new constant, which we will call K_a, the **ionization constant** of the acid:

$$K_{eq}[H_2O] = K_a = \frac{[H_3O^+][A^-]}{[HA]}$$

(Another name for ionization constant is **dissociation constant.**) Acid ionization constants have units of molarity, but the units are seldom shown.

An acid ionization constant indicates the position of equilibrium for the dissociation of a particular acid. **Strong acids** ionize almost totally, and thus the position of equilibrium for their dissociations lies far to the right. Although the word *strong* is not very precise, we usually classify strong acids as acids having ionization constants of 1 or greater. The common strong acids are listed in Table 14-3.

TABLE 14-3	Common Strong Acids
Acid	Formula
Hydrochloric acid	HCl
Hydrobromic acid	HBr
Hydroiodic acid	HI
Nitric acid	HNO_3
Perchloric acid	$HClO_4$
Chromic acid	H_2CrO_4
Sulfuric acid	H_2SO_4

Weak acids are acids having ionization constants of less than 1. These acids are only slightly ionized in water and are usually less harmful to tissue than strong acids. A number of weak acids are listed in Table 14-4.

Note that acid strength refers to the extent of acid ionization, not to the concentration of an acid in a solution. Thus, strong acids can have high concentrations or low concentrations, but they are always strong acids. The same is true for weak acids; regardless of their concentrations, they are still weak acids.

TABLE 14-4	Common Weak Acids
Acid	Formula
Acetic acid	$HC_2H_3O_2$
Boric acid	H_3BO_3
Carbonic acid	H_2CO_3
Phosphoric acid	H_3PO_4
Sulfurous acid	H_2SO_3

EXAMPLE 14.2

The following equilibrium concentrations existed in a solution of hydrofluoric acid:

$$[HF] = 0.0921 \, M$$
$$[H_3O^+] = 7.9 \times 10^{-3} \, M$$
$$[F^-] = 7.9 \times 10^{-3} \, M$$

Calculate K_a for hydrofluoric acid.

Solution

Since hydrofluoric acid ionizes according to the following equation,

$$HF(aq) + H_2O(l) \rightleftharpoons H_3O^+(aq) + F^-(aq)$$

the expression for the ionization constant is

$$K_a = \frac{[H_3O^+][F^-]}{[HF]}$$

Substituting concentration values into the expression, we obtain

$$K_a = \frac{(7.9 \times 10^{-3})(7.9 \times 10^{-3})}{0.0921} = 6.8 \times 10^{-4}$$

MATH TIP

Since units of K_a are seldom shown, it is not necessary to include concentration units for the reactants and products. To solve this problem using a calculator, Press 7.9 [EE] 3 [+/−] [÷] 0.0921 [×] 7.9 [EE] 3 [+/−] [=]. The display reads 6.77633 −04. The answer should be written 6.8×10^{-4}.

EXERCISE 14.2

In a solution of a certain weak acid, HA, the following equilibrium concentrations existed:

$$[HA] = 0.10 \, M$$
$$[H_3O^+] = 2.0 \times 10^{-3} \, M$$
$$[A^-] = 2.0 \times 10^{-3} \, M$$

Calculate K_a for the acid. ■

EXAMPLE 14.3

Hydrofluoric acid has a K_a of 6.8×10^{-4}. If the equilibrium concentration of HF is 0.20 M, what is the concentration of H_3O^+?

Solution

Hydrofluoric acid ionizes as follows:

$$HF(aq) + H_2O(l) \rightleftharpoons H_3O^+(aq) + F^-(aq)$$

Since each mole of HF that dissociates forms one mole of H_3O^+ and one mole of F^-, we know that at equilibrium,

$$[H_3O^+] = [F^-]$$

Then, substituting into the expression for K_a,

$$K_a = \frac{[H_3O^+][F^-]}{[HF]}$$

$$= \frac{[H_3O^+][H_3O^+]}{[HF]}$$

$$= \frac{[H_3O^+]^2}{[HF]}$$

Now, solving for $[H_3O^+]$,

$$[H_3O^+]^2 = K_a[HF]$$

or,

$$[H_3O^+] = \sqrt{K_a\,[HF]}$$

$$= \sqrt{(6.8 \times 10^{-4})(0.20)}$$

$$= \sqrt{1.36 \times 10^{-4}}$$

$$= 1.2 \times 10^{-2}\ M$$

MATH TIP

Solving $\sqrt{(6.8 \times 10^{-4})(0.20)}$ on the calculator, Press 6.8 $\boxed{EE}$ 4 $\boxed{+/-}$ $\boxed{\times}$ 0.20 $\boxed{=}$ $\boxed{\sqrt{x}}$. The display reads 1.16619 -02. The answer is rounded off to two significant figures and reported as $1.2 \times 10^{-2}\ M$.

EXERCISE 14.3   Hydrocyanic acid, HCN(aq), has a K_a of 4.9×10^{-10}. If the equilibrium concentration of HCN is 0.15 M, what is the concentration of H_3O^+? ■

Table 14-5 gives examples of acid-base conjugate pairs. Consider one such pair, hydrofluoric acid and fluoride ion:

$$HF(aq) + H_2O(l) \rightleftharpoons H_3O^+(aq) + F^-(aq)$$

$$K_a = 6.8 \times 10^{-4}$$

Hydrofluoric acid is a weak acid, as indicated by its K_a. Thus, in the forward reaction, only a few molecules of HF transfer protons to water, in turn forming only a few F^- ions. But the reverse reaction, in which F^- acts as a proton acceptor, goes nearly to completion; thus, F^- is a stronger base

TABLE 14-5   Acid-Base Conjugate Pairs

Acid	Formula	Conjugate Base	K_a (25° C)
Acetic acid	$HC_2H_3O_2$	$C_2H_3O_2^-$	1.8×10^{-5}
Carbonic acid	H_2CO_3	HCO_3^-	4.3×10^{-7}
Formic acid	$HCHO_2$	CHO_2^-	1.7×10^{-4}
Hydrocyanic acid	HCN	CN^-	4.9×10^{-10}
Hydrofluoric acid	HF	F^-	6.8×10^{-4}
Nitrous acid	HNO_2	NO_2^-	4.5×10^{-4}
Phosphoric acid	H_3PO_4	$H_2PO_4^-$	7.5×10^{-3}
Sulfurous acid	H_2SO_3	HSO_3^-	1.3×10^{-2}

Base	Formula	Conjugate Acid	K_b (25° C)
Ammonia	NH_3	NH_4^+	1.8×10^{-5}
Carbonate ion	CO_3^{2-}	HCO_3^-	2.1×10^{-4}
Hypochlorite ion	ClO^-	HClO	3.1×10^{-7}

than H_2O, because more HF exists in the equilibrium mixture than H_3O^+. We can also say that HF is a weaker acid than H_3O^+, because most HF molecules do not donate protons to H_2O. By considering relative acid strengths, we can make the following general statement: The weaker an acid, the stronger its conjugate base, and the stronger an acid, the weaker its conjugate base. These relative strengths and weaknesses are indicated by K_a values in Table 14-5.

14.5 POLYPROTIC ACIDS

Tables 14-3 and 14-4 list several acids that are capable of donating more than one proton from each of their formula units; such acids are called **polyprotic acids.** They donate protons in a stepwise manner, and each step has its own K_a. Two ionization steps are associated with sulfuric acid, for example:

$$H_2SO_4(aq) + H_2O(l) \rightleftharpoons H_3O^+(aq) + HSO_4^-(aq) \qquad K_a = 1 \times 10^3$$
$$HSO_4^-(aq) + H_2O(l) \rightleftharpoons H_3O^+(aq) + SO_4^{2-}(aq) \qquad K_a = 1.2 \times 10^{-2}$$

Note that the first step has a K_a corresponding to a strong acid, but the second step has a value characteristic of a weak acid. This means that the second step occurs less readily than the first and is thus less complete. Because of the high K_a of the first step, sulfuric acid is considered a strong acid.

The H_2SO_4 molecule loses a proton to water readily because the sulfur and oxygen atoms draw bonding electrons away from the hydrogen atoms (remember that S and O are quite electronegative), causing the O—H bonds to be highly polarized:

Sulfuric acid

A proton can be lost from either OH group in sulfuric acid with equal probability. However, after HSO_4^- is formed, its negative charge partially satisfies the electronegativities of the sulfur and oxygen atoms, and the remaining O—H bond is only weakly polar:

Hydrogen sulfate anion

Thus, HSO_4^- is a weak acid. Another way of looking at the situation is this: adding a second negative charge to HSO_4^- requires a large amount of energy, and thus SO_4^{2-} forms only to a small extent in the reversible reaction with water.

Another common polyprotic acid is phosphoric acid, H_3PO_4. In this case, proton dissociation occurs in three steps:

$$H_3PO_4(aq) + H_2O(l) \rightleftharpoons H_3O^+(aq) + H_2PO_4^-(aq) \qquad K_a = 7.5 \times 10^{-3}$$

$$H_2PO_4^-(aq) + H_2O(l) \rightleftharpoons H_3O^+(aq) + HPO_4^{2-}(aq) \qquad K_a = 6.2 \times 10^{-8}$$

$$HPO_4^{2-}(aq) + H_2O(l) \rightleftharpoons H_3O^+(aq) + PO_4^{-3}(aq) \qquad K_a = 2.2 \times 10^{-13}$$

The K_a of the first step indicates that H_3PO_4 is a much weaker acid than H_2SO_4. Phosphorus is less electronegative than sulfur; thus, the O—H bonds in H_3PO_4 are less polar than those in H_2SO_4 and other strong oxyacids. When the first proton dissociates from H_3PO_4, it can come from any one of the three OH groups with equal probability. The resulting negative charge on $H_2PO_4^-$ depolarizes the remaining O—H bonds, making this anion a weaker acid than H_3PO_4. Similarly, the two negative charges on HPO_4^{2-} make it an even weaker acid than $H_2PO_4^-$.

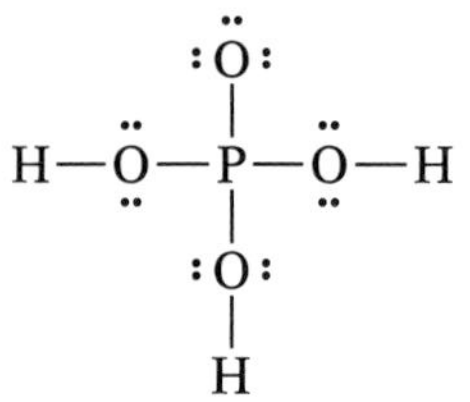

Phosphoric acid

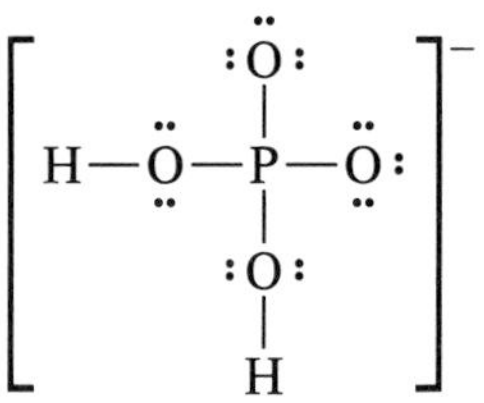

Dihydrogen phosphate anion

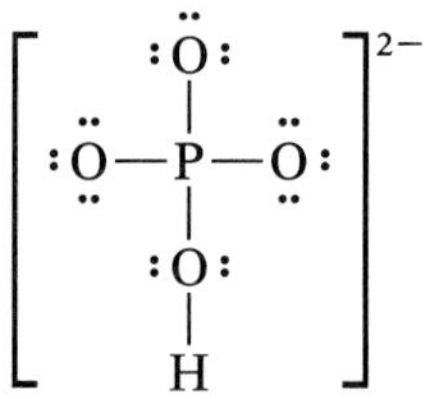

Hydrogen phosphate anion

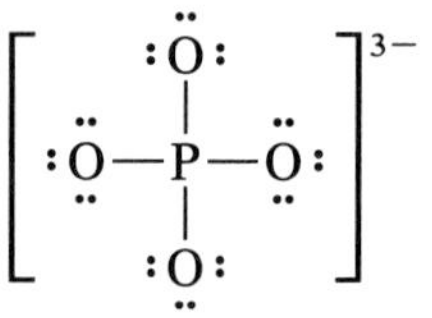

Phosphate anion

Each of the anions formed is a conjugate base and is thus capable of accepting a proton from an acid.

Carbonic acid, H_2CO_3, is another example of a polyprotic acid. The central atom, carbon, is less electronegative than sulfur and phosphorus; thus, H_2CO_3 is a weaker acid than H_2SO_4 and H_3PO_4.

$$H_2CO_3(aq) + H_2O(l) \rightleftharpoons H_3O^+(aq) + HCO_3^-(aq) \qquad K_a = 4.3 \times 10^{-7}$$

$$HCO_3^-(aq) + H_2O(l) \rightleftharpoons H_3O^+(aq) + CO_3^{2-}(aq) \qquad K_a = 5.6 \times 10^{-11}$$

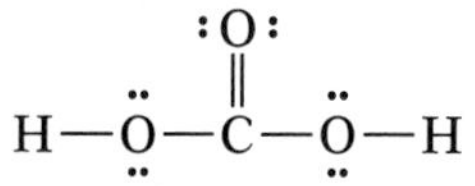

Carbonic acid

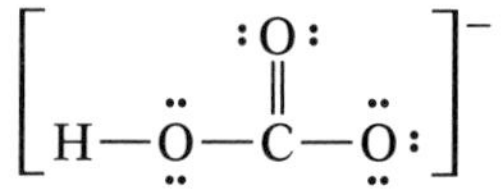

Bicarbonate anion or hydrogen carbonate anion

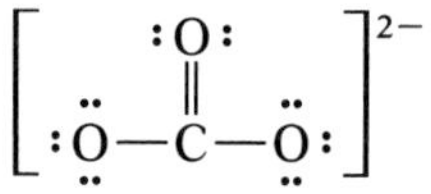

Carbonate anion

14.6 BASE STRENGTH

Bases that contain hydroxide ions can also be classified as strong or weak, depending on their abilities to provide hydroxide ions in aqueous solution. The soluble metal hydroxides NaOH and KOH are ionic solids that dissociate completely in water to form basic solutions:

$$NaOH(s) \xrightarrow{H_2O} Na^+(aq) + OH^-(aq)$$

$$KOH(s) \xrightarrow{H_2O} K^+(aq) + OH^-(aq)$$

Thus, NaOH and KOH are strong bases. Few other metal hydroxides are very soluble in water; for example, $Mg(OH)_2$ dissolves in water only to the extent of 0.011 g/L. For this reason, suspensions of $Mg(OH)_2$ (sold as milk of magnesia) are taken to alleviate discomfort caused by high levels of hydrochloric acid in the stomach (frequently referred to as hyperacidity). The $Mg(OH)_2$ dissolves slowly as it neutralizes stomach acid, and thus very few hydroxide ions are in solution in the stomach at any time.

Aside from the metal hydroxides, there are a great many other bases, most of which do not contain hydroxide ions. Ammonia is an example. It is a base because it is a proton acceptor:

$$NH_3(g) + H_2O(l) \rightleftharpoons NH_4^+(aq) + OH^-(aq)$$

Since this reaction does not proceed very far to the right, ammonia is a weak base.

Bases with structures similar to that of ammonia do not contain hydroxide ions and are usually weak bases. For these, we can write the following general equation for acceptance of a proton by such bases:

$$B + H_2O(l) \rightleftharpoons BH^+(aq) + OH^-(aq)$$

The expression for the equilibrium constant is

$$K_{eq} = \frac{[BH^+][OH^-]}{[B][H_2O]}$$

and the ionization constant, K_b, for the weak base is

$$K_b = K_{eq}[H_2O] = \frac{[BH^+][OH^-]}{[B]}$$

The proton acceptor (B) can be a molecule, such as NH_3, or it can be an anion.

EXAMPLE 14.4

In a solution of ammonium hydroxide, the following equilibrium concentrations existed:

$$NH_3(g) + H_2O(l) \rightleftharpoons NH_4^+(aq) + OH^-(aq)$$
$$[NH_3] = 0.10\ M$$
$$[NH_4^+] = 1.35 \times 10^{-3}\ M$$
$$[OH^-] = 1.35 \times 10^{-3}\ M$$

Calculate K_b for ammonium hydroxide.

Solution   The expression for K_b is

$$K_b = \frac{[NH_4^+][OH^-]}{[NH_3]}$$

Substituting the equilibrium concentrations, we obtain

$$K_b = \frac{(1.35 \times 10^{-3})(1.35 \times 10^{-3})}{0.10} = 1.8 \times 10^{-5}$$

EXERCISE 14.4   A weak base, B, reacts with water as follows:

$$B(aq) + H_2O(l) \rightleftharpoons BH^+(aq) + OH^-(aq)$$

From the following equilibrium concentrations, calculate K_b.

$$[B] = 0.40\ M$$
$$[BH^+] = 7.7 \times 10^{-8}\ M$$
$$[OH^-] = 7.7 \times 10^{-8}\ M$$

The bicarbonate ion, HCO_3^-, is an example of an anion that is a weak base; it reacts with water as follows:

$$HCO_3^-(aq) + H_2O(l) \rightleftharpoons H_2CO_3(aq) + OH^-(aq)$$

Thus, an aqueous solution of a bicarbonate compound will actually be basic. Sodium bicarbonate, $NaHCO_3$ (baking soda), is often an ingredient in antacids. These nonprescription medications are used to combat hyperacidity in the stomach, because the hydroxide ions, formed when sodium bicarbonate dissolves in water, neutralize the excess stomach acid. The characteristic belching following ingestion of sodium bicarbonate is caused by formation of carbon dioxide from carbonic acid:

$$H_2CO_3(aq) \rightleftharpoons H_2O(l) + CO_2(g)$$

Bases with ionization constants of 1 or greater are classified as strong bases, while those with ionization constants of less than 1 are classified as weak bases. As mentioned earlier, soluble metal hydroxides are strong bases. Hydroxides of Group IA elements are soluble in water and thus are strong bases. Most other metal hydroxides dissolve in water to such small extents that they are not thought of as strong bases. Examples of weak bases are given in Table 14-6.

TABLE 14-6   Common Weak Bases

Base	Formula	K_b(25° C)
Ammonia	NH_3	1.8×10^{-5}
Carbonate ion	CO_3^-	2.1×10^{-4}
Cyanide ion	CN^-	1.6×10^{-5}
Triethylamine	$(C_2H_5)_3N$	5.2×10^{-4}
Trimethylamine	$(CH_3)_3N$	6.3×10^{-5}

14.7 ACID AND BASIC ANHYDRIDES

Most (but not all) oxides of nonmetals react with water to form acids; such substances are called **acid anhydrides.** The word *anhydride* means "without hydrogen;" thus, acid anhydrides are "acids without hydrogen." A familiar acid anhydride is carbon dioxide, CO_2. When CO_2 dissolves in water, 99% exists as hydrated CO_2, but the other 1% forms carbonic acid:

$$CO_2(aq) + H_2O(l) \rightleftharpoons H_2CO_3(aq)$$

Thus, CO_2, like other acid anhydrides, is converted to an acid by reaction with water, in which the hydrogen of water becomes the hydrogen of the acid. The dissociation of carbonic acid,

$$H_2CO_3(aq) + H_2O(l) \rightleftharpoons H_3O^+(aq) + HCO_3^-(aq)$$

imparts a sour taste to the solution, accounting for the tangy taste of carbonated beverages. Other common acid anhydrides and their reactions with water are:

$$SO_2(g) + H_2O(l) \longrightarrow H_2SO_3(aq)$$
$$SO_3(g) + H_2O(l) \longrightarrow H_2SO_4(aq)$$
$$N_2O_5(g) + H_2O(l) \longrightarrow 2\ HNO_3(aq)$$
$$P_4O_{10}(s) + 6\ H_2O(l) \longrightarrow 4\ H_3PO_4(aq)$$

All of these reactions are used for the commercial production of acids.

Acid anhydrides form oxyacids in which the oxidation number of the nonmetal is the same as that of the nonmetal in the acid anhydride. Thus, it is possible to derive formulas of acid anhydrides from formulas of the corresponding acids.

EXAMPLE 14.5   Give the formula of the anhydride of the acid HNO_2.

Solution   First, find the oxidation number of N in HNO_2.

Atoms	Individual Oxidation Numbers	Sum of Oxidation Numbers = 0
H	+1	+1
2 O	−2	−4
N	+3	+3
		0

Then, find the formula of the oxide of nitrogen in which N has an oxidation number of +3.

Atoms	Individual Oxidation Numbers	Sum of Oxidation Numbers = 0
2 N	+3	+6
3 O	−2	−6
		0

Therefore, the acid anhydride of HNO_2 must be N_2O_3.

PERSPECTIVE ACID RAIN

In 1959, a Norwegian fisheries inspector connected the decline in Scandinavian fish populations to the increasing acidity of rain and snow. The source of the so-called acid rain was traced to Europe's heavily industrialized areas, and the situation has worsened to the point that hundreds of Scandinavian lakes no longer support fish. By 1972, acid rain had spread to North America, and fish populations were being wiped out in Canada and the eastern United States. The westward spread of acid rain has continued, and streams in many of the far western states are now tainted with acids.

The airborne acidity comes from automobile exhausts and from burning fossil fuels such as coal, oil, and natural gas. The high temperatures in automobile engines and other combustion processes cause oxidation of atmospheric nitrogen, forming nitrogen dioxide, NO_2. In addition, fossil fuels contain sulfur compounds, which are oxidized to sulfur dioxide, SO_2, and sulfur trioxide, SO_3, during combustion. All of these oxides are acid anhydrides, and they react with moisture in the atmosphere to form their corresponding acids:

Burning Fossil Fuels Increases the Acidity of Rain.

Falling rain and snow bring the acids to earth, raising the acid levels of lakes, rivers, and streams and harming crops and building exteriors.

$$3\ NO_2(g) + H_2O(l) \longrightarrow 2\ HNO_3(aq) + NO(g)$$
$$SO_2(g) + H_2O(l) \longrightarrow H_2SO_3(aq)$$
$$SO_3(g) + H_2O(l) \longrightarrow H_2SO_4(aq)$$

EXERCISE 14.5 Give the formula of the acid anhydride of the acid $HClO_4$. ■

Ionic metal oxides react with water to form bases; these metal oxides are known as **basic anhydrides.** Basic anhydrides are converted to metal hydroxides by reaction with water; thus, the hydrogen of water becomes the hydrogen of the base, as shown in the following examples:

$$Na_2O(s) + H_2O(l) \longrightarrow 2\ NaOH(aq)$$
$$K_2O(s) + H_2O(l) \longrightarrow 2\ KOH(aq)$$
$$MgO(s) + H_2O(l) \longrightarrow Mg(OH)_2(s)$$

Since basic anhydrides are ionic, the metal ion in the anhydride has the same charge (and thus the same oxidation number) as the metal ion in the base. It is thus possible to derive formulas of basic anhydrides from formulas of the corresponding bases.

EXAMPLE 14.6 Give the formula of the anhydride of the base $Al(OH)_3$.

Solution First, find the charge of Al from the periodic table. Since Al is in Group

IIIA, it forms Al^{3+}. Then, find the formula of the oxide of Al, remembering that it contains Al^{3+} and O^{2-}:

$$Al \overset{\text{③+}}{\underset{}{\rightleftarrows}} O^{\text{②−}} = Al_2O_3$$

Thus, Al_2O_3 is the anhydride of $Al(OH)_3$.

EXERCISE 14.6   Give the formula of the anhydride of the base $Fe(OH)_2$.   ■

Since acid and base anhydrides are simply acids and bases without hydrogen, the anhydrides react with each other just as acids and bases do. However, because there is no hydrogen in the anhydrides, no water can be formed. Some examples of reactions between acid and base anhydrides are

$$SO_3(g) + CaO(s) \longrightarrow CaSO_4(s)$$
$$CO_2(g) + Na_2O(s) \longrightarrow Na_2CO_3(s)$$

14.8 ACID-BASE TITRATIONS

Acid-base neutralization reactions can be used to determine the concentration of aqueous acids and bases by the technique known as titration. A **titration** is a procedure in which we measure the volume of a standard solution required for complete reaction with another solution of unknown concentration. The titration of an acid with a base, illustrated in Figure 14-3, is performed as follows:

1. A carefully measured volume of an acid of unknown concentration is placed in a flask. A few drops of acid-base indicator are then added to the acid in the flask. (In Figure 14-3, the indicator is phenolphthalein.)
2. A solution of base of known concentration (called the standard base) is added to the acid through a buret. (A buret is a long transparent tube, calibrated in units of mL, from which a solution, the titrant, can be added to another solution.)
3. Standard base is added slowly to the acid until one drop of base causes a change in the color of the indicator. At this point, the **end point,** all of the acid has been neutralized, and there is just the very slightest excess of base in the flask, causing the indicator to change color.

At the end of titration, just enough base has been added to neutralize the acid. Since the concentration of the standard base is known, the concentration of the acid can be calculated. Titrations are routinely performed to determine acid levels in such substances as vinegar, wine, and waste water. Concentrations of bases can also be determined by titration, using a solution of standard acid as the titrant.

EXAMPLE 14.7   A 25.00-mL sample of hydrochloric acid required 29.72 mL of 0.1025 M NaOH to reach the end point of a titration. What was the concentration of the hydrochloric acid?

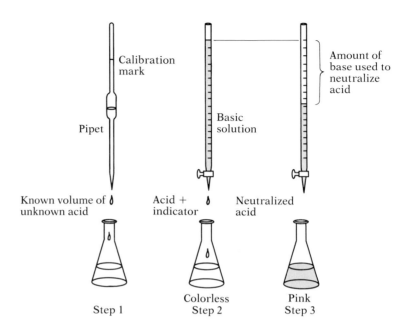

FIGURE 14-3 Titration of an Acid with a Base

Solution   First, write the equation for the neutralization reaction:

$$HCl(aq) + NaOH(aq) \longrightarrow NaCl(aq) + H_2O(l)$$

Next, calculate moles of NaOH used in the titration:

$$(29.72\ \text{mL})\left(\frac{1\ \text{L}}{1000\ \text{mL}}\right)\left(\frac{0.1025\ \text{mol NaOH}}{\text{L}}\right) = 0.003046\ \text{mol NaOH}$$

Then, convert moles of NaOH to moles of HCl:

$$(0.003046\ \text{mol NaOH})\left(\frac{1\ \text{mol HCl}}{1\ \text{mol NaOH}}\right) = 0.003046\ \text{mol HCl}$$

Finally, calculate the molarity of the hydrochloric acid:

$$M = \frac{0.003046\ \text{mol HCl}}{0.02500\ \text{L}} = 0.1218\ \text{mol HCl/L}$$

Or,

$$M = \frac{(29.72\ \text{mL})\left(\dfrac{1\ \text{L}}{1000\ \text{mL}}\right)\left(\dfrac{0.1025\ \text{mol NaOH}}{\text{L}}\right)\left(\dfrac{1\ \text{mol HCl}}{1\ \text{mol NaOH}}\right)}{0.02500\ \text{L}}$$

$$= 0.1218\ \text{mol HCl/L}$$

MATH TIP

For review consult the mole map, Figure 12-11. It can be used for titration problems, which involve solution stoichiometry.

g NaOH → mol NaOH → mol HCl → g HCl

↑
vol of
NaOH solution

When mol HCl are found, then molarity of HCl can be calculated from the relationship

$$M = \frac{\text{moles HCl}}{\text{L HCl solution}}$$

The volume of HCl was given, 25.00 mL = 0.02500 L.

EXERCISE 14.7   A 20.00-mL sample of sulfuric acid required 18.27 mL of 0.1586 *M* NaOH to reach the end point of a titration. What was the molarity of the sulfuric acid?   ∎

14.9 NORMALITY *(optional)*

In the reaction between HCl and NH_3, one mole of each reacts to form one mole of NH_4Cl:

$$HCl(g) + NH_3(g) \longrightarrow NH_4Cl(s)$$

Similarly, one mole of hydrochloric acid reacts with one mole of sodium hydroxide to produce one mole each of sodium chloride and water:

$$HCl(aq) + NaOH(aq) \longrightarrow NaCl(aq) + H_2O(l)$$

In both cases, one mole of acid neutralized one mole of base.

Let's examine the reaction of a polyprotic acid, H_2SO_4, with NaOH. In this case, one mole of H_2SO_4 reacts with two moles of NaOH:

$$H_2SO_4(aq) + 2\ NaOH(aq) \longrightarrow Na_2SO_4(aq) + 2\ H_2O(l)$$

Thus, one mole of H_2SO_4 has the same neutralizing power as two moles of HCl, or we can say that one mole of H_2SO_4 is equivalent to two moles of HCl in neutralizing power. In acid-base reactions, it is sometimes more useful to think in terms of equivalent amounts instead of molar amounts. An **equivalent of an acid** is the weight of acid that neutralizes one mole of hydroxide ions, and an **equivalent of a base** is the weight of base that neutralizes one mole of hydronium ions. From these definitions, it is possible to calculate the number of equivalents in one mole of acid or base. For acids having one dissociable proton (monoprotic acids), one mole of acid will neutralize one mole of OH^-, and thus one mole of a monoprotic acid contains one equivalent of the acid. But for polyprotic acids, one mole contains more than one equivalent of the acid. A mole of a diprotic acid (an acid having two dissociable protons) will neutralize two moles of hydroxide ion, and thus one mole of a diprotic acid contains two equivalents of the acid. Similarly, a mole of a triprotic acid (an acid having three dissociable protons) will neutralize three moles of hydroxide ions and thus contains three equivalents of the acid.

The number of equivalents in a mole of base is calculated by considering the number of moles of hydronium ions that can be neutralized by one mole of the base. For example, one mole of NaOH will neutralize one mole of H_3O^+, and therefore one mole of NaOH contains one equivalent of NaOH. For a base like $Ca(OH)_2$, one mole of base neutralizes two moles of H_3O^+, and thus one mole of $Ca(OH)_2$ contains two equivalents of $Ca(OH)_2$. These relationships are summarized in Table 14-7.

A solution containing one equivalent of a substance in one liter of solution is called a one-normal (1 N) solution. **Normality** is concentration expressed as the number of equivalents per liter of solution (eq/L); it is defined as

$$N = \frac{\text{equivalents solute}}{\text{liters of solution}}$$

Thus, a solution having 2.00 equivalents per liter is a 2.00 N solution, and 500.0 mL of a 2.00 N solution contains 1.00 equivalent of solute.

TABLE 14-7 Equivalents of Common Acids and Bases

Acid	Moles of Hydroxide Ions Neutralized by One Mole of Acid	Equivalents/Mol
HCl	1	1
$HC_2H_3O_2$	1	1
H_2SO_4	2	2
H_3PO_4	3	3

Base	Moles of Hydronium Ions Neutralized by One Mole of Base	Equivalents/Mol
NaOH	1	1
KOH	1	1
$Ca(OH)_2$	2	2

EXAMPLE 14.8 What is the normality of a hydrochloric acid solution containing 50.0 g of HCl in 1.00 L?

Solution Since normality is number of equivalents per liter, we must first find the number of equivalents in one mole of HCl. Since there is only one dissociable proton,

$$1 \text{ mol HCl} = 1 \text{ eq HCl} = 36.5 \text{ g HCl}$$

Then we can calculate the number of equivalents of HCl in 50.0 g:

$$(50.0 \text{ g HCl})\left(\frac{1 \text{ eq HCl}}{36.5 \text{ g HCl}}\right) = 1.37 \text{ eq HCl}$$

Last, we calculate normality:

$$N = \frac{\text{equivalents solute}}{\text{liters of solution}}$$

$$= \frac{1.37 \text{ eq HCl}}{1.00 \text{ L}} = 1.37 \text{ eq HCl/L}$$

Or,

$$N = \frac{(50.0 \text{ g HCl})\left(\dfrac{1 \text{ eq HCl}}{36.5 \text{ g HCl}}\right)}{1.00 \text{ L}} = 1.37 \text{ eq HCl/L}$$

Thus, the solution has a concentration of 1.37 N.

EXERCISE 14.8 What is the normality of a solution containing 125 g of NaOH in 1.50 L? ■

EXAMPLE 14.9 How many grams of H_2SO_4 are in 1.00 L of a 1.50 N solution?

Solution First, find the number of equivalents in one mole of H_2SO_4. Since there are two dissociable protons,

$$1 \text{ mol } H_2SO_4 = 2 \text{ eq } H_2SO_4$$

MATH TIP
Like molarity, normality can be used in calculations as a conversion factor to convert from volume of solution to the number of equivalents of solute in solution.

Then, multiply the given information, 1.00 L H_2SO_4, by the appropriate conversion factors to achieve grams of H_2SO_4:

$$(1.00 \text{ L } H_2SO_4)\left(\frac{1.50 \text{ eq } H_2SO_4}{1 \text{ L } H_2SO_4}\right)\left(\frac{1 \text{ mol } H_2SO_4}{2 \text{ eq } H_2SO_4}\right)\left(\frac{98.1 \text{ g } H_2SO_4}{\text{mol } H_2SO_4}\right)$$

$$= 73.6 \text{ g } H_2SO_4$$

EXERCISE 14.9 How many grams of phosphoric acid, H_3PO_4, are in 500.0 mL of a 6.00 N solution?

For very dilute solutions of acids and bases, it is sometimes useful to express concentrations as *milliequivalents per liter*, meq/L, where 1 meq = 0.001 eq.

EXAMPLE 14.10 How many meq are in 1.00 mL of a 1.00 N solution?

Solution
$$\left(\frac{1.00 \text{ eq}}{\text{L}}\right)\left(\frac{1 \text{ L}}{1000 \text{ mL}}\right)\left(\frac{\text{meq}}{0.001 \text{ eq}}\right) = 1.00 \text{ meq/mL}$$

This example illustrates that a 1.00 N solution contains 1.00 eq/L and 1.00 meq/mL.

EXERCISE 14.10 How many meq are in 20.00 mL of a 0.1000 N solution?

For a solution of acid or base whose molarity is known, the normality can easily be calculated. The normality is simply the molarity multiplied by the number of equivalents per mole of acid or base:

$$N = \left(\frac{\text{mol}}{\text{L}}\right)\left(\frac{\text{eq}}{\text{mol}}\right) = \text{eq/L}$$

Thus, a 1 M solution of HCl would also be a 1 N solution, but a 1 M solution of H_2SO_4 would be a 2 N solution. Similarly, a 1 M solution of NaOH would also be a 1 N solution, but a 1 M solution of $Ca(OH)_2$ would be a 2 N solution.

EXAMPLE 14.11 What is the normality of a 4.00 M sulfuric acid solution?

Solution Since there are two equivalents of H_2SO_4 per mole,

$$N = \left(\frac{4.00 \text{ mol } H_2SO_4}{\text{L}}\right)\left(\frac{2 \text{ eq } H_2SO_4}{\text{mol } H_2SO_4}\right) = 8.00 \text{ eq } H_2SO_4/\text{L}$$

EXERCISE 14.11 What is the normality of a 0.0200 M solution of $Ca(OH)_2$?

In acid-base reactions, one equivalent of acid neutralizes one equivalent of base:

$$H_3O^+(aq) + OH^-(aq) \longrightarrow 2 \text{ } H_2O(l)$$

Consider the titration of an acid of unknown concentration with a standard base whose concentration is expressed as normality. At the end point of the titration, just enough standard base has been added to neutralize the acid, and the number of equivalents of base added are then exactly equal to the number of equivalents of acid originally in the flask:

$$\text{eq base} = \text{eq acid}$$

and

$$\text{eq base} = (\text{L})\left(\frac{\text{eq}}{\text{L}}\right) = V_bN_b$$

$$\text{eq acid} = (\text{L})\left(\frac{\text{eq}}{\text{L}}\right) = V_aN_a$$

or,

$$V_bN_b = V_aN_a \qquad (1)$$

where V_b is volume of base, N_b is the normality of the base, V_a is the volume of acid, and N_a is the normality of the acid. Rearranging equation (1), we get

$$N_a = N_b\left(\frac{V_b}{V_a}\right) \qquad (2)$$

Thus, the normality of the acid (N_a) is easily calculated from the normality of the base (N_b), volume of the base added (V_b), and volume of acid used for the titration (V_a). Volume can be expressed as mL or L, as long as the same units are used for both volumes.

MATH TIP

Since the units of V_a and V_b will cancel, the volume units can be expressed in several different ways; liters are not required. The only requirement is that both volumes have identical units.

For titration of a base with standard acid, equation (1) is rearranged to give

$$N_b = N_a\left(\frac{V_a}{V_b}\right) \qquad (3)$$

EXAMPLE 14.12 A 25.00-mL sample of acid required 37.33 mL of 0.2000 *N* standard base to reach the end point of a titration. What was the normality of the acid?

Solution

$$N_a = N_b\left(\frac{V_b}{V_a}\right)$$

$$= 0.2000\ \text{eq/L}\left(\frac{37.33\ \cancel{\text{mL}}}{25.00\ \cancel{\text{mL}}}\right)$$

$$= 0.2986\ \text{eq/L}$$

EXERCISE 14.12 A 20.00-mL sample of acid required 26.21 mL of 0.1048 *N* standard base to reach the end point of a titration. What was the normality of the acid? ■

14.10 SALTS

When acids and bases neutralize each other, one product is always an ionic compound called a **salt.** Salts are among the most abundant compounds found in nature. Most rocks and minerals are salts. As water flows over rocks and soil, it dissolves salts to the extent allowed by their solubilities. Hard water contains dissolved calcium, magnesium, and iron salts, and large quantities of dissolved salts are found in the oceans. Salts are typical ionic compounds, existing in the pure state as crystalline solids with very high melting points.

Hydrogen chloride and ammonia produce the salt ammonium chloride:

$$HCl(g) + NH_3(g) \longrightarrow NH_4Cl(s)$$

Notice that this is a combination reaction. If the base contains hydroxide ion, then water is formed as a product along with the salt, as illustrated by the reaction of hydrochloric acid and sodium hydroxide:

$$HCl(aq) + NaOH(aq) \longrightarrow NaCl(aq) + H_2O(l)$$

Thus, the reaction between an acid and a base that contains hydroxide ions is a double replacement reaction. These two examples illustrate that a salt is always formed in acid-base neutralization; the salt is composed of cations from the base and anions from the acid.

The two equations we have just written for acid-base neutralization are **nonionic equations,** because they do not give formulas for the ions that exist individually in solution. We can rewrite the last equation with the formulas of the dissolved ions:

$$H^+(aq) + Cl^-(aq) + Na^+(aq) + OH^-(aq) \longrightarrow$$
$$Na^+(aq) + Cl^-(aq) + H_2O(l)$$

This is now a **total ionic equation;** it contains formulas for all ions and molecules as they actually exist in a reaction mixture.

When solutes react with each other in aqueous solution, the products will remain dissolved if they are soluble, or they will leave the solution if they are insoluble. Those remaining in solution can exist as ions, as molecules, or as a mixture of both, according to the following general rules.

1. *Ions* Soluble ionic compounds and strong acids and bases exist as hydrated ions in aqueous solution; in ionic equations they are represented by formulas of their hydrated ions.
2. *Molecules* Most molecular compounds that are soluble in water exist as molecules in aqueous solution and are represented in ionic equations by formulas of their molecules. Some molecular compounds, such as weak acids and bases, exist as mixtures of ions and molecules in aqueous solution. Because these exist to a greater extent as molecules than as ions, they are represented in ionic equations by formulas of their molecules.

Notice that we used a single arrow in the preceding neutralization equations; neutralization reactions of strong acids and bases go to completion. But notice also that sodium ions and chloride ions did not actually participate in the chemical reaction. The only actual reaction was the combination of $H^+(aq)$ and $OH^-(aq)$ to form $H_2O(l)$. (Note that $H^+(aq)$ is actually $H_3O^+(aq)$; $H^+(aq)$ is often used to represent $H_3O^+(aq)$ for the sake of simplicity.) The other ions in the reaction mixture (Na^+ and Cl^-) did not participate in the actual reaction; these ions are called **spectator ions.**

Formulas of spectator ions appear on both sides of an equation, and we can cancel them to write a **net ionic equation**, which contains only the formulas of ions and molecules that actually participate in a reaction. We can write the net ionic equation for the neutralization of HCl(aq) by NaOH(aq) as follows:

Complete ionic equation: $H^+(aq) + \cancel{Cl^-(aq)} + \cancel{Na^+(aq)} + OH^-(aq) \longrightarrow$
$$\cancel{Na^+(aq)} + \cancel{Cl^-(aq)} + H_2O(l)$$

Net ionic equation: $H^+(aq) + OH^-(aq) \longrightarrow H_2O(l)$

We use net ionic equations when we want to emphasize only the substances undergoing change in a chemical reaction.

Up until now, you have balanced only nonionic equations. Balancing ionic equations is very much like balancing nonionic equations, but there are two differences. One is that subscripts are used in ionic equations only for polyatomic ions; the subscripts of monatomic ions in nonionic equations become coefficients in ionic equations. The second difference is that the electrical charges, as well as the elements, must be in balance in an ionic equation.

EXAMPLE 14.13   From the following nonionic equation,

$$2\ AgNO_3(aq) + MgCl_2(aq) \longrightarrow 2\ AgCl(s) + Mg(NO_3)_2(aq)$$

derive the total ionic equation and the net ionic equation.

Solution   Total ionic equation:

$$2\ Ag^+(aq) + 2\ \cancel{NO_3^-(aq)} + \cancel{Mg^{2+}(aq)} + 2\ Cl^-(aq) \longrightarrow$$
$$2\ AgCl(s) + \cancel{Mg^{2+}(aq)} + 2\ \cancel{NO_3^-(aq)}$$

Net ionic equation:

$$Ag^+(aq) + Cl^-(aq) \longrightarrow AgCl(s)$$

EXERCISE 14.13   From the following nonionic equation,

$$BaCl_2(aq) + Na_2SO_4(aq) \longrightarrow BaSO_4(s) + 2\ NaCl(aq)$$

derive the total ionic equation and the net ionic equation.   ■

14.11   SOLUBILITIES OF SALTS AND OTHER IONIC COMPOUNDS

Sodium chloride is very soluble in water, but some salts are not. An example is the salt formed by neutralizing barium hydroxide, $Ba(OH)_2$, with sulfuric acid, H_2SO_4:

$$Ba(OH)_2(aq) + H_2SO_4(aq) \longrightarrow BaSO_4(s) + 2\ H_2O(l)$$

If we write the total ionic equation for this reaction,

$$Ba^{2+}(aq) + 2\ OH^-(aq) + 2\ H^+(aq) + SO_4^{2-}(aq) \longrightarrow$$
$$BaSO_4(s) + 2\ H_2O(l)$$

we see that all substances participated in the reaction, so the total ionic equation is the same as the net ionic equation.

Salts that are not soluble in water have such strong ionic forces holding their crystals together that water molecules cannot pull many of the ions apart. When ions of such salts are together in the same aqueous solution, they immediately form an insoluble solid called a **precipitate** and the solution is then saturated with the salt. In a chemical equation we sometimes use an arrow pointing down ($\downarrow$) to indicate formation of a precipitate. For example, if we mix a barium chloride solution with a sodium sulfate solution, insoluble barium sulfate will form:

$$BaCl_2(aq) + Na_2SO_4(aq) \longrightarrow BaSO_4(s) \downarrow + 2\ NaCl(aq)$$

The total ionic equation is

$$Ba^{2+}(aq) + 2\ Cl^-(aq) + 2\ Na^+(aq) + SO_4^{2-}(aq) \longrightarrow$$
$$BaSO_4(s) \downarrow + 2\ Cl^-(aq) + 2\ Na^+(aq)$$

and the net ionic equation is

$$Ba^{2+}(aq) + SO_4^{2-}(aq) \longrightarrow BaSO_4(s) \downarrow$$

The aqueous solubilities of a great many ionic compounds are known, so there are some general guidelines that can be used to make rough predictions of solubilities:

1. Compounds containing Group IA cations, NH_4^+, NO_3^-, ClO_3^-, ClO_4^-, or $C_2H_3O_2^-$ are usually water soluble.
2. Binary compounds containing halogen anions are usually water soluble, except for the halides of Ag^+, Hg_2^{2+}, and Pb^{2+}.
3. Compounds containing SO_4^{2-} are usually water soluble, except for $CaSO_4$, $SrSO_4$, $BaSO_4$, Hg_2SO_4, $HgSO_4$, $PbSO_4$, and Ag_2SO_4.
4. The oxides and hydroxides of Group IA elements, Ba^{2+}, Sr^{2+}, and NH_4^+ are water soluble.
5. Most other ionic compounds are not soluble in water.

These solubility guidelines can be used to predict when insoluble compounds will be formed by double replacement reactions occurring in aqueous solution. The reaction of $Ba(OH)_2$ with H_2SO_4 is an example, as is the reaction of $BaCl_2$ with Na_2SO_4; as illustrated earlier, both reactions form insoluble $BaSO_4$. Another example is the formation of insoluble $CaCO_3$ in the reaction of Na_2CO_3 with $CaCl_2$:

$$Na_2CO_3(aq) + CaCl_2(aq) \longrightarrow CaCO_3(s) + 2\ NaCl(aq)$$

EXAMPLE 14.14

An aqueous solution of $AgNO_3$ is mixed with an aqueous solution of NaCl. Will there be a reaction? If so, write the balanced nonionic equation, the total ionic equation, and the net ionic equation.

Solution

The possible products from a double replacement reaction are AgCl and $NaNO_3$. Since AgCl is insoluble, a reaction will occur.

Nonionic equation: $AgNO_3(aq) + NaCl(aq) \longrightarrow AgCl(s) + NaNO_3(aq)$

Total ionic equation: $Ag^+(aq) + \cancel{NO_3^-(aq)} + \cancel{Na^+(aq)} + Cl^-(aq) \longrightarrow$
$AgCl(s) + \cancel{Na^+(aq)} + \cancel{NO_3^-(aq)}$

Net ionic equation: $Ag^+(aq) + Cl^-(aq) \longrightarrow AgCl(s)$

EXERCISE 14.14   An aqueous solution of H_2SO_4 is mixed with an aqueous solution of $Mg(NO_3)_2$. Will there be a reaction? If so, write the balanced nonionic equation, the total ionic equation, and the net ionic equation. ◾

A reaction will also occur if at least one product is a molecular compound. Typically, these compounds are gases, H_2O, or any other molecular compound that does not ionize appreciably in water.

14.12 SOLUBILITY PRODUCT CONSTANTS *(optional)*

Compounds such as barium sulfate ($BaSO_4$) are commonly said to be insoluble in water, but since solubility is a matter of degree, it is more precise to speak of such compounds as being slightly soluble in water. Barium sulfate, like all slightly soluble ionic compounds, exists in water as an equilibrium mixture of dissolved ions and undissolved solid. We can describe the equilibrium by the following equation:

$$BaSO_4(s) \rightleftharpoons Ba^+(aq) + SO_4^{2-}(aq)$$

The expression for the equilibrium constant is

$$K_{eq} = \frac{[Ba^{2+}][SO_4^{2-}]}{[BaSO_4]}$$

However, $[BaSO_4]$ corresponds to the concentration of a pure substance and thus is a constant. Knowing this, we can combine K_{eq} and $[BaSO_4]$ to define a new constant, K_{sp}:

$$K_{eq}[BaSO_4] = K_{sp} = [Ba^{2+}][SO_4^{2-}]$$

K_{sp} is called the **solubility product constant;** it is the product of the concentrations of dissolved ions, with the concentrations raised to powers corresponding to the coefficients in the equation for the saturation equilibrium. Units of K_{sp} values are not usually shown. K_{sp} values are quite small, due to the extremely low concentrations of dissolved ions in solutions saturated with slightly soluble compounds.

EXAMPLE 14.15   Write the K_{sp} expression for Ag_2CrO_4, a slightly soluble salt.

Solution   First, write the equation that describes the saturation equilibrium:

$$Ag_2CrO_4(s) \rightleftharpoons 2\,Ag^+(aq) + CrO_4^{2-}(aq)$$

Then, derive the K_{sp} expression:

$$K_{sp} = [Ag^+]^2[CrO_4^{2-}]$$

EXERCISE 14.15 Write the K_{sp} expression for $Mn(OH)_2$, a slightly soluble base. ■

The K_{sp} value for a particular compound is related to solubility—for formula units containing the same number of ions, the higher the K_{sp} value, the greater the solubility. Table 14-8 gives K_{sp} values for a variety of slightly soluble compounds. The product of the ion concentrations of a slightly soluble compound (when raised to their appropriate powers) cannot exceed the K_{sp} value at a given temperature. For example, AgCl has a K_{sp} of 1.6×10^{-10} at 25° C. The equation for the saturation equilibrium is

$$AgCl(s) \rightleftharpoons Ag^+(aq) + Cl^-(aq)$$

and

$$K_{sp} = [Ag^+][Cl^-]$$

Thus, if the product of $[Ag^+]$ and $[Cl^-]$ were to exceed 1.6×10^{-10} at 25° C, AgCl would precipitate.

TABLE 14-8 Solubility Product Constants (K_{sp}) at 25° C

Name	Formula	K_{sp}
Aluminum hydroxide	$Al(OH)_3$	3.7×10^{-15}
Barium carbonate	$BaCO_3$	8.1×10^{-9}
Barium fluoride	BaF_2	1.7×10^{-6}
Barium sulfate	$BaSO_4$	1.1×10^{-10}
Bismuth sulfide	Bi_2S_3	1.6×10^{-72}
Cadmium sulfide	CdS	1.4×10^{-28}
Calcium carbonate	$CaCO_3$	8.7×10^{-9}
Calcium fluoride	CaF_2	4.0×10^{-11}
Chromium hydroxide	$Cr(OH)_3$	3×10^{-29}
Cobalt sulfide	CoS	3.1×10^{-26}
Cupric sulfide	CuS	1.0×10^{-44}
Cuprous bromide	Cu_2Br_2	4.2×10^{-8}
Cuprous iodide	Cu_2I_2	5.1×10^{-12}
Ferric hydroxide	$Fe(OH)_3$	1.1×10^{-36}
Ferrous hydroxide	$Fe(OH)_2$	1.6×10^{-14}
Ferrous sulfide	FeS	3.7×10^{-19}
Lead carbonate	$PbCO_3$	3.3×10^{-14}
Lead chloride	$PbCl_2$	2.4×10^{-4}
Lead fluoride	PbF_2	4.1×10^{-8}
Lead sulfide	PbS	3.4×10^{-28}
Magnesium hydroxide	$Mg(OH)_2$	1.2×10^{-11}
Manganese sulfide	MnS	1.4×10^{-15}
Mercurous chloride	Hg_2Cl_2	3.5×10^{-18}
Mercuric sulfide	HgS	4.0×10^{-54}
Nickel sulfide	NiS	1.4×10^{-24}
Silver bromide	$AgBr$	7.7×10^{-13}
Silver chloride	$AgCl$	1.6×10^{-10}
Silver sulfide	Ag_2S	1.8×10^{-49}
Stannous sulfide	SnS	1×10^{-26}
Strontium carbonate	$SrCO_3$	1.6×10^{-9}
Zinc hydroxide	$Zn(OH)_2$	1.8×10^{-14}
Zinc sulfide	ZnS	1.2×10^{-23}

EXAMPLE 14.16   If enough $AgNO_3$ and $NaCl$ were dissolved in the same solution at 25° C to make $[Ag^+]$ 1.2×10^{-4} M and $[Cl^-]$ 2.0×10^{-3} M, would $AgCl$ precipitate?

Solution   Compare the product of the proposed concentrations of dissolved ions to the K_{sp} value:

$$[Ag^+][Cl^-] = (1.2 \times 10^{-4})(2.0 \times 10^{-3}) = 2.4 \times 10^{-7}$$
$$K_{sp} = 1.6 \times 10^{-10}$$

MATH TIP
$2.4 \times 10^{-7} > 1.6 \times 10^{-10}$ since the larger the negative exponent, the closer the number's value is to zero. See Table 2-1.

Since the product of $[Ag^+]$ and $[Cl^-]$ in the proposed solution would be higher than the K_{sp} value, $AgCl$ would precipitate.

EXERCISE 14.16   The K_{sp} for $BaSO_4$ is 1.1×10^{-10} at 25° C. If enough $BaCl_2$ and Na_2SO_4 were dissolved to make $[Ba^{2+}]$ 3.5×10^{-5} M and $[SO_4{}^{2-}]$ 7.4×10^{-2} M, would $BaSO_4$ precipitate?   ■

For slightly soluble compounds, the sources of the ions in solution are not important. Thus, for $AgCl$ to precipitate, Ag^+ ions could be provided by any soluble silver compound and Cl^- ions could be provided by any soluble chloride compound. The addition of either ion to a solution containing the other ion will shift the saturation equilibrium in the direction of undissolved compound:

$$AgCl(s) \rightleftharpoons Ag^+(aq) + Cl^-(aq)$$
$$\longleftarrow \text{ increased } [Ag^+]$$
$$\longleftarrow \text{ increased } [Cl^-]$$

Thus, adding Ag^+ ions to a solution containing Cl^- ions can precipitate $AgCl$ and reduce $[Cl^-]$. The process of increasing the concentration of one of the ions of a slightly soluble compound and causing the saturation equilibrium to shift to the left, thus decreasing the concentration of the other ion, is called the **common ion effect.** The common ion effect is just another application of LeChatelier's principle. The common ion effect results in a change in the relative concentrations of ions, but K_{sp} remains constant at a given temperature.

EXAMPLE 14.17   If $[Ag^+]$ and $[Cl^-]$ are each 3.0×10^{-6} M initially, what will $[Ag^+]$ be if $[Cl^-]$ is raised to 0.050 M at 25° C?

Solution
$$K_{sp} = [Ag^+][Cl^-]$$
$$[Ag^+] = \frac{K_{sp}}{[Cl^-]}$$
$$= \frac{1.6 \times 10^{-10}}{0.050}$$
$$= 3.2 \times 10^{-9} \, M$$

EXERCISE 14.17   The K_{sp} for CaF_2 is 4.0×10^{-11} at 25° C. If $[Ca^{2+}]$ and $[F^-]$ are each 6.0×10^{-5} M initially, what will $[F^-]$ be if $[Ca^{2+}]$ is raised to 0.10 M at 25° C?   ■

SUMMARY

The **Arrhenius theory** defines acids as substances that produce H^+ in water and bases as substances that produce OH^- in water. The more general **Bronsted-Lowry theory** defines acids as proton donors and bases as proton acceptors. An acid forms its **conjugate base** by losing a proton, and a base forms its **conjugate acid** by gaining a proton. In aqueous solution, acids have a sour taste, turn blue litmus red, neutralize bases, and dissolve certain metals. Aqueous bases have a bitter taste and slippery feeling, turn red litmus blue, and neutralize acids. Strong acids and bases have **ionization constants** of 1 or greater, while weak acids and bases have ionization constants of less than 1. **Polyprotic acids** are capable of donating more than one proton from each formula unit and ionize in steps. The acid strength of a polyprotic acid is related to the electronegativity of the central nonmetallic atom. An **acid anhydride** is a nonmetallic oxide that reacts with water to form an acid. A **basic anhydride** is a metallic oxide that reacts with water to form a base.

Concentrations of acids and bases can be determined by **titration** and are often expressed as **normality,** the number of equivalents per liter. An **equivalent of an acid** is the weight of acid that neutralizes one mole of OH^-, while an **equivalent of a base** is the weight of base that neutralizes one mole of H_3O^+. Normality can be calculated from molarity or determined by titration.

A **salt** is an ionic compound formed when an acid neutralizes a base. The salt is composed of cations from the base and anions from the acid. Ionic equations can be used to illustrate formation of salts. If ions of an insoluble salt are together in the same solution, they will form a **precipitate.** General solubility guidelines are used to predict formation of insoluble compounds by double replacement reactions in aqueous solution.

Insoluble compounds can be characterized by their **solubility product constants,** K_{sp}. The K_{sp} value of a compound is related to solubility; the higher the K_{sp} value, the greater the solubility. The process of increasing the concentration of one of the ions of an insoluble compound and causing the saturation equilibrium to shift toward formation of precipitate is called the **common ion effect.**

STUDY QUESTIONS AND PROBLEMS

(More difficult questions and problems are marked with an asterisk.)

CONCEPTS OF ACIDS AND BASES

1. Define the following terms:
   a. Arrhenius theory
   b. Bronsted-Lowry theory
   c. Hydronium ion
   d. Conjugate acid
   e. Conjugate base
   f. Amphoteric
2. What are the major differences between Arrhenius theory and Bronsted-Lowry theory?

3. Write the chemical equation for the reaction of each acid with water, and identify each acid-base conjugate pair.
   a. HCl   b. HNO_3 c. HBr   d. $HClO_4$

4. For each of the following acids and bases, write the formula of the conjugate base or acid.

ACID	BASE
a. HI	e. HSO_4^-
b. $HClO_3$	f. OH^-
c. $H_2PO_4^-$	g. NH_3
d. NH_4^+	h. CO_3^{2-}

PROPERTIES OF ACIDS AND BASES

5. Summarize the major properties of acids and bases.
6. What is an acid-base indicator? Give an example of one.
7. Write a chemical equation to illustrate acid-base neutralization.

ACID STRENGTH

8. Explain what *acid strength* means.
9. Distinguish between strong and weak acids, and give an example of each.
10. What is $[H_3O^+]$ in a 6.0 M solution of hydrochloric acid, HCl(aq)?
11. In an aqueous solution of acetic acid, $HC_2H_3O_2$, the following equilibrium concentrations exist.

$$[HC_2H_3O_2] = 0.0987 \ M$$
$$[H_3O^+] = 1.3 \times 10^{-3} \ M$$
$$[C_2H_3O_2^-] = 1.3 \times 10^{-3} \ M$$

Calculate K_a for acetic acid.

12. Hydrogen cyanide, a toxic gas, dissolves in water to form hydrocyanic acid, a weak acid:

$$HCN(aq) + H_2O(l) \rightleftharpoons H_3O^+(aq) + CN^-(aq)$$

If the following equilibrium concentrations exist,

$$[HCN] = 0.20 \ M$$
$$[H_3O^+] = 9.9 \times 10^{-6} \ M$$
$$[CN^-] = 9.9 \times 10^{-6} \ M$$

calculate K_a for hydrocyanic acid.

13. Hydrofluoric acid is a weak acid because it contains fluorine, the most electronegative element. Fluorine atoms attract bonding electrons and hydrogen atoms in HF so strongly that little dissociation occurs:

$$HF(aq) + H_2O(l) \rightleftharpoons H_3O^+(aq) + F^-(aq)$$
$$K_a = 6.8 \times 10^{-4}$$

If the equilibrium concentration of HF is 0.099 M, what is the equilibrium concentration of H_3O^+?

14. Fill in the blanks in the following statements.
   a. A weak acid has a ___strong___ conjugate base.
   b. A strong acid has a ___weak___ conjugate base.
   c. A weak base has a ___strong___ conjugate acid.
   d. A strong base has a ___weak___ conjugate acid.

POLYPROTIC ACIDS

15. What does the term *polyprotic acid* mean?
16. Write chemical equations for the stepwise ionization of sulfuric acid, H_2SO_4. Identify all acid-base conjugate pairs.
17. Write chemical equations for the stepwise ionization of phosphoric acid, H_3PO_4. Identify all acid-base conjugate pairs.
18. Why is phosphoric acid, H_3PO_4, a weaker acid than sulfuric acid, H_2SO_4?
19. For a polyprotic acid, why does K_a decrease with each successive loss of a proton?

BASE STRENGTH

20. Distinguish between strong bases and weak bases, and give an example of each.
21. What is $[OH^-]$ in a 2.0 M solution of KOH?
22. In an aqueous solution of ammonia, the following equilibrium concentrations existed:

$$[NH_3] = 0.148\ M$$
$$[NH_4^+] = 1.6 \times 10^{-3}\ M$$
$$[OH^-] = 1.6 \times 10^{-3}\ M$$

What is K_b for the following reaction?

$$NH_3(aq) + H_2O(l) \rightleftharpoons NH_4^+(aq) + OH^-(aq)$$

23. Write a chemical equation to illustrate why a solution of $NaHCO_3$ is basic.

ACID AND BASIC ANHYDRIDES

24. What are acid and basic anhydrides?
25. Derive the formula of the anhydride of each of the following.
   a. $Ca(OH)_2$ **c.** H_2SiO_3 **e.** H_3BO_3
   b. $HClO$ **d.** $Zn(OH)_2$
*26. Write the chemical equation for the reaction of each anhydride with water.
   a. SrO **c.** CoO **e.** SeO_3
   b. Br_2O **d.** Cl_2O_5
*27. New automobiles come equipped with catalytic converters that reduce NO_2 in exhaust to N_2. How is this beneficial? (*Hint:* Consider the reaction of NO_2 with moisture in the atmosphere.)

ACID-BASE TITRATIONS

28. How is a titration performed?

29. A 25.00-mL sample of H_2SO_4 required 32.15 mL of 0.6000 M NaOH to reach the end point of a titration. What was the molarity of the H_2SO_4?

30. A 20.00-mL sample of aqueous $Ba(OH)_2$ required 15.27 mL of 0.1000 M hydrochloric acid to reach the end point of a titration. What was the molarity of the $Ba(OH)_2$?

31. When dissolved in water, potassium hydrogen phthalate ($KHC_8H_4O_4$) behaves like an acid, donating one proton per formula unit. A solution containing a 0.6135 g sample of pure potassium hydrogen phthalate required 26.72 mL of a solution of NaOH to reach the end point of a titration. What was the molarity of the NaOH?

$$KHC_8H_4O_4(aq) + NaOH(aq) \longrightarrow KNaC_8H_4O_4(aq) + H_2O(l)$$

*32. A 1.263 g sample of impure $Mg(OH)_2$ required 30.64 mL of 0.6000 M HCl to reach the end point of a titration. What was the %(w/w) of $Mg(OH)_2$ in the sample?

NORMALITY

33. Define the following terms:
   a. Equivalent of an acid   d. Diprotic acid
   b. Equivalent of a base   e. Triprotic acid
   c. Monoprotic acid   f. Normality

34. Listed below are several acids. Classify each as monoprotic, diprotic, or triprotic.
   a. H_2SO_4 b. H_3PO_4 c. HCl   d. H_2CO_3

35. Calculate the normality of a solution containing 392.4 g of H_2SO_4 in 2.00 L of solution.

36. What is the normality of a solution containing 20.0 g of NaOH in 250.0 mL of solution?

37. How many grams of KOH are in 600.0 mL of a 6.00 N solution?

38. How many meq are in 100.0 mL of a 0.01000 N solution?

39. How many meq are in 50.0 mL of a 0.150 N solution?

40. What is the normality of a 2.00 M solution of H_3PO_4?

41. What is the normality of a 1.50 M solution of H_2SO_4?

42. What is the normality of a 6.00 M solution of NaOH?

43. In a titration, 40.00 mL of an acid solution required 20.05 mL of 1.502 N KOH to reach the end point. What was the normality of the acid?

44. Calculate the volume of 0.4000 N base required to titrate 15.00 mL of 1.000 N HNO_3.

SALTS

45. Define the following terms:
   a. Salt   d. Spectator ions
   b. Nonionic equation   e. Net ionic equation
   c. Total ionic equation

46. a. Write the total ionic equation for the complete neutralization of NaOH by H_2SO_4 in water.
   b. Write the net ionic equation for the reaction in part a.

47. a. Write the total ionic equation for the complete neutralization of $Ca(OH)_2$ by H_2SO_4 in water; the salt formed is insoluble in water.
    b. Write the net ionic equation for the reaction in part a.

SOLUBILITIES OF SALTS AND OTHER IONIC COMPOUNDS

48. In chemistry, what is meant by a *precipitate?*
49. Using the solubility guidelines, determine which of the following ionic compounds are not soluble in water.
    a. Rb_2SO_4 c. $PbSO_4$ e. $PbCO_3$
    b. Ag_2O d. $HgBr_2$
50. Write the balanced nonionic equation for any reactions that occur when aqueous solutions of the following are mixed.
    a. $AlCl_3$ and $NaOH$ c. $AgNO_3$ and $NaClO_3$
    b. $Ba(C_2H_3O_2)_2$ and K_2SO_4 d. $FeCl_2$ and $LiOH$
51. Write the total ionic equation for any reaction that would occur in question 50.
52. Write the net ionic equation for any reaction that would occur in question 50.

SOLUBILITY PRODUCT CONSTANTS

53. Write the K_{sp} expression for each of the following.
    a. $Al(OH)_3$ b. Hg_2CrO_4 c. BaC_2O_4 d. $Sr_3(PO_4)_2$
54. The solubility of $Cd(OH)_2$ at $25°$ C is 1.7×10^{-5} mol/L. Calculate the K_{sp} of $Cd(OH)_2$ at $25°$ C.
55. The solubility of $Ce(OH)_3$ at $25°$ C is 5.2×10^{-6} mol/L. Calculate the K_{sp} of $Ce(OH)_3$ at $25°$ C.
56. Which of the following two compounds has the lower solubility in mol/L?

$$Ag_2CO_3 \quad K_{sp} = 8.2 \times 10^{-12}$$
$$CuCO_3 \quad K_{sp} = 2.5 \times 10^{-10}$$

57. The K_{sp} for $Cr(OH)_3$ is 3×10^{-29} at $25°$ C. If $[Cr^{3+}]$ and $[OH^-]$ are each 1.00×10^{-8} M initially, what will $[Cr^{3+}]$ be if $[OH^-]$ is raised to 0.15 M at $25°$ C?

GENERAL EXERCISES

58. Write chemical equations to illustrate how H_2O can act as an acid and a base.
59. For the weak acid $HCHO_2$ (formic acid), $K_a = 1.7 \times 10^{-4}$. Calculate $[H_3O^+]$ in a solution that contains 0.20 M $HCHO_2$ at equilibrium.

$$HCHO_2(aq) + H_2O(l) \rightleftharpoons CHO_2^-(aq) + H_3O^+(aq)$$

60. Write chemical equations for the stepwise ionization of carbonic acid, H_2CO_3. Identify all acid-base conjugate pairs.
61. Urea, a metabolic product found in urine, is a weak base:

$$CON_2H_4(aq) + H_2O(l) \rightleftharpoons CON_2H_5^+(aq) + OH^-(aq)$$

A typical urine sample contains the following concentrations:

$$[CON_2H_4] = 0.40\ M$$
$$[CON_2H_5^+] = 7.7 \times 10^{-8}\ M$$
$$[OH^-] = 7.7 \times 10^{-8}\ M$$

What is K_b for urea?

*62. Give the product of each reaction.
   a. $CaO(s) + SiO_2(s)$ c. $B_2O_3(s) + 3\ Na_2O(s)$
   b. $BaO(s) + SO_3(g)$

*63. The catalytic converters mentioned in question 27 have an unfortunate disadvantage in that they oxidize SO_2 in exhaust to SO_3. Why is this undesirable? (*Hint:* Consider the reactions of both oxides with moisture in the atmosphere.)

64. Potatoes are peeled commercially by soaking them in NaOH solution for a short time and then spraying off the softened peel with a jet of water. If 42.74 mL of 1.000 M H_2SO_4 was required to reach the end point in a titration of 20.00 mL of the NaOH solution used for the potatoes, what was the molarity of the NaOH?

65. A 0.2981 g sample of impure oxalic acid, $H_2C_2O_4$, required 27.56 mL of 0.1845 M NaOH to reach the end point of a titration. What was the %(w/w) of $H_2C_2O_4$ in the sample?

66. How many grams of H_3PO_4 are in 5.00 L of a 2.00 N solution?

67. A volume of 50.0 mL of 0.100 N NaOH was added to 15.0 mL of 0.400 N HCl. Was the resulting solution neutral, acidic, or basic? Explain your answer.

*68. Write the balanced nonionic equation for the acid-base neutralization that would produce each salt.
   a. $CaBr_2(aq)$ c. $NaClO_4(aq)$
   b. $KC_2H_3O_2(aq)$ d. $Mg_3(PO_4)_2(s)$

*69. At 25° C, the K_{sp} of CaF_2 is 4.0×10^{-11}. What is the solubility of CaF_2 in mol/L?

70. The K_{sp} for Bi_2S_3 is 1.6×10^{-72} at 25° C. If $[Bi^{3+}]$ and $[S^{2-}]$ are each 2.5×10^{-16} M initially, what will $[Ba^{3+}]$ be if $[S^{2-}]$ is raised to 0.10 M at 25° C?

ELECTROLYTES, pH, AND BUFFERS

CHAPTER 15

OUTLINE

15.1 Electrolytes
15.2 The Ionization of Water
15.3 pH
15.4 The pH Scale
15.5 Measurement of pH
15.6 Additional pH Calculations
(optional)
15.7 Hydrolysis *(optional)*
15.8 Buffers
Perspective: The Acid-Base
Balance of Blood
Summary
Study Questions and Problems

Electrolytes are substances that conduct electricity when they are melted or dissolved in water. They can be acids, bases, or salts, but they must produce ions to conduct electricity. Our bodies contain many electrolytes, but the principal ions produced by electrolytes in the body are K^+, Na^+, and Cl^-. On the average, the concentration of K^+ is 40 times higher inside cells than outside. The opposite is true for Na^+ and Cl^-; their concentrations inside cells are about 1/2 and 1/20, respectively, of their concentrations in body fluids that surround the cells. Electrolytes also have important purposes outside of the human body; for example, your automobile battery uses electrolytes to store energy.

The level of acid in aqueous solution is often measured in units of pH. At one time farmers checked the pH of soil simply by tasting it. If the soil tasted sour, it was acidic, with a pH of less than 7. A sweet taste meant that the soil was alkaline and had a pH greater than 7. Modern farmers might use meters to measure soil pH, and simpler devices are available for measuring the pH of swimming pool water and other common substances. The human body is so sensitive to pH changes that it possesses chemical means of responding to even slight pH changes in blood. The chemical systems responsible for maintaining a relatively constant pH in blood and other solutions are called *buffers*. In this chapter you will learn how such buffers work. Before going ahead with this chapter, you should review ionization constants (chapter 14), logarithms (appendix A), and LeChatelier's principle (chapter 13).

15.1   ELECTROLYTES

When you think of electricity, you might visualize metal wires that carry electric current. In general, metals are good conductors of electricity because the outer electrons of their atoms are held rather loosely and thus can be made to move from one atom to another. Most nonmetals do not conduct electricity, because the outer electrons of their atoms are either held very tightly or shared by other nearby atoms. For the same reason, molecular compounds are usually not electrical conductors. Solid ionic compounds do not conduct electricity because their ions are unable to move from their fixed positions. However, if an ionic compound is melted or dissolved in water, it will conduct electricity because the ions are free to move and they can carry electrons to charged electrodes (see Figure 15-1). Substances that conduct electricity when they are melted or dissolved in water are called **electrolytes**.

We can find out whether a substance is an electrolyte by dissolving the substance in water and measuring the electrical conductivity of the solution. For example, pure water just barely conducts electricity, but tap water is a better conductor because it contains dissolved minerals that exist as ions in solution. (For this reason, we are cautioned not to use electric appliances in tap water unless they are designed for such use; without proper insulation of the appliance, tap water could carry the electric current to our bodies, causing severe electrical shock.)

Electrode

A conductor through which an electric current enters or leaves.

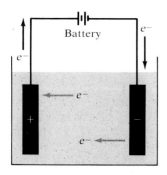

FIGURE 15-1   Movement of Electrons as Electricity Flows through a Liquid. Electrons are carried through the liquid by ions.

TABLE 15-1 Examples of Strong Electrolytes, Weak Electrolytes, and Nonelectrolytes

Substance	Formula	Type of Substance	Classification
Hydrochloric acid	HCl	Strong acid	Strong electrolyte
Sulfuric acid	H_2SO_4	Strong acid	Strong electrolyte
Sodium hydroxide	NaOH	Strong base	Strong electrolyte
Sodium chloride	NaCl	Salt	Strong electrolyte
Calcium nitrate	$CaNO_3$	Salt	Strong electrolyte
Acetic acid	$HC_2H_3O_2$	Weak acid	Weak electrolyte
Formic acid	$HCHO_2$	Weak acid	Weak electrolyte
Ammonia	NH_3	Weak base	Weak electrolyte
Methylamine	CH_3NH_2	Weak base	Weak electrolyte
Glucose	$C_6H_{12}O_6$	Sugar (molecular solid)	Nonelectrolyte
Sucrose	$C_{12}H_{22}O_{11}$	Sugar (molecular solid)	Nonelectrolyte
Ethanol	C_2H_5OH	Alcohol (molecular liquid)	Nonelectrolyte

Ions in solution can come from acids, bases, or salts. **Strong electrolytes** are compounds that dissociate completely into ions when dissolved in water. All strong acids and strong bases are strong electrolytes, and so are salts that are water soluble. Aqueous solutions of these substances have high electrical conductivity. **Weak electrolytes** are substances that dissociate incompletely into ions when dissolved in water. Aqueous solutions of weak electrolytes show poor electrical conductivity. Weak acids and weak bases are examples of weak electrolytes, as are salts with low water solubility.

Some water-soluble substances produce no ions when dissolved in water and thus do not conduct electricity; they are called **nonelectrolytes.** They are usually molecular compounds in which the covalent bonds are not polar enough to be separated by water molecules. Most sugars fit this category, as well as many alcohols. Table 15-1 gives examples of each class of electrolytes.

15.2 THE IONIZATION OF WATER

As we mentioned earlier, pure water shows a trace of conductivity, indicating that a very small number of ions are present. If pure water is subjected to a strong electric current, hydrogen and oxygen are produced as a result of the following reactions, which occur at the electrodes as the water conducts electricity:

$$2\ H^+(aq) + 2\ e^-(aq) \longrightarrow H_2(g)$$
$$2\ OH^-(aq) - 4\ e^-(aq) \longrightarrow O_2(g) + 2\ H^+(aq)$$

(H^+(aq) is a simplified representation of a hydronium ion.) These reactions confirm the presence of ions in pure water, indicating that pure water must ionize to some extent to form $H_3O^+(aq)$ and $OH^-(aq)$. Thus, water is a very weak acid and a very weak base. It is capable of donating protons and of accepting protons, and ionization of water occurs according to the following equation:

$$H_2O(l) + H_2O(l) \rightleftharpoons H_3O^+(aq) + OH^-(aq)$$

or,

$$2\,H_2O(l) \rightleftharpoons H_3O^+(aq) + OH^-(aq)$$

Measurements show that only about one out of every 10^9 water molecules is ionized at any one instant at room temperature.

The expression for the ionization constant of water is derived below:

$$K_{eq} = \frac{[H_3O^+][OH^-]}{[H_2O]^2}$$

$$K_{eq}[H_2O]^2 = [H_3O^+][OH^-]$$

$$K_w = K_{eq}[H_2O]^2 = [H_3O]^+[OH^-]$$

The ionization constant for water has a special name—the **ion-product constant**, K_w. From measurements of the extremely low electrical conductivity of water, the value of K_w at 25° C has been determined to be 1.00×10^{-14}.

Like all ionization constants, K_w is a true constant. Thus, if $[H_3O^+]$ is increased by adding acid to water, $[OH^-]$ decreases and the solution is acidic. If $[OH^-]$ is increased by the addition of base to water, $[H_3O^+]$ decreases and the solution is basic. When $[H_3O^+] = [OH^-]$, the solution is said to be neutral. Pure water is neutral.

15.3   pH

pH

From the French *pouvoir hydrogene*, meaning "power of hydrogen."

Because the product of $[H_3O^+]$ and $[OH^-]$ in water is constant at a particular temperature, the total acid-base balance in any aqueous solution can be expressed by stating the concentration of H_3O^+. It is convenient to express $[H_3O^+]$ in terms of **pH**, which is defined as the negative logarithm of $[H_3O^+]$:

$$pH = -\log[H_3O^+]$$

(If you need to review logarithms, see appendix A.)

Although the idea of taking a negative logarithm may seem complicated, we can simplify the process by remembering that the log of a number expressed as 1×10^n is simply n. Thus, the log of 1×10^3 is 3, and the log of 1×10^{-3} is -3. pH is defined as a negative logarithm, so if $[H_3O^+] = 1 \times 10^{-3}$ M, then the pH is 3, the value of the exponent with its sign changed. Let's see how this works by calculating the pH of pure water:

$$K_w = [H_3O^+][OH^-]$$

$\sqrt{1.00 \times 10^{-14}}$ can also be written as $(\sqrt{1.00})(\sqrt{10^{-14}})$ since $\sqrt[n]{ab} = (\sqrt[n]{a})(\sqrt[n]{b})$. Since the square root of 1.00 is 1.00, this problem simplifies to $(1.00)(\sqrt{10^{-14}})$. To proceed further,

$$\sqrt{10^{-14}} = 10^{-14/2}$$
$$= 10^{-7}$$

since

$$\sqrt[n]{a^x} = a^{x/n}$$

This problem can also be solved on the calculator by the following use of the square root key $\boxed{\sqrt{x}}$:

Press 1.00 $\boxed{EE}$ 14 $\boxed{+/-}$ $\boxed{\sqrt{x}}$.

The display reads 1. -07. The answer is written 1.00×10^{-7}.

Since, in pure water,

$$[H_3O^+] = [OH^-]$$

then, at 25° C,

$$K_w = [H_3O^+][H_3O^+] = [H_3O^+]^2 = 1.00 \times 10^{-14}$$

Thus,

$$[H_3O^+]^2 = 1.00 \times 10^{-14}$$
$$[H_3O^+] = \sqrt{1.00 \times 10^{-14}} = 1.00 \times 10^{-7}$$

Now we must take the negative logarithm of $[H_3O^+]$:

$$pH = -\log [H_3O^+]$$
$$= -\log (1.00 \times 10^{-7})$$
$$= -(-7)$$
$$= 7$$

Thus, pure water at 25° C has a pH of 7. In fact, all neutral aqueous solutions at 25° C have pH values of 7. If a solution is acidic, $[H_3O^+]$ at 25° C will be greater than $1.00 \times 10^{-7} M$, and the pH will be less than 7. At 25° C, a basic solution will have $[H_3O^+]$ less than 1.00×10^{-7}, and its pH will be greater than 7. *Notice that pH decreases as a solution becomes more and more acidic, and it increases as a solution becomes more and more basic.*

As another example, consider the acid level in the human stomach, whose lining secretes hydrochloric acid to aid digestion. Although $[H_3O^+]$ is somewhat variable in the stomach, a typical value is 0.01 M. Let's calculate the pH of stomach acid corresponding to this $[H_3O^+]$:

$$pH = -\log [H_3O^+]$$
$$= -\log (0.01) = -(\log 1 \times 10^{-2})$$
$$= -(-2)$$
$$= 2$$

Thus, the pH of stomach acid in this example is 2. The pH of stomach acid is normally in the range of 1–2.

EXAMPLE 15.1

Calculate the pH of a solution in which $[H_3O^+]$ is $1.0 \times 10^{-4} M$. Is the solution acidic or basic?

Solution

$$pH = -\log [H_3O^+]$$
$$= -\log (1.0 \times 10^{-4})$$
$$= -(-4)$$
$$= 4$$

The solution is acidic, because its pH is less than 7.

EXERCISE 15.1

Calculate the pH of a solution in which $[H_3O^+]$ is $1.0 \times 10^{-9} M$. Is the solution acidic or basic? ■

EXAMPLE 15.2 Calculate $[H_3O^+]$ in a solution whose pH is 8.0.

Solution

$$pH = -\log [H_3O^+]$$
$$\log [H_3O^+] = -pH$$
$$= -8.0$$

Taking the antilogarithm of -8.0,

$$[H_3O^+] = 1 \times 10^{-8}\ M$$

EXERCISE 15.2 Calculate $[H_3O^+]$ in a solution whose pH is 6.0. ■

Because K_w for water is a constant at a given temperature, $[OH^-]$ is automatically indicated by pH. We can show the relationship between $[OH^-]$ and pH by starting with the definition of K_w:

$$K_w = [H_3O^+][OH^-] = 1.00 \times 10^{-14}$$

When numbers are multiplied, their logarithms are added; thus,

$$\log K_w = \log [H_3O^+] + \log [OH^-] = -14$$

and the sum of logarithms of $[H_3O^+]$ and $[OH^-]$ at 25° is -14. Let's go one step further by defining **pOH** as the negative logarithm of $[OH^-]$:

$$pOH = -\log [OH^-]$$

Then,

$$\log [H_3O^+] + \log [OH^-] = -14$$

and,

$$-\log [H_3O^+] - \log [OH^-] = 14$$
$$pH + pOH = 14$$

TABLE 15-2 Concentrations, pH, and pOH

$[H_3O^+]$	pH	$[OH^-]$	pOH	
10^{0}	0	10^{-14}	14	acidic
10^{-1}	1	10^{-13}	13	acidic
10^{-2}	2	10^{-12}	12	acidic
10^{-3}	3	10^{-11}	11	acidic
10^{-4}	4	10^{-10}	10	acidic
10^{-5}	5	10^{-9}	9	acidic
10^{-6}	6	10^{-8}	8	acidic
10^{-7}	7	10^{-7}	7	neutral
10^{-8}	8	10^{-6}	6	basic
10^{-9}	9	10^{-5}	5	basic
10^{-10}	10	10^{-4}	4	basic
10^{-11}	11	10^{-3}	3	basic
10^{-12}	12	10^{-2}	2	basic
10^{-13}	13	10^{-1}	1	basic
10^{-14}	14	10^{-0}	0	basic

Now you see that the sum of pH and pOH is always 14 at 25° C. (Since K_w varies only slightly with temperature, this relationship holds true for temperatures within a few degrees of 25° C, i.e., normal room temperature.)

Returning to the stomach acid example, the pH of 2 gives a pOH of 12 and $[OH^-]$ of 1×10^{-12} M. Thus, when pH is less than 7, $[H_3O^+]$ must be greater than 1×10^{-7} M; there are more H_3O^+ ions in solution than there are OH^- ions, and the solution is acidic. Similarly, when the pH is greater than 7, $[H_3O^+]$ must be less than 1×10^{-7} M, $[OH^-]$ must be greater than 1×10^{-7} M, and the solution is basic. In a neutral solution, the pH is 7, and $[H_3O^+]$ and $[OH^-]$ are both 1×10^{-7} M. Thus, in a neutral solution $[H_3O^+]$ and $[OH^-]$ are balanced. These relationships are summarized in Table 15-2.

EXAMPLE 15.3   If the pH of a solution is 3, what is the pOH?

Solution   Since

$$pH + pOH = 14$$

then,

$$pOH = 14 - pH$$
$$= 14 - 3$$
$$= 11$$

EXERCISE 15.3   If the pH of a solution is 8, what is the pOH?   ■

EXAMPLE 15.4   If the pH of a solution is 9, what is $[OH^-]$?

Solution   Since

$$pH + pOH = 14$$

then,

$$pOH = 14 - pH$$
$$= 14 - 9$$
$$= 5$$

And then,

$$pOH = -\log [OH^-]$$
$$\log [OH^-] = -pOH = -5$$
$$[OH^-] = 1 \times 10^{-5} \, M$$

EXERCISE 15.4   If the pH of a solution is 5, what is $[OH^-]$?   ■

15.4 THE pH SCALE

The **pH scale** has a range 0–14; it is used for expressing levels of acidity in aqueous solutions. The pH scale has become popular because pH values of

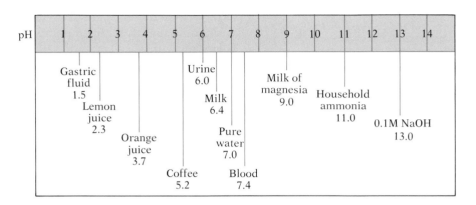

FIGURE 15-2   The pH Scale

common aqueous solutions are usually simple positive numbers within the range of the scale. The pH scale can be used for all kinds of aqueous solutions — lakes and rivers, stomach acid, blood, urine, rainfall, even shampoos — and the use of simple positive numbers simplifies data tabulation and record-keeping.

The pH scale is shown in Figure 15-2, along with pH values for some familiar solutions. A difference of one pH unit anywhere along the scale represents a tenfold change in $[H_3O^+]$. Notice that pH values for several common solutions are not whole numbers. For example, the pH of milk is 6.4, so its $[H_3O^+]$ is between $1 \times 10^{-6} M$ and $1 \times 10^{-7} M$. In many cases, it is sufficient to know just the range of $[H_3O^+]$ corresponding to a decimal pH value. The calculation of decimal pH values will be discussed later.

EXAMPLE 15.5   Find the pH of a 0.1 M solution of hydrochloric acid.

Solution   Since hydrochloric acid is a strong acid, it ionizes completely, producing a solution in which $[H_3O^+]$ is 0.1 M. Thus,

$$\begin{aligned} pH &= -\log [H_3O^+] \\ &= -\log 0.1 = -\log (1 \times 10^{-1}) \\ &= -(-1) \\ &= 1 \end{aligned}$$

EXERCISE 15.5   Find the pH of a 0.001 M solution of hydrochloric acid.   ■

EXAMPLE 15.6   Find the pH of a 0.001 M aqueous solution of NaOH.

Solution   Since NaOH is a strong base, it ionizes completely in water, producing a solution in which $[OH^-]$ is 0.001 M. Thus,

$$[OH^-] = 0.001 \ M = 1 \times 10^{-3} \ M$$
$$pOH = -\log [OH^-]$$
$$= -\log (1 \times 10^{-3})$$
$$= -(-3)$$
$$= 3$$
$$pH + pOH = 14$$
$$pH = 14 - pOH$$
$$= 11$$

EXERCISE 15.6   Find the pH of a 0.01 M aqueous solution of KOH.   ■

15.5   MEASUREMENT OF pH

The pH of a solution can be estimated fairly accurately by using acid-base indicators. An indicator is a weak acid or base, often derived from plant material, that changes color when it is neutralized. Let's examine an indicator acid, which we'll call HInd. HInd ionizes like any other weak acid, and its ions are in equilibrium with undissociated indicator in aqueous solution:

$$HInd(aq) + H_2O(l) \rightleftharpoons H_3O^+(aq) + Ind^-(aq)$$

The HInd and Ind$^-$ forms have different colors; the amount of each in a solution depends on the $[H_3O^+]$ (and thus the pH) of the solution, according to LeChatelier's principle. By adding a small amount of indicator to a solution and noting the color, it is possible to tell which of the two forms of indicator predominates in the mixture and thus estimate pH. The many indicators known at present change colors at various pH values between 1 and 14. Some of these are listed in Table 15-3. pH paper is paper tape

TABLE 15-3   Examples of Acid-Base Indicators

Indicator	Color Change (increasing pH)	pH Range
Thymol blue	red to yellow	1.2–2.8
Bromphenol blue	yellow to blue	3.0–4.6
Methyl orange	red to yellow	3.1–4.4
Bromcresol green	yellow to blue	3.8–5.4
Methyl red	red to yellow	4.2–6.2
Litmus	red to blue	4.5–8.3
Bromthymol blue	yellow to blue	6.0–7.6
Phenol red	yellow to red	6.8–8.4
Phenolphthalein	colorless to red	8.3–10.0
Alizarin yellow	yellow to violet	10.1–12.0
1,3,5-Trinitrobenzene	colorless to yellow	12.0–14.0

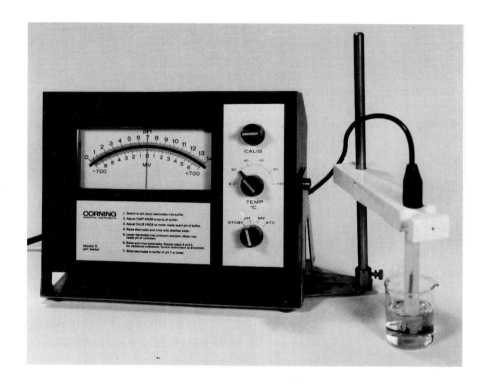

FIGURE 15-3 A pH Meter. The single probe (right) is called a combination electrode, because it combines both electrodes into a single unit.

impregnated with various indicators; it is routinely used for determination of approximate pH. The advantages of using acid-base indicators for estimating pH are the low cost and the short time required.

pH can also be measured with an instrument called a **pH meter,** an electronic device with two sensing probes (electrodes) (see Figure 15-3). When the electrodes are immersed in an aqueous solution, the hydronium ions in the solution generate an electric signal which is registered on the meter. The strength of the signal is directly proportional to $[H_3O^+]$, and the meter is calibrated in units of pH. Thus, pH is read directly from the meter. Measuring pH in this way is not as convenient as using pH paper, but the pH meter is much more precise and can be used in colored solutions that might interfere with the observation of indicator color.

In the past several years, extremely small pH electrodes have been developed, and some are so small that they can be inserted directly into living cells. In addition to biological and medical uses, pH meters are used routinely in environmental monitoring and agricultural work.

15.6   ADDITIONAL pH CALCULATIONS *(optional)*

It is easy to calculate pH when $[H_3O^+] = 1 \times 10^n\, M$, where n is zero or a negative whole number; in these cases, $pH = -n$. But there are other calculations of pH that are not so straightforward. For example, how would you calculate the pH of a solution in which $[H_3O^+] = 2.0 \times 10^{-4}\, M$?

MATH TIP

To find the logarithm of 2.0 using the calculator,

Press 2.0 $\boxed{\log}$.

The display reads 0.301029996. With logarithms, the number of significant digits in the given number dictates the number of significant digits to the right of the decimal point in the logarithm. The log of 2.0 is written as 0.30. The log (2.0×10^{-4}) can be found using a calculator without separating the numerical and the exponential part of the number, finding separate logs and then adding. To do this,

Press 2.0 $\boxed{EE}$ 4 $\boxed{+/-}$ $\boxed{\log}$.

The display reads $-3.69897\ 00$. The answer is reported as -3.70 with the correct number of significant figures.

In this case, you need to take the logarithms of both 2.0 and 10^{-4}, because

$$\log [H_3O^+] = \log (2.0 \times 10^{-4}) = \log 2.0 + \log 10^{-4}$$

You can find the logarithm of 2.0 with a calculator or a table of logarithm (appendix B contains a table of logarithms). Then,

$$\begin{aligned} \log [H_3O^+] &= \log (2.0 \times 10^{-4}) \\ &= 0.30 + (-4) \\ &= -3.70 \end{aligned}$$

Since

$$pH = -\log [H_3O^+]$$

then,

$$pH = -(-3.70) = 3.70$$

Thus, a solution in which $[H_3O^+] = 2.0 \times 10^{-4}$ M has a pH of 3.70. This calculation illustrates that when $[H_3O^+]$ has any value other than 1×10^n M, the corresponding pH will be a decimal number.

EXAMPLE 15.7   Find the pH of a solution in which $[H_3O^+] = 4.2 \times 10^{-5}$ M.

Solution

$$\begin{aligned} pH &= -\log [H_3O^+] \\ &= -\log (4.2 \times 10^{-5}) \\ &= -(\log 4.2 + \log 10^{-5}) \\ &= -[0.62 + (-5)] \\ &= -[-4.38] \\ &= 4.38 \end{aligned}$$

EXERCISE 15.7   Find the pOH of a solution in which $[OH^-] = 6.2 \times 10^{-3}$ M.   ∎

To find the $[H_3O^+]$ corresponding to a given decimal pH, you must use antilogarithms, the numbers that logarithms represent. You can find antilogarithms by using a calculator or a table of logarithms. The following example illustrates finding the $[H_3O^+]$ of a solution of known decimal pH.

EXAMPLE 15.8   Find the $[H_3O^+]$ of a solution whose pH is 5.72.

Solution

$$\begin{aligned} pH &= -\log [H_3O^+] \\ \log [H_3O^+] &= -pH \\ &= -5.72 \\ [H_3O^+] &= \text{antilog} (-5.72) \\ &= \text{antilog} (0.28 - 6) \\ &= [\text{antilog } 0.28][\text{antilog} (-6)] \\ &= 1.9 \times 10^{-6} \ M \end{aligned}$$

EXERCISE 15.8   Find the $[OH^-]$ of a solution whose pH is 7.85.   ∎

If the pH of a solution of a weak acid and the concentration of the weak acid are known, the K_a of the acid can be calculated. Thus, the K_a for a weak acid can be determined in the laboratory simply by measuring the pH of a solution of the weak acid of known concentration. Example 15.9 demonstrates this calculation.

EXAMPLE 15.9

A 0.10 M aqueous solution of a weak acid, HA, has a pH of 3.20. What is the K_a of the acid?

Solution

First, write the equation for ionization of the weak acid in water:

$$HA(aq) + H_2O(l) \rightleftharpoons H_3O^+(aq) + A^-(aq)$$

Then, convert pH to $[H_3O^+]$:

$$pH = -\log [H_3O^+]$$
$$\log [H_3O^+] = -pH = -3.20$$
$$[H_3O^+] = \text{antilog} (-3.20)$$
$$= \text{antilog} (0.80 - 4)$$
$$= [\text{antilog } 0.80][\text{antilog} (-4)]$$
$$= 6.3 \times 10^{-4} \, M$$

Next, calculate [HA] at equilibrium, keeping in mind that $[HA]_{dissociated} = [H_3O^+]$, because each mole of HA that dissociates produces one mole of H_3O^+:

$$[HA]_{equilibrium} = [HA]_{initial} - [HA]_{dissociated}$$
$$= 0.10 \, M - 6.3 \times 10^{-4} \, M$$
$$= 0.10 \, M \text{ (In other words, } 6.3 \times 10^{-4} \, M \text{ is negligible compared to } 0.10 \, M.)$$

MATH TIP

To square 6.3×10^{-4}, square each part of the number separately:

$$(6.3 \times 10^{-4})^2 = (6.3)^2(10^{-4})^2$$
$$= (40)(10^{-8})$$
$$= 4.0 \times 10^{-7}$$

[Note that when an exponential number, i.e., 10^{-4}, is raised to a power, the original exponent is multiplied by the power to which the number is being raised to obtain the new exponent. Thus, $(10^{-4})^2 = 10^{-8}$.]

Finally, calculate K_a, keeping in mind that $[H_3O^+] = [A^-]$:

$$K_a = \frac{[H_3O^+][A^-]}{[HA]}$$
$$= \frac{(6.3 \times 10^{-4})^2}{0.10}$$
$$= 4.0 \times 10^{-6}$$

EXERCISE 15.9

A 0.25 M aqueous solution of a weak acid, HA, has a pH of 2.16. What is the K_a of the acid?   ■

EXAMPLE 15.10

A 0.20 M aqueous solution of a weak acid, HA, is 1.42% ionized. Calculate the pH of the solution, and find K_a for the acid.

Solution

First, write the equation for ionization of the weak acid in water:

$$HA(aq) + H_2O(l) \rightleftharpoons H_3O^+(aq) + A^-(aq)$$

Then, calculate $[H_3O^+]$, keeping in mind that $[HA]_{dissociated} = [H_3O^+]$, because each mole of HA that dissociates produces one mole of H_3O^+:

$$[HA]_{dissociated} = (0.20 \, M)(0.0142) = 0.0028 \, M = 2.8 \times 10^{-3} \, M = [H_3O^+]$$

MATH TIP
1.42% ionized means that a fraction of the original quantity of HA dissociated. This fraction, 0.0142, is the percentage converted to a decimal. Multiplication by the decimal

(0.20 *M*)(0.0142)

yields the molar quantity of HA that dissociated.

Next, calculate pH:

$$pH = -\log [H_3O^+]$$
$$= -\log (2.8 \times 10^{-3})$$
$$= -(\log 2.8 + \log 10^{-3})$$
$$= -[0.45 - 3]$$
$$= 2.55$$

Finally, calculate K_a:

$$[HA]_{equilibrium} = [HA]_{initial} - [HA]_{dissociated}$$
$$= 0.20 \, M - (2.8 \times 10^{-3} \, M)$$
$$= 0.20 \, M$$
$$K_a = \frac{[H_3O^+][A^-]}{[HA]}$$
$$= \frac{(2.8 \times 10^{-3})^2}{0.20}$$
$$= 3.9 \times 10^{-5}$$

(Note that pH is not needed for the calculation of K_a; thus, K_a could have been calculated before pH.)

EXERCISE 15.10 A 0.15 *M* solution of a weak acid, HA, is 2.03% ionized. Calculate the pH of the solution, and find K_a for the acid. ■

15.7 HYDROLYSIS *(optional)*

Hydrolysis
(high-DRAH-lih-sis): from the Greek *hydro-*, meaning "water," and *-lysis*, meaning "breaking down or decomposition."

When an acid and a base react, the reaction is referred to as neutralization, but this statement is not entirely accurate—the solution formed after an acid reacts with a base may not be exactly neutral, that is, it may not have a pH of 7. The pH of such a solution will be different from 7 if any of the ions of the salt can react with water. An example of a salt whose ions can react with water is sodium bicarbonate, $NaHCO_3$. When $NaHCO_3$ dissolves in water, Na^+ ions and HCO_3^- ions are formed immediately. However, HCO_3^- is the conjugate base of the weak acid H_2CO_3. In general, the conjugate base of a weak acid is itself a fairly strong base, and it will accept protons from water. Thus, HCO_3^- reacts with water as follows:

$$HCO_3^-(aq) + H_2O(l) \rightleftharpoons H_2CO_3(aq) + OH^-(aq)$$

Thus, when $NaHCO_3$ dissolves in water, the weak acid H_2CO_3 and OH^- ions are formed, in addition to Na^+ ions. The reaction is called **hydrolysis** because molecules of water are split apart; H^+ associates with HCO_3^-, and OH^- is left. The solution then contains the weak acid and OH^- ions. Because the weak acid ionizes to such a small extent, the number of OH^- ions exceeds the number of H_3O^+ ions, and the solution is basic. This example illustrates a general principle regarding hydrolysis: *if a salt is composed of cations from a strong base and anions from a weak acid, the salt will form a basic solution when dissolved in water.*

Now let's consider another type of salt, one composed of cations from a weak base and anions from a strong acid. Ammonium chloride, NH_4Cl, is an example of such a salt. When NH_4Cl dissolves in water, it immediately ionizes completely into NH_4^+ ions and Cl^- ions. But NH_4^+ is the conjugate acid of a weak base, and thus is a fairly strong acid. Hence, NH_4^+ will react with water by donating protons to water molecules,

$$NH_4^+(aq) + H_2O(l) \rightleftharpoons NH_3(aq) + H_3O^+(aq)$$

and the resulting solution contains a weak base and H_3O^+ ions. Since the weak base ionizes only slightly, there are more H_3O^+ ions in solution than there are OH^- ions, and the solution is acidic. *Thus, when a salt composed of cations from a weak base and anions from a strong acid dissolves in water, the solution will be acidic.*

Perhaps you are wondering about a salt composed of cations from a strong base and anions from a strong acid. In this case, an aqueous solution of the salt does not contain the conjugate base of a weak acid or the conjugate acid of a weak base, so hydrolysis does not occur, and the solution is neutral. On the other hand, if a salt is composed of cations from a weak base and anions from a weak acid, the ions will undergo hydrolysis when dissolved in water, but the solution can be acidic or basic, depending on the particular ions involved.

Since reaction of an acid with a base may produce ions that hydrolyze, the resulting solution may not be neutral, but the examples discussed above can be used to help predict whether the solution will be acidic, basic, or neutral.

EXAMPLE 15.11   Predict whether a solution of NaCN will be acidic, basic, or neutral. If you think the solution will be acidic or basic, write an equation for the reaction that occurs.

Solution   To determine whether hydrolysis will occur, we must consider both the cation and the anion of the salt. If one is the conjugate of a strong acid or base and the other is not, the solution will not be neutral. If the cation is the conjugate acid of a weak base, the solution will be acidic. On the other hand, if the anion is the conjugate base of a weak acid, the solution will be basic. In this case, we have a salt composed of a cation that is the conjugate acid of a strong base (NaOH) and an anion that is the conjugate base of a weak acid (HCN). Thus, the solution will be basic.

$$CN^-(aq) + H_2O(l) \rightleftharpoons HCN(aq) + OH^-(aq)$$

EXERCISE 15.11   Predict whether a solution of KNO_3 will be acidic, basic, or neutral. If you think the solution will be acidic or basic, write an equation for the reaction that occurs.   ∎

15.8 BUFFERS

Blood contains substances that help it maintain a rather constant pH despite the entry of acids and bases. Solutions that are able to resist pH change when moderate amounts of acid or base are added are called **buffer solutions** or **buffers.**

Buffer solutions contain either a weak acid and a salt of that weak acid or a weak base and a salt of that weak base. The two components of a buffer solution are called a buffer pair; Table 15-4 lists some common buffer pairs. Let's examine the buffer pair consisting of acetic acid (the weak acid) and sodium acetate (the salt of the weak acid). Since acetic acid is a weak acid, it ionizes only slightly in water:

$$HC_2H_3O_2(aq) + H_2O(l) \rightleftharpoons H_3O^+(aq) + C_2H_3O_2^-(aq)$$
$$K_a = 1.8 \times 10^{-5}$$

In contrast, the salt sodium acetate is a strong electrolyte and ionizes completely in water:

$$NaC_2H_3O_2(s) \xrightarrow{H_2O} Na^+(aq) + C_2H_3O_2^-(aq)$$

Thus, $HC_2H_3O_2$ provides the solution with a small number of H_3O^+ ions and $C_2H_3O_2^-$ ions and a relatively large amount of un-ionized $HC_2H_3O_2$. In contrast, $NaC_2H_3O_2$ provides the solution with a large number of Na^+ ions and $C_2H_3O_2^-$ ions. If a moderate quantity of H_3O^+ ions is added from any source, the equilibrium adjusts itself, according to LeChatelier's principle, by forming more un-ionized $HC_2H_3O_2$:

$$HC_2H_3O_2(aq) + H_2O(l) \rightleftharpoons H_3O^+(aq) + C_2H_3O_2^-(aq)$$
$$\longleftarrow \text{increase in } [H_3O^+]$$

On the other hand, if a moderate amount of OH^- ions is added from any source, they are removed due to neutralization by the H_3O^+ ions that are present. The equilibrium responds to this loss of H_3O^+ ions by forming more H_3O^+ ions from $HC_2H_3O_2$:

$$HC_2O_3H_2(aq) + H_2O(l) \rightleftharpoons H_3O^+(aq) + C_2H_3O_2^-(aq)$$
$$\text{decrease in } [H_3O^+] \longrightarrow$$

Hence, moderate amounts of either acid or base can be absorbed by the buffer solution without significant alteration of pH. The weak acid component of the buffer pair is a source of un-ionized acid that can provide more H_3O^+ ions when needed, and the salt of the weak acid provides an abundance of acid anions that can react with excess H_3O^+ ions.

Another example of a buffer pair is one composed of aqueous ammonia (NH_3) and ammonium chloride (NH_4Cl), a weak base and its salt. The salt

TABLE 15-4 Common Buffer Pairs

Buffer Pair	Formula	Useful pH Range
Formic acid/sodium formate	$HCHO_2/NaCHO_2$	2.8–4.8
Acetic acid/sodium acetate	$HC_2H_3O_2/NaC_2H_3O_2$	3.8–5.8
Sodium dihydrogen phosphate/ sodium hydrogen phosphate	NaH_2PO_4/Na_2HPO_4	6.2–8.2
Tetraboric acid/sodium tetraborate	$H_2B_4O_7/Na_2B_4O_7$	8.0–10.0
Sodium bicarbonate/sodium carbonate	$NaHCO_3/Na_2CO_3$	9.2–11.2

Under normal conditions, the buffer systems in blood restrict the pH to a very narrow range, usually about 7.35–7.45, but occasionally illness or other unusual conditions create situations that overload the buffering capacity of blood. The condition in which the pH of blood falls below 7.35 is referred to as *acidosis*. If the pH of blood rises above 7.45, the condition is called *alkalosis*. An interesting aspect of the human body is that it can respond to offset the potential effects of acidosis and alkalosis. We are indeed fortunate to have such a response mechanism, because if blood pH varies outside of its normal range by a very large extent, the condition can be fatal.

Acidosis can occur if breathing is impaired, as in pneumonia, for example; the shallow breathing that is characteristic of pneumonia does not allow for proper ventilation of the lungs, and the CO_2 produced as a waste product in the cells is not properly expelled. This causes a buildup of CO_2 dissolved in the blood and thus an increase in the level of H_2CO_3, accounting for a lower blood pH. The natural response by the central nervous system is to increase the rate of breathing to clear the lungs and the blood of excess CO_2. However, the lungs do not respond well if they are prevented from functioning normally, so another system serves as

a backup. The kidneys respond to such a condition by excreting more H_3O^+ in the urine than they normally do, thus helping to minimize the severity of the acidosis. Acidosis can also occur as a result of overingestion of acidic substances, starvation, dehydration, or prolonged intestinal upset. In all cases, the kidneys and respiratory system play important roles in offsetting the potentially disastrous results of severe acidosis.

Alkalosis may occur if breathing is too fast, as a result of excitement, trauma, or high body temperature. In such instances, too much CO_2 is released by the lungs, thus reducing the amount of H_2CO_3 in the blood and making it too basic. Other causes of alkalosis are loss of stomach acid by prolonged vomiting or overingestion of antacids (which are basic substances). The condition is registered by the central nervous system, which signals the lungs to slow down the rate of breathing. However, in cases of hysteria and similar situations, the breathing rate does not slow enough, and once again the kidneys play a backup role. They excrete urine that contains a lower amount of H_3O^+ than normal, thus conserving acid in the blood to neutralize the excess base.

---

ionizes completely in aqueous solution into NH_4^+ and Cl^-, but because NH_3 is a weak base, the following equilibrium exists:

$$NH_3(aq) + H_2O(l) \rightleftharpoons NH_4^+(aq) + OH^-(aq)$$
$$K_b = 1.8 \times 10^{-5}$$

Thus, addition of moderate amounts of H_3O^+ ions from any source neutralizes some of the OH^- ions, and the equilibrium shifts to form more OH^-:

$$NH_3(aq) + H_2O(l) \rightleftharpoons NH_4^+(aq) + OH^-(aq)$$
decrease in $[OH^-] \longrightarrow$

Addition of a moderate amount of OH^- ions from any source causes the equilibrium to shift to the left, and the excess OH^- ions are removed:

$$NH_3(aq) + H_2O(l) \rightleftharpoons NH_4^+(aq) + OH^-(aq)$$
$\longleftarrow$ increase in $[OH^-]$

As we mentioned earlier, the pH of blood is maintained by buffers. In fact, there are at least three buffer systems in blood, and their presence emphasizes the necessity of restricting blood pH to the range of 7.35–7.45. This is necessary because enzymes are sensitive to pH, and most of them carry out their catalytic roles best in the pH range of 7.35–7.45. The buffer pair that exists in highest concentration in blood is composed of carbonic acid (H_2CO_3) and the bicarbonate ion (HCO_3^-). The H_2CO_3 is formed when CO_2 given off by cells dissolves in blood, and the HCO_3^- ion is the principal anion formed when H_2CO_3 ionizes:

$$CO_2(g) + H_2O(l) \rightleftharpoons H_2CO_3(aq)$$

$$H_2CO_3(aq) + H_2O(l) \rightleftharpoons H_3O^+(aq) + HCO_3^-(aq)$$
$$K_a = 4.3 \times 10^{-7}$$

(The ionization constant for the second dissociation of H_2CO_3 is so small that, for all practical purposes, no further ionization occurs.) Carbonic acid and bicarbonate ion function as a buffer pair composed of a weak acid and its salt, but their buffering capability is especially good near neutral pH.

The second buffer system in blood is composed of dihydrogen phosphate ion, $H_2PO_4^-$, and hydrogen phosphate ion, HPO_4^{2-}. These ions are formed from phosphate compounds that the body continually synthesizes. They respond to the entry of acidic and basic substances like a buffer pair composed of a weak acid and its salt. The third buffer in blood is a mixture of proteins, which are very large molecules. Proteins are able to remove excess hydronium ions from blood, and the protonated protein molecules can also provide hydronium ions when necessary to maintain a rather constant pH.

SUMMARY

Electrolytes are substances that conduct electricity when melted or dissolved in water. **Strong electrolytes** ionize completely in water, while **weak electrolytes** ionize only partially in water. **Nonelectrolytes,** although water soluble, do not form ions in water.

Pure water ionizes to a very small extent to produce equal quantities of hydronium ions and hydroxide ions. The ionization constant for water is called the **ion-product constant** (K_w); it has a value of 1.00×10^{-14} at $25°$ C. **pH** is defined as $-\log [H_3O^+]$. The **pH scale** is a convenient means of expressing the level of acidity of aqueous solutions. Values of pH are usually in the range of 0–14 for common solutions. pH can be estimated by use of colored indicators, or it can be measured precisely with a pH meter. This instrument has two electrodes that are immersed in the solution. The electrical signal generated by hydronium ions registers on the meter, which is calibrated in units of pH.

A solution formed after an acid neutralizes a base will not be neutral if any of the ions of the salt can react with water by **hydrolysis,** in which molecules of water are split apart. A salt composed of cations from a strong base and anions from a weak acid forms a basic solution, while a salt composed of cations from a weak base and anions from a strong acid forms an acidic solution. A salt composed of cations from a strong base and anions from a strong acid forms a neutral solution.

Buffers are solutions that have the ability to resist pH changes that would be caused by the addition of moderate amounts of acid or base. A buffer consists of a weak acid and its salt or a weak base and its salt. Blood contains three buffer systems: carbonic acid/bicarbonate ion, dihydrogen phosphate ion/hydrogen phosphate ion, and a mixture of protein molecules.

STUDY QUESTIONS AND PROBLEMS

(More difficult questions and problems are marked with an asterisk.)

ELECTROLYTES

1. Define the following terms:
   a. Electrolyte   c. Weak electrolyte
   b. Strong electrolyte   d. Nonelectrolyte
2. Why are most metals good conductors of electricity?
3. Why are most nonmetals and molecular compounds poor conductors of electricity?
4. Why do ionic compounds conduct electricity when melted or dissolved in water but not in the solid state?
5. Distinguish between strong electrolytes and weak electrolytes, and give an example of each.
6. Distinguish between weak electrolytes and nonelectrolytes, and give an example of each.

THE IONIZATION OF WATER

7. Explain why tap water is usually a good conductor of electricity, but pure water is only a weak conductor.
8. Write the equation for the ionization of water.
9. What is the name of the ionization constant of water?
10. Derive the expression for K_w.

pH

11. Define pH and pOH.
12. If $[H_3O^+]$ is $1.00 \times 10^{-3}\ M$ in an aqueous solution, what is $[OH^-]$?
13. Calculate the pH for each of the following hydronium ion concentrations.
    a. $0.100\ M$ c. $1.00 \times 10^{-4}\ M$ e. $1.00\ M$
    b. $1.00 \times 10^{-12}\ M$ d. $1.00 \times 10^{-9}\ M$ f. $1.00 \times 10^{-6}\ M$
14. Calculate the pOH for each of the hydronium ion concentrations in question 13.

THE pH SCALE

15. Why do you think the pH scale has an upper limit of 14?
16. For a $0.00100\ M$ KOH solution,
    a. What is the $[OH^-]$?   c. What is the pH?
    b. What is the $[H_3O^+]$?   d. What is the pOH?
17. If you needed 500.0 mL of hydrochloric acid, pH 1, how would you prepare it from a stock solution of $12.0\ M$ hydrochloric acid?
18. Water that is exposed to the atmosphere dissolves significant amounts

of CO_2. On the basis of this information, explain why purified water has a pH slightly below 7 after standing exposed to air for some time.

MEASUREMENT OF pH

19. Explain how pH can be estimated with indicators.
20. What are the advantages of pH paper over a pH meter?
21. What are the advantages of a pH meter over pH paper?

ADDITIONAL pH CALCULATIONS

22. What is the pH of each of the following solutions?
    a. $[H_3O^+] = 0.25\ M$ c. $[OH^-] = 4.3 \times 10^{-6}\ M$
    b. $[OH^-] = 0.35\ M$ d. $[H_3O^+] = 2.6 \times 10^{-3}\ M$
23. What is the pH of each of the following solutions?
    a. $[H_3O^+] = 6.8 \times 10^{-4}\ M$ c. $[H_3O^+] = 0.015\ M$
    b. $[OH^-] = 3.7 \times 10^{-2}\ M$ d. $[OH^-] = 6.5 \times 10^{-12}\ M$
24. What is the $[H_3O^+]$ of each of the following solutions?
    a. pH = 3.05   c. pH = 6.12
    b. pOH = 7.33   d. pOH = 11.45
25. What is the $[OH^-]$ of each of the following solutions?
    a. pH = 0.37   c. pH = 10.56
    b. pOH = 3.42   d. pOH = 8.91
26. A 0.35 M solution of a weak acid, HA, has a pH of 3.47. What is K_a for the acid?
27. Cocaine (represented by B) is a weak base, and it ionizes as follows:

$$B(aq) + H_2O(l) \rightleftharpoons BH^+(aq) + OH^-(aq)$$

A $5.0 \times 10^{-3}\ M$ solution of cocaine has a pH of 10.04. What is K_b for cocaine?
28. Lactic acid, $HC_3H_5O_3$, forms in muscles during vigorous exercise. If a 0.25 M aqueous solution of lactic acid ionizes to the extent of 2.4%, calculate the pH of the solution, and find the K_a for lactic acid.

HYDROLYSIS

29. What does *hydrolysis* mean?
30. Predict whether an aqueous solution of each of the following will be acidic, basic, or neutral. If you think the solution will be acidic or basic, write an equation for the reaction that occurs. (Refer to chapter 14 for weak acids and weak bases.)
    a. KNO_2 b. NH_4Br c. $NaC_2H_3O_2$ d. $LiHSO_4$

BUFFERS

31. Explain what is meant by *buffer* and *buffer pair*.
32. Explain how a buffer composed of a weak acid and its salt resists change in pH.
33. Explain how a buffer composed of a weak base and its salt resists change in pH.
34. Can a solution of HCl and NaCl act as a buffer? Explain your answer.
35. What are the three buffer systems in blood? Which is most important? Why?

36. Explain how the carbonic acid/bicarbonate ion buffer system controls blood pH.

GENERAL EXERCISES

37. Both $Ba(OH)_2$ and H_2SO_4 are strong electrolytes, but when equimolar amounts of these are mixed together in water, the solution shows almost no electrical conductivity. Write a chemical equation to explain this observation.

38. The normal pH of urine is 6.0.
   a. What is the $[H_3O^+]$?   **c.** What is the pOH?
   b. What is the $[OH^-]$?

39. At 37° C (normal body temperature for a human), the ion-product constant of water is 2.42×10^{-14}. What is the concentration of H_3O^+ in pure water at this temperature? What is the concentration of OH^-?

40. A sample of blood has a pH of 7.40. What is $[H_3O^+]$ in the blood? What is $[OH^-]$?

41. A 0.24 M solution of a weak acid, HA, has a pH of 2.65. What is K_a for the acid?

42. Dichloracetic acid, $HC_2HO_2Cl_2$, is a weak acid with one dissociable proton. A 0.200 M solution of dichloroacetic acid is 33% ionized. Calculate the pH of the solution, and find the K_a of dichloroacetic acid.

43. Benzoic acid, $HC_7H_5O_2$, is a weak acid with one dissociable proton. Its K_a is 6.2×10^{-5} at 25° C. What is the pH of a 0.20 M solution of benzoic acid? What is the percent ionization of the acid?

44. Predict whether an aqueous solution of each of the following will be acidic, basic, or neutral. If you think the solution will be acidic or basic, write an equation for the reaction that occurs. (Refer to chapter 14 for weak acids and weak bases.)
   a. NaI   **b.** RbF   **c.** KH_2PO_4 **d.** Na_2SO_3

45. Indicate which of the following pairs of substances form buffer solutions when equimolar quantities are dissolved together in water. Explain each of your answers. (Refer to chapter 14 for weak acids and weak bases.)
   a. HNO_2 and KNO_2 **c.** NH_4NO_3 and HNO_3
   b. HCN and KClO   **d.** KH_2PO_4 and K_2HPO_4

OXIDATION AND REDUCTION

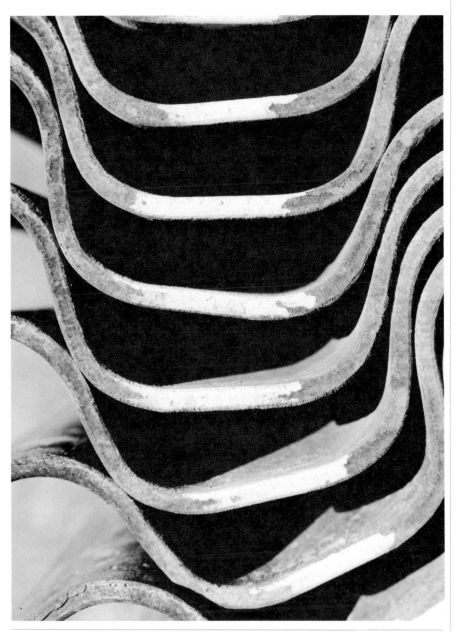

OUTLINE

16.1 Concepts of Oxidation and
        Reduction
16.2 Balancing Redox Equations: The
        Oxidation State Method
16.3 Balancing Redox Equations: The
        Ion-Electron Method
        Reactions in Acidic Solution
        Reactions in Basic Solution
16.4 The Activity Series of Metals
16.5 Voltaic Cells
16.6 Batteries
        Lead Storage Battery
        Dry Cell
        Nickel-Cadmium Battery
16.7 Electrolytic Cells
        Perspective: Charles Martin Hall
            and the Production of Aluminum
        Summary
        Study Questions and Problems

Oxidation-reduction reactions are abundant in the world about us. All combustion reactions are oxidation-reduction reactions, and so is photosynthesis. Burning dead leaves oxidizes the covalent carbon compounds in the leaves to carbon dioxide (carbon dioxide then serves as the fuel for living plants to make the covalent carbon compounds that they need). The oxidation-reduction reactions that make up the metabolic cycles in our cells allow us to derive energy from food. One of the more practical ways in which oxidation-reduction reactions are used is in the construction and operation of batteries that provide electricity to automobiles, flashlights, and even cardiac pacemakers. In this chapter, you will explore the basic principles of oxidation-reduction reactions and some uses that society has found for them. Before proceeding, you will find it helpful to review the introduction to oxidation and reduction (chapter 6), oxidation numbers (chapter 7), balancing equations (chapter 9), and ionic equations (chapter 14).

16.1 CONCEPTS OF OXIDATION AND REDUCTION

In chapter 6 you were introduced to oxidation and reduction in the context of ionic compounds, and, in chapter 7, the idea of oxidation numbers was expanded to include molecular compounds. Recall that oxidation and reduction always go together; when one substance is oxidized, another must become reduced. Since *oxidation-reduction reaction* is a bit of a mouthful, this term is often shortened to *redox reaction*. **Redox reactions** are reactions that involve changes in oxidation states. All reactions with O_2 are redox reactions. Consider the formation of iron(III) oxide from iron and oxygen:

$$4 \underset{0}{Fe} + 3 \underset{0}{O_2} \longrightarrow 2 \underset{+3 \ -2}{Fe_2O_3}$$

The oxidation numbers are shown below the formulas in the equation. The increase in the oxidation number of Fe from 0 to +3 represents a loss of electrons by Fe when it forms the product. **Oxidation** is defined as a loss of electrons and therefore an increase in oxidation number. The decrease in oxidation number of O from 0 to −2 represents a gain of electrons by O when it forms the product. **Reduction** is defined as a gain of electrons and therefore a decrease in oxidation number. The changes in oxidation number corresponding to oxidation and reduction are summarized in Figure 16-1.

In redox reactions, the substance that causes oxidation is called the **oxidizing agent** or the **oxidant.** The oxidizing agent is also the substance that is reduced. In the preceding example, the oxidizing agent is oxygen. The substance that causes reduction is called the **reducing agent** or the **reductant.** The reducing agent is oxidized in the course of a redox reaction. In the formation of Fe_2O_3, iron is the reducing agent, because it reduces oxygen.

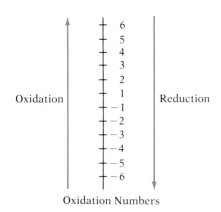

FIGURE 16-1 Oxidation Is an Increase in Oxidation Number, and Reduction Is a Decrease in Oxidation Number.

Consider another example of a redox reaction, the combustion of carbon:

$$\underset{0}{C} + \underset{0}{O_2} \longrightarrow \underset{+4\ -2}{CO_2}$$

In this reaction, there are no ions, yet oxidation and reduction have occurred, since the oxidation number of C increases from 0 to $+4$, while the oxidation number of O decreases from 0 to -2. Thus, carbon is oxidized and oxygen is reduced. This example illustrates that ions need not be present in redox reactions; *the only requirement for a redox reaction is that there be an increase in oxidation number at the same time that there is a decrease in oxidation number.* There are occasions when the same element undergoes both oxidation and reduction.

You can see that oxidation and reduction are always opposites of each other. The general equation for a redox reaction is

$$A(ox) + B(red) \longrightarrow C(red) + D(ox)$$

where reactant A is in an oxidized form and reactant B is in a reduced form. During the reaction, A is changed to a reduced form (C) and B is changed to an oxidized form (D). Thus, A is reduced by B, and B is oxidized by A; A is the oxidizing agent, and B is the reducing agent. Redox reactions need not involve O_2, as the following example illustrates.

EXAMPLE 16.1   Identify the substance that is oxidized, the substance that is reduced, the oxidizing agent, and the reducing agent in the following reaction.

$$2\ Na + Cl_2 \longrightarrow 2\ NaCl$$

Solution   First, assign oxidation numbers to elements in reactants and product, using the rules given in chapter 7:

$$\underset{0}{2\ Na} + \underset{0}{Cl_2} \longrightarrow \underset{+1\ -1}{2\ NaCl}$$

Then, find the element whose oxidation number increases; that element is Na, so Na is the substance that is oxidized, and it also is the reducing agent.

Next, find the element whose oxidation number decreases; that element is Cl, so Cl is the substance that is reduced, and it is also the oxidizing agent.

EXERCISE 16.1   Identify the substance that is oxidized, the substance that is reduced, the oxidizing agent, and the reducing agent in the following reaction.

$$H_2 + Br_2 \longrightarrow 2\ HBr \qquad \blacksquare$$

16.2 BALANCING REDOX EQUATIONS: THE OXIDATION STATE METHOD

Simple redox equations can be balanced by the trial-and-error method given in chapter 9, but complex redox equations require a more systematic approach. One approach is the **oxidation state method.** In this method, the nonionic equation is used, and it is not necessary to know whether the substances are ionic or covalent. The method is based on the changes in the oxidation states of the elements that are oxidized and reduced. The key to the method is that the total increase in oxidation state must equal the total decrease in oxidation state. As an example, let's balance the following redox equation.

$$HCl + K_2Cr_2O_7 \longrightarrow KCl + CrCl_3 + Cl_2 + H_2O$$

Five steps are needed:

1. Write the complete formulas of the reactants, followed by an arrow and the complete formulas of the products. Then, using oxidation numbers, identify the elements that undergo changes in oxidation state, and balance their atoms.

   In our example, Cl and Cr exhibit changes in oxidation state, and we must balance these two elements before going on. Since there are 2 Cl's in the Cl_2 molecule, we must put a 2 in front of HCl; similarly, since there are 2 Cr's in $K_2Cr_2O_7$, we must put a 2 in front of $CrCl_3$. Notice that Cl in KCl and $CrCl_3$ did not change oxidation state.

$$\underset{-1}{2\,HCl} + \underset{+6}{K_2Cr_2O_7} \longrightarrow KCl + \underset{+3}{2\,CrCl_3} + \underset{0}{Cl_2} + H_2O$$

2. Draw a bridge between reactant and product to indicate the element that gets oxidized; draw another bridge to indicate the element that gets reduced. On each bridge, note the *total* change in oxidation state between reactant and product.

$$\underset{-1}{2\,HCl} + \underset{+6}{K_2Cr_2O_7} \longrightarrow KCl + \underset{+3}{2\,CrCl_3} + \underset{0}{Cl_2} + H_2O$$

(bridges: $+2$ and -6)

3. Balance the total increase in oxidation state with the total decrease in oxidation state by placing a coefficient in front of one or both changes in oxidation state. (This is done by finding the lowest common multiple of the two changes in oxidation state.)

$$\underset{-1}{2\,HCl} + \underset{+6}{K_2Cr_2O_7} \longrightarrow KCl + \underset{+3}{2\,CrCl_3} + \underset{0}{Cl_2} + H_2O$$

(bridges: $3(+2)$ and -6)

   The 3 in front of $+2$ makes the total increase in oxidation state, $+6$, equal to the total decrease in oxidation state, -6.

4. Use the coefficient(s) derived in step 3 as coefficients of the respective compounds or elements in the equation.

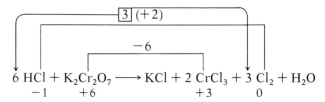

$$\underset{-1}{6\,HCl} + \underset{+6}{K_2Cr_2O_7} \longrightarrow KCl + \underset{+3}{2\,CrCl_3} + \underset{0}{3\,Cl_2} + H_2O$$

(bridges: $\boxed{3}(+2)$ and -6)

Note that the 3 is multiplied by both the 2 in front of HCl and the 1 that is understood to be in front of Cl_2.

5. Balance the rest of the equation by inspection. Since the Cr's are already balanced, we must now balance the other elements. There are 2 K's on the left, so we must place a 2 in front of KCl. Next, changing the 6 in front of HCl to 14 balances the Cl's, and placing a 7 in front of H_2O balances the H's. Thus, the balanced equation is

$$14 \, HCl + K_2Cr_2O_7 \longrightarrow 2 \, KCl + 2 \, CrCl_3 + 3 \, Cl_2 + 7 \, H_2O$$

EXAMPLE 16.2   Balance the following equation by the oxidation state method.

$$HNO_3 + Sn \longrightarrow SnO_2 + NO_2 + H_2O$$

Solution   Balancing oxidation with reduction (steps 1–4), we get

$$\boxed{4}\,(-1)$$
$$+4$$
$$4 \, HNO_3 + Sn \longrightarrow SnO_2 + 4 \, NO_2 + H_2O$$
$$\;+5 \qquad 0 \qquad\quad +4 \qquad +4$$

Then, balancing the rest of the equation by inspection (step 5) we get

$$4 \, HNO_3 + Sn \longrightarrow SnO_2 + 4 \, NO_2 + 2 \, H_2O$$

EXERCISE 16.2   Balance the following equation by the oxidation state method.

$$HNO_3 + H_2S \longrightarrow S + NO + H_2O$$ ∎

16.3   BALANCING REDOX EQUATIONS: THE ION-ELECTRON METHOD

When redox reactions occur in aqueous solution, it is simpler to write net ionic equations. Then changes in the oxidation states of monatomic ions are obvious, because they correspond to changes in the charges of the monatomic ions. For a polyatomic ion or a molecule, the entire unit undergoes change, not individual elements in the polyatomic ion or molecule.

We use the **ion-electron method** to balance equations for redox reactions in aqueous solution. This method makes use of the fact that each redox equation is actually the sum of two equations, one for the oxidation and the other for the reduction. Each of these two equations is called a **half-reaction.** The steps in balancing redox equations by the ion-electron method are slightly different for reactions in acidic solution than for those in basic solution. Applying the ion-electron method to these two types of solutions is discussed in the following paragraphs.

Reactions in Acidic Solution   Let's work through the ion-electron method by balancing the equation for the following redox reaction, which occurs in an acidic solution (in the unbalanced equation, H_2O and H^+ are not shown, but the proper number of H_2O molecules and H^+ ions will be determined in the course of balancing the equation):

$$Cr_2O_7{}^{2-}(aq) + Cl^-(aq) \longrightarrow Cr^{3+}(aq) + Cl_2(g)$$

To balance the equation for this reaction and others in acidic solutions, six steps are needed:

1. Write skeleton equations for the oxidation half-reaction and the reduction half-reaction, and balance the elements undergoing oxidation and reduction. (We recognize the reduction reaction by the decrease in oxidation number of Cr, and we recognize the oxidation reaction by the increase in oxidation number of Cl.)

$$\underset{+6}{Cr_2O_7{}^{2-}} \longrightarrow \underset{+3}{2\ Cr^{3+}} \qquad \text{(reduction)}$$

$$\underset{-1}{2\ Cl^-} \longrightarrow \underset{0}{Cl_2} \qquad \text{(oxidation)}$$

2. Balance oxygen in the half-reactions. Since the reaction occurs in aqueous acid, H_2O and H^+ can be added as needed. For each O atom needed, add one H_2O to the side of the equation that is deficient in oxygen.

$$Cr_2O_7{}^{2-} \longrightarrow 2\ Cr^{3+} + 7\ H_2O \qquad \text{(reduction)}$$

$$2\ Cl^- \longrightarrow Cl_2 \qquad \text{(oxidation)}$$

3. Balance hydrogen by adding H^+ to the side of the equation that is deficient in hydrogen.

$$14\ H^+ + Cr_2O_7{}^{2-} \longrightarrow 2\ Cr^{3+} + 7\ H_2O \qquad \text{(reduction)}$$

$$2\ Cl^- \longrightarrow Cl_2 \qquad \text{(oxidation)}$$

4. Balance the charges in the half-reactions by adding as many electrons, e^-, as needed. (Notice that electrons (e^-) appear on the left in the reduction reaction and on the right in the oxidation reaction.)

$$6\ e^- + 14\ H^+ + Cr_2O_7{}^{2-} \longrightarrow 2\ Cr^{3+} + 7\ H_2O \qquad \text{(reduction)}$$

$$2\ Cl^- \longrightarrow Cl_2 + 2\ e^- \qquad \text{(oxidation)}$$

5. The electrons added represent electrons gained or lost in each half-reaction. Since the number of electrons gained must be equal to the number of electrons lost, the number of e^- must be the same in each half-reaction. Thus, one or both equations must be multiplied by the smallest number that balances the electron transfer. In our example, the oxidation half-reaction must be multiplied by 3 to balance electron transfer:

$$3\ [2\ Cl^- \longrightarrow Cl_2 + 2\ e^-] \qquad \text{(oxidation)}$$

or

$$6\ Cl^- \longrightarrow 3\ Cl_2 + 6\ e^- \qquad \text{(oxidation)}$$

6. The two half-reactions have now been balanced for elements, charge,

and electron transfer. Add the two half-reactions together and cancel any identical species appearing on opposite sides of the total equation.

$$6\ e^- + 14\ H^+ + Cr_2O_7^{2-} \longrightarrow 2\ Cr^{3+} + 7\ H_2O \quad \text{(reduction)}$$
$$6\ Cl^- \longrightarrow 3\ Cl_2 + 6\ e^- \quad \text{(oxidation)}$$

$$\cancel{6\ e^-} + 14\ H^+ + Cr_2O_7^{2-} + 6\ Cl^- \longrightarrow$$
$$2\ Cr^{3+} + 7\ H_2O + 3\ Cl_2 + \cancel{6\ e^-}$$

Now the equation is balanced, and we can rewrite it in final form:

$$Cr_2O_7^{2-}(aq) + 14\ H^+(aq) + 6\ Cl^-(aq) \longrightarrow$$
$$2\ Cr^{3+}(aq) + 7\ H_2O(l) + 3\ Cl_2(g)$$

As confirmation, check the balance of elements and charges.

Notice that our example is simply the net ionic equation for the reaction that was used in section 16.2 to explain the oxidation state method of balancing redox equations. Thus, both methods can be used to balance equations for oxidation-reduction reactions occurring in aqueous solution. However, the ion-electron method starts with the net ionic equation, while the oxidation state method starts with the nonionic equation for the reaction.

EXAMPLE 16.3   Use the ion-electron method to balance the equation for the following reaction, which occurs in acidic solution.

$$MnO_4^-(aq) + As_4O_6(s) \longrightarrow Mn^{2+}(aq) + H_3AsO_4(aq)$$

(Note that H_3AsO_4 is a weak acid and thus exists predominantly in the un-ionized form in water.)

Solution   Step 1:

$$\underset{+3}{As_4O_6} \longrightarrow 4\ \underset{+5}{H_3AsO_4} \quad \text{(oxidation)}$$

$$\underset{+7}{MnO_4^-} \longrightarrow \underset{+2}{Mn^{2+}} \quad \text{(reduction)}$$

Step 2:
$$10\ H_2O + As_4O_6 \longrightarrow 4\ H_3AsO_4 \quad \text{(oxidation)}$$
$$MnO_4^- \longrightarrow Mn^{2+} + 4\ H_2O \quad \text{(reduction)}$$

Step 3:
$$10\ H_2O + As_4O_6 \longrightarrow 4\ H_3AsO_4 + 8\ H^+ \quad \text{(oxidation)}$$
$$8\ H^+ + MnO_4^- \longrightarrow Mn^{2+} + 4\ H_2O \quad \text{(reduction)}$$

Step 4:
$$10\ H_2O + As_4O_6 \longrightarrow 4\ H_3AsO_4 + 8\ H^+ + 8\ e^- \quad \text{(oxidation)}$$
$$5\ e^- + 8\ H^+ + MnO_4^- \longrightarrow Mn^{2+} + 4\ H_2O \quad \text{(reduction)}$$

Step 5:
$$5[10\ H_2O + As_4O_6 \longrightarrow 4\ H_3AsO_4 + 8\ H^+ + 8\ e^-] \quad \text{(oxidation)}$$
$$8[5\ e^- + 8\ H^+ + MnO_4^- \longrightarrow Mn^{2+} + 4\ H_2O] \quad \text{(reduction)}$$

Step 6:

$$50\ H_2O + 5\ As_4O_6 \longrightarrow 20\ H_3AsO_4 + 40\ H^+ + 40\ e^- \quad \text{(oxidation)}$$
$$40\ e^- + 64\ H^+ + 8\ MnO_4^- \longrightarrow 8\ Mn^{2+} + 32\ H_2O \quad \text{(reduction)}$$

$$\overset{18}{\cancel{50}}\ H_2O + 5\ As_4O_6 + \cancel{40\ e^-} + \overset{24}{\cancel{64}}\ H^+ + 8\ MnO_4^- \longrightarrow$$
$$20\ H_3AsO_4 + \cancel{40\ H^+} + \cancel{40\ e^-} + 8\ Mn^{2+} + \cancel{32\ H_2O}$$

Thus, the final equation is

$$18 \, H_2O(l) + 5 \, As_4O_6(s) + 24 \, H^+(aq) + 8 \, MnO_4^-(aq) \longrightarrow$$
$$20 \, H_3AsO_4(aq) + 8 \, Mn^{2+}(aq)$$

EXERCISE 16.3 Use the ion-electron method to balance the equation for the following reaction, which occurs in acidic solution.

$$Mn^{2+}(aq) + BiO_3^-(aq) \longrightarrow MnO_4^-(aq) + Bi^{3+}(aq)$$ ∎

Reactions in Basic Solution The ion-electron method for reactions in basic solution is similar to the method for acidic solutions, except that OH^- and H_2O are added as needed. Thus, the major difference occurs in step 3. As an example, let's balance the equation for the following reaction, which occurs in basic solution.

$$Bi_2O_3(s) + ClO^-(aq) \longrightarrow BiO_3^-(aq) + Cl^-(aq)$$

1. Write the skeleton equations for the half-reactions and balance the elements undergoing oxidation and reduction.

$$\underset{+3}{Bi_2O_3} \longrightarrow 2 \, \underset{+5}{BiO_3^-} \qquad \text{(oxidation)}$$

$$\underset{+1}{ClO^-} \longrightarrow \underset{-1}{Cl^-} \qquad \text{(reduction)}$$

2. Balance oxygen in the half-reactions. For each O atom needed, add *two* OH^- ions to the side of the equation that is deficient in oxygen.

$$6 \, OH^- + Bi_2O_3 \longrightarrow 2 \, BiO_3^- \qquad \text{(oxidation)}$$
$$ClO^- \longrightarrow Cl^- + 2 \, OH^- \qquad \text{(reduction)}$$

3. Balance hydrogen by adding H_2O to the side of the equation that is deficient in hydrogen. (The idea in steps 2 and 3 is that 2 OH^- ions are equivalent to $O + H_2O$.)

$$6 \, OH^- + Bi_2O_3 \longrightarrow 2 \, BiO_3^- + 3 \, H_2O \qquad \text{(oxidation)}$$
$$H_2O + ClO^- \longrightarrow Cl^- + 2 \, OH^- \qquad \text{(reduction)}$$

4. Balance charges in the half-reactions by adding e^- where needed.

$$6 \, OH^- + Bi_2O_3 \longrightarrow 2 \, BiO_3^- + 3 \, H_2O + 4 \, e^- \qquad \text{(oxidation)}$$
$$2 \, e^- + H_2O + ClO^- \longrightarrow Cl^- + 2 \, OH^- \qquad \text{(reduction)}$$

5. Balance electron transfer.

$$6 \, OH^- + Bi_2O_3 \longrightarrow 2 \, BiO_3^- + 3 \, H_2O + 4 \, e^- \qquad \text{(oxidation)}$$
$$2[2 \, e^- + H_2O + ClO^- \longrightarrow Cl^- + 2 \, OH^-] \qquad \text{(reduction)}$$

6. Add together the two half-reactions, cancelling any identical species appearing on both sides of the total equation.

$$6 \, OH^- + Bi_2O_3 \longrightarrow 2 \, BiO_3^- + 3 \, H_2O + 4 \, e^- \qquad \text{(oxidation)}$$
$$\underline{4 \, e^- + 2 \, H_2O + 2 \, ClO^- \longrightarrow 2 \, Cl^- + 4 \, OH^-} \qquad \text{(reduction)}$$

$$\overset{2}{\cancel{6} \, OH^-} + Bi_2O_3 + \cancel{4 \, e^-} + \cancel{2 \, H_2O} + 2 \, ClO^- \longrightarrow$$
$$\overset{1}{} 2 \, BiO_3^- + \cancel{3 \, H_2O} + \cancel{4 \, e^-} + 2 \, Cl^- + \cancel{4 \, OH^-}$$

Thus, the final equation is

$$2\ OH^-(aq) + Bi_2O_3(s) + 2\ ClO^-(aq) \longrightarrow$$
$$2\ BiO_3{}^-(aq) + H_2O(l) + 2\ Cl^-(aq)$$

EXAMPLE 16.4   Use the ion-electron method to balance the equation for the following reaction, which occurs in basic solution.

$$MnO_4{}^-(aq) + C_2O_4{}^{2-}(aq) \longrightarrow MnO_2(s) + CO_2(g)$$

Solution   Step 1:

$$\underset{+3}{C_2O_4{}^{2-}} \longrightarrow 2\ \underset{+4}{CO_2} \qquad \text{(oxidation)}$$

$$\underset{+7}{MnO_4{}^-} \longrightarrow \underset{+4}{MnO_2} \qquad \text{(reduction)}$$

Step 2:

$$C_2O_4{}^{2-} \longrightarrow 2\ CO_2 \qquad \text{(oxidation)}$$
$$MnO_4{}^- \longrightarrow MnO_2 + 4\ OH^- \qquad \text{(reduction)}$$

Step 3:

$$C_2O_4{}^{2-} \longrightarrow 2\ CO_2 \qquad \text{(oxidation)}$$
$$2\ H_2O + MnO_4{}^- \longrightarrow MnO_2 + 4\ OH^- \qquad \text{(reduction)}$$

Step 4:

$$C_2O_4{}^{2-} \longrightarrow 2\ CO_2 + 2\ e^- \qquad \text{(oxidation)}$$
$$3\ e^- + 2\ H_2O + MnO_4{}^- \longrightarrow MnO_2 + 4\ OH^- \qquad \text{(reduction)}$$

Step 5:

$$3[C_2O_4{}^{2-} \longrightarrow 2\ CO_2 + 2\ e^-] \qquad \text{(oxidation)}$$
$$2[3\ e^- + 2\ H_2O + MnO_4{}^- \longrightarrow MnO_2 + 4\ OH^-] \qquad \text{(reduction)}$$

Step 6:

$$3\ C_2O_4{}^{2-} \longrightarrow 6\ CO_2 + 6\ e^- \qquad \text{(oxidation)}$$
$$\underline{6\ e^- + 4\ H_2O + 2\ MnO_4{}^- \longrightarrow 2\ MnO_2 + 8\ OH^-} \qquad \text{(reduction)}$$
$$3\ C_2O_4{}^{2-} + \cancel{6\ e^-} + 4\ H_2O + 2\ MnO_4{}^- \longrightarrow$$
$$6\ CO_2 + \cancel{6\ e^-} + 2\ MnO_2 + 8\ OH^-$$

Thus, the final equation is

$$3\ C_2O_4{}^{2-}(aq) + 4\ H_2O(l) + 2\ MnO_4{}^-(aq) \longrightarrow$$
$$6\ CO_2(g) + 2\ MnO_2(s) + 8\ OH^-(aq)$$

EXERCISE 16.4   Use the ion-electron method to balance the equation for the following reaction, which occurs in basic solution.

$$S^{2-}(aq) + I_2(s) \longrightarrow SO_4{}^{2-}(aq) + I^-(aq) \qquad \blacksquare$$

16.4   THE ACTIVITY SERIES OF METALS

In chapter 5, we pointed out that metals have low ionization energies compared to nonmetals, and thus atoms of metals tend to lose electrons and form cations. We observe this behavior when certain metals react with water to liberate hydrogen. For example, the equation for the reaction of sodium with water is

$$2\ Na(s) + 2\ H_2O(l) \longrightarrow 2\ NaOH(aq) + H_2(g)$$

Sodium, like all alkali metals, reacts vigorously with cold water. Alkaline earth metals react less vigorously with water. For example, calcium reacts slowly with cold water,

$$Ca(s) + 2 H_2O(l) \longrightarrow Ca(OH)_2(aq) + H_2(g)$$

and magnesium does not react with cold water but will react with hot water vapor

$$Mg(s) + 2 H_2O(g) \longrightarrow Mg(OH)_2(s) + H_2(g)$$

Thus, the tendency for metals to produce hydrogen from water decreases as metallic character of the metals decreases. This trend is related to the ease with which electrons are lost by atoms of metals.

It was mentioned in chapter 14 that many, but not all, metals react with acids to produce hydrogen. The reaction of aluminum with acid is

$$2 Al(s) + 6 H^+(aq) \longrightarrow 2 Al^{3+}(aq) + 3 H_2(g)$$

where $H^+(aq)$ represents a hydrated hydrogen ion (hydronium ion). Notice that the hydrogen gas actually comes from hydrogen ions. In fact, it is the hydrogen ions in water that allow the production of hydrogen from water by active metals. When a metal produces hydrogen from water or acids, atoms of the metal spontaneously lose electrons to hydrogen ions, converting them to hydrogen. Thus, the production of hydrogen from water or acids by metals is actually a spontaneous redox reaction. The oxidation process corresponds to loss of electrons by the metal atoms

$$M \longrightarrow M^+ + e^- \qquad \text{(oxidation)}$$

(where M represents an alkali metal), and the reduction process corresponds to a gain of electrons by the hydrated protons:

$$2 H^+ + 2 e^- \longrightarrow H_2 \qquad \text{(reduction)}$$

If we multiply the oxidation equation by 2 to balance electron transfer, we can add the two equations together:

$$
\begin{array}{ll}
2 M \longrightarrow 2 M^+ + 2 e^- & \text{(oxidation)} \\
\underline{2 H^+ + 2 e^- \longrightarrow H_2} & \text{(reduction)} \\
2 M + 2 H^+ + 2 e^- \longrightarrow 2 M^+ + 2 e^- + H_2 &
\end{array}
$$

or,

$$2 M(s) + 2 H^+(aq) \longrightarrow 2 M^+(aq) + H_2(g)$$

The final equation describes the redox reaction between an alkali metal and acid.

The metals that react spontaneously with acids are said to displace hydrogen from the acids. Atoms of these metals have a greater tendency to exist as cations than do atoms of H_2; hence these metals are said to be more active than H_2. On the other hand, there are some metals that are less active than hydrogen; copper, silver, and mercury are examples. These metals do not react spontaneously with acids. We can conclude that atoms of these metals have less tendency to exist as cations than do atoms of hydrogen. By studying the spontaneous reactions of metals with acids, chemists have been able to list metals according to their activities in this

TABLE 16-1	The Activity Series of Selected Metals and Hydrogen

Ease of oxidation ↑

K	→	K$^+$	+ e^-
Ba	→	Ba^{2+}	+ 2 e^-
Ca	→	Ca^{2+}	+ 2 e^-
Na	→	Na$^+$	+ e^-
Mg	→	Mg^{2+}	+ 2 e^-
Al	→	Al^{3+}	+ 3 e^-
Zn	→	Zn^{2+}	+ 2 e^-
Cr	→	Cr^{3+}	+ 3 e^-
Fe	→	Fe^{2+}	+ 2 e^-
Ni	→	Ni^{2+}	+ 2 e^-
Sn	→	Sn^{2+}	+ 2 e^-
Pb	→	Pb^{2+}	+ 2 e^-
H$_2$	→	2 H$^+$	+ 2 e^-
Cu	→	Cu^{2+}	+ 2 e^-
As	→	As^{3+}	+ 3 e^-
Ag	→	Ag$^+$	+ e^-
Hg	→	Hg^{2+}	+ 2 e^-
Au	→	Au^{3+}	+ 3

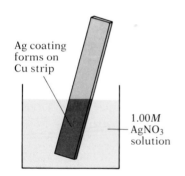

Ag coating forms on Cu strip

1.00M AgNO$_3$ solution

FIGURE 16-2 The Displacement of Silver by Copper

reaction. This listing is called the **activity series of metals.** Table 16-1 illustrates the series with some of the more common metals. Metals that displace hydrogen from acids are said to be more active than hydrogen. Note that the activities of metals decrease going down the list. Metals that do not displace hydrogen from acids are said to be less active than hydrogen; they are listed below hydrogen in the table.

The activity series of metals actually corresponds to a listing of metals by their abilities to become oxidized (lose electrons) relative to hydrogen. Atoms of any metal above hydrogen in the series will spontaneously lose electrons to protons

$$2\ M + 2\ H^+ \longrightarrow 2\ M^+ + H_2$$

(where M represents an alkali metal), while atoms of any metal below hydrogen will undergo the reverse reaction:

$$2\ M^+ + H_2 \longrightarrow 2\ M + 2\ H^+$$

In fact, *atoms of a metal will spontaneously lose electrons to cations of any other metal (or hydrogen) listed below it in the series.* Such reactions are called **metal displacements.** Thus, potassium atoms will lose electrons to sodium ions

$$K + Na^+ \longrightarrow K^+ + Na$$

and chromium atoms will lose electrons to tin ions,

$$2\ Cr + 3\ Sn^{2+} \longrightarrow 2\ Cr^{3+} + 3\ Sn$$

(In the first reaction, potassium was oxidized and sodium was reduced; in the second reaction, chromium was oxidized and tin was reduced.) If a strip of copper is placed in a solution of silver nitrate, AgNO$_3$, some of the copper will dissolve (how much depends on the concentration of AgNO$_3$), and silver metal will form.

$$Cu + 2\ Ag^+ \longrightarrow Cu^{2+} + 2\ Ag$$

Figure 16-2 illustrates this reaction; it is sometimes used to make sparkling "silver trees" from copper wire in demonstrations.

EXAMPLE 16.5 Will there be a reaction if calcium metal is placed in an aqueous solution of zinc sulfate, ZnSO$_4$? If so, write the net ionic equation.

Solution Ca is listed above Zn in the activity series, so there will be a reaction. The net ionic equation is derived as follows:

$$Ca + Zn^{2+} + \cancel{SO_4^{2-}} \longrightarrow Ca^{2+} + Zn + \cancel{SO_4^{2-}}$$
$$Ca + Zn^{2+} \longrightarrow Ca^{2+} + Zn$$

EXERCISE 16.5 Will there be a reaction if an iron nail is placed in an aqueous solution of lead nitrate, Pb(NO$_3$)$_2$? If so, write the net ionic equation. ∎

16.5 VOLTAIC CELLS

Spontaneous reactions like those discussed in the preceding section release energy. An interesting feature of such reactions is that the released energy can be converted to electrical energy. Chemical reactions capable of producing electrical energy are called **electrochemical reactions.** One example of an electrochemical reaction is the reaction between zinc metal and aqueous copper(II) sulfate:

$$Zn(s) + Cu^{2+}(aq) \longrightarrow Zn^{2+}(aq) + Cu(s)$$

Electrons are transferred spontaneously from Zn to Cu^{2+}; during the reaction, the zinc metal dissolves and copper metal forms. The half-reactions for this redox system are

$$Zn \longrightarrow Zn^{2+} + 2\ e^- \qquad \text{(oxidation)}$$
$$Cu^{2+} + 2\ e^- \longrightarrow Cu \qquad \text{(reduction)}$$

Each half-reaction can be run in a separate container, provided there is a way for electrons to flow from the oxidation reaction to the reduction reaction, as there is in the apparatus in Figure 16-3.

The apparatus in which an electrochemical reaction is run is called an **electrochemical cell.** In the electrochemical cell illustrated in Figure 16-3, one vessel contains aqueous $CuSO_4$ and the other contains aqueous $ZnSO_4$. Both $CuSO_4$ and $ZnSO_4$ are strong electrolytes and are thus completely ionized. Immersed in each solution is an **electrode,** a solid rod made of an electrical conductor. In this case, one electrode is made of zinc metal and the other of copper metal. When the two electrodes are connected by a metal wire, there is a tendency for the zinc electrode to lose electrons, which pass through the wire to the copper electrode. There they are released to the solution of $CuSO_4$. In this way, electrons can flow from

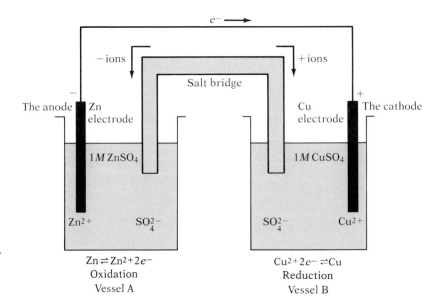

FIGURE 16-3 A Voltaic (Galvanic) Cell. Electrons flow from the zinc electrode (anode) through the wire to the copper electrode (cathode).

the zinc metal in one container to the Cu^{2+} ions in the other container, allowing oxidation-reduction to occur. One half-reaction takes place at each electrode. The electrode where oxidation occurs is called the **anode.** In this example, the anode is made of zinc metal, and the reaction that takes place is

$$Zn \longrightarrow Zn^{2+} + 2\ e^-$$

The electrode where reduction occurs is called the **cathode.** In this example the reaction at the cathode is

$$Cu^{2+} + 2\ e^- \longrightarrow Cu$$

Thus, the anode conducts electrons away from its solution, while the cathode provides electrons to its solution. In the process, electrons flow through the connecting wire — this is electricity, which can then be used for any suitable purpose.

As the current flows, the anode (Zn) dissolves, and the cathode (Cu) builds up mass. Since Zn atoms become Zn^{2+} ions in vessel A and Cu^{2+} ions become Cu atoms in vessel B, the number of positive charges in vessel A increases while the number of positive charges in vessel B decreases. There must be some means of conducting current between the two vessels and allowing ions to move from vessel B to vessel A to balance the charges. The salt bridge accomplishes this purpose. It is a glass tube filled with a gel containing aqueous KCl. The gel allows movement of ions but not mixing of the two solutions it connects. Thus, ions can move from vessel B through the salt bridge to vessel A and maintain electrical neutrality.

Electrochemical cells that provide electricity, like the one just described, are called **voltaic cells** (after Alessandro Volta, an eighteenth century Italian scientist) or **galvanic cells** (after Luigi Galvani, another eighteenth century Italian scientist). Although there are many different voltaic cells, the one described above was invented by the English scientist John F. Daniell in 1836 and is known as the Daniell cell. It was used to provide electricity for early telegraphs.

16.6 BATTERIES

The invention of voltaic cells in the eighteenth century set the stage for the development of batteries. **Batteries** are commercial voltaic cells that are used as sources of electricity. The three most common types of batteries are the lead storage battery (used in automobiles), the dry cell (used in flashlights and portable radios), and the rechargeable nickel-cadmium battery (often used in electronic calculators).

Lead Storage Battery   The lead storage battery (see Figure 16-4) is used in automobiles to provide electricity for ignition and for using the lights and radio when the automobile is not running. The typical lead storage battery contains three or six cells joined together as a series to produce 6 or 12 volts of electricity. Each cell has a lead anode, shaped as a grid, and a cathode made of lead dioxide (PbO_2), also shaped as a grid (lead dioxide is

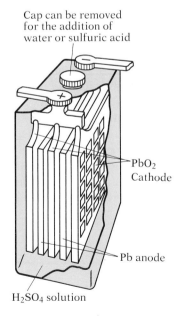

Cap can be removed for the addition of water or sulfuric acid

PbO_2 Cathode

Pb anode

H_2SO_4 solution

FIGURE 16-4   A Lead Storage Battery. Under the usual operating conditions, the aqueous solution of H_2SO_4 has a concentration of about 38%(w/w).

not soluble in water). The two electrodes are separated by an inert spacer, and both are immersed in an aqueous solution of sulfuric acid (called "battery acid"). When the two electrodes are connected by an electrical conductor, the following half-reactions occur spontaneously:

$$\text{Anode:} \quad Pb(s) + H_2SO_4(aq) \longrightarrow PbSO_4(s) + 2\ H^+(aq) + 2\ e^-$$

$$\text{Cathode:} \quad 2\ e^- + 2\ H^+(aq) + PbO_2(s) + H_2SO_4(aq) \longrightarrow$$
$$PbSO_4(s) + 2\ H_2O(l)$$

$$\text{Net Reaction:} \quad Pb(s) + PbO_2(s) + 2\ H_2SO_4(aq) \longrightarrow$$
$$2PbSO_4(s) + 2\ H_2O(l)$$

Thus, electrons flow from the Pb anode to the PbO_2 cathode, and the battery is said to *discharge*. During discharge, $PbSO_4$ collects on the grids of both electrodes, and the amount of H_2SO_4 decreases.

Automobile batteries would fully discharge and thus become "dead" after short use if they did not have the capability of being recharged. This capability is part of the automobile design. After the engine is started, the alternator is activated; it generates an electric current that passes through the battery in the opposite direction of spontaneous electron flow. This action forces the reverse, nonspontaneous reaction to occur

$$2\ PbSO_4(s) + 2\ H_2O(l) \longrightarrow Pb(s) + PbO_2(s) + 2\ H_2SO_4(aq)$$

thus recharging the battery. As soon as the battery is fully recharged, the alternator shuts off, and the battery is ready for a fresh start. However, some energy is lost each time the battery is discharged and recharged, and eventually the battery cannot be recharged. Since the density of battery acid goes from 1.35 g/mL in a fully charged battery to 1.05 g/mL in a dead battery, the density of battery acid indicates the extent of discharge of a battery.

Dry Cell The dry cell was invented in 1865 by Georges Leclanche, a French chemist. These batteries contain a gel or paste instead of a solution, and they are discarded when they become fully discharged, rather than being recharged. Their advantages are their small size and low cost.

The most common dry cell is illustrated in Figure 16-5. It contains a zinc anode as an outer shell and a carbon (graphite) cathode, which provides a surface on which the cathode reaction can occur but does not itself undergo change. Between the anode and cathode is a paste containing MnO_2, graphite, NH_4Cl, and water. The half-reactions of the dry cell are

$$\text{Anode:} \quad Zn(s) \longrightarrow Zn^{2+}(aq) + 2\ e^-$$
$$\text{Cathode:} \quad 2\ NH_4^+(aq) + 2\ MnO_2(s) + 2\ e^- \longrightarrow$$
$$Mn_2O_3(s) + 2\ NH_3(aq) + H_2O(l)$$

$$\text{Net Reaction:} \quad Zn(s) + 2\ NH_4^+(aq) + 2\ MnO_2(s) \longrightarrow$$
$$Zn^{2+}(aq) + Mn_2O_3(s) + 2\ NH_3(aq) + H_2O(l)$$

A salt bridge is not needed because the paste prevents mixing of reactants from the two compartments.

Two other popular dry cells are the alkaline battery, which also contains MnO_2 and Zn, and the mercury battery, which contains mercury(II) oxide,

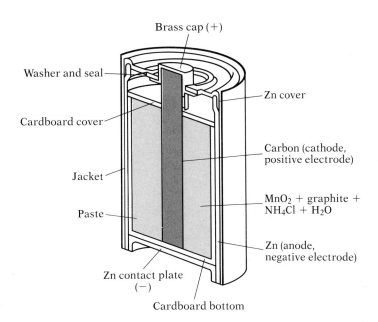

FIGURE 16-5 A Dry-Cell Battery

HgO, as its oxidizing agent and Zn as its reducing agent. Both are actually alkaline batteries because they use KOH in place of NH_4Cl, and both are constructed so that they can deliver more current than the common zinc cell. These batteries stand up better under heavy use and last longer than ordinary dry cells.

Nickel-Cadmium Battery The nickel-cadmium battery consists of a nickel(IV) oxide cathode and a cadmium anode immersed in an aqueous solution of KOH. Since these batteries can be recharged by attaching them to an electrical outlet and reversing their flow of electrons, they have longer lifetimes than dry cells. However, some energy is lost during the charge-discharge cycle, and they eventually wear out.

16.7 ELECTROLYTIC CELLS

Electrochemical reactions that are not spontaneous can be made to occur by providing electrical energy to them. As we saw in the preceding section, this process allows certain batteries to be recharged. The electrical energy can come from another electrochemical source, such as a battery, or it can be provided by an electrical generator. The process of forcing non-spontaneous electrochemical reactions to occur by providing electrical energy from an outside source is called **electrolysis;** the electrochemical cell in which electrolysis is carried out is called an **electrolytic cell.**

When electricity is conducted through a solution, as in an electrolytic cell, ions carry electrons from one electrode to the other (see Figure 16-6). The negatively charged electrode is called the cathode; it takes on a negative charge because it receives electrons (negative charges) from the electrical source. The other electrode, the anode, loses electrons to the elec-

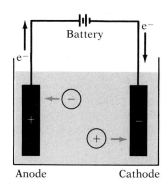

FIGURE 16-6 The Movement of Ions and Electrons during Electrolysis

Electrolysis

(e-lec-TRAH-lih-sis): from *electro-*, meaning "electricity," and *-lysis*, meaning "breaking down or decomposition."

○R **PERSPECTIVE**   CHARLES MARTIN HALL AND THE PRODUCTION OF ALUMINUM

Charles Martin Hall was a chemistry student at Oberlin College in Ohio when one of his professors remarked that a fortune was to be made by the person who could discover an inexpensive way to produce aluminum. Hall was so impressed by the professor's remarks that he set out to devise such a process. He worked hard in the chemistry laboratories at Oberlin, even past graduation, and in 1886, at the age of 21, Charles Martin Hall discovered what is now called the *Hall process* for producing aluminum. At about the same time, Paul Heroult, a Frenchman the same age as Hall, made the same discovery. The success of the Hall process is based on the discovery that aluminum ore, Al_2O_3, dissolves in molten cryolite, Na_3AlF_6, producing a solution that conducts electricity. Electrolysis of the solution produces aluminum metal of 99.0–99.9% purity at the cathode:

$$Al^{3+} + 3\ e^- \longrightarrow Al$$

At first Hall's discovery did not meet with great favor among industrialists, but after several failures to obtain financial backing, Hall persuaded the Mellon family to establish the Pittsburgh Reduction Company in 1888. By 1890, Hall had become vice president of the company, and in 1907, its name was changed to Aluminum Company of America (ALCOA). In the years that followed, Hall did indeed make a fortune for himself, but remained deeply grateful to Oberlin

Charles Martin Hall

College for giving him the education and opportunity to begin his work. He expressed his gratitude by leaving more than five million dollars to Oberlin College in his will.

Cations migrate to the cathode, and anions migrate to the anode.

trical source and thus becomes positively charged. The electrical source provides the driving force for electron flow, and electric current passes through the solution and then back to the battery. Thus, the system is a closed circuit through which electricity flows, provided there are sufficient numbers of ions in the solution. Positively charged ions (cations) are attracted to the negatively charged electrode (cathode), and negatively charged ions (anions) are attracted to the positively charged electrode (anode). During electrolysis, anions lose electrons to the anode, and thus become oxidized. Cations receive electrons from the cathode and become reduced. The net result is that electrons are carried through the solution by ions of the electrolyte.

Electrolysis can be performed with molten NaCl (see Figure 16-7), since its ions are free to move about and it is thus a good conductor of electricity. At the anode, Cl^- ions lose electrons and form neutral atoms, which quickly unite to produce Cl_2; at the same time, Na^+ ions pick up electrons from the cathode to form Na atoms.

$$2\ Cl^- \longrightarrow Cl_2 + 2\ e^-$$
$$2\ Na^+ + 2\ e^- \longrightarrow 2\ Na$$

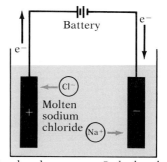

FIGURE 16-7 Electrolysis of Molten Sodium Chloride

Anode, where chlorine gas is formed

Cathode, where metallic sodium is formed

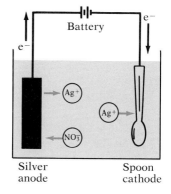

Silver anode

Spoon cathode

FIGURE 16-8 Diagram of an Electroplating Apparatus

In this way, molten NaCl is decomposed by electrical energy into its constituent elements:

$$2\ NaCl(l) + \text{electrical energy} \longrightarrow 2\ Na(l) + Cl_2(g)$$

As long as anions and cations are present in the melt, current will flow. However, when all of the NaCl has decomposed, the circuit will be broken, and the electric current will stop flowing.

Electrolysis has numerous applications. It is used extensively in extraction and purification of elements from ores or other compounds. Electrolysis of molten NaCl is used for the commercial production of sodium metal and chlorine. Electrolysis is also the basis for electroplating, the process in which one metal is coated with another. An example of electroplating is shown in Figure 16-8, where a spoon made of an inexpensive metal is being coated with a thin layer of silver to make it more attractive and more resistant to corrosion, yet the silver-plated spoon is much less expensive than a spoon made entirely of silver. A fairly recent application of electrolysis is the removal of unwanted hair. Small needles are used to transmit electric current to individual hair roots, and the resulting electrolysis kills the hair follicle. Warts and other small skin growths can also be removed in this way.

SUMMARY

Redox reactions are reactions that involve changes in oxidation states. **Oxidation** is a loss of electrons and therefore an increase in oxidation state. **Reduction** is a gain of electrons and therefore a decrease in oxidation state. The substance that causes oxidation is the **oxidizing agent,** and the substance that causes reduction is the **reducing agent.** In the course of a redox reaction, the oxidizing agent is reduced and the reducing agent is oxidized.

Two methods are used for balancing redox equations. The **oxidation state method** is used for nonionic equations to balance the total increase in oxidation state with the total decrease in oxidation state. The steps used in the oxidation state method are given on pp. 388–389. The **ion-electron method** is used for net ionic equations, and it balances the numbers of electrons transferred in the two **half-reactions.** The steps used in the ion-electron method depend on whether the reaction takes place in an acidic or a basic solution. For reactions in acidic solutions, H^+ and H_2O can be added to balance the equation; the complete steps are given on pp. 390–391. For reactions in basic solution, OH^- and H_2O can be added to balance the equation; the steps used for basic solutions are given on pp. 392–393.

The **activity series of metals** lists metals in decreasing order of ability to displace hydrogen from acids and thus become oxidized. Atoms of a metal will spontaneously lose electrons to cations of any other metal (or hydrogen) listed below it in the series. Such reactions are called **metal displacements.**

Chemical reactions capable of producing electrical energy are called **electrochemical reactions.** An apparatus in which an electrochemical reaction is run is called an **electrochemical cell.** An electrochemical cell contains two **electrodes.** Oxidation takes place at the **anode,** and reduction takes place at the **cathode.** Electrochemical cells that provide electricity are called **voltaic cells** or **galvanic cells. Batteries** are commercial voltaic cells used as sources of electricity. The three most common types of batteries are the **lead storage battery,** the **dry cell,** and the rechargeable **nickel-cadmium battery.**

The process of forcing nonspontaneous electrochemical reactions to occur by providing electrical energy from an outside source is called **electrolysis;** the electrochemical cell in which electrolysis is carried out is called an **electrolytic cell.** In electrolysis, electrons flow from an electrical source to the cathode. The electric current is transmitted from the cathode to the anode by ions, and the cathode sends electrons back to the electrical source. In the process, anions are oxidized and cations are reduced.

STUDY QUESTIONS AND PROBLEMS

(More difficult questions and problems are marked with an asterisk.)

CONCEPTS OF OXIDATION AND REDUCTION

1. Define the following terms:
    a. Redox reaction
    b. Oxidation
    c. Reduction
    d. Oxidizing agent (oxidant)
    e. Reducing agent (reductant)

2. Identify the redox reaction(s) in the following list.
    a. $CO_2 + H_2O \rightarrow H_2CO_3$
    b. $H_2S + 4 Br_2 + 4 H_2O \rightarrow H_2SO_4 + 8 HBr$
    c. $H_3PO_4 + Al(OH)_3 \rightarrow AlPO_4 + 3 H_2O$
    d. $2 AgNO_3 + H_2S \rightarrow Ag_2S + 2 HNO_3$

3. For each reaction, identify the substance that is oxidized, the substance that is reduced, the oxidizing agent, and the reducing agent.
    a. $3 P + 5 HNO_3 + 2 H_2O \rightarrow 5 NO + 3 H_3PO_4$
    b. $Sn + 4 HNO_3 \rightarrow SnO_2 + 4 NO_2 + 2 H_2O$
    c. $I_2 + 5 Cl_2 + 6 H_2O \rightarrow 2 HIO_3 + 10 HCl$
    d. $K_2Cr_2O_7 + 6 FeCl_2 + 14 HCl \rightarrow 2 CrCl_3 + 2 KCl + 6 FeCl_3 + 7 H_2O$

BALANCING REDOX EQUATIONS: THE OXIDATION STATE METHOD

4. Use the oxidation state method to balance the following equations.
    a. $KMnO_4 + KCl + H_2SO_4 \rightarrow MnSO_4 + K_2SO_4 + H_2O + Cl_2$
    b. $H_2O + P_4 + HClO \rightarrow H_3PO_4 + HCl$
    c. $Sb + HNO_3 \rightarrow Sb_4O_6 + NO + H_2O$
    d. $PbO_2 + HI \rightarrow PbI_2 + I_2 + H_2O$

5. Use the oxidation state method to balance the following equations.
    a. $Cu + H_2SO_4 \rightarrow CuSO_4 + SO_2 + H_2O$
    b. $Pb + PbO_2 + H_2SO_4 \rightarrow PbSO_4 + H_2O$
    c. $MnO_2 + HI \rightarrow MnI_2 + I_2 + H_2O$
    d. $NF_3 + AlCl_3 \rightarrow N_2 + Cl_2 + AlF_3$

BALANCING REDOX EQUATIONS: THE ION-ELECTRON METHOD

6. Explain what is meant by a *half-reaction*.
7. Use the ion-electron method to balance the following equations. All reactions occur in acidic solution.
   a. $AsH_3(g) + Ag^+(aq) \rightarrow As_4O_6(s) + Ag(s)$
   b. $Zn(s) + H_2MoO_4(aq) \rightarrow Zn^{2+}(aq) + Mo^{3+}(aq)$
   c. $Se(s) + BrO_3^-(aq) \rightarrow H_2SeO_3(aq) + Br^-(aq)$
   d. $H_3AsO_3(aq) + MnO_4^-(aq) \rightarrow H_3AsO_4(aq) + Mn^{2+}(aq)$
8. Use the ion-electron method to balance the following equations. All reactions occur in acidic solution.
   a. $ClO^-(aq) + I^-(aq) \rightarrow Cl^-(aq) + I_2(s)$
   b. $Cr(s) + H^+(aq) \rightarrow Cr^{3+}(aq) + H_2(g)$
   c. $H_2SeO_3(aq) + H_2S(aq) \rightarrow Se(s) + HSO_4^-(aq)$
   d. $Fe^{2+}(aq) + Cr_2O_7^{2-}(aq) \rightarrow Fe^{3+}(aq) + Cr^{3+}(aq)$
9. Use the ion-electron method to balance the following equations. All reactions occur in basic solution.
   a. $S^{2-}(aq) + I_2(s) \rightarrow SO_4^{2-}(aq) + I^-(aq)$
   b. $Si(s) + OH^-(aq) \rightarrow SiO_3^{2-}(aq) + H_2(g)$
   c. $Cr(OH)_3(s) + BrO^-(aq) \rightarrow CrO_4^{2-}(aq) + Br^-(aq)$
   d. $S_2O_3^{2-}(aq) + ClO^-(aq) \rightarrow SO_4^{2-}(aq) + Cl^-(aq)$

THE ACTIVITY SERIES OF METALS

10. What is the activity series of metals? How can it be used to predict metal displacements?
11. Use the activity series of metals (Table 16-1) to identify which of the following metals will displace hydrogen from acids, and write balanced equations to illustrate the displacement.
    a. Ba   b. Ag   c. Cr   d. Pb
12. From the activity series of metals (Table 16-1), predict which of the following reactions will occur.
    a. $Zn(s) + Ni^{2+}(aq) \rightarrow Zn^{2+}(aq) + Ni(s)$
    b. $Pb(s) + Zn^{2+}(aq) \rightarrow Pb^{2+}(aq) + Zn(s)$
    c. $Fe(s) + 2 Ag^+(aq) \rightarrow Fe^{2+}(aq) + 2 Ag(s)$
    d. $Hg(l) + Pb^{2+}(aq) \rightarrow Hg^{2+}(aq) + Pb(s)$

VOLTAIC CELLS

13. Define the following terms:
    a. Electrochemical reaction   e. Cathode
    b. Electrochemical cell   f. Voltaic (galvanic) cell
    c. Electrode   g. Daniell cell
    d. Anode
14. Could the following reaction be carried out in a voltaic cell? Why or why not?

$$Al^{3+}(aq) + As(s) \longrightarrow Al(s) + As^{3+}(aq)$$

15. Why must the two electrodes of a voltaic cell be connected by an electrical conductor?
16. What is the purpose of a salt bridge in a voltaic cell?

17. Describe how a voltaic cell produces electricity.

18. Write the equation for the reaction that occurs in the Daniell cell; then write equations for the half-reactions. Which half-reaction occurs at the anode? Which half-reaction occurs at the cathode?

19. Sketch a diagram of a Daniell cell, and label the anode and cathode. Indicate the direction of electron flow.

BATTERIES

20. What is a battery? What are the three common types? What are their similarities and differences?

21. In the net reaction that occurs during discharge of a lead storage battery, what is oxidized? What is reduced? What are the oxidizing and reducing agents?

22. Explain why the density of battery acid decreases during discharge.

23. In lead storage batteries, why is it important that $PbSO_4$ is not soluble in water?

24. In the net reaction that occurs during discharge of an ordinary dry cell, what is oxidized? What is reduced? What are the oxidizing and reducing agents?

25. Why is a salt bridge not necessary in a dry cell?

ELECTROLYTIC CELLS

26. What does electrolysis mean? What is an electrolytic cell?

27. Sketch a diagram of an electrolytic cell in which electrolysis of molten $CuCl_2$ could be carried out. Label the anode and cathode, and indicate the directions of electron flow and ion movement.

28. Distinguish between a voltaic cell and an electrolytic cell.

29. Electrolysis of water can be performed if an electrolyte is dissolved in the water. The electrolyte does not undergo change, but the water decomposes to its elements. Write the net equation for the electrolysis of water.

GENERAL EXERCISES

30. For each reaction, identify the substance that is oxidized, the substance that is reduced, the oxidizing agent, and the reducing agent.
   a. $2\ Fe^{2+}(aq) + Br_2(l) \rightarrow 2\ Fe^{3+}(aq) + 2\ Br^-(aq)$
   b. $3\ Ag(s) + 4\ H^+(aq) + NO_3^-(aq) \rightarrow 3\ Ag^+(aq) + NO(g) + 2\ H_2O(l)$
   c. $Cu(s) + 2\ Ag^+(aq) \rightarrow Cu^{2+}(aq) + 2\ Ag(s)$
   d. $2\ MnO_4^-(aq) + 5\ S^{2-}(aq) + 16\ H^+(aq) \rightarrow 2\ Mn^{2+}(aq) + 5\ S(s) + 8\ H_2O(l)$

31. Use the ion-electron method to balance the following equations. All reactions occur in basic solution.
   a. $NiO_2(s) + Fe(s) \rightarrow Ni(OH)_2(s) + Fe(OH)_3(s)$
   b. $MnO_4^-(aq) + I^-(aq) \rightarrow MnO_4^{2-}(aq) + IO_4^-(aq)$
   c. $S_2O_3^{2-}(aq) + I_2(s) \rightarrow SO_4^{2-}(aq) + I^-(aq)$
   d. $As(s) + OH^-(aq) \rightarrow AsO_3^{3-}(aq) + H_2(g)$

*32. A tarnished silver spoon (tarnish is Ag_2S) was left in water in a galvanized iron container overnight. The next day, all signs of tarnish were gone, and the spoon was shiny. What happened? (*Hint:* Galvanized iron is iron coated with zinc.)

33. Suppose you wanted to make a voltaic cell in which the following reaction occurs:

$$Mg(s) + 2\, Ag^+(aq) \longrightarrow Mg^{2+}(aq) + 2\, Ag(s)$$

a. Write equations for the expected half-reactions.
b. What would be the composition of your anode and cathode?
c. Sketch a diagram of your proposed cell.
d. On the sketch, indicate the direction of electron flow.

34. Why must electric current be applied to recharge a lead storage battery?

35. Why is a salt bridge not necessary in a lead storage battery?

*36. Bumping and shaking dislodges $PbSO_4$ from the electrode grids of a lead storage battery and shortens its lifetime. Why?

*37. After prolonged discharge, a dry cell appears to be dead, but if allowed to rest, the battery often "comes back to life." What causes this behavior? (*Hint:* Consider the diffusion of ions in the paste.)

38. What is the major reason that nickel-cadmium batteries can be recharged?

39. Explain how an electrolytic cell can be used for electroplating.

RADIOACTIVITY AND NUCLEAR PROCESSES

CHAPTER
17

OUTLINE

17.1 Radioactivity
17.2 Alpha Decay
17.3 Beta Decay
    Electron Emission
    Positron Emission
    Electron Capture
17.4 Gamma Decay
    Perspective: Madame Curie
17.5 Nuclear Reactions
17.6 Natural and Artificial Radioactivity
17.7 Detection and Measurement of
    Nuclear Radiation
    Detecting Nuclear Radiation
    Measurements of Nuclear
      Radiation
17.8 Radiation Safety
17.9 Applications of Radiochemistry
    Diagnostic X Rays
    Archeological Dating
    Isotopic Tracers
    Radiation Therapy
    Medical Diagnosis
17.10 Nuclear Energy
17.11 Nuclear Fission
    Nuclear Power Plants
    The Atomic Bomb
    Perspective: The Development of
      the Atomic Bomb
17.12 Nuclear Fusion
    Summary
    Study Questions and Problems

The discovery of radioactivity by the French physicist Henri Becquerel in 1896 came only a few months after the discovery of X rays in Germany. These two events led to an unprecedented period of exciting advances in the understanding of atomic structure: J. J. Thomson discovered the electron in 1897, the three different types of radioactive decay were first observed in 1900, and in 1902 it was recognized that nuclear changes accompany radioactive decay. One of the most significant developments, Einstein's theory of the equivalence of mass and energy ($E = mc^2$), occurred in 1905. The twentieth century has witnessed the unraveling of the structure of the atom and the unleashing of nuclear fury at Hiroshima and Nagasaki.

Before beginning your study of radioactivity and nuclear processes, you should review subatomic particles (chapter 3), atomic number (chapter 3), mass number (chapter 3), isotopes (chapter 3), conservation of mass (chapter 1), and Einstein's equation (chapter 1).

17.1   RADIOACTIVITY

Radioactivity is the process in which unstable atomic nuclei are transformed into more stable nuclei. Stabilization can be achieved by emission of particles or energy, or by capturing electrons that exist outside the nucleus. As the atoms of a radioactive element undergo transformations that help them achieve stability, their nuclei are changed into nuclei of another element. The process of radioactive change might seem to fulfill the dreams of alchemists who searched for the secret of transmutation, except that the products of radioactivity are usually not gold. Even so, a number of radioactive isotopes do have very important medical uses (see Table 17-1).

All isotopes of elements beyond bismuth (atomic number 83) are radioactive. For most of these elements, the mass number of the most stable or best known isotope appears in parentheses in the position for atomic weight in the periodic table. Some elements with atomic numbers smaller than 84 also have **radioisotopes** (short for *radioactive isotopes*); these

TABLE 17-1   Radioisotopes in Medicine

Radioisotope	Use
Radium-226	Destruction of tumors
Cobalt-60	Destruction of tumors
Phosphorus-32	Treatment of leukemia
Iodine-131	Treatment of overactive thyroid gland
Technetium-99m	Various organ scans
Xenon-133	Lung scans
Copper-64	Diagnosis of Wilson's disease (copper storage disease)
Selenium-75	Pancreas scan
Barium-131	Detection of bone tumors

TABLE 17-2 The Types of Radioactive Decay

Type of Decay	Nuclear Change	Particle Mass (amu)	Particle Charge	Particle Description
Alpha	Emission of $_2^4\alpha$	4	+2	Helium nucleus
Beta	Emission of $_{-1}^0\beta$	0	−1	Electron
	Emission of $_1^0\beta$	0	+1	Positron
	Capture of $_{-1}^0\beta$	0	−1	Electron
Gamma	Emission of γ-rays	no particle	no particle	High-energy radiation

include hydrogen-3 (tritium), carbon-14, and cobalt-60, where the numbers 3, 14, and 60 are the mass numbers of the isotopes.

Since all atoms of atomic number greater than 83 are radioactive, scientists have studied the nuclear composition of these elements in attempting to understand radioactivity. One of their important observations is that the nuclei of atoms of low atomic number have approximately the same number of neutrons as protons; thus the neutron-to-proton ratio, n/p, in most of these atoms is very close to 1. In contrast, atoms of atomic number greater than 83 have n/p ratios approaching 1.5, that is, radioisotopes tend to have 1.5 times as many neutrons as protons in their nuclei. It is thought that neutrons help to stabilize the nucleus by partially insulating the positively charged protons so that they do not repel each other strongly enough to leave the nucleus. However, atoms of atomic number 84 or greater have so many protons in their nuclei that the number of neutrons present is not great enough to provide stability, and radioactive decay occurs. **Radioactive decay** is the process in which the unstable nucleus of a radioisotope forms a new nucleus as it attempts to achieve stability. If the product nucleus is stable, no further change will occur; however, if the product nucleus is also unstable, additional decay occurs until a stable nucleus is achieved.

When radioactive decay was first discovered, three types were found. They were named alpha (α), beta (β), and gamma (γ), for the first three letters of the Greek alphabet. The types of radioactive decay are summarized in Table 17-2 and discussed in the sections that follow.

17.2 ALPHA DECAY

Alpha decay occurs when unstable isotopes attempt to stabilize themselves by emitting alpha particles. An **alpha particle** contains two protons and two neutrons; it is identical to a helium-4 nucleus. Thus, an alpha particle has an atomic number of 2 and a mass number of 4. Its symbol is $_2^4\alpha$, where the subscript is the atomic number and the superscript is the mass number. Emission of an alpha particle reduces the atomic number of the unstable nucleus by 2 and the mass number by 4. The process of alpha decay was first observed by Ernest Rutherford in 1902, when he found that radium-226 spontaneously changes to radon-222:

$$_{88}^{226}\text{Ra} \longrightarrow {}_2^4\alpha + {}_{86}^{222}\text{Rn}$$

Note that the sum of the atomic numbers on the left side of the arrow is equal to the sum of the atomic numbers on the right side; the same is true for mass numbers. These equalities reflect the law of conservation of mass. Thus, the total number of protons in the products of the nuclear process is the same as the total number of protons in the starting material. The same is true for the neutrons and thus for the sum of the protons and neutrons.

17.3 BETA DECAY

There are three types of beta decay. All involve either electrons or similar particles called positrons.

Electron Emission   One form of beta decay occurs when electrons are emitted by the nucleus. In a nuclear reaction, emitted electrons are called **beta particles,** and their symbol is $_{-1}^{0}\beta$. (The atomic number of a beta particle or an electron is -1, because an electron has a charge that is opposite to that of a proton but of the same magnitude.) Since there are no electrons in a nucleus, you are probably wondering how a nucleus can emit them: *the electron is created at the instant of emission by conversion of a neutron to an electron and a proton.* Emission of a beta particle increases the atomic number of the unstable nucleus by one but has no effect on its mass number. In the following example, carbon-14 is converted to nitrogen-14 by electron emission:

$$^{14}_{6}C \longrightarrow {}_{-1}^{0}\beta + {}^{14}_{7}N$$

Notice again that the sum of atomic numbers and mass numbers is the same on both sides of the arrow.

Positron Emission   A **positron** ($_{1}^{0}\beta$) is very similar to an electron except that the positron has a positive charge. A positron is created when a proton loses a small, positively charged particle, the positron, and becomes a neutron. Positron emission reduces the atomic number of the unstable nucleus by one but has no effect on its mass number. A positron emission occurs when antimony-116 is converted to tin-116:

$$^{116}_{51}Sb \longrightarrow {}_{1}^{0}\beta + {}^{116}_{50}Sn$$

Electron Capture   Some radioisotopes are unstable because of an excessive number of positive charges in their nuclei. The number of positive charges is reduced when the nucleus captures an electron from outside. Electron capture causes a decrease in the number of positive charges in a nucleus but has no effect on mass number; in other words, entry of an electron into a nucleus changes a proton into a neutron. An example of electron capture is the decay of gold-195 to platinum-195,

$$^{195}_{79}Au + {}_{-1}^{0}e \longrightarrow {}^{195}_{78}Pt$$

where $_{-1}^{0}e$ represents an electron.

Marie Sklodowski Curie (1867–1934) was a pioneer in the study of radioactivity. She was a brilliant high school student in her native Poland, and by working as a governess for several years, she saved enough money to begin studies at the Sorbonne, the famous Paris university. Marie often studied late into the night in her garret in the students' quarter, where she lived on a diet of mostly bread, butter, and tea. After two years of study, her academic devotion was rewarded with a master's degree in physics, and after another year she received a master's degree in mathematics.

Near the end of her studies at the Sorbonne, Marie met a young French physicist, Pierre Curie, and they were married in 1895. As Marie searched for a research project for her doctorate, she became intrigued with Henri Becquerel's newly discovered "radioactivity," a term originated by Marie. Pierre joined her in the study of radioactive minerals, and in 1898 they announced their discovery of two new elements, polonium and radium. In 1903, the Curies shared the Nobel Prize with Becquerel for investigations of radioactivity. Marie was awarded her doctorate that same year.

One year after the joint Nobel Prize, Pierre was run down and killed by a horse-drawn wagon in a tragic accident on the streets of Paris. Marie then devoted all of her energy to completing the scientific work they had started. In 1911, she was awarded a second Nobel

Marie Curie in Her Laboratory

Prize, an unprecedented achievement, for her earlier discovery of polonium and radium. She is remembered as Madame Curie, a woman of insatiable curiosity and selfless dedication to science.

17.4 GAMMA DECAY

Alpha and beta decay frequently produce unstable products that release energy in the form of gamma rays. **Gamma rays** (γ rays) are a highly penetrating form of energy. In most cases, gamma rays are given off only when other types of decay occur. Gamma decay causes no change in atomic number or mass number, but other accompanying types of decay can affect these quantities. Gamma emission occurs along with beta emission in the decay of cobalt-60, a radioisotope used in cancer treatment:

$$^{60}_{27}\text{Co} \longrightarrow {}^{60}_{28}\text{Ni} + {}^{0}_{-1}\beta + \gamma$$

17.5 NUCLEAR REACTIONS

The atomic product of radioactive decay can be predicted if the radioisotope and the type of decay are known. For example, a tritium atom ($^{3}_{1}\text{H}$) decays by giving off one beta particle. Thus, we can write the following incomplete equation for the reaction:

$$\,^3_1\text{H} \longrightarrow \,^{0}_{-1}\beta + ?$$

Since the sum of atomic numbers on both sides of the equation must be the same, 1 in this case, then the atomic product must have an atomic number of 2 ($-1 + 2 = 1$). Similarly, we deduce the mass number of the atomic product to be 3 ($0 + 3 = 3$). Since the element with atomic number 2 is helium, the atomic product must be an isotope of helium. The complete equation is

$$\,^3_1\text{H} \longrightarrow \,^{0}_{-1}\beta + \,^3_2\text{He}$$

EXAMPLE 17.1   An atom of potassium-40 emits a beta particle and gamma radiation. Write the complete equation for the decay.

Solution   By looking up the atomic number of potassium, we can write the following incomplete equation:

$$\,^{40}_{19}\text{K} \longrightarrow \,^{0}_{-1}\beta + \gamma + ?$$

Since the sum of atomic numbers on both sides of the equation must be the same, 19 in this case, then the atomic product must have an atomic number of 20 ($-1 + 20 = 19$). Similarly, we deduce the mass number of the atomic product to be 40. Since the element with atomic number 20 is calcium, the complete equation is

$$\,^{40}_{19}\text{K} \longrightarrow \,^{0}_{-1}\beta + \gamma + \,^{40}_{20}\text{Ca}$$

EXERCISE 17.1   Write the complete equation for positron emission by nitrogen-13.   ■

If the type of decay and the atomic product are known, the starting material can be deduced. For example, the element polonium, discovered by Madame Curie and named for her native Poland, is produced by alpha decay. In the production of $\,^{218}_{84}\text{Po}$, the incomplete equation is

$$? \longrightarrow \,^4_2\alpha + \,^{218}_{84}\text{Po}$$

Because the atomic number of the starting material must be 86 ($2 + 84$), the starting element must be the isotope of radon with mass number 222. Thus, the complete equation is

$$\,^{222}_{86}\text{Rn} \longrightarrow \,^4_2\alpha + \,^{218}_{84}\text{Po}$$

EXAMPLE 17.2   Write the complete equation for the positron emission by the radioisotope that produces indium-109.

Solution   By looking up the atomic number of indium, we can write the following incomplete equation:

$$? \longrightarrow \,^0_1\beta + \,^{109}_{49}\text{In}$$

Since the atomic number of the starting radioisotope must be 50 ($1 + 49$), then the starting radioisotope must be the isotope of tin with mass number 109. The complete equation is

$$\,^{109}_{50}\text{Sn} \longrightarrow \,^0_1\beta + \,^{109}_{49}\text{In}$$

EXERCISE 17.2 Write the complete equation for the alpha decay of the radioisotope that produces osmium-186. ■

It is also possible to predict the type of radioactive decay if the starting material and atomic product are identified. For example, nitrogen-13 decays to produce carbon-13. The incomplete equation is

$$^{13}_{7}\text{N} \longrightarrow ? + {}^{13}_{6}\text{C}$$

The reaction occurs by emission of a particle of atomic number 1 and mass number 0, corresponding to a positron. Therefore, the complete equation is

$$^{13}_{7}\text{N} \longrightarrow {}^{0}_{1}\beta + {}^{13}_{6}\text{C}$$

EXAMPLE 17.3 Uranium-238, the predominant radioisotope in uranium ore, decays to produce thorium-234. Write the complete equation for this process.

Solution The incomplete equation is

$$^{238}_{92}\text{U} \longrightarrow ? + {}^{234}_{90}\text{Th}$$

It indicates that the nuclear decay produces a particle of atomic number 2 (92 − 90) and mass number 4 (238 − 234). This corresponds to an alpha particle; therefore, the complete equation is

$$^{238}_{92}\text{U} \longrightarrow {}^{4}_{2}\alpha + {}^{234}_{90}\text{Th}$$

EXERCISE 17.3 Write the complete equation for the formation of rhodium-99 from palladium-99. ■

17.6 NATURAL AND ARTIFICIAL RADIOACTIVITY

Approximately one-third of the elements have natural radioisotopes. Most of these are very slow to decay and have existed since the earth was formed. Natural radioisotopes stabilize themselves by the three radioactive decay processes described in the preceding sections. Some natural radioisotopes with faster decay rates are continually formed by bombardment from cosmic rays. **Cosmic rays** are streams of particles coming into the earth's atmosphere from the sun and outer space. The incoming particles are mostly protons, with some electrons, alpha particles, and nuclei of atoms larger than helium. These streams are called **primary cosmic rays.** When the particles of primary cosmic rays collide with atoms and molecules of gases in the upper atmosphere, the collisions produce **secondary cosmic rays** composed of all of the known elementary particles —electrons, protons, neutrons, positrons, and other less familiar particles. The larger nuclei of primary cosmic rays are broken up by the collisions, generating such radioisotopes as hydrogen-3 (tritium) and carbon-14.

When a radioisotope undergoes decay, the product is often radioactive, that is, the decay process did not produce a stable product. Instead, the

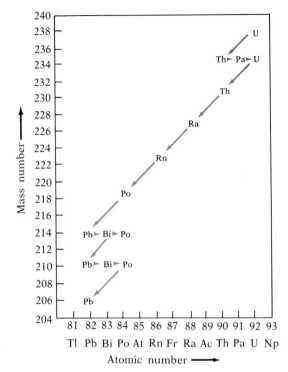

FIGURE 17-1 The Uranium-238 Disintegration Series. $^{238}_{92}U$ first decays to $^{234}_{90}Th$, and subsequent decay processes ultimately form $^{206}_{82}Pb$. Diagonal arrows represent loss of an alpha particle and horizontal arrows represent loss of a beta particle.

product decays to another product, which may or may not be stable. Thus a whole succession of radioactive disintegrations may take place, ultimately forming a stable isotope. Naturally occurring radioisotopes of atomic number greater than 81 decay by one of three **radioactive disintegration series.** One of these is illustrated in Figure 17-1. All of the radioactive disintegration series produce stable isotopes of lead as end products.

Nuclear reactions can be used to produce artificial radioisotopes. If a stable nucleus is bombarded with alpha particles, neutrons, or other subatomic particles, the first atom may be changed into a new atom that is unstable. The resulting radioactive decay is called **artificial radioactivity.** This technique has been used to synthesize elements beyond uranium in the periodic table (the **transuranium elements**), as illustrated by the following equations:

$$^{238}_{92}U + ^{1}_{0}n \longrightarrow ^{239}_{93}Np + ^{0}_{-1}\beta$$
$$^{238}_{92}U + ^{2}_{1}H \longrightarrow ^{238}_{93}Np + 2^{1}_{0}n$$
$$\longrightarrow ^{238}_{94}Pu + ^{0}_{-1}\beta$$

As these equations illustrate, nuclear bombardment brings about **transmutation,** the conversion of one element to another. Natural radioactivity also results in transmutation. However, it was through artificial transmutation that Ernest Rutherford proved the existence of protons in 1919:

$$^{14}_{7}N + ^{4}_{2}\alpha \longrightarrow ^{17}_{8}O + ^{1}_{1}H$$

Another artificial transmutation led to the discovery of the neutron by James Chadwick in 1932:

$$^{9}_{4}\text{Be} + {}^{4}_{2}\alpha \longrightarrow {}^{12}_{6}\text{C} + {}^{1}_{0}n$$

17.7 DETECTION AND MEASUREMENT OF NUCLEAR RADIATION

Radiation emitted specifically by radioactive matter is called nuclear radiation.

Radiation is energy emitted by matter, including the kinetic energy of moving particles. Radioactive matter gives off radiation in the form of alpha particles, beta particles, or gamma rays, and other types of radiation (protons, neutrons, and X rays) can be created artificially.

Some forms of radiation can cause ionization in substances they strike; these types of radiation are called **ionizing radiation.** Table 17-3 lists the major types of ionizing radiation, along with the properties and sources of each. (X rays are a form of ultraviolet light; they will be discussed further in section 17.9).

Detecting Nuclear Radiation   Ionizing radiation can be detected by the **Geiger-Muller counter** (Geiger counter is a shortened version of the name). This device contains an unreactive gas in a closed chamber (see Figure 17-2); the window of the chamber allows high-energy radiation, such as beta particles, gamma particles, and X rays, to pass through into the chamber. Once inside, the radiation produces ions from some of the gas molecules, and the ions move to electrodes, completing an electric circuit. The intensity of the radiation is registered by a meter and a clicking sound. The speed of the clicking indicates the intensity of the radiation, and the intensity is registered on the meter.

For detection of low-energy nuclear radiation, such as alpha and some beta radiation, a device called a **scintillation counter** is used. In this instrument, special compounds that emit light (scintillations) when they absorb nuclear radiation are exposed to the material being tested. The emitted light pulses can then be counted by extremely sensitive detectors,

TABLE 17-3   The Major Types of Ionizing Radiation

Type	Symbol	Mass (amu)	Charge	Source
Alpha	${}^{4}_{2}\alpha$	4	+2	Spontaneous radioactive decay (primarily from heavy atoms)
Beta	${}^{0}_{-1}\beta$	0	−1	Spontaneous radioactive decay
Beta (positron)	${}^{0}_{1}\beta$	0	+1	Spontaneous radioactive decay
Proton	${}^{1}_{1}\text{H}$	1	+1	Artificially produced by nuclear reactors
Neutron	${}^{1}_{0}n$	1	0	Artificially produced by nuclear reactors
Gamma	γ	0	0	Spontaneous radioactive decay
X ray	—	0	0	X ray machines

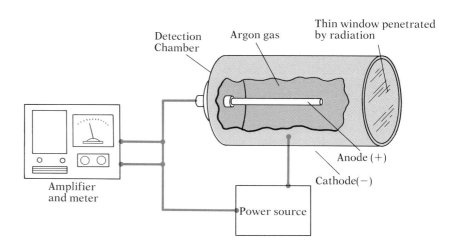

FIGURE 17-2 Schematic Diagram of a Geiger Counter

FIGURE 17-3 A Radiation Film Badge. On each film badge, the worker's name is written on the white plastic holder (lower left), and the date (upper center) is printed on paper that encloses the radiation-sensitive film.

converted to electrical signals, and registered on a meter. In this way, very low levels of nuclear radiation can be detected and measured. This method is particularly useful for detecting the low-energy beta radiation of carbon-14, a radioisotope widely used in research.

The radiation produced by radioactivity has many of the same qualities as ordinary light. For example, nuclear radiation will expose photographic film. Thus, people who work around nuclear radiation often wear **film badges** (Figure 17-3), sometimes called **dosimeters,** to record the amount of radiation exposure they receive. These badges contain several layers of film, each shielded from light but not from radiation. The amount of radiation exposure received by the wearer can be determined from the darkness of the film. In most cases, cumulative records are kept to ensure that workers are not exposed to unsafe accumulated amounts of radiation over long periods of time.

Measurements of Nuclear Radiation   Each different radioisotope decays at its own characteristic rate. It takes some radioisotopes a long time to decay to a significant degree, whereas others decay rapidly. In order to compare rates of decay, the quantity called **half-life** is used. The half-life is the length of time required for half of the atoms of a radioisotope

TABLE 17-4 The Radioactive Decay of Mercury-203	
Time (number of half-lives)	Percent of Hg-203 Remaining
0	100
1 × 47 days (1 half-life)	50
2 × 47 days (2 half-lives)	25
3 × 47 days (3 half-lives)	12.5
4 × 47 days (4 half-lives)	6.25
5 × 47 days (5 half-lives)	3.125
10 × 47 days (10 half-lives)	0.098

to decay. For example, mercury-203 has a half-life of 47 days, while carbon-14 has a half-life of 5730 years. This means that half of the Hg-203 atoms will have decayed after just 47 days, but it will take almost 6,000 years for half of the C-14 atoms to decay. Furthermore, even though more radiation is emitted when a large number of radioactive atoms is present, the half-life of a radioisotope is always the same, regardless of how much of the radioisotope is present initially. Thus, after the first 47 days, the number of Hg-203 atoms will be reduced by one-half; after the second 47 days, the number will be reduced by one-half again, and so on. Table 17-4 illustrates how the relative amount of Hg-203 diminishes with each half-life. Note that the Hg-203 will never completely disappear, but its amount will diminish more and more, approaching zero. This behavior is illustrated in Figure 17-4. Note the constant value for the half-life.

Nuclear radiation measurements are reported in a variety of units, depending on the purpose of the measurement. For example, a hospital purchasing a source of radiation would want to know the **activity** of the sample (the number of radioactive disintegrations in a given period of time), the **dosage** (the amount of radiation delivered by the radioactive material), and the **energy level** of the radiation (the energy of radiation emitted by radioactive materials or X-ray generators). Nuclear radiation measurements can be made and expressed for all three purposes. The types of nuclear radiation measurements and their units are summarized in Table 17-5.

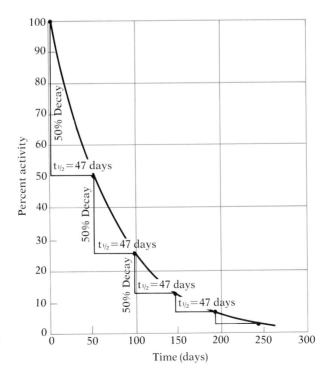

FIGURE 17-4 The Radioactive Decay of Mercury-203

TABLE 17-5 Nuclear Radiation Measurements and Units

Measurement	Radiation Type	Effect Measured	Units and Definitions
Activity	All types	Nuclear disintegrations	Curie (Ci): 3.7×10^7 disintegrations/sec.
			Becquerel (Bq): 1 disintegration/sec.
Dosage	X or γ rays	Ionization	roentgen (R), the amount of radiation that produces 2.1×10^{-9} coulombs of charge in 1 cm³ of dry air at 0° C and 1 atm pressure.
	All types	Energy absorbed	rad: 100 ergs of energy absorbed/g of tissue.
			gray (Gy): 1 J of energy absorbed/kg of tissue.
	All types	Biological damage	rem: damage equivalent to the absorption of 1 roentgen.
Energy	All types	Energy emitted	electron-volt (eV): energy absorbed by an electron accelerating through a voltage change of 1 volt.

17.8 RADIATION SAFETY

We are all aware of the need to avoid excessive dosages of nuclear radiation. Nuclear radiation is harmful because of the chemical changes it causes in living tissue. It strips electrons from atoms to form ions and breaks apart molecules to form fragments called **free radicals.** Free radicals are highly reactive structures and will react almost indiscriminately with anything nearby. Figure 17-5 illustrates the effects of radiation on water molecules, the most abundant compound in living tissue.

Generally speaking, the more massive a particle is, the less deeply it penetrates into tissue. Table 17-6 gives the penetrating abilities of common forms of radiation. Alpha particles travel only short distances through tissue, on the order of 0.05 mm. Beta particles, having smaller mass, can penetrate up to about 4 mm into tissue. X and gamma rays have no mass, so they are the most penetrating forms of radiation, passing through as much as 50 cm of tissue. The least penetrating forms of nuclear radiation usually cause the most biological damage; because the large particles do not travel very far in the tissue, they transfer all of their energy to a rather small volume of tissue. The smaller particles, as well as X and gamma rays, travel over longer distances inside the tissue, and they release their energy all along the path, delivering less energy per unit volume and doing less harm.

Free radical

A highly reactive structure having one or more unpaired electrons.

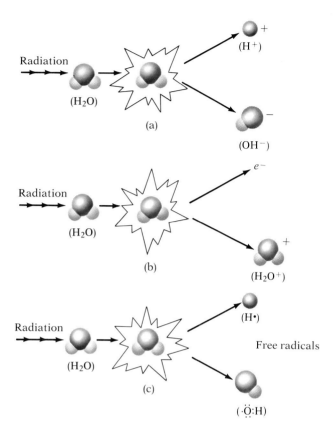

FIGURE 17-5 The Effects of Ionizing Radiation on Water Molecules. After the radiation is absorbed, a water molecule can (a) split into H^+ and OH^-; (b) lose an electron and become a positively charged ion, H_2O^+; or (c) split into two free radicals.

The ions and free radicals formed by the interaction of radiation and water in living tissue react with nearby biological molecules to change their structures. The most significant biological molecule affected is DNA (deoxyribonucleic acid), the carrier of genetic information in cells. Exposure to very high levels of nuclear radiation (200–600 rems) damages the DNA of cells so severely that a large number of the exposed cells die. If the entire body is exposed to this level of nuclear radiation, severe illness

TABLE 17-6 Penetrating Abilities of Common Forms of Nuclear Radiation

Radiation Type	Approximate Depth of Penetration of Radiation into:		
	Dry Air	*Tissue*	*Lead*
Alpha	4 cm	0.05 mm	0
Beta	6–300 cm	0.06–4 mm	0.005–0.3 mm
	Thickness Needed to Reduce Initial Intensity by 10%		
Gamma	400 meters	50 cm	30 mm
X rays	120–240 meters	15–30 cm	0.3–1.5 mm

TABLE 17-7 Probable Short-Term Effects of Exposure to Nuclear Radiation

Dose (rems)	Probable Effect
0–25	No noticeable effects
25–100	Slight blood changes
100–200	Some vomiting within hours, with fatigue and loss of appetite; except for the blood-forming system, recovery will be complete within a few weeks.
200–600	For doses of 300 rems and more, vomiting will occur in two hours or less; severe blood changes will be accompanied by hemorrhage and infection; recovery will occur in 20–100% of all cases, usually within one month to one year.
600–1000	Vomiting within one hour; severe blood changes, hemorrhaging, and infection; loss of hair; 80–100% of exposed individuals will die within two months; survivors will be convalescent over a long period of time.

occurs within hours, and death is likely. Exposure to lower levels of nuclear radiation gives less dramatic short-term symptoms, or none at all, depending on the extent of exposure, but death is not usually imminent for exposures of less than 200 rems. Table 17-7 summarizes the short-term effects of exposure to nuclear radiation.

Brief exposures to low-level nuclear radiation (0–25 rems) gives humans no noticeable short-term effects, but the cumulative effects of such exposure are not yet fully understood. Medical evidence suggests that protracted exposure to low-level nuclear radiation causes weakness and may lead to the development of cancer. Since nuclear radiation alters DNA, it is thought that the altered DNA sometimes lacks the ability to control normal cellular processes, thus allowing the uncontrolled growth and proliferation characteristic of cancer cells. This theory helps to explain the premature deaths of a number of early workers in radioactivity and X-ray technology. The altered DNA may be passed along by reproduction to future generations, possibly causing birth defects and increased susceptibility to cancer.

It should be noted that everyone is exposed to small amounts of nuclear radiation from artificial sources (medical procedures, nuclear power plants, research laboratories, industrial wastes, radioactive fallout, and so forth), as well as from cosmic rays and natural deposits of radioactive minerals. The level of such "background" radiation is thought to average about 0.2 rems, but the level may be increasing. The effects of background radiation on any one generation may not be significant, but the possible genetic effects on future generations are not known.

In light of the hazards of nuclear radiation, we have learned to treat it with respect. Safety precautions are based on the properties of nuclear radiation. Radiation moves in straight lines from its origin, streaming out in all directions, much as light moves from the sun in all directions. For sufficient protection from nuclear radiation, a shield must be inserted between the exposed person and the emitting source. Since alpha and beta particles have the least penetrating abilities, these can be stopped by paper, wood, or cardboard. However, protection against gamma and X rays requires use of very dense materials; lead is the most common substance used. Thus, radiologists spread lead-impregnated cloths over parts of the body not intended to receive radiation. It is also important that radiologists and associated workers take extra precautions to shield themselves from nuclear radiation, as they have far more opportunities for exposure than most people.

A second means of nuclear radiation protection is to stay as far away from the radioactive source as possible. The rays spread out in the shape of a cone, and less radiation will strike a given amount of surface area as it moves away from the source.

In addition to proper shielding and distance from the source, low-intensity sources are used whenever possible to keep the exposure to a minimum. Above all, radioactive materials should not be taken into the body except for medical treatment under the direction of a physician; workers should never eat, drink, or smoke in the vicinity of radioactive substances.

17.9 APPLICATIONS OF RADIOCHEMISTRY

Many practical uses have been found for radioisotopes and X rays. Some of the major ones are discussed in this section.

Diagnostic X Rays **X rays** are high-energy ultraviolet light resulting from electron transitions within atoms. For example, when a nucleus captures an electron, energy is released as an outer electron falls into the space created by the electron capture, as illustrated in Figure 17-6. Most useful X rays are generated by a high-voltage electrical discharge in a vacuum tube; high-speed electrons bombard a target metal, knocking out inner shell electrons and thus producing X rays. X rays have a high energy content and pass through soft tissues to varying degrees. They are used in medicine and dentistry to examine internal organs and the structure of bone and teeth. After passing through the soft tissue, the X rays darken a photographic film placed on the other side of the body. Bones and other hard substances absorb the X rays, leaving white images on the photographic film.

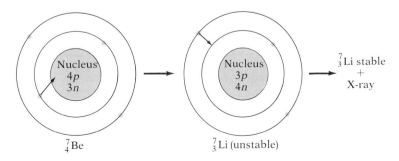

FIGURE 17-6 X ray Emission Following Electron Capture

Archeological Dating   As we mentioned earlier, secondary cosmic radiation is responsible for producing carbon-14 in our atmosphere:

$$^{14}_{7}N + ^{1}_{0}n \longrightarrow ^{14}_{6}C + ^{1}_{1}H$$

This radioisotope makes up only a small fraction of a percent of all carbon on earth. Although new C-14 is continuously being formed by cosmic radiation, some is lost by radioactive decay; its beta radiation is easily detected.

$$^{14}_{6}C \longrightarrow ^{14}_{7}N + ^{0}_{-1}\beta$$

Because of its simultaneous formation and decomposition, the amount of C-14 in the environment was reasonably constant up until atmospheric testing of nuclear weapons was begun.

Carbon-14 makes its way into living plants and animals as they exchange carbon compounds with their environment. Thus, any living creature begins at conception to give off and take in C-14; an equilibrium is soon reached, and then the level of C-14 within the organism remains constant and equal to the level of C-14 in the surroundings. When death occurs, exchange of C-14 with the environment ceases, but the decay of the previously incorporated C-14 continues. After 5,730 years (the half-life of C-14), organic matter that was once part of a living organism will contain one-half of its original concentration of C-14. By knowing the half-life of C-14 and measuring relative amounts of C-14 in organic matter, archeologists can estimate the age of a sample. In practice, the carbon content of a sample is converted to carbon dioxide, CO_2, by burning the sample, and the level of beta radiation is measured. Some examples of objects dated with C-14 are given in Table 17-8. Unfortunately, nuclear weapons testing has altered the C-14 content of the atmosphere to the extent that currently living species cannot be dated this way by future archeologists.

Other radioisotopes also can be used for archeological dating. For example, uranium-238, with a half-life of 4.5×10^9 years, can be used to determine the age of rocks containing uranium. Potassium-40 and rubidium-87 are also used for dating minerals and rocks.

TABLE 17-8 Approximate Ages of Some Objects Dated with Carbon-14	
Object	Age (years)
Franchthi Cave artifacts (Greece)	22,000
Lescaux Cave wall paintings (France)	15,000
Bristlecone pine trees (U.S. southwest)	7,000
Crater Lake, Oregon	6,500
Egyptian tombs	4,900
Stonehenge (England)	3,700
California giant sequoia	2,900
Dead Sea Scrolls	1,900

Isotopic Tracers Chemical compounds containing relatively large amounts of radioisotopes are said to be *tagged* or *labelled*. Because labelled compounds are easy to detect, they can be used to monitor chemical processes that might otherwise be too complicated to study. Biochemical research has depended heavily on the use of labelled compounds to study cellular reactions that occur in complex sequences called *metabolic pathways*. For example, the use of carbon dioxide labelled with C-14 allowed Melvin Calvin, who won the Nobel Prize in chemistry in 1961, to explain the process of photosynthesis. He shined light on blue-green algae in the presence of $^{14}_{6}CO_2$ for short time periods and then analyzed the cells for the presence of C-14 in other compounds. By isolating and analyzing the compounds containing C-14, Calvin was able to deduce the chemical route by which CO_2 is used by plants to synthesize glucose and then cellulose. In essence, Calvin allowed the algae to do the chemistry for him and then analyzed the results. Similar studies have been done on many major metabolic pathways. Other uses of labelled compounds include monitoring the accumulation and use of iodine by the thyroid gland, studying the incorporation of iron into the hemoglobin of blood, and measuring the rates of chemical reactions.

Radiation Therapy Nuclear radiation can be used to treat cancer and other abnormal growths, because rapidly multiplying cells are more sensitive to radiation damage. Alpha and beta radiation are not normally used for therapy because they have such low penetrating abilities. However, gamma and X rays readily penetrate tissues, and they are used extensively. The most common examples are machine-generated X rays and gamma rays from cobalt-60 and cesium-137. These radiations can be shielded and guided so that they are directed at the cancer site (see Figure 17-7). In this way, cancer cells are preferentially destroyed; unfortunately, some normal cells are also damaged or destroyed, and the patient may suffer the

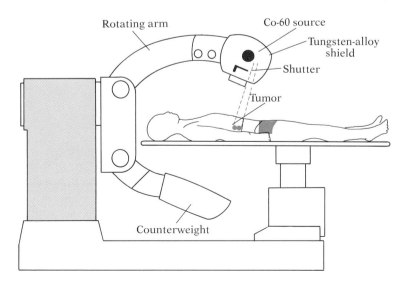

FIGURE 17-7 Radiation Therapy for Cancer with Cobalt-60

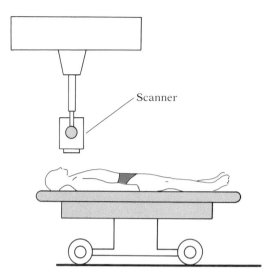

FIGURE 17-8   A Scanning Scintillation Camera

effects of radiation sickness during therapy. However, nuclear radiation is one of the more effective ways of treating cancer.

Medical Diagnosis   Nuclear medicine is the rapidly growing field in which radioisotopes are used to diagnose disease. Radioisotopes can be introduced into the body by injection or ingestion. As they move through the body, they continue to emit radiation, usually gamma rays, which can be detected by scanning with a scintillation camera (Figure 17-8). This device is similar to a scintillation counter in that tiny flashes of light are produced when radiation strikes crystals inside the camera. These light flashes are then used to expose photographic film. The developed film resembles an X ray photograph, except that the images produced are images of the particular body parts from which radiation is coming. In this way, organ and bone scans are obtained. The radioactive compounds used are those that have a tendency to accumulate in certain body parts or tumors. For example, if compounds containing radioactive iodine (I-131) are taken into the body, the I-131 is concentrated in the thyroid gland. Compounds containing radioactive strontium (Sr-85) accumulate in bones; because strontium and calcium are in the same chemical family in the periodic table, both are used for making bone tissue.

17.10   NUCLEAR ENERGY

It is a well-known fact that nuclear changes release energy. The theoretical basis for the energy release lies in the equation worked out by Albert Einstein in 1905:

$$E = mc^2$$

This now famous equation states that energy (in ergs) is equal to mass (in grams) multiplied by the square of the speed of light (in cm/sec). Since the speed of light is such a large number (3×10^{10} cm/sec), a very small amount of matter can release quite a large amount of energy.

Einstein's equation does not tell the whole story about the origin of nuclear energy. We can learn more by examining the actual mass of a nucleus, for example, the nucleus of helium-4. If we add up the masses of the component parts of the helium nucleus, we get a total of 4.031882 amu:

Two protons	2×1.007276 amu = 2.014552 amu
Two neutrons	2×1.008665 amu = <u>2.017330 amu</u>
	Total mass = 4.031882 amu

But when the mass of the helium nucleus is actually measured, it is found to be 4.001506 amu; that is, some of the mass calculated to be in the helium nucleus is not actually there. This deficiency in mass is called the **mass defect;** for helium, the mass defect is 0.030376 amu. If we use Einstein's equation to calculate the energy equivalent to the mass defect of helium, we get a value of 28 million electron-volts (MeV). The energy equivalent of the mass defect is called the **nuclear binding energy,** which is defined as the energy released in the combination of particles to form a nucleus. All atomic nuclei weigh less than the sum of masses of their subatomic particles; this mass defect is considered to be equivalent to the energy holding particles of the nuclei together.

The Einstein equation can be used to calculate nuclear binding energy for all atomic nuclei. The nuclear binding energy of a particular nucleus divided by the number of neutrons and protons in the nucleus is called the **average binding energy.** A plot of average binding energy as a function of mass number is shown in Figure 17-9. As you can see, average binding energy is very low in the region of low mass number; then it rises to a maximum around mass number 56, corresponding to iron. Iron is one of the more abundant elements in the universe; its high average binding

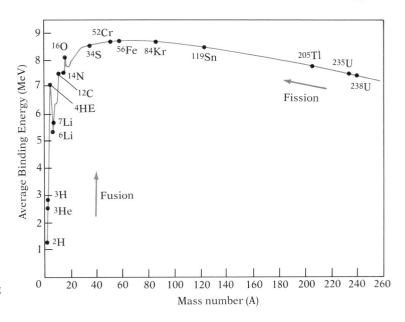

FIGURE 17-9 A Plot of Average Binding Energy as a Function of Mass Number

FIGURE 17-10   A Nuclear Power Plant

energy and, consequently, its high nuclear stability are factors in determining its abundance.

As mass number increases beyond iron, average binding energy begins to fall again. Thus, isotopes having intermediate mass numbers are the most stable. It would appear that converting isotopes of small mass number to isotopes of intermediate mass number would result in release of energy and greater stability. Indeed, this is what happens in **nuclear fusion,** the process of combining two light nuclei to produce a heavier nucleus of intermediate mass number. Similarly, splitting very heavy nuclei into lighter ones should also cause release of energy and greater stability. Such processes do occur; **nuclear fission** is the name given to the process of splitting a heavy nucleus into lighter nuclei of intermediate mass number. In both fusion and fission, a mass decrease occurs and energy is released.

In radioactive decay, small amounts of mass are converted to energy. The energy released appears mostly as the kinetic energy of the particles and nuclei produced by the reaction. These particles and nuclei collide with atoms and molecules in their surroundings, gradually losing their excess energy by transfer to other particles. Most of this lost energy is ultimately converted to heat.

17.11   NUCLEAR FISSION

When certain large atoms are bombarded with particles emitted by a nuclear source, smaller atoms are produced. This process is known as **nuclear fission.** The binding energies of the parent nucleus and the smaller fragments are sufficiently different that large amounts of energy are liberated. One of the most common examples of fission is the fission of uranium-235 that results from bombardment with neutrons:

$$^{235}_{92}\text{U} + ^{1}_{0}n \longrightarrow ^{135}_{53}\text{I} + ^{97}_{39}\text{Y} + 4\,^{1}_{0}n$$

$$^{235}_{92}\text{U} + ^{1}_{0}n \longrightarrow ^{139}_{56}\text{Ba} + ^{94}_{36}\text{Kr} + 3\,^{1}_{0}n$$

$$^{235}_{92}\text{U} + ^{1}_{0}n \longrightarrow ^{103}_{42}\text{Mo} + ^{131}_{50}\text{Sn} + 2\,^{1}_{0}n$$

$$^{235}_{92}\text{U} + ^{1}_{0}n \longrightarrow ^{139}_{54}\text{Xe} + ^{95}_{38}\text{Sr} + 2\,^{1}_{0}n$$

The equations given above show only a few of the many decomposition routes for fission of U-235. In each case, more neutrons are produced than are consumed. Thus, fission of U-235 can become self-sustaining, since the neutrons produced can serve as bombarding neutrons for fission of more U-235. Two well-known applications of nuclear fission are found in nuclear power plants and the atomic bomb.

Nuclear Power Plants   The rate of the chain reaction of nuclear fission can be controlled so that energy can be produced slowly enough to be used constructively. Controlled fission is carried out in **nuclear reactors.**

Most nuclear power plants, such as the one shown in Figure 17-10, use fission reactors that require U-235 as fuel. However, since this isotope accounts for only 0.7% of natural uranium, there is a limit to the amount of fuel that will be available for nuclear power generation in the future. Furthermore, naturally occurring uranium must be enriched so that the percentage of U-235 is increased to 3% of the total. The enrichment process is both costly and energy-consuming.

As the United States has become more dependent on nuclear power, the

FIGURE 17-11   The Atomic Explosion at Nagasaki, Japan, in 1945

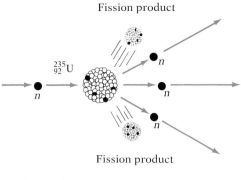

(a) subcritical

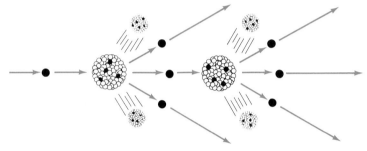

(b) critical chain reaction

FIGURE 17-12 Mass Considerations in a Nuclear Chain Reaction Initiated by Neutron Bombardment of U-235. (a) In a subcritical mass, the number of neutrons produced that cause further fission averages less than one per reaction, and a chain reaction will not be sustained. (b) In a critical mass, the number of neutrons produced that cause further fission averages one per reaction, so the chain reaction will be self-sustaining. (c) In a supercritical mass, the number of neutrons produced that cause further fission averages more than one per reaction, and the accelerating chain reaction can lead to an explosion.

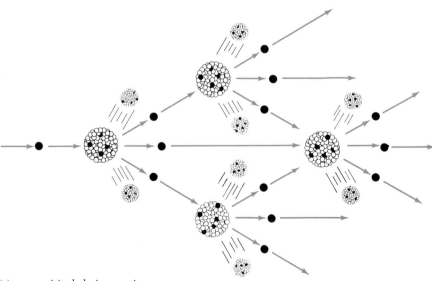

(c) supercritical chain reaction

amount of nuclear waste has built up. At present, there is no permanent repository for the waste, but three sites for possible permanent storage are under consideration. One is the government-owned Hanford Reservation in Washington, where a layer of basalt at a depth of 1,000 meters might serve as a future permanent storage container. Another possibility is the

Nevada Test Site, federal land in southern Nevada used for weapons testing. The volcanic rock under the earth's surface and the arid climate are factors in favor of the Nevada location as a permanent storage site. The third candidate, a salt site, will be selected from Gulf Coast salt domes and bedded salt basins in Utah and Texas.

The Atomic Bomb   The violent release of energy is what distinguishes an atomic explosion (Figure 17-11) from a nuclear reactor. The release of large quantities of energy in such a short time period is a result of two factors. One is that the fissionable material (either uranium-235 or plutonium-239) is highly concentrated, on the order of 97% of the total fuel mass. The other factor is related to the **critical mass,** the smallest amount of fissionable isotope required for a self-sustaining nuclear chain reaction. If the fissionable isotope is present in amounts less than the critical mass, too many of the neutrons produced will escape to the surroundings and not be available for sustaining the chain reaction (see Figure 17-12).

Fissionable isotopes must be handled in amounts less than their critical masses during construction and transportation of atomic bombs. Thus, it is necessary to assemble a supercritical mass at the time that the bomb is to be exploded. Two designs have been used to do this. In one (see Figure 17-13), a small nonnuclear explosive propels a subcritical mass at one end of a barrel into a subcritical mass at the other end, producing a supercritical mass that will explode upon exposure to a neutron source. This is called the **gun method** of assembly. In the second method, called the **implosion method,** a number of nonnuclear explosive charges are placed on the inner surface of a sphere. These fire many subcritical pieces of fissionable material into one common ball at the center of the sphere, thus

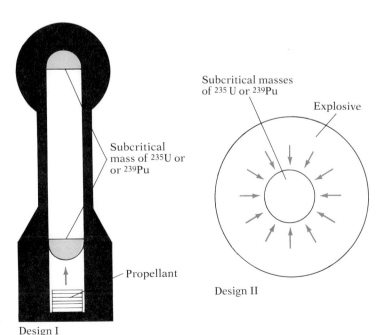

Subcritical masses of ^{235}U or ^{239}Pu

Explosive

Subcritical mass of ^{235}U or or ^{239}Pu

Propellant

Design I

Design II

FIGURE 17-13 Designs of Atomic Bombs. Design I is the "gun" design, and Design II is the implosion design.

Our present knowledge of nuclear power is the result of developments in the 1930s in Europe. The Italian physicist Enrico Fermi was the first to study the bombardment of uranium with neutrons, and two Austrians, Lise Meitner and her nephew Otto Frisch, calculated the energy that could be derived from U-235 fission. When Nazi Germany invaded Austria in 1938, Meitner fled to Sweden and conveyed her research findings to the Danish physicist Neils Bohr. Bohr then brought the information to the attention of Albert Einstein, who had already taken refuge in the United States. Einstein convinced President Franklin D.

Roosevelt of the importance of the work, and in 1942 the United States began a massive research project on atomic energy under the code name Manhattan Project.

Some of the most illustrious scientists the world has known gathered to work on the Manhattan Project. On December 2, 1942, Enrico Fermi achieved the first self-sustaining nuclear chain reaction, accomplishing this feat in a laboratory hastily constructed under the bleachers at the football field of the University of Chicago. In less than three years time, knowledge gained from this work resulted in the atomic bombs used at Hiroshima and Nagasaki.

---

creating the supercritical mass by implosion. The bomb can then be triggered by a neutron source. Both atomic bomb designs were used in 1945, the gun assembly at Hiroshima and the implosion assembly at Nagasaki.

17.12 NUCLEAR FUSION

The sun is the only known continuous nuclear fusion reactor. In fact, the fusion of hydrogen atoms to produce helium atoms is the main source of the sun's energy. The reaction below is a summary of a series of nuclear reactions in the sun:

$$_1^1H + {}_1^2H \longrightarrow {}_2^3He$$

Although nuclear fusion produces large quantities of energy, very high temperatures are required to sustain the process, since the positively charged nuclei must overcome their electrical repulsions for each other in order to fuse together. At temperatures in the sun (about 15 million degrees C), the light nuclei are moving fast enough that they collide with sufficient impact to stay fused.

It once seemed impossible that the temperatures needed for fusion reactions would ever be achieved on earth, but by 1952, scientists had developed the **hydrogen bomb,** a fission bomb surrounded by fuel that can undergo fusion. The fusion fuel is a compound formed by combination of hydrogen-2 (deuterium) and lithium-6; its formula is LiH. When the fission bomb releases its large quantities of heat and neutrons, the LiH atoms undergo fusion, as illustrated by the following sequence of reactions:

$$_3^6Li + {}_0^1n \longrightarrow {}_2^4He + {}_1^3H$$
$$_1^2H + {}_1^3H \longrightarrow {}_2^4He + {}_0^1n$$

As a result of worldwide energy shortages, much interest has developed in the possibility of putting nuclear fusion to practical use. The reaction between deuterium (H-2) and tritium (H-3) could be used as a source of power, since both reactants can be obtained rather easily. However, two major problems remain to be solved. One is the matter of containment. Fusion occurs in the sun because the reactants are held close together by the sun's large gravitational forces. Such powerful forces do not exist naturally on earth, and there are no materials known that can withstand the temperatures of fusion and thus serve as a container. A second problem is that the temperatures needed for fusion have been achieved in hydrogen bombs but not in the controlled systems necessary for useful energy production. However, much progress has been made toward solving these problems, and there is optimism that the fusion process can be harnessed for the benefit of humanity.

SUMMARY

Radioactivity occurs when an unstable atomic nucleus attempts to stabilize itself by emitting particles or energy or by capturing an electron. Nuclear instability results from proton-proton repulsions; all elements with atomic numbers greater than 83 are radioactive, as are some isotopes of lower atomic number.

The three kinds of **nuclear decay** are alpha decay, beta decay, and gamma decay. **Alpha decay** occurs when a nucleus emits a particle of atomic number 2 and mass number 4; an alpha particle is identical to a nucleus of He-4. **Beta decay** can be nuclear emission of an electron (a beta particle) or a positron, or capture of an electron by the nucleus. **Gamma decay** is release of gamma rays, high-energy radiation, by a nucleus. All three types of nuclear radiation can be described by equations. Since each radioisotope decays at its own rate, the **half-life** is used to compare rates of decay.

Most of the naturally occurring radioisotopes decay very slowly and have existed since formation of the earth. Others are formed by bombardment from **secondary cosmic rays.** Naturally occurring radioisotopes of atomic number greater than 81 decay by one of three **radioactive disintegration series.** Stable nuclei can be bombarded with subatomic particles to produce new, unstable nuclei. In this way, nuclear bombardment brings about **transmutation.**

Nuclear radiation and X rays form ions and free radicals that react with nearby biological molecules in living tissues. Cells can withstand small radiation dosages, but high levels of damage to DNA can cause cell death. Possible long-range effects of nuclear radiation exposure include cancer and genetic defects. Those working with radiation should observe all safety precautions. Nuclear radiation can be detected and measured by Geiger counters, scintillation counters, and film badges.

There are numerous applications of radiochemistry. Knowledge of the half-life of C-14 allows estimation of the age of archeological specimens. Isotopic tracers are used to study metabolic pathways and other complex chemical systems. Radiation therapy is a powerful treatment for cancer; normally gamma and X rays are used for this purpose. Radioisotopes taken into the body can be used to diagnose organ and bone abnormalities.

The Einstein equation indicates that matter and energy are interchangeable. This relationship can be used to calculate **nuclear binding energy** from **mass defect.** In **nuclear fusion,** two light atomic nuclei are combined to produce a heavier nucleus of greater binding energy; the excess energy is released to the surroundings. Energy is also released by **nuclear fission,** the splitting of a heavy atomic nucleus into lighter nuclei. Nuclear power plants use the fission process. Continuous nuclear fusion occurs in the sun, and the process is the basis of the hydrogen bomb. Fusion power is one possible solution to the worldwide energy shortage.

STUDY QUESTIONS AND PROBLEMS

(More difficult questions and problems are marked with an asterisk.)

RADIOACTIVITY

1. Explain what is meant by *radioactivity* and *radioactive decay.*
2. How do radioisotopes attempt to achieve stability?
3. How does radioactivity bring about transmutation?
4. Explain the significance of the neutron-to-proton ratio in radioactivity.

ALPHA DECAY

5. What is an alpha particle? How does emission of an alpha particle help to stabilize a radioactive atom?
6. How does the law of conservation of mass apply to alpha decay?
7. Describe the changes in atomic number and mass number resulting from alpha decay.

BETA DECAY

8. What is a beta particle? Should emission of a beta particle help to stabilize a radioactive atom? Explain your answer.
9. How are beta particles created?
10. What is a positron? How does emission of a positron help to stabilize a radioactive atom?
11. What is meant by electron capture? How does the process help to stabilize a radioactive atom?
12. Summarize the processes involved in beta decay, indicating how each affects atomic number and mass number.

GAMMA DECAY

13. What are gamma rays? What is their origin?
14. How does gamma emission affect atomic number and mass number?
15. Name the three types of radioactive decay, and give the distinguishing features of each.

NUCLEAR REACTIONS

16. Write the complete equation for each nuclear reaction.
  a. Alpha decay by polonium-198
  b. Electron emission to form bismuth-210
  c. Formation of osmium-188 from platinum-192
  d. Alpha decay of polonium-205
17. Write the complete equation for each nuclear reaction.
  a. Electron emission by neptunium-238
  b. Electron capture by plutonium-237
  c. Electron emission by helium-6
  d. Formation of beryllium-8 from lithium-8

NATURAL AND ARTIFICIAL RADIOACTIVITY

18. Define the following terms:
  a. Cosmic rays
  b. Primary cosmic rays
  c. Secondary cosmic rays
  d. Radioactive disintegration series
  e. Artificial radioactivity
  f. Transuranium elements
  g. Transmutation
*19. In theory, each of the following transformations can be accomplished by bombardment of a single atom with a single particle. Write the complete equation for each transformation.
  a. Formation of $^{240}_{95}Am$ from $^{239}_{94}Pu$
  b. Formation of $^{242}_{96}Cm$ from $^{239}_{94}Pu$
  c. Formation of $^{243}_{97}Bk$ from $^{241}_{95}Am$
  d. Formation of $^{245}_{98}Cf$ from $^{242}_{96}Cm$
20. Irene and Frederick Joliot-Curie (Irene was the daughter of Marie Curie) first observed induced radioactivity in 1934 by bombarding $^{27}_{13}Al$ with alpha particles. The products from a single atom of $^{27}_{13}Al$ were a neutron and $^{30}_{15}P$, which subsequently decayed by positron emission. Write complete equations for this two-step process.

DETECTION AND MEASUREMENT OF NUCLEAR RADIATION

21. Define the following terms:
  a. Radiation
  b. Ionizing radiation
  c. X rays
  d. Geiger counter
  e. Scintillation counter
  f. Film badge
  g. Half-life
  h. Activity of a radioactive sample
  i. Dosage of nuclear radiation
  j. Energy level of radiation
22. Describe how X rays are generated by machines.
23. Explain how a Geiger counter works.
24. Explain how a scintillation counter works.
25. What information is conveyed by each of the following units for measuring nuclear radiation?
  a. Curie
  b. Roentgen
  c. Rad
  d. Rem
  e. Electron-volt

RADIATION SAFETY

26. What are free radicals? How are they harmful inside cells?
27. If a person receives 10 rems of nuclear radiation each year for 60 years, there might not be any noticeable effects. However, if the same person receives the same total of 600 rems in one dose, the exposure is likely to be fatal. Explain why the large single dose appears to be so much more harmful than the multiple smaller doses.
28. Both Marie Curie and her daughter Irene died of leukemia, probably caused by years of exposure to nuclear radiation. Explain how nuclear radiation might produce cancer. (Leukemia is a form of cancer characterized by proliferation of white blood cells.)
29. Give four ways of minimizing exposure for those who work with nuclear radiation.

APPLICATIONS OF RADIOCHEMISTRY

30. Explain the basis of archeological dating with C-14.
31. How did Melvin Calvin use radioisotope tracers to learn about photosynthesis?
32. Why does radiation sickness sometimes accompany radiation treatment for cancer?
33. Explain how radioisotopes are used in performing organ scans.

NUCLEAR ENERGY

34. Define the following terms:
    a. Mass defect
    b. Nuclear binding energy
    c. Average binding energy
    d. Nuclear fusion
    e. Nuclear fission
35. How are mass defect and nuclear binding energy related to the energy produced by nuclear processes?
36. Distinguish between nuclear fission and nuclear fusion.

NUCLEAR FISSION

37. Why does nuclear fission release energy?
38. What are some advantages and disadvantages of nuclear power.
39. Give two major differences between a nuclear reactor in a power plant and an atomic bomb.
40. What is a critical mass? What role does it play in nuclear chain reactions?
41. Describe two methods of assembling an atomic bomb.

NUCLEAR FUSION

42. Give a summary equation for the nuclear fusion processes that provide the sun's energy.
43. Distinguish between an atomic bomb and a hydrogen bomb.
44. What are the two major problems yet to be solved in developing nuclear fusion as a practical energy source?

GENERAL EXERCISES

45. What is the role of neutrons in an atomic nucleus?

46. How are positrons created?

47. Write the complete equation for each nuclear reaction.

    a. Positron emission and gamma decay by cobalt-56

    b. Decay of radium-223 to produce radon-219

    c. Formation of boron-10 by positron emission

    d. Formation of oxygen-18 by electron emission

    e. Formation of osmium-190 by electron capture

*48. Suggest how tritium (H-3) might be formed by cosmic rays. Write the complete equation to illustrate.

49. What percentage of original radioactive atoms will remain in a sample of radioisotope after a period of four half-lives?

50. In the early 1920s, a number of women were employed to paint coatings of a radium compound on watch dials to make the dials glow in the dark. The radium compound contained Ra-226, which emits weakly penetrating alpha particles and is thus considered safe to handle. The women had a habit of using their lips to put fine points on their paint brushes. Many of the women later died of cancer. Why do you think the incidence of cancer among the women was high?

51. It is ironic that nuclear radiation can be both a cause and a cure for cancer. Suggest an explanation for this seeming contradiction.

52. How can a nuclear reaction become a chain reaction?

INTRODUCTION TO ORGANIC CHEMISTRY

CHAPTER 18

OUTLINE

18.1 The Nature of Organic Chemistry
18.2 The Hydrocarbons
18.3 Alkanes
18.4 Naming Alkanes
     Perspective: Petroleum Refining
18.5 Alkenes and Alkynes
18.6 Naming Alkenes and Alkynes
18.7 Cyclic Aliphatic Hydrocarbons
18.8 Benzene and Aromatic
     Hydrocarbons
18.9 Classification of Organic
     Compounds by Functional
     Group
18.10 Orbital Hybridization in
     Hydrocarbon Molecules
     (optional)
     Bonding in Methane
     Bonding in Ethene
     Bonding in Benzene
     Bonding in Ethyne
     Summary
     Study Questions and Problems

Until the nineteenth century, it was thought that certain substances could only be produced by living organisms. These substances were thus referred to as *organic,* and it was believed that a mysterious "vital force" present only in living matter was responsible for their synthesis. However, the vital force theory was disproved in 1828 by a German chemist, Friedrich Wöhler. Wöhler accidentally prepared urea, an organic compound found in urine, by heating ammonium cyanate, an inorganic compound:

$$NH_4OCN \xrightarrow{\Delta} H_2N-\overset{\overset{\displaystyle O}{\|}}{C}-NH_2$$
$$\text{ammonium cyanate} \qquad\qquad \text{urea}$$

Since that time, organic chemistry has come to mean the study of most covalent carbon compounds. Inorganic chemistry, which includes all the preceding material in this book, was once the study of gases, rocks, and ores; it now includes all chemical compounds except most covalent carbon compounds. In this chapter we will concentrate on the binary organic compounds called hydrocarbons. In preparation, you should review shapes of molecules (chapter 6), covalent bonds (chapter 6) and multiple covalent bonds (chapter 6).

18.1   THE NATURE OF ORGANIC CHEMISTRY

Some covalent carbon compounds, such as carbonic acid (H_2CO_3) and its salts and hydrocyanic acid (HCN) and its salts, are traditionally classified as inorganic compounds.

Organic chemistry is the study of most covalent carbon compounds. Perhaps the main reason that organic and inorganic chemistry still exist as separate fields is the enormous number of covalent carbon compounds. Of the approximately 3 million known chemical compounds, about 2.7 million (90%) contain covalent carbon. Organic compounds are usually weak electrolytes or nonelectrolytes, with relatively low melting and boiling points. The burning process in which a substance reacts with oxygen to produce carbon dioxide, water, heat, and light, is characteristic of organic compounds.

The unique chemical character of carbon accounts for the multitude of organic compounds. Carbon, the first element in Group IVA, has neither a strong tendency to lose electrons nor a strong tendency to gain electrons. It prefers to share pairs of electrons and thus form covalent bonds. A carbon atom can form single, double, or triple bonds to other carbon atoms, or it can bond to atoms of other elements. Carbon atoms form single bonds to atoms of hydrogen, phosphorus, sulfur, and the halogens, and single or multiple bonds to atoms of nitrogen and oxygen. Because of its versatility in forming covalent bonds, carbon can bond to itself and to other nonmetallic elements to form a seemingly limitless number of compounds. One of the simplest types of organic compounds is the group known as the hydrocarbons.

18.2   THE HYDROCARBONS

Hydrocarbons are binary molecular compounds composed of hydrogen and carbon. In hydrocarbon molecules, carbon atoms form the framework by linking together in chains or rings. Hydrogen atoms are also attached to the carbon atoms, to give each carbon atom four bonds. The carbon atoms are arranged in unbranched chains, in branched chains, in rings, or in more complicated structures. Theoretically, the number of carbon atoms in a hydrocarbon molecule can be infinite, but, except for polymers, the largest hydrocarbon to be synthesized contains 110 carbon atoms.

Most hydrocarbons occur naturally as constituents of petroleum and natural gas. In addition, some hydrocarbons are found in trees and other plants. The major component of turpentine, which is obtained from pine trees, is a hydrocarbon called α-pinene, $C_{10}H_{16}$. Hydrocarbons are also present in insect waxes such as beeswax. Lycopene, $C_{40}H_{56}$, is the red pigment of tomatoes and watermelon, and the group of hydrocarbons called the carotenes give carrots their yellow-orange color. Crude plantation rubber is composed almost entirely of a hydrocarbon polymer whose molecules contain approximately 100,000 carbon atoms each. Because hydrocarbons have been so abundant in the past, we have grown dependent on them as fuel and as raw material for the manufacture of plastics, synthetic rubber, solvents, explosives, lubricants, alcohols, synthetic fabrics, and many other useful products.

Even though the hydrocarbons have very diverse structures, they have several physical properties in common. Most hydrocarbons are colorless, and as nonpolar molecules, they are insoluble in water but usually soluble in one another. They are gases, liquids, or solids at room temperature, depending on formula weight. Hydrocarbons with less than 5 carbon atoms are gases at room temperature, while those with 6–18 carbon atoms are liquids at room temperature. The liquid hydrocarbons are less dense than water, and they float on the surface of water. Hydrocarbons with more than 18 carbon atoms are solids at room temperature.

Hydrocarbons are classified as aliphatic or aromatic. The term **aliphatic** comes from the Greek word for *fat*, and it was used originally to designate hydrocarbons that could be derived from animal fat. These hydrocarbon molecules are usually unbranched or branched chains, but *aliphatic* now means all hydrocarbons that are not aromatic. The word **aromatic** was first used to refer to the pleasant aroma of certain hydrocarbons, but it is now used for hydrocarbons that have special ring structures containing multiple covalent bonds. When we study aromatic hydrocarbons later in this chapter, the distinction between aliphatic and aromatic hydrocarbons will become clearer. But first, let's focus on the aliphatic hydrocarbons. These hydrocarbons are divided into three major subclasses: alkanes, alkenes, and alkynes. We will start with the alkanes.

18.3   ALKANES

Alkanes are hydrocarbons that contain only single bonds. Each carbon atom is bonded to four other atoms; this arrangement corresponds to maximum bonding to other atoms, and the carbon atoms are said to be saturated. Thus, alkanes are known as **saturated hydrocarbons.** The car-

TABLE 18-1   The First Ten Linear Alkanes

Number of Carbon Atoms	Structure	Name
1	CH_4	methane
2	CH_3CH_3	ethane
3	$CH_3CH_2CH_3$	propane
4	$CH_3CH_2CH_2CH_3$	butane
5	$CH_3CH_2CH_2CH_2CH_3$	pentane
6	$CH_3CH_2CH_2CH_2CH_2CH_3$	hexane
7	$CH_3CH_2CH_2CH_2CH_2CH_2CH_3$	heptane
8	$CH_3CH_2CH_2CH_2CH_2CH_2CH_2CH_3$	octane
9	$CH_3CH_2CH_2CH_2CH_2CH_2CH_2CH_2CH_3$	nonane
10	$CH_3CH_2CH_2CH_2CH_2CH_2CH_2CH_2CH_2CH_3$	decane

bon atoms can be bonded together in an unbranched chain, a branched chain, or even a ring.

All open-chain (noncyclic) alkanes have the general formula C_nH_{2n+2}, where n is the number of carbon atoms in the molecule. The alkanes whose carbon atoms are joined in unbranched chains are referred to as linear alkanes. Table 18-1 gives the first ten linear alkanes. Notice that as the number of carbon atoms increases from one alkane to the next, the structures differ only by a —CH_2— group. A series of compounds in which each member differs from the preceding member only by a single, constant group is called a **homologous series.**

The simplest alkane is methane, CH_4. We can use ball-and-stick models to represent the three-dimensional structure of methane, as illustrated in Figure 18-1. The hydrogen atoms of methane are considered to occupy the corners of a regular tetrahedron, with angles of 109.5° between them.

Ethane (C_2H_6), propane (C_3H_8), and butane (C_4H_{10}) are also represented by ball-and-stick models in Figure 18-1. In these compounds also, the

Homologous

Of the same chemical type, but differing by a fixed increment in certain constituents.

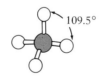

Methane, CH_4

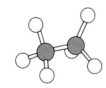

Ethane, C_2H_6

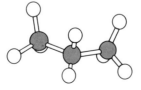

Propane, C_3H_8

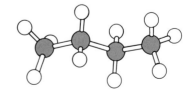

Butane, C_4H_{10}

FIGURE 18-1   Ball-and-Stick Models of Methane, Ethane, Propane, and Butane

bonds to carbon are separated by angles of 109.5°. In general, when there are four single bonds to a carbon atom, the bond angles will be 109.5°.

The alkanes having no branches are called linear alkanes. This term is a bit misleading, because unbranched alkanes do not actually have a linear structure. Instead, the carbon atoms occupy positions in space corresponding to the corners of a zigzag line. We usually show linear alkanes as having extended structures, like the structure of butane in Figure 18-1. However, groups of atoms rotate freely about single bonds, so linear alkanes having four or more carbon atoms are quite flexible. They twist and turn continuously in space, and at one instant, a linear alkane might have quite a different shape than at another instant.

Although models are helpful in visualizing the three-dimensional aspects of alkane structure, we must use formulas to represent alkanes on paper. Three kinds of formulas are commonly used. The **molecular formula** shows the number and kinds of atoms present in a molecule, such as

$$\text{Ethane} \quad C_2H_6$$
$$\text{Propane} \quad C_3H_8$$
$$\text{Butane} \quad C_4H_{10}$$

The **condensed structural formula** gives major structural features of a molecule:

$$\text{Ethane} \quad CH_3CH_3 \text{ or } CH_3—CH_3$$
$$\text{Propane} \quad CH_3CH_2CH_3 \text{ or } CH_3—CH_2—CH_3$$
$$\text{Butane} \quad CH_3CH_2CH_2CH_3 \text{ or } CH_3—CH_2—CH_2—CH_3$$

The **structural formula** provides as many structural features as possible in two dimensions:

$$\text{Ethane} \quad \begin{array}{c} \text{H} \quad \text{H} \\ | \quad | \\ \text{H}—\text{C}—\text{C}—\text{H} \\ | \quad | \\ \text{H} \quad \text{H} \end{array}$$

$$\text{Propane} \quad \begin{array}{c} \text{H} \quad \text{H} \quad \text{H} \\ | \quad | \quad | \\ \text{H}—\text{C}—\text{C}—\text{C}—\text{H} \\ | \quad | \quad | \\ \text{H} \quad \text{H} \quad \text{H} \end{array}$$

$$\text{Butane} \quad \begin{array}{c} \text{H} \quad \text{H} \quad \text{H} \quad \text{H} \\ | \quad | \quad | \quad | \\ \text{H}—\text{C}—\text{C}—\text{C}—\text{C}—\text{H} \\ | \quad | \quad | \quad | \\ \text{H} \quad \text{H} \quad \text{H} \quad \text{H} \end{array}$$

In an alkane containing four or more carbon atoms, the atoms can be arranged in more than one way, yet all of the arrangements give the same molecular formula. Compounds having the same molecular formula but different arrangements of atoms are called **isomers.** Butane is the simplest alkane to have isomers. (Methane, ethane, and propane molecules do not contain enough carbon atoms to allow a different arrangement of

Isomer

From the Greek words *iso*, meaning "equal," and *mer*, meaning "parts."

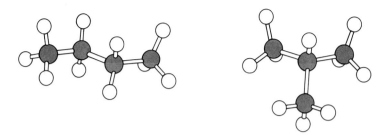

FIGURE 18-2   Ball-and-Stick Models of
the Isomers of Butane, C_4H_{10}

their atoms.) There are two isomers of butane:

$$CH_3-CH_2-CH_2-CH_3 \qquad CH_3-CH-CH_3$$
$$\qquad\qquad\qquad\qquad\qquad\qquad\qquad | $$
$$\qquad\qquad\qquad\qquad\qquad\qquad\quad CH_3$$

linear isomer   branched isomer

boiling point $0°$ C   boiling point $-12°$ C

Each is an individual compound with its own set of physical properties. Ball-and-stick models of the butane isomers are shown in Figure 18-2.

The larger the number of carbon atoms in an alkane molecule, the more isomers the alkane will have. There are two isomers of butane, three of pentane, and nine of heptane. Table 18-2 gives the structures of the isomers of the first seven open-chain alkanes.

Most alkanes are isolated from petroleum. They are considered highly desirable energy sources because of the heat produced in combustion. Methane is the main component of natural gas, which is used for cooking and heating. Propane, because of its higher boiling point, can easily be liquefied by moderate pressure, and it is sold as liquefied propane gas (LPG) for use in camping and in homes not served by natural gas. Butane is also easily liquefied, and it is the fuel of butane cigarette lighters. Inhalation of any of these gases in large amounts can produce sleepiness, unconsciousness, and even death. Since they are colorless and odorless, it is difficult to detect their presence. For this reason, strong-smelling sulfur compounds are added to natural gas, so that leaks can be detected.

Alkanes with 5–12 carbon atoms are liquids at room temperature; gasoline for automobiles is a blend of these alkanes. The alkanes having 12–18 carbon atoms are also liquids; they are used for diesel fuel, jet fuel, and heating oil. Most alkanes with 19 or more carbon atoms are solids at room temperature and are used as lubricants and greases. They are often present in skin and hair lotions as replacements for natural oils; petroleum jelly is an example of one formulation used as a skin softener or protective coating. Paraffin, the wax sealant for homemade jams and jellies, is a mixture of solid alkanes.

18.4   NAMING ALKANES

Through the years, alkanes and other organic compounds have acquired numerous common names. Since common names can be ambiguous as well as difficult to remember, a nomenclature system has been devised by

TABLE 18-2 Isomers of the First Seven Open-Chain Alkanes

Number of Carbon Atoms	Isomeric Structures	Number of Isomers	Boiling Point (°C)
1	CH_4	1	-162
2	CH_3-CH_3	1	-89
3	$CH_3-CH_2-CH_3$	1	-42
4	$CH_3-CH_2-CH_2-CH_3$	2	0
	$CH_3-\underset{\underset{CH_3}{\displaystyle\vert}}{CH}-CH_3$		-12
5	$CH_3-CH_2-CH_2-CH_2-CH_3$	3	36
	$CH_3-\underset{\underset{CH_3}{\displaystyle\vert}}{CH}-CH_2-CH_3$		28
	$CH_3-\underset{\underset{CH_3}{\displaystyle\vert}}{\overset{\overset{CH_3}{\displaystyle\vert}}{C}}-CH_3$		10
6	$CH_3-CH_2-CH_2-CH_2-CH_2-CH_3$	5	69
	$CH_3-\underset{\underset{CH_3}{\displaystyle\vert}}{CH}-CH_2-CH_2-CH_3$		60
	$CH_3-CH_2-\underset{\underset{CH_3}{\displaystyle\vert}}{CH}-CH_2-CH_3$		63
	$CH_3-\underset{\underset{CH_3}{\displaystyle\vert}}{CH}-\underset{\underset{CH_3}{\displaystyle\vert}}{CH}-CH_3$		58
	$CH_3-\underset{\underset{CH_3}{\displaystyle\vert}}{\overset{\overset{CH_3}{\displaystyle\vert}}{C}}-CH_2-CH_3$		50
7	$CH_3-CH_2-CH_2-CH_2-CH_2-CH_2-CH_3$	9	98
	$CH_3-\underset{\underset{CH_3}{\displaystyle\vert}}{CH}-CH_2-CH_2-CH_2-CH_3$		90
	$CH_3-CH_2-\underset{\underset{CH_3}{\displaystyle\vert}}{CH}-CH_2-CH_2-CH_3$		92
	$CH_3-\underset{\underset{CH_3}{\displaystyle\vert}}{CH}-\underset{\underset{CH_3}{\displaystyle\vert}}{CH}-CH_2-CH_3$		90
	$CH_3-\underset{\underset{CH_3}{\displaystyle\vert}}{CH}-CH_2-\underset{\underset{CH_3}{\displaystyle\vert}}{CH}-CH_3$		80
	$CH_3-\underset{\underset{CH_3}{\displaystyle\vert}}{\overset{\overset{CH_3}{\displaystyle\vert}}{C}}-CH_2-CH_2-CH_3$		79
	$CH_3-CH_2-\underset{\underset{CH_3}{\displaystyle\vert}}{\overset{\overset{CH_3}{\displaystyle\vert}}{C}}-CH_2-CH_3$		86
	$CH_3-\underset{\underset{CH_3}{\displaystyle\vert}}{\overset{\overset{CH_3}{\displaystyle\vert}}{C}}-\underset{\underset{CH_3}{\displaystyle\vert}}{CH}-CH_3$		81
	$CH_3-CH_2-\underset{\underset{CH_2-CH_3}{\displaystyle\vert}}{CH}-CH_2-CH_3$		93

the International Union of Pure and Applied Chemistry (IUPAC). This system uses the following rules for naming open-chain alkanes.

1. The names of all alkanes end in -*ane*.
2. The name is based on the longest continuous chain of carbon atoms; this *main chain* is considered to be the parent compound of the alkane. Table 18-1 lists the first ten parent compounds. If there are no branches in the alkane molecule, it is simply named by using Table 18-1.
3. If there are branches in the alkane molecule, a number is assigned to each carbon atom in the longest continuous chain, beginning at the end of the chain nearest a branch. In the example below, the longest continuous chain has seven carbon atoms. Do not be misled by the bend in the main chain; remember, there is free rotation about single bonds, and thus the main chain can have many different shapes in three-dimensions.

$$
\begin{array}{c}
\overset{1}{C}H_3 - \overset{2}{C}H - \overset{3}{C}H_2 - \overset{4}{C}H - CH_2 - CH_3 \\
\quad\quad | \quad\quad\quad\quad\quad | \\
\quad\quad CH_3 \quad\quad\quad \overset{5}{C}H_2 \\
\quad\quad\quad\quad\quad\quad\quad\quad | \\
\quad\quad\quad\quad\quad\quad\quad\quad \overset{6}{C}H_2 \\
\quad\quad\quad\quad\quad\quad\quad\quad | \\
\quad\quad\quad\quad\quad\quad\quad\quad \overset{7}{C}H_3
\end{array}
$$

4. Saturated hydrocarbon substituents, or branches, are called **alkyl groups.** The name of an alkyl group is derived from the corresponding alkane by changing the -*ane* ending of the alkane to -*yl*. For example,

$-CH_3$ is *methyl* (derived from *methane*)

$-CH_2CH_3$ is *ethyl* (derived from *ethane*)

$-CH_2CH_2CH_3$ is *propyl* (derived from *propane*)

5. The names of any alkyl groups attached to the main chain are written before the name of the parent compound, in alphabetical order. The Greek prefixes *di-, tri-, tetra-, penta-,* and so on, are used to indicate two or more identical alkyl group attachments. The position of each substituent on the main chain is designated by the numbers assigned in rule 3. Commas separate numbers, and hyphens separate numbers from words. The name of the compound is written as one word, with the name of the main chain appearing last in the word.

$$
\begin{array}{c}
\overset{1}{C}H_3 - \overset{2}{C}H - \overset{3}{C}H_2 - \overset{4}{C}H - CH_2 - CH_3 \\
\quad\quad | \quad\quad\quad\quad\quad | \\
\quad\quad CH_3 \quad\quad\quad \overset{5}{C}H_2 \\
\quad\quad\quad\quad\quad\quad\quad\quad | \\
\quad\quad\quad\quad\quad\quad\quad\quad \overset{6}{C}H_2 \\
\quad\quad\quad\quad\quad\quad\quad\quad | \\
\quad\quad\quad\quad\quad\quad\quad\quad \overset{7}{C}H_3
\end{array}
$$

4-ethyl-2-methylheptane

$$CH_3 - \underset{\underset{|}{\overset{|}{\underset{2}{C}}}}{\overset{\overset{CH_3}{|}}{}} - \underset{3}{CH_2} - \underset{4}{CH_2} - \underset{5}{\overset{|}{CH}} - CH_2 - CH_3$$

CH₃ ... ⁶CH₂ ... ⁷CH₂ ... ⁸CH₃

2,2-dimethyl-5-ethyloctane

The following examples will give you some practice in naming alkanes.

EXAMPLE 18.1 Derive the IUPAC name of the following alkane.

$$CH_3 - \underset{\underset{CH_3}{|}}{CH} - CH_3$$

Solution First, locate the main chain. In this case, it has three carbon atoms; thus, the parent compound is propane.

$$CH_3 - \underset{\underset{CH_3}{|}}{CH} - CH_3$$

Next, number the carbon atoms in the main chain. In this case, we can start at either end, since each is the same distance from the branch.

$$\underset{1}{CH_3} - \underset{2}{\underset{\underset{CH_3}{|}}{CH}} - \underset{3}{CH_3}$$

There is only one substituent, and its name is methyl. However, the only possible location for the methyl group in this compound is on carbon 2, and although it is tempting to name the compound *2-methylpropane*, this is one of the few times that a number is not needed in the name. Thus the compound's name is simply *methylpropane*.

EXERCISE 18.1 Derive the IUPAC name for the following alkane.

$$CH_3 - CH_2 - \underset{\underset{\underset{CH_3}{|}}{\underset{CH_2}{|}}}{CH} - CH_3$$

EXAMPLE 18.2 Derive the IUPAC name of the following alkane.

$$CH_3 - \underset{\overset{|}{CH_3}}{\overset{\overset{CH_3}{|}}{CH}} - \underset{\overset{|}{CH_3}}{\overset{\overset{CH_3}{|}}{CH}} - \underset{\underset{\underset{CH_3}{|}}{\underset{CH_2}{|}}}{CH} - CH_3$$

PERSPECTIVE   PETROLEUM REFINING

Petroleum is crude oil found in natural underground deposits. It is a complicated mixture of alkanes, small amounts of other hydrocarbons, and compounds of sulfur, nitrogen, and oxygen. Petroleum deposits are the result of spontaneous degradation of plant and animal remains that have accumulated through the ages. Converting the complex mixture of chemicals in petroleum to separate, useful products is the process called *petroleum refining*.

Fractional distillation is the primary refinery process. In fractional distillation, crude oil is heated in a distilling tower containing several openings along the side (sidestreams) where vapors are withdrawn. The lower formula weight compounds have relatively low boiling points and come off near the top of the column. Higher formula weight compounds have relatively high boiling points and are taken off from the lower openings. In this way, the petroleum is separated into fractions of molecules grouped by boiling range and thus roughly by formula weight. The remaining solid residue, called asphalt, is composed of high formula weight hydrocarbons that were not volatilized under the operating conditions.

A Petroleum Refinery

Solution   First, locate the main chain. The main chain is bent, but the bend is not intended to convey any information about the molecule's three-dimensional structure. The main chain has six carbon atoms, and thus its name is hexane.

$$
\begin{array}{c}
\qquad\quad\; CH_3 \quad CH_3 \\
\qquad\quad\;\; | \qquad\;\; | \\
CH_3-CH-CH-CH-CH_3 \\
\qquad\qquad\qquad\quad | \\
\qquad\qquad\qquad\; CH_2 \\
\qquad\qquad\qquad\quad | \\
\qquad\qquad\qquad\; CH_3
\end{array}
$$

Next, number the carbon atoms in the main chain, starting at the end nearest a branch.

$$
\begin{array}{c}
\qquad\quad\; CH_3 \quad CH_3 \\
\qquad\quad\;\; | \qquad\;\; | \\
\underset{1}{CH_3}-\underset{2}{CH}-\underset{3}{CH}-\underset{4}{CH}-CH_3 \\
\qquad\qquad\qquad\quad | \\
\qquad\qquad\qquad\; \underset{}{^{5}CH_2} \\
\qquad\qquad\qquad\quad | \\
\qquad\qquad\qquad\; ^{6}CH_3
\end{array}
$$

All three substituents are —CH_3, or methyl groups. Thus, the name is 2,3,4-trimethylhexane. Note that all of the following compounds are also 2,3,4-trimethylhexane. They are just drawn differently.

$$
\begin{array}{cccc}
 & CH_3 & CH_3 \\
 & | & | \\
CH_3{-}CH{-} & CH{-}CH{-} & CH_2{-}CH_3 \\
 & | \\
 & CH_3
\end{array}
\qquad
\begin{array}{ccc}
CH_3 & CH_3 & CH_3 \\
| & | & | \\
CH_3{-}CH{-} & CH{-} & CH \\
 & & | \\
 & & CH_2 \\
 & & | \\
 & & CH_3
\end{array}
$$

$$
\begin{array}{cc}
CH_3 & CH_3 \\
| & | \\
CH{-}CH{-}CH{-} & CH_2{-}CH_3 \\
| & | \\
CH_3 & CH_3
\end{array}
$$

EXERCISE 18.2   Derive the IUPAC name of the following alkane.

$$
\begin{array}{ccccc}
CH_3{-}CH{-}CH_2{-}CH{-}CH_2{-}CH_3 \\
\quad\quad | \quad\quad\quad\quad | \\
\quad\quad CH_2 \quad\quad\quad CH_3 \\
\quad\quad | \\
\quad\quad CH_3
\end{array}
$$

18.5   ALKENES AND ALKYNES

Alkyne
(al-KINE)

Unsaturated hydrocarbons are hydrocarbons whose molecules have one or more multiple bonds. There are two types of unsaturated aliphatic hydrocarbons: alkenes and alkynes. Molecules of the **alkenes** have one or more double bonds between carbon atoms. Because of the double bonds, an alkene molecule does not contain as many hydrogen atoms as an alkane molecule with the same number of carbon atoms, so we say it is *unsaturated*. The carbon atoms in an alkene molecule, like those in an alkane, can be bonded together consecutively in an unbranched chain, a branched chain, or a ring.

The open-chain alkenes with one double bond have the general formula C_nH_{2n}, where n is the number of carbon atoms in the molecule. Thus, the alkenes, like the alkanes, form a homologous series of compounds. Table 18-3 lists representative alkenes with their names and formulas.

The simplest alkene is ethene; Figure 18-3 shows three-dimensional features of the ethene molecule. All of the atoms lie in the same plane. In fact, in any molecule containing a double bond, the atoms connected by the double bond and the atoms bonded to them occupy the same plane. Groups connected by a double bond cannot rotate about the double bond. The bond angles around the carbon atoms connected by a double bond are 120°.

Like the alkanes, the alkenes can have isomers. However, the presence of a double bond allows alkenes to display an additional type of isomerism called *geometric isomerism*. **Geometric isomers** are isomers of alkenes in

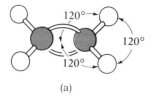

(a)

(b)

FIGURE 18-3 The Ethene Molecule, C_2H_4. (a) Ball-and-Stick Model (b) Space-Filling Model

TABLE 18-3 Examples of Alkenes		
Structure	Molecular Formula	Name
$CH_2\!=\!CH_2$	C_2H_4	ethene (ethylene)
$CH_3CH\!=\!CH_2$	C_3H_6	propene (propylene)
$CH_3CH_2CH\!=\!CH_2$	C_4H_8	1-butene (butylene)
$CH_3CH_2CH_2CH\!=\!CH_2$	C_5H_{10}	1-pentene
$CH_3CH_2CH_2CH_2CH\!=\!CH_2$	C_6H_{12}	1-hexene
$CH_3\!-\!\overset{\displaystyle |}{\underset{\displaystyle CH_3}{C}}\!=\!CH_2$	C_4H_8	2-methylpropene
$CH_3\!-\!\overset{\displaystyle |}{\underset{\displaystyle CH_3}{CH}}\!-\!CH\!=\!CH_2$	C_5H_{10}	3-methyl-1-butene
$CH_3CH_2\!-\!\overset{\displaystyle |}{\underset{\displaystyle CH_3}{C}}\!=\!CH_2$	C_5H_{10}	2-methyl-1-butene

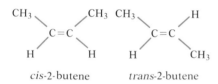

cis-2-butene   trans-2-butene

FIGURE 18-4 The Geometric Isomers of Butene

which identical substituents on each of the carbon atoms connected by a double bond can be on the same side of the double bond or on opposite sides of the double bond. If the identical substituents are on the same side of the double bond, the isomer is designated as *cis*. If the identical substituents are on opposite sides of the double bond, the isomer is designated as *trans*. Butene is the simplest alkene to have geometric isomers, as shown in Figure 18-4. The isomers of butene are thus *cis*-2-butene and *trans*-2-butene; they are different compounds, each having its own set of physical properties. Only those alkenes having two different groups on each of the double-bonded carbon atoms can exist as geometric isomers. Thus, alkene molecules having double bonds at the end of the carbon chain, including ethene and propene, do not have geometric isomers.

The low formula weight alkenes (1–4 carbon atoms) are produced commercially by removal of hydrogen from alkanes in natural gas or petroleum. Ethene is produced in greatest abundance and is considered to be the most important alkene. It is used to make polyethylene, antifreeze (ethylene glycol), ethyl alcohol, and polystyrene. Propene and the butenes

TABLE 18-4 Examples of Alkynes		
Structure	Molecular Formula	Name
$H\!-\!C\!\equiv\!C\!-\!H$	C_2H_2	ethyne
$CH_3\!-\!C\!\equiv\!C\!-\!H$	C_3H_4	propyne
$CH_3CH_2\!-\!C\!\equiv\!C\!-\!H$	C_4H_6	1-butyne
$CH_3CH_2CH_2\!-\!C\!\equiv\!C\!-\!H$	C_5H_8	1-pentyne
$CH_3CH_2CH_2CH_2\!-\!C\!\equiv\!C\!-\!H$	C_6H_{10}	1-hexyne
$CH_3\overset{\displaystyle |}{\underset{\displaystyle CH_3}{CH}}\!-\!C\!\equiv\!C\!-\!H$	C_5H_8	3-methyl-1-butyne
$CH_3CH_2\overset{\displaystyle |}{\underset{\displaystyle CH_3}{CH}}\!-\!C\!\equiv\!C\!-\!H$	C_6H_{10}	3-methyl-1-pentyne

are also manufactured in large quantity, and they serve as raw materials for making other compounds that are used as solvents or in the manufacture of plastics, detergents, synthetic rubber, food preservatives, and many other products.

The **alkynes** are aliphatic hydrocarbons having one or more triple bonds. Because the alkynes contain even less hydrogen per carbon atom than alkenes, the alkynes are more unsaturated (or less saturated) than the alkenes. As with alkanes and alkenes, the carbon atoms of alkyne molecules can join together to make unbranched chains, branched chains, or rings. The open-chain alkynes with one triple bond have the general formula C_nH_{2n-2}. Table 18-4 gives examples of alkynes.

The simplest alkyne is ethyne, commonly known as acetylene. The three-dimensional structure of ethyne is shown in Figure 18-5. In ethyne, all atoms lie in a straight line; thus, we say that ethyne has linear geometry. The bond angles of ethyne are 180°. For all alkynes, the two carbon atoms connected by the triple bond and the atoms attached to them lie in a straight line.

Ethyne is the only alkyne produced commercially in large amounts. It is derived from soft (bituminous) coal and from the high-temperature conversion of methane to ethyne:

$$2\ CH_4 \xrightarrow{1600°\ C} H-C{\equiv}C-H + 3\ H_2$$

About half of the ethyne produced is used in oxyacetylene torches; the remainder is used as raw material for synthesizing other organic compounds.

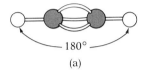

FIGURE 18-5 The Ethyne Molecule, C_2H_2. (a) Ball-and-Stick Model (b) Space-Filling Model

18.6 NAMING ALKENES AND ALKYNES

For reasons lost in history, ethene was once known as olefiant gas, and the alkenes were called *olefins*. This term is still used as an alternative name for the alkenes. In addition, several of the simple alkenes have common names ending in *-ylene;* for example, the common name of ethene is ethylene. However, because of ambiguities in common names, the following IUPAC rules should be used to derive systematic names for open-chain alkenes.

1. The names of all alkenes end in *-ene.*
2. Alkene names are based on the longest continuous chain of carbon atoms containing the double bond. This main chain is considered the parent compound of the alkene. The main chain is named by selecting the alkane with the same number of carbon atoms in a chain and changing the *-ane* ending of the alkane name to *-ene.*
3. The carbon atoms in the main chain are numbered, starting at the end nearest the double bond. This gives the first carbon of the double bond the lowest possible number. (This is the carbon atom used to locate the bond. Note that numbers are not needed to locate the double bond in

ethene and propene.) For example, the alkene below is named 1-butene, not 3-butene or 4-butene.

$$CH_3—CH_2—CH=CH_2$$

4. Groups attached to the main chain are identified individually by numbers and names, as with alkanes.

5. If the alkene contains more than one double bond, the location of each is identified by a number. When there are two double bonds, the name of the parent compound ends in *-diene* instead of *-ene;* when there are three double bonds, the name of the parent compound ends in *-triene,* and so on.

$$CH_2=CH—CH=CH_2 \qquad CH_3—CH=CH—CH=CH—CH=CH_2$$
1,3-butadiene   1,3,5-heptatriene

$$\begin{array}{c} CH_2=C—CH=CH_2 \\ | \\ CH_3 \end{array} \qquad \begin{array}{c} CH_3—C=CH—CH=C—CH=CH_2 \\ | \qquad\qquad\qquad | \\ CH_3 \qquad\qquad\quad CH_3 \end{array}$$
2-methyl-1,3-butadiene 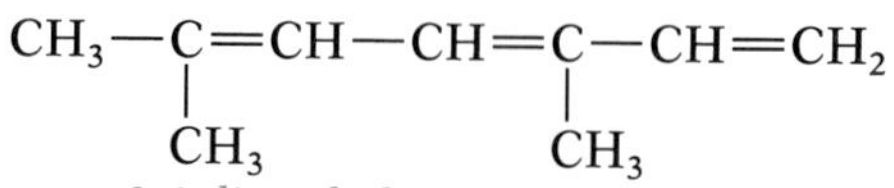   3,6-dimethyl-1,3,5-heptatriene

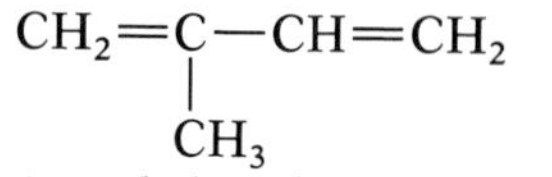

EXAMPLE 18.3   Derive the IUPAC name of the following alkene.

$$CH_3—CH=CH—CH_2—CH_3$$

Solution   This alkene is linear, with five carbon atoms and one double bond; hence, its name must end in *-pentene.* We number the carbon atoms starting at the end nearest the double bond,

$$\underset{1}{C}H_3—\underset{2}{C}H=\underset{3}{C}H—\underset{4}{C}H_2—\underset{5}{C}H_3$$

Thus, the correct name is 2-pentene.

EXERCISE 18.3   Derive the IUPAC name of the following alkene.

$$CH_3—CH=CH—CH_3$$ ∎

EXAMPLE 18.4   Derive the IUPAC name of the following alkene.

$$\begin{array}{c} CH_3—CH_2—CH_2—CH—CH_3 \\ | \\ CH \\ \| \\ CH_2 \end{array}$$

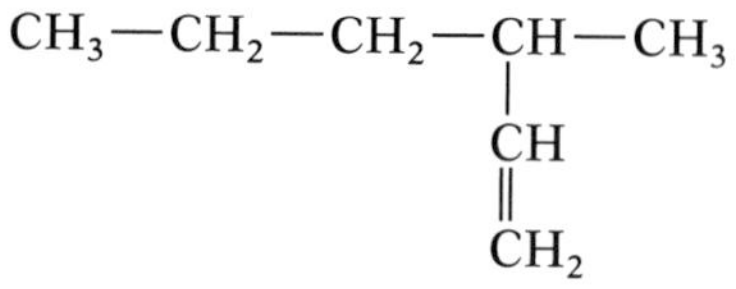

Solution   We first select the main chain and number it:

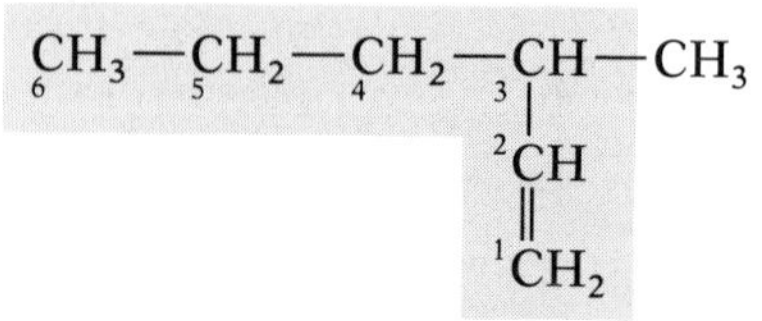

$$\begin{array}{c} \underset{6}{C}H_3—\underset{5}{C}H_2—\underset{4}{C}H_2—\underset{3}{C}H—CH_3 \\ | \\ \underset{2}{C}H \\ \| \\ \underset{1}{C}H_2 \end{array}$$

Since the only substituent is a methyl group at carbon 3, the correct name is 3-methyl-1-hexene.

EXERCISE 18.4 Derive the IUPAC name of the following alkene.

$$CH_3-CH-CH=CH-CH-CH_3$$
$$\qquad\quad\;| \qquad\qquad\quad\; |$$
$$\qquad\quad CH_3 \qquad\qquad CH_2$$
$$\qquad\qquad\qquad\qquad\qquad\; |$$
$$\qquad\qquad\qquad\qquad\qquad CH_3$$

IUPAC names of open-chain alkynes are derived much the same way as for open-chain alkenes, except that the name ending for alkynes is *-yne*. Alkyne names are based on the longest continuous chain containing the triple bond. The chain is numbered so that the first carbon atom of the triple bond has the lowest possible number.

EXAMPLE 18.5 Derive the IUPAC name of the following alkyne.

$$CH_3-CH_2-C\equiv C-H$$

Solution We number the main chain as follows,

$$\underset{4}{CH_3}-\underset{3}{CH_2}-\underset{2}{C}\equiv\underset{1}{C}-H$$

and thus the correct name is 1-butyne.

EXERCISE 18.5 Derive the IUPAC name of the following alkyne.

$$CH_3-CH_2-CH-C\equiv C-H$$
$$\qquad\qquad\qquad | $$
$$\qquad\qquad\quad CH_3$$

18.7 CYCLIC ALIPHATIC HYDROCARBONS

As we have mentioned, the carbon atoms of aliphatic hydrocarbons can be joined together in rings; some examples are shown in Figure 18-6. Such compounds are named by adding the prefix *cyclo-* to the name of the corresponding open-chain alkane, alkene, or alkyne. Cyclic molecules are conveniently represented by geometric figures, as shown below,

cyclopentane   cyclohexane   cyclohexene

where each line stands for a C—C bond and each corner of the figure stands for a carbon atom with the appropriate number of hydrogen atoms to satisfy carbon's valence of 4.

Cyclic aliphatic hydrocarbons are colorless compounds obtained from petroleum and by chemical synthesis. These compounds have physical properties similar to those of open-chain aliphatic hydrocarbons, and they have many of the same uses. Cyclopropane was once a popular general anesthetic, due to its fast action, but because of its flammability, its use has been discontinued as nonflammable replacements have been developed.

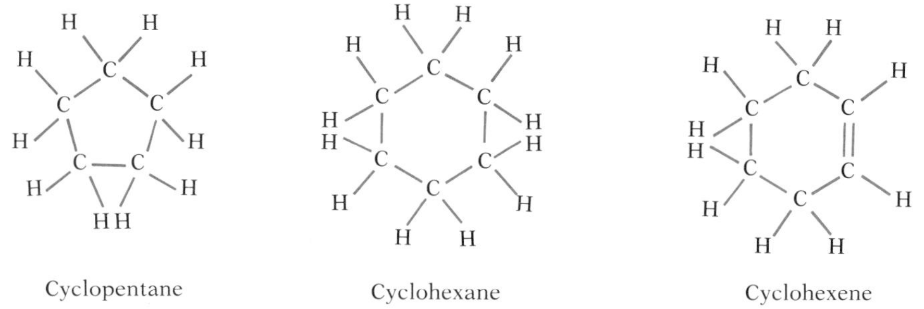

FIGURE 18-6 Examples of Cyclic Aliphatic Hydrocarbon Molecules

Cyclopentane   Cyclohexane   Cyclohexene

18.8 BENZENE AND AROMATIC HYDROCARBONS

During the middle of the nineteenth century, chemists recognized a group of hydrocarbons with distinctly different chemical properties than the aliphatic hydrocarbons. These hydrocarbons have a high degree of unsaturation, yet they do not show the chemical behavior expected of alkenes or alkynes. Many of their derivatives have pleasant aromas, and some are responsible for the odors of cloves, cinnamon, anise, wintergreen, and other naturally occurring substances. Because of the pleasant odors of many compounds of this group, they became known as aromatic compounds. It was noticed in 1861 that many aromatic hydrocarbons have formulas suggesting that they are derived from the aromatic hydrocarbon known as benzene, C_6H_6. Since then, the word **aromatic** has come to mean any compound that is structurally related to benzene. In modern times, benzene has come to be regarded as the model for all aromatic hydrocarbons.

Friedrich August Kekulé, a German chemist, made the first sound suggestion about the structure of benzene in 1865. He proposed that benzene has two structures that, for reasons he did not understand, could not be isolated from each other. Kekulé's structures were

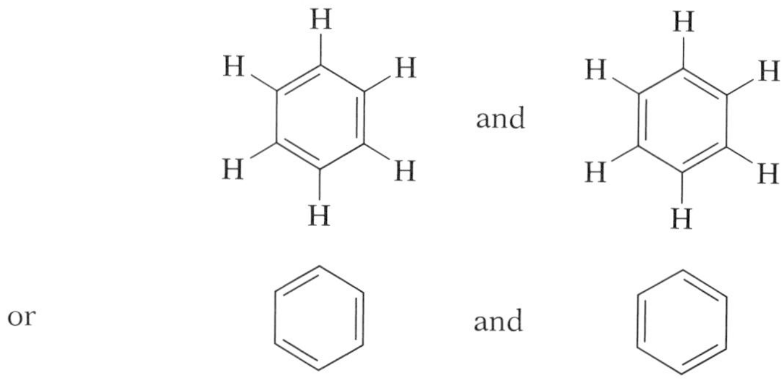

Kekulé
(KEH-ku-lay)

Notice that they differ from each other only in the placement of double bonds, which is essentially the placement of electrons.

We now recognize that the Kekulé structures are simply two alternative representations of the true structure of the benzene molecule. Each of the double bonds shown in Kekulé's structures is not actually localized between two carbon atoms; instead, the three double bonds are spread over the entire ring. This means that the attractive force between each two carbon atoms corresponds to 1½ covalent bonds. In fact, the benzene molecule is actually a *hybrid* of the two Kekulé structures. The double bonds shown to be in alternating positions in the Kekulé structures are unusual in their ability to spread over all the carbon atoms in the ring. This feature is the unique characteristic of all aromatic hydrocarbons. Modern chemistry defines an **aromatic hydrocarbon** as a cyclic hydrocarbon having one or more double bonds spread over all the carbon atoms in the ring. The structures of aromatic molecules that show the alternative placement of electrons are called **resonance structures.** To indicate bond delocalization in benzene, we often draw the benzene molecule as a hexagon with a circle inside, as shown below.

In general, aromatic hydrocarbons are unusually stable and therefore much less reactive than would be expected. There are no true double bonds in the molecules, and thus aromatic hydrocarbons do not resemble the alkenes in chemical behavior. Like molecules of all aromatic hydrocarbons, the benzene molecule is completely flat, with all twelve atoms lying in the same plane. It is a colorless liquid, and like other hydrocarbons, it is insoluble in water but soluble in nonpolar liquids. Benzene is obtained from the gases released when coal is heated in the absence of air. It is of major importance in the manufacture of plastics, dyes, detergents, and insecticides.

A compound made (derived) from another compound is referred to as a **derivative** of the parent compound. Benzene derivatives are formed when one or more hydrogen atoms on the ring are replaced by other atoms or groups. Examples of benzene derivatives are given in Table 18-5.

Most aromatic hydrocarbons have more complex structures than benzene. For example, the **polycyclic aromatic hydrocarbons** consist of molecules containing two or more aromatic rings fused together. The simplest of these is naphthalene; it is a white solid at room temperature

TABLE 18-5	Examples of Benzene Derivatives (common names are given in parentheses)
Structure	Name
⬡—CH₃	methylbenzene (toluene)
⬡—OH	hydroxybenzene (phenol)
⬡—NH₂	aminobenzene (aniline)

and is often the ingredient responsible for the odor of mothballs. The structures of naphthalene and other examples of polycyclic aromatic hydrocarbons are shown below.

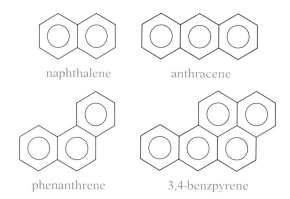

naphthalene   anthracene

phenanthrene   3,4-benzpyrene

Some polycyclic aromatic hydrocarbons whose molecules contain four or more fused rings have been found to cause cancer in laboratory animals. For example, 3,4-benzpyrene causes gene mutations and lung cancer in test animals; it is a product of incomplete combustion of tobacco, coal, and petroleum and is found in cigarette smoke and automobile exhaust.

18.9 CLASSIFICATION OF ORGANIC COMPOUNDS BY FUNCTIONAL GROUP

Because of the enormous number of organic compounds, chemists have found ways to simplify the study of their structures and reactions. One way is to classify the compounds by functional group. A **functional group** is a distinctive group of atoms that is found in molecules of many different organic compounds but always reacts the same way. In general, alkyl groups are much less reactive than other parts of organic molecules and are considered to be the skeleton to which one or more functional groups may be attached. The functional group is the part of the organic molecule that is more likely to undergo change; that is, the functional group functions, but it is usually attached to an inert structural support (the alkyl group). Thus, an organic compound can be classified by the functional group it contains. Table 18-6 illustrates the common functional groups. Note that R is used to indicate that any alkyl group can be attached to the functional group, and H means that a hydrogen atom can be attached to the functional group in place of an alkyl group.

Functional groups provide a very convenient way to classify and recognize types of organic compounds. The double bonds of alkenes and the triple bonds of alkynes are considered functional groups, since they are more reactive than the single bonds of alkyl groups. Thus, any alkene will undergo roughly the same kinds of reactions as any other alkene. The same is true of alkynes. Similarly, any aliphatic molecule containing a hydroxyl group, —OH, is an alcohol, and it will show approximately the

TABLE 18-6 The Functional Groups of Organic Molecules

Functional Group	Functional Group Name	Class Name of Compounds	General Formula of Compounds	Condensed Formula of Compounds	Examples
$>C=C<$	double bond	alkene	H H | | R—C=C—R (H) (H)	R—CH=CH—R	$H_2C=CH_2$ ethene
—C≡C—	triple bond	alkyne	R—C≡C—R (H) (H)	R—C≡C—R	H—C≡C—H ethyne
—OH	hydroxyl	alcohol	R—OH	ROH	CH_3OH methyl alcohol
—C—O—C—	ether	ether	R—O—R	ROR	H_3C—O—CH_3 dimethyl ether
O ‖ C	carbonyl	aldehyde	O ‖ R—C—H (H)	RCHO	H_3C—C=O | H acetaldehyde
		ketone	O ‖ R—C—R	RCOR	O ‖ H_3C—C—CH_3 acetone
—C⟨O OH	carboxyl	carboxylic acid	O ‖ R—C (H) OH	RCOOH	O ‖ H_3C—C OH acetic acid
—C⟨O O—C—	ester	ester	O ‖ R—C (H) O—R	RCOOR	O ‖ H_3C—C OCH_2CH_3 ethyl acetate
—N—	amino	amine	R—N—H (primary) | H	RNH_2	H_3C—NH_2 methyl amine
			R—N—R (secondary) | H	R_2NH	H_3C—N—CH_3 | H dimethyl amine
			R—N—R (tertiary) | R	R_3N	H_3C—N—CH_3 | CH_3 trimethyl amine

—C⟨O NH₂	amide	amide	O H
			 ‖ /
			 R—C—N
			 (H) \
			      H
				$RCONH_2$	O
					 ‖
					 H_3C—C
					    NH_2
					 acetamide

TABLE 18-6 *Continued*

Functional Group	Functional Group Name	Class Name of Compounds	General Formula of Compounds	Condensed Formula of Compounds	Examples
$-NO_2$	nitro	nitro	$R-NO_2$	RNO_2	H_3C-NO_2 nitromethane
$-SH$	thiol	thiol	$R-SH$	RSH	H_3C-SH methanethiol
$-S-S-$	disulfide	disulfide	$R-S-S-R$	$RSSR$	$H_3C-S-S-CH_3$ dimethyl disulfide
$-X$ ($-F, -Cl,$ $-Br, -I$)	halo- (fluoro-, chloro-, bromo-, iodo-)	halide	$R-X$	RX	H_3CCl chloromethane

same chemical behavior as all other alcohol molecules. Examples of some common alcohols are given below, with common names in parentheses.

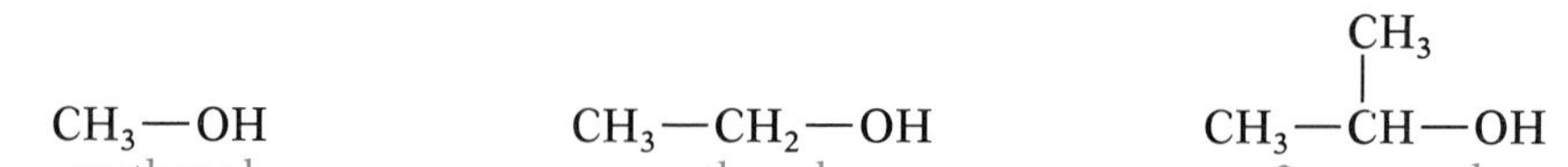

$$CH_3-OH$$
methanol
(methyl alcohol; wood alcohol)

$$CH_3-CH_2-OH$$
ethanol
(grain alcohol)

$$CH_3-\overset{\overset{\displaystyle CH_3}{|}}{CH}-OH$$
2-propanol
(isopropyl alcohol; rubbing alcohol)

Other commonly occurring functional groups are the carboxyl group, $-CO_2H$, of carboxylic acids and the amino group, $-\overset{|}{N}-$, of amines. Carboxylic acids are typical weak acids, and you have already encountered a familiar one, acetic acid. Carboxylic acids are known for their disagreeable odors; butyric acid, shown below,

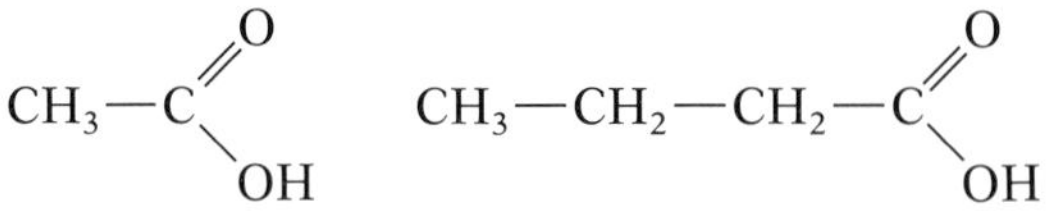

$$CH_3-C\overset{\displaystyle O}{\underset{\displaystyle OH}{}}$$
acetic acid

$$CH_3-CH_2-CH_2-C\overset{\displaystyle O}{\underset{\displaystyle OH}{}}$$
butyric acid

is the odor of rancid butter. Amines also have distinctive odors resembling ammonia. The similarity between odors of amines and the odor of ammonia is attributed to their structural similarities.

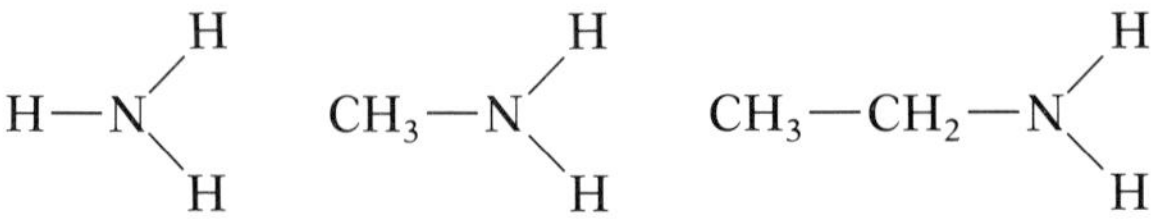

$$H-N\overset{\displaystyle H}{\underset{\displaystyle H}{}}$$
ammonia

$$CH_3-N\overset{\displaystyle H}{\underset{\displaystyle H}{}}$$
methylamine

$$CH_3-CH_2-N\overset{\displaystyle H}{\underset{\displaystyle H}{}}$$
ethylamine

Many organic molecules contain more than one functional group. For example, all of the amino acids, which are the components of proteins, contain an amino group and a carboxyl group, and some contain additional functional groups, such as the hydroxyl group in serine and the thiol group, —SH, in cysteine:

$$CH_3-CH-CO_2H \qquad HO-CH_2-CH-CO_2H \qquad HS-CH_2-CH-CO_2H$$

$$\underset{\text{alanine}}{\overset{|}{NH_2}} \qquad\qquad \underset{\text{serine}}{\overset{|}{NH_2}} \qquad\qquad \underset{\text{cysteine}}{\overset{|}{NH_2}}$$

The chemical behavior of molecules containing more than one functional group is a composite of the individual behaviors of each functional group.

18.10 ORBITAL HYBRIDIZATION IN HYDROCARBON MOLECULES *(optional)*

In chapter 6, covalent bonds were described as regions in space where single atomic orbitals of two atoms overlap; since each of the two atomic orbitals contained one electron, the covalent bond consists of two electrons shared by two atoms. Thus, covalent bond formation corresponds to the merging of two atomic orbitals. In the case of the hydrogen molecule (see Figure 18-7), two H atoms come together, their $1s$ orbitals overlap, and the H_2 molecule forms. The resulting electron cloud is then spread over the entire molecule. This new electron cloud, formed by merged atomic orbitals, is called a **molecular orbital.** Like an atomic orbital, a molecular orbital can contain a maximum of two electrons.

Each single covalent bond in a molecule is a molecular orbital called a **sigma molecular orbital.** A sigma molecular orbital is symmetrical about its axis, as illustrated in Figure 18-8. This means that a cross-section of the electron cloud, taken at a right angle to the bonding axis, has the shape of a circle. Since all single covalent bonds are sigma molecular orbitals, single covalent bonds are also called *sigma bonds.*

Molecular orbitals play an important role in explaining bonding patterns in organic molecules. We will discuss the fundamental bonding patterns of organic molecules in the following paragraphs.

Bonding in Methane Before applying the concept of molecular orbitals to the methane molecule, CH_4, we must first analyze the atomic orbitals of the carbon atom, whose electron configuration is $1s^2 2s^2 2p^2$. The distribution of electrons in the carbon atom is shown in Figure 18-9. The valence shell of the carbon atom contains four electrons; two are paired in the $2s$ orbital and two exist singly in $2p$ orbitals. This arrangement suggests that only two $2p$ orbitals could overlap with orbitals from other atoms to form

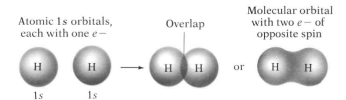

Atomic $1s$ orbitals, each with one $e-$ Overlap   Molecular orbital with two $e-$ of opposite spin

FIGURE 18-7 Formation of the Sigma Molecular Orbital of the H_2 Molecule

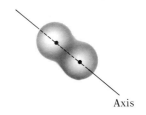

Axis

FIGURE 18-8 The Sigma Molecular Orbital of the H_2 Molecule Is Symmetrical about the Axis Joining the Two Atoms.

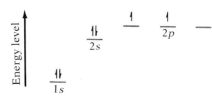

FIGURE 18-9 The Distribution of Electrons in the Atomic Orbitals of a Lone Carbon Atom. The small arrows represent individual electrons and their direction of spin.

sigma molecular orbitals and thus sigma bonds. (Remember, a covalent bond forms by overlap of two atomic orbitals from different atoms, with each atomic orbital usually containing one electron.) The $2s$ orbital cannot overlap with another atomic orbital containing an electron, because the $2s$ orbital already contains two electrons.

How, then, does the carbon atom form four sigma bonds to hydrogen in the methane molecule? The answer is that the orbitals of a carbon atom in a molecule are different from those of a lone carbon atom. In forming CH_4 and other molecules containing four sigma bonds to carbon, the atomic orbitals of carbon's valence shell form four new atomic orbitals to be used in sigma bonds. The valence shell of a lone carbon atom contains a $2s$ orbital and three $2p$ orbitals (one of which is empty), but the valence shell of a carbon atom forming methane contains four new orbitals, each containing a single electron. Since the new orbitals are considered to be a mixture of an s orbital and three p orbitals, they are called **hybrid orbitals,** and they are designated as sp^3 orbitals. (The sp^3 notation refers to the mixture of one s orbital, $s^{(1)}$, and three p orbitals, p^3.)

The hypothetical formation of sp^3 hybrid orbitals is illustrated in Figure 18-10. Note that all of the sp^3 hybrid orbitals are at the same energy level, intermediate between the $2s$ orbital and the $2p$ orbitals. Thus, the sp^3 orbitals are of equal energy, and each contains a single electron. The shape of a single sp^3 orbital is shown in Figure 18-11(a). Note that an sp^3 orbital consists of a major lobe and a minor lobe. It is the major lobe that overlaps with an atomic orbital of another atom to form a sigma bond.

Figure 18-11(b) illustrates the shape of the four sp^3 orbitals of a single carbon atom; the minor lobes have been omitted for clarity. Note that the sp^3 orbitals point toward the corners of a regular tetrahedron, with angles of 109.5° between orbitals. The carbon atom is said to be sp^3-hybridized, and its orbitals are now ready for overlap with the atomic orbitals of four hydrogen atoms to form methane. Figure 18-12(a) illustrates the overlap of atomic orbitals in methane. Each pair of overlapping atomic orbitals forms a sigma molecular orbital and thus a sigma bond.

Formation of sigma bonds using sp^3 orbitals is not restricted to methane; it occurs in any molecule containing a carbon atom bonded to four other atoms. Thus, carbon tetrachloride, CCl_4, contains a sigma bond between the carbon atom and each chlorine atom. Ethane, C_2H_6, contains seven sigma bonds, as illustrated in Figure 18-12(b). Two sp^3 orbitals overlap to form the C—C sigma bond, and an sp^3 orbital overlaps with the $1s$ orbital of hydrogen in each of the six C—H sigma bonds.

Bonding in Ethene In ethene, CH_2=CH_2, each carbon atom is bonded to another carbon atom and two hydrogen atoms. The atomic orbitals of a lone carbon atom do not allow bonding to three atoms, and thus the

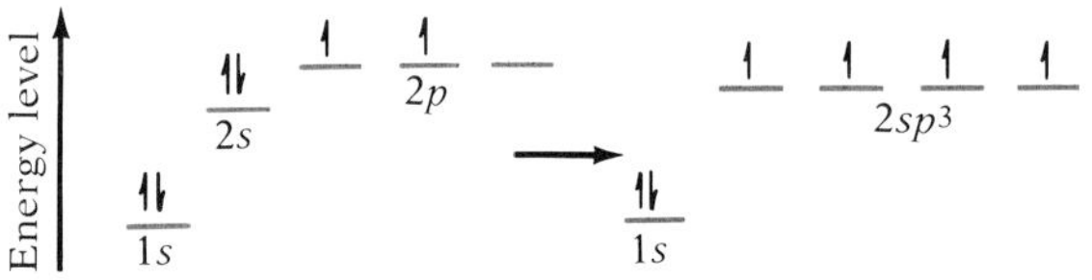

FIGURE 18-10 Hypothetical Formation of sp^3 Hybrid Orbitals

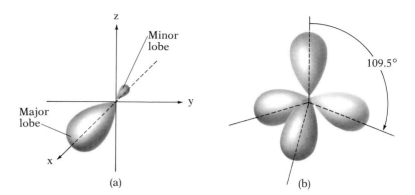

FIGURE 18-11   (a) A Single *sp³* Orbital (b) The Major Lobes of the Four *sp³* Orbitals of a Carbon Atom

atomic orbitals of carbon must be hybridized to permit formation of ethene. In this case, only three of the four atomic orbitals in carbon's valence shell become hybridized. The 2*s* orbital and two of the 2*p* orbitals combine to form three new orbitals, as illustrated in Figure 18-13. The resulting hybrid orbitals are designated as ***sp²* orbitals.** *sp²* orbitals are shaped much like *sp³* orbitals, but the three *sp²* orbitals of a carbon atom lie at 120° angles from one another in the same plane and point toward the corners of an equilateral triangle (see Figure 18-14(a)). Thus, a carbon atom and its three *sp²* orbitals are said to have a trigonal (trianglelike) shape. However, let's not forget about the leftover *p* orbital containing one electron that was not used in hybridization. Its dumbbell is perpendicular to the plane of the three *sp²* orbitals, as illustrated in Figure 18-14(b).

When two *sp²*-hybridized carbon atoms are brought together, as in the formation of ethene, an *sp²* orbital of one atom overlaps with an *sp²* orbital of the other atom to form a sigma bond. Then the parallel *p* orbitals (see Figure 18-15) of each atom merge above and below the sigma bond to form a new molecular orbital, called a pi (π) molecular orbital. A **pi molecular orbital** is formed by side-to-side overlap of *p* orbitals. The pi molecular orbital corresponds to a second bond, a pi bond, between the two carbon atoms. Thus, the double bond in ethene, $CH_2{=}CH_2$, is composed of one

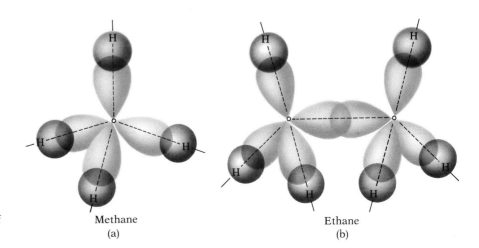

FIGURE 18-12   (a) Overlap of Atomic Orbitals in Methane, CH_4 (b) Overlap of Atomic Orbitals in Ethane, C_2H_6

Methane
(a)

Ethane
(b)

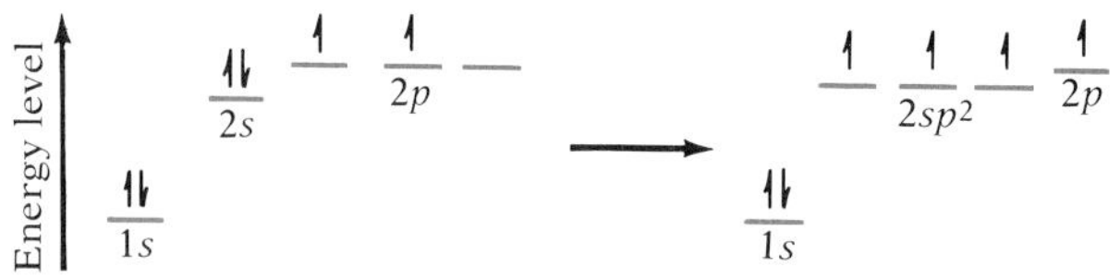

Energy level

$\frac{\uparrow\downarrow}{2s}$ $\frac{\uparrow}{}$ $\frac{\uparrow}{2p}$ $\frac{}{}$ ⟶ $\frac{}{}$ $\frac{\uparrow\ \ \uparrow\ \uparrow}{2sp^2}$ $\frac{\uparrow}{2p}$

$\frac{\uparrow\downarrow}{1s}$ $\frac{\uparrow\downarrow}{1s}$

FIGURE 18-13 Hypothetical Formation of sp^2 Hybrid Orbitals

sigma bond and one pi bond. The four overlapping lobes of the p orbitals in the pi bond prevent rotation about the double bond and lock the four hydrogen atoms of ethene into the same plane. The bond angles about each carbon atom are 120°. Note that the sigma bond of a double bond is formed by end-to-end overlap of sp^2 hybrid orbitals, while the pi bond is formed by side-to-side overlap of unhybridized p orbitals. The double bonds of all alkenes are formed in the same way.

Bonding in Benzene   The molecular orbital picture of carbon-carbon double bonds is helpful in understanding the electron delocalization of benzene, C_6H_6, (see Figure 18-16) and other aromatic hydrocarbons. In the benzene ring, sigma bonds are formed by overlapping sp^2 orbitals. Since each carbon atom has a p orbital with one electron in it, the six p orbitals can overlap above and below the ring, allowing delocalization of p electrons among all the carbon atoms. These electrons are considered to occupy a pi cloud that has the shape of a doughnut above and below the plane of the ring.

Bonding in Ethyne   In ethyne, CH≡CH, each carbon atom is bonded to another carbon atom and to a hydrogen atom. As you might expect, the carbon atoms are hybridized in ethyne. However, only two of the four atomic orbitals in carbon's valence shell are used for hybridization, as illustrated in Figure 18-17. The $2s$ orbital combines with a $2p$ orbital to form two **sp hybrid orbitals.** The two sp orbitals have shapes similar to those of sp^3 and sp^2 orbitals; they lie in a straight line, pointing in opposite directions (see Figure 18-18(a)), with an angle of 180° between them. Thus, the shape of a carbon atom with its two sp orbitals is linear. The two p orbitals that were not hybridized are perpendicular to each other (see Figure 18-18(b)) and to the axes of the sp orbitals.

When two sp-hybridized carbon atoms are brought together, as in the formation of ethyne, an sp orbital of one atom overlaps end-to-end with an sp orbital of the other carbon atom to form a sigma bond. Then the two sets of parallel p orbitals on each carbon atom overlap side-to-side, and two pi

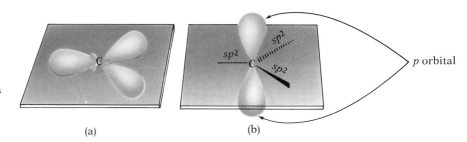

FIGURE 18-14   (a) Trigonal sp^3 Orbitals of a Carbon Atom (b) The p Orbital Perpendicular to the Plane of the sp^2 Orbitals

(a)   (b)

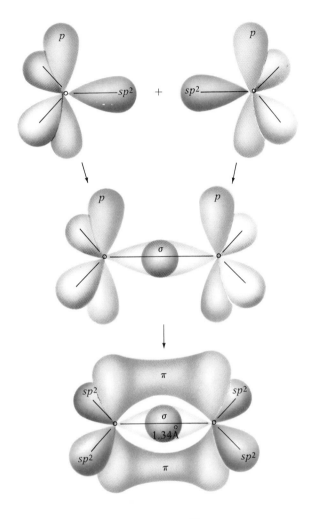

FIGURE 18-15 Formation of the Double Bond of Ethene. Two sp^2-hybridized carbon atoms form a sigma bond and a pi bond.

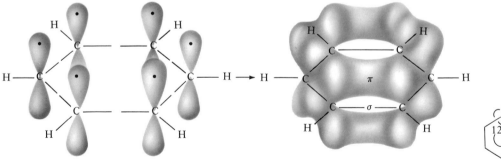

FIGURE 18-16 Orbital Representation of Bonding in Benzene, C_6H_6. Sigma bonds are formed between carbon atoms by overlap of sp^2 orbitals, and electron delocalization occurs by overlap of p orbitals. (The orbitals of the C—C and C—H sigma bonds have been omitted for clarity.)

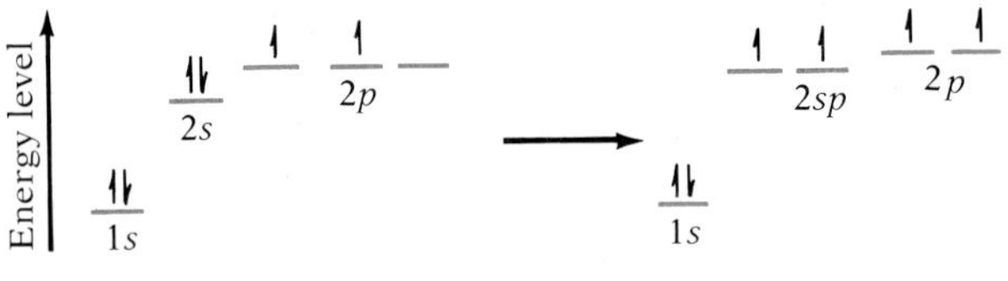

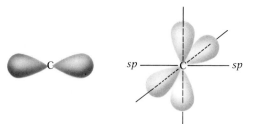

FIGURE 18-17 Hypothetical Formation of *sp* Hybrid Orbitals

FIGURE 18-18 (a) The *sp* Orbitals of a Carbon Atom (b) The *p* Orbitals Are Perpendicular to Each Other and to the Linear *sp* Orbitals.

Two *sp* orbitals   Two *p* orbitals

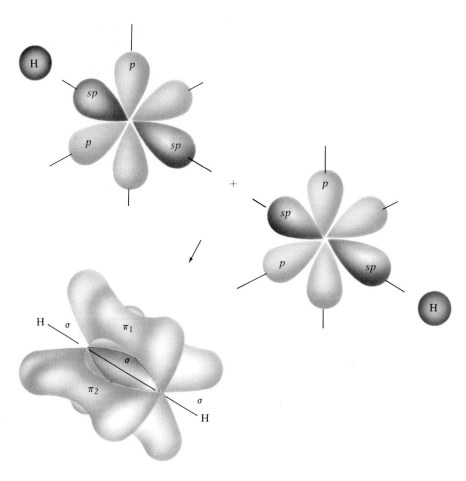

FIGURE 18-19 Formation of the Triple Bond of Ethyne, CH≡CH. Two *sp*-hybridized carbon atoms form a sigma bond and two pi bonds. (The orbitals comprising the C—H bonds have been omitted for clarity.)

bonds form (see Figure 18-19). Thus, the triple bond of ethyne is composed of a sigma bond and two pi bonds. The triple bonds of all alkynes are formed in the same way.

SUMMARY

Organic chemistry is the study of most covalent carbon compounds; about 90% of the known chemical compounds are organic. Carbon atoms form single or multiple bonds to other carbon atoms and single bonds to atoms of hydrogen, nitrogen, oxygen, phosphorus, sulfur, and the halogens. Carbon atoms also form multiple bonds to atoms of nitrogen and oxygen.

Hydrocarbons are organic compounds composed of carbon and hydrogen; carbon atoms form the skeleton of the molecule and hydrogen atoms are attached to the carbon atoms. Hydrocarbons are insoluble in water, less dense than water, and soluble in one another. Hydrocarbons are classified as **aliphatic** or **aromatic.**

Alkanes are aliphatic hydrocarbons containing only single bonds. The open-chain alkanes have the general formula C_nH_{2n+2}. Alkanes have bond angles of 109.5°, and the open-chain alkanes can be linear or branched. Since there is free rotation about single bonds, linear alkanes can exist in the extended conformation as long, zigzag arrangements of carbon atoms, or they can adopt more compact shapes. Alkanes having four or more carbon atoms exist as isomers. The lower formula weight alkanes are used as gaseous fuels for cooking and heating. Liquid alkanes are blended to make gasoline, diesel fuel, jet fuel, and heating oil. Solid alkanes are used as lubricants, greases, and sealants. Rules for naming alkanes are given on pp. 444–445.

Alkenes are aliphatic hydrocarbons having one or more double bonds. Open-chain alkenes with one double bond have the general formula C_nH_{2n}. The four atoms surrounding a double bond lie in the same plane, with bond angles of 120°. Alkenes with four or more carbon atoms and internal double bonds exist as **geometric isomers** because double bonds do not allow rotation of attached groups. **Alkynes** are hydrocarbons having one or more triple bonds. Open-chain alkynes with one triple bond have the general formula C_nH_{2n-2}. The triple bond has linear geometry, with bond angles of 180°. Rules for naming alkenes are given on pp. 449–450, and rules for naming alkynes are given on p. 451. **Cyclic aliphatic hydrocarbons** are aliphatic hydrocarbons whose molecules are closed rings. They are named by adding *cyclo-* to the beginning of the name of the corresponding open-chain hydrocarbon.

Aromatic hydrocarbons are unsaturated cyclic compounds that are structurally related to benzene. Benzene is a hybrid of two resonance structures and, like all aromatic compounds, is unusually stable. The bonds in the benzene ring are intermediate between single and double bonds. Benzene **derivatives** are formed when one or more hydrogen atoms are replaced by other atoms or groups. Aromatic hydrocarbons containing two or more benzene rings fused together are called **polycyclic aromatic hydrocarbons.**

Organic compounds can be classified by functional group. A **functional group** is a distinctive group of atoms that is present in many different organic compounds and that always reacts the same way. Alkyl groups are the skeletons to which functional groups are attached. An organic molecule can contain more than one functional group.

Covalent bond formation corresponds to the merging of two atomic orbitals of different atoms. The new electron cloud formed by merged atomic orbitals is called a **molecular orbital.** Each single covalent bond in a molecule is a **sigma molecular orbital.** When a carbon atom bonds to four atoms, the atomic orbitals of carbon's valence shell form four hybrid orbitals called *sp³* **orbitals.** *sp³* orbitals are of equal energy and each contains a major and a minor lobe. The major lobe overlaps with an atomic orbital of another atom to form a **sigma bond.** When a carbon atom bonds to three atoms, the *s* orbital and two *p* orbitals of its valence shell form three hybrid orbitals called *sp²* **orbitals.** The *sp²* orbitals overlap with atomic orbitals of other atoms to form three sigma bonds; the remaining *p* orbital of the carbon atom can overlap side-to-side with the *p* orbital of another carbon atom to form a **pi molecular orbital** and thus a **pi bond.** A double bond is composed of a sigma bond and a pi bond. In the benzene molecule, electron delocalization occurs by side-to-side overlap of the *p* orbitals of all six *sp²*-hybridized carbon atoms. When a carbon atom bonds to two other atoms, the *s* orbital and one *p* orbital of its valence shell form two hybrid orbitals called *sp* **orbitals.** The *sp* orbitals overlap with atomic orbitals of other atoms to form two sigma bonds. The remaining *p* orbitals can overlap side-to-side with *p* orbitals of another carbon atom to form two pi bonds. A triple bond is composed of a sigma bond and two pi bonds.

STUDY QUESTIONS AND PROBLEMS

THE NATURE OF ORGANIC CHEMISTRY

1. Distinguish between organic and inorganic chemistry.
2. Give the type of covalent bond (single, double, or triple) that a carbon atom can form to each atom. (There may be more than one answer for some atoms.)
   a. C   c. P   e. Cl   g. O
   b. H   d. S   f. N

THE HYDROCARBONS

3. Define the following terms:
   a. Hydrocarbon   c. Aromatic hydrocarbon
   b. Aliphatic hydrocarbon
4. In what ways can the carbon atoms be arranged in hydrocarbon molecules?
5. What are the major natural sources of the hydrocarbons?
6. Summarize the physical properties of the hydrocarbons.

ALKANES

7. Define the following terms:
   a. Alkane   e. Condensed structural formula
   b. Saturated hydrocarbon   f. Structural formula
   c. Homologous series   g. Isomers
   d. Molecular formula

8. What is the general molecular formula of the open-chain alkanes? Which of the following hydrocarbons is (are) open-chain alkane(s)?
   a. $C_{12}H_{24}$ c. C_6H_{10} e. C_7H_{16}
   b. C_4H_{10} d. C_5H_{10} f. $C_{10}H_{18}$

9. Draw the structural formulas of the following linear alkanes.
   a. C_5H_{12} b. C_8H_{18} c. C_3H_8

10. Without looking at Table 18-2, draw structural formulas of all the isomers of C_6H_{14}.

11. What is the natural source of alkanes? Give some specific uses of alkanes.

NAMING ALKANES

12. Derive the IUPAC name of each of the following alkanes.

a. 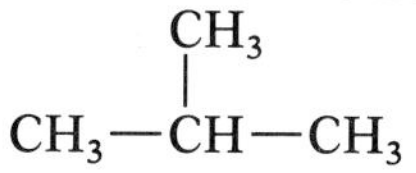
$$CH_3-\overset{\overset{\displaystyle CH_3}{|}}{CH}-CH_3$$

b. 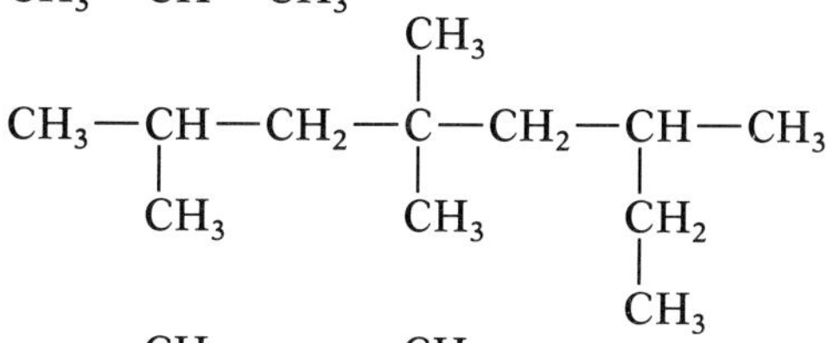
$$CH_3-\overset{\overset{\displaystyle CH_3}{|}}{CH}-CH_2-\overset{\overset{\displaystyle CH_3}{|}}{\underset{\underset{\displaystyle CH_3}{|}}{C}}-CH_2-\overset{\overset{\displaystyle CH_3}{|}}{\underset{\underset{\displaystyle \underset{\displaystyle CH_2}{|}}{}}{CH}}-CH_3$$
$$CH_3$$

c.
$$CH_3-\overset{\overset{\displaystyle CH_3}{|}}{CH}-CH_2-\overset{\overset{\displaystyle CH_3}{|}}{CH_2}$$

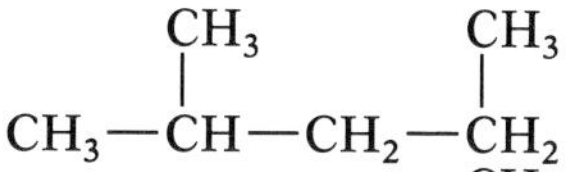

d. 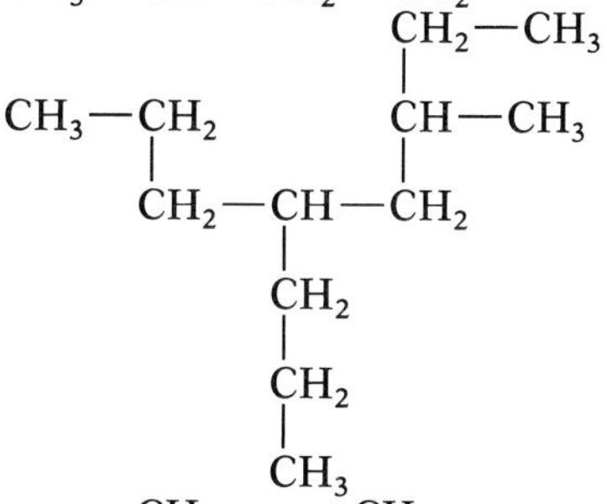
$$CH_3-CH_2$$
$$\underset{\underset{\displaystyle CH_2}{|}}{CH_2}-\underset{\underset{\displaystyle CH_2}{|}}{CH}-\underset{\underset{\displaystyle \underset{\displaystyle CH_2}{|}}{CH_2}}{CH_2}$$

e. 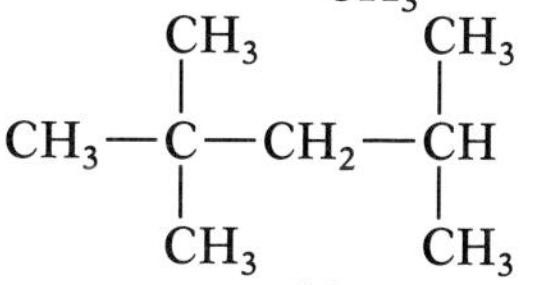
$$CH_3-\overset{\overset{\displaystyle CH_3}{|}}{\underset{\underset{\displaystyle CH_3}{|}}{C}}-CH_2-\overset{\overset{\displaystyle CH_3}{|}}{\underset{\underset{\displaystyle CH_3}{|}}{CH}}$$

13. Draw structural formulas for the three isomers of pentane and derive the IUPAC name of each.

14. Draw the structural formula of each alkane.
   a. 2-Methylpentane   d. 3,4-Diethyl-5-methylnonane
   b. 2,2-Dimethylbutane   e. 4-Methyl-6-propyldecane
   c. 3,3-Dimethyl-5-ethyldecane

ALKENES AND ALKYNES

15. Define the following terms:
    a. Unsaturated hydrocarbon   c. Geometric isomers
    b. Alkene   d. Alkyne
16. Distinguish between saturated and unsaturated hydrocarbons.
17. What is the general molecular formula of the open-chain alkenes that contain one double bond? Which of the hydrocarbons in question 8 is (are) open-chain alkene(s)?
18. Describe the geometry of the ethene molecule, C_2H_4.
19. Designate each alkene as *cis*, *trans*, or neither.

    a.
    $$CH_3 \diagdown \qquad \diagup CH_3$$
    $$\qquad C=C$$
    $$H \diagup \qquad \diagdown H$$

    c.
    $$CH_3-CH_2 \diagdown \qquad \diagup H$$
    $$\qquad\qquad C=C$$
    $$H \diagup \qquad \diagdown CH_3$$

    b.
    $$CH_3-CH_2-CH_2 \diagdown \qquad \diagup H$$
    $$\qquad\qquad\qquad C=C$$
    $$H \diagup \qquad \diagdown H$$

    d.
    $$CH_3 \diagdown \qquad \diagup CH_2-CH_3$$
    $$\qquad C=C$$
    $$H \diagup \qquad \diagdown H$$

20. How are alkenes produced? Give some specific uses of alkenes.
21. What is the general molecular formula of the open-chain alkynes that contain one triple bond? Which of the hydrocarbons in question 8 is (are) open-chain alkyne(s)?
22. Describe the geometry of the ethyne molecule, C_2H_2.
23. Give the common name of ethyne. What are its sources and major uses?

NAMING ALKENES AND ALKYNES

24. Derive the IUPAC name of each of the following.

    a. $CH_2{=}CH$
    $\qquad\quad |$
    $\qquad\quad CH_3$

    b. $CH_3-CH-CH_2-C{\equiv}C-H$
    $\qquad\qquad\;\; |$
    $\qquad\qquad\;\; CH_2-CH_3$

    c. $CH_2{=}CH-CH_2-CH-CH_3$
    $\qquad\qquad\qquad\qquad |$
    $\qquad\qquad\qquad\qquad CH_3$

    d. $CH_3-CH{=}C-CH_3$
    $\qquad\qquad\quad |$
    $\qquad\qquad\quad CH_2$
    $\qquad\qquad\quad |$
    $\qquad\qquad\quad CH_3$

    e. $H-C{\equiv}C-CH-CH_2-CH-CH_2-CH_3$
    $\qquad\qquad\quad\; | \qquad\qquad\; |$
    $\qquad\qquad\quad\; CH_3 \qquad\quad CH_2-CH_3$

    f. $CH_3 \qquad\qquad CH_2-CH_3$
    $\;\;\; | \qquad\qquad\qquad |$
    $\;\;\; CH-CH{=}CH-CH_2$

    g. $CH_3-CH-CH_2-CH{=}CH-CH_2-CH_3$
    $\qquad\qquad |$
    $\qquad\qquad CH_3$

CYCLIC ALIPHATIC HYDROCARBONS

25. What are cyclic aliphatic hydrocarbons?
26. Draw the structural formula and the geometric figure of each molecule.
    a. Cyclopentene   b. Cyclopropane   c. Cycloheptane

BENZENE AND AROMATIC HYDROCARBONS

27. Define each of the following terms:
    a. Aromatic hydrocarbon   c. Derivative
    b. Resonance structures   d. Polycyclic aromatic hydrocarbon
28. How did the term *aromatic* first arise?
29. What are the Kekulé structures of benzene? Are they actual compounds? Explain your answer.
30. What are the chemical characteristics of aromatic hydrocarbons?
31. Draw the structural formulas of two specific polycyclic aromatic hydrocarbons.

CLASSIFICATION OF ORGANIC COMPOUNDS BY FUNCTIONAL GROUP

32. What is a functional group? How are functional groups helpful in classifying organic compounds?
33. Why are alkyl groups not considered functional groups?
34. Circle and name each functional group in the following molecules.

    a. vitamin A

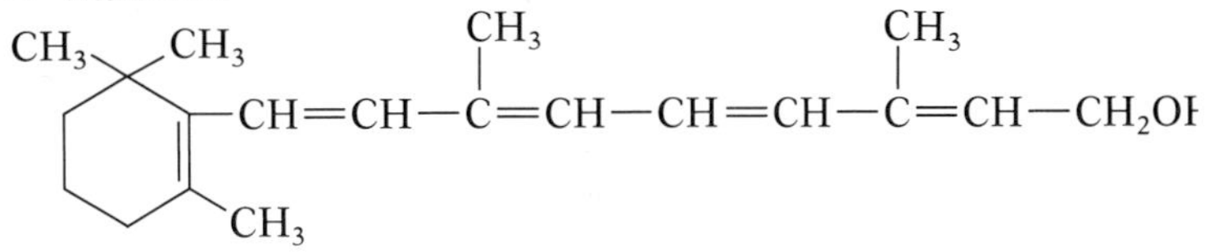

    b. niacin
       (a vitamin)

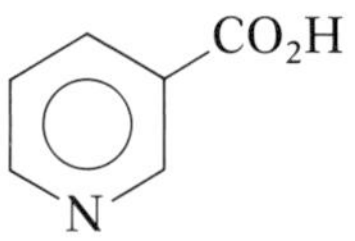

    c. thyroxine
       (a hormone)

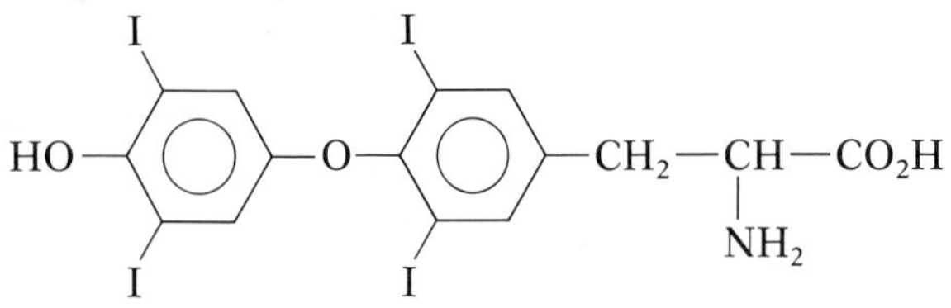

d. testosterone
(a hormone)

e. epinephrine
(a hormone)

f. butyl acetate
(odor of bananas)

$$CH_3-\overset{\overset{\displaystyle O}{\|}}{C}-O-CH_2-CH_2-CH_2-CH_3$$

ORBITAL HYBRIDIZATION IN HYDROCARBON MOLECULES

35. Define the following terms:
   a. Molecular orbital   **e.** sp^2 orbital
   b. Sigma molecular orbital   **f.** Pi molecular orbital
   c. Hybrid orbital   **g.** sp orbital
   d. sp^3 orbital
36. How is a molecular orbital formed?
37. Without looking at Figure 18-10, draw a diagram showing energy levels to illustrate the hypothetical formation of sp^3 orbitals.
38. Without looking at Figure 18-12, sketch the orbitals in the methane molecule, CH_4.
39. Without looking at Figure 18-13, draw a diagram showing energy levels to illustrate the hypothetical formation of sp^2 orbitals.
40. Describe electron delocalization in the benzene ring in terms of orbitals.
41. Without looking at Figure 18-17, draw a diagram showing energy levels to illustrate the hypothetical formation of sp orbitals.
42. Without looking at Figure 18-19, sketch the orbitals in the ethyne molecule, C_2H_2.
43. Describe the molecular orbital composition of each of the following.
   a. Single bond   **b.** Double bond   **c.** Triple bond

GENERAL EXERCISES

44. What is the basis of carbon's tendency to form covalent bonds?
45. Describe the geometry of the methane molecule, CH_4.

46. Why is there only one isomer of C_3H_8?

47. Gasolines are blended to achieve desired octane ratings. The octane scale is based on a rating of 100 for 2,2,4-trimethylpentane (an octane isomer) and 0 for heptane. This means that 2,2,4-trimethylpentane burns very smoothly in an automobile engine, and heptane burns very unevenly, giving rough engine performance. Draw the structural formulas of 2,2,4-trimethylpentane and heptane.

48. Why is it that butene can have geometric isomers but propene cannot?

49. Draw the structural formula of each of the following.
   a. 3-Heptene
   b. 2-Methyl-3-octyne
   c. 2-Methyl-2-hexene
   d. 3-Propyl-2-octene
   e. 5-Ethyl-3,5-dimethyl-1-heptene
   f. 2,4-Dimethyl-3-nonene
   g. 2,4,6-decatriene

50. What is the actual bonding pattern in the benzene ring? Explain your answer.

51. Without looking at Figure 18-15, sketch the orbitals in the ethene molecule C_2H_4.

REVIEW OF BASIC MATHEMATICS

APPENDIX A

A.1 DIAGNOSTIC TEST FOR BASIC CALCULATIONS

Take the following short test to check your skill in basic calculations. Work through the test carefully, and then check your work against the answers given at the end of the test. If all your answers were correct, then your skills in basic calculations are probably sufficient for introductory chemistry, and you can move on to the next diagnostic test (section A.3, Diagnostic Test for Basic Algebra, pp. 485–486). If you got at least 80% of the answers correct, go back through your work and find your mistakes. If you made only minor errors and can redo the calculations correctly, then proceed to the next diagnostic test. However, if you cannot make the corrections or if you got less than 80% of the answers correct, you should work through section A.2, Review of Basic Calculations.

DIAGNOSTIC TEST FOR BASIC CALCULATIONS

A. Give the absolute value of each number.
   1. $+7$ **2.** -2 **3.** -13 **4.** $+10$
B. Complete the following additions without using a calculator.
   1. $4.61 + 17.5$ **3.** $(-24.9) + 14.3$
   2. $(-65.7) + (-31.4)$ **4.** $51.6 + (-47.9) + (-11.3)$
C. Complete the following subtractions without using a calculator.
   1. $24.7 - 16.2$ **3.** $(-64.35) - 75.6$
   2. $17.62 - 11.4$ **4.** $(-11.23) - (-14.19)$
D. Complete the following multiplications.
   1. $(41)(4.9)(3.2)$ **3.** $(-47)(5)(2.3)$
   2. $(-2)(3)(-15)$ **4.** $(-4.2)(3.6)(7.9)(-2.15)$
E. Find the following whole-number roots.
   1. $\sqrt{81}$ **2.** $\sqrt[3]{64}$ **3.** $\sqrt{36}$ **4.** $\sqrt[3]{343}$
F. Solve the following problems.
   1. The circumference (C) of a circle is given by the formula $C = 2\pi r$, where $\pi = 3.14$ and r is the radius of the circle. Find the circumference of a circle having a radius of 4 inches.
   2. The volume (V) of a cube is given by the formula $V = l^3$, where l is the length of each side of the cube. Find the volume of a cube having sides 6 cm long.
G. Complete the following calculations.
   1. $\dfrac{(-3)(8)}{12}$ **3.** $\dfrac{(-6)(-2)(3)(5)}{9}$

   2. $\dfrac{(-2)(5)(-7)}{10}$ **4.** $\dfrac{(8\ \text{in}^2)(6\ \text{in})}{24\ \text{in}}$

471

H. Complete the following calculations to give answers that are fractions.
1. $\frac{1}{4} + \frac{2}{8}$　　3. $\frac{1}{3} + \frac{1}{4}$　　5. $\frac{2}{3} \times \frac{4}{9}$
2. $\frac{1}{6} - \frac{1}{12}$　　4. $\frac{7}{9} - \frac{2}{3}$　　6. $\frac{3}{10} \div 5$

I. Complete the following operations.
1. $0.06 =$ _____ %　　3. $70\% =$ _____ (decimal number)
2. $\frac{3}{7} =$ _____ %　　4. $105\% =$ _____ (decimal number)

Answers

Section A.　1. 7　2. 2　3. 13　4. 10
Section B.　1. 22.11　2. -97.1　3. -10.6　4. -7.6
Section C.　1. 8.5　2. 6.22　3. -139.95　4. 2.96
Section D.　1. 642.88　2. 90　3. -540.5　4. 256.813
Section E.　1. 9　2. 4　3. 6　4. 7
Section F.　1. 25.12 in　2. 216 cm³
Section G.　1. -2　2. 7　3. 20　4. 2 in²
Section H.　1. $\frac{4}{8}$ or $\frac{1}{2}$　2. $\frac{1}{12}$　3. $\frac{7}{12}$　4. $\frac{1}{9}$　5. $\frac{8}{27}$　6. $\frac{3}{50}$
Section I.　1. 6%　2. 42.86%　3. 0.70　4. 1.05

A.2　REVIEW OF BASIC CALCULATIONS

Positive and Negative Numbers　A **whole number** is any number having a value of 0, 1, 2, 3, 4, 5, 6, 7, 8, 9, 10, 11, 12, ... Whole numbers are also called **integers.** A **decimal number** is a number whose value is between two whole numbers. For example, 2.5 is a decimal number. It has a value greater than 2 but less than 3. A decimal number with a value greater than 0 but less than 1 is written with a zero to the left of the decimal point and the decimal value to the right of the decimal point. Examples of decimal numbers greater than 0 but less than 1 are 0.59, 0.0036, and 0.00027.

Whole numbers (except zero) and decimal numbers can be **positive,** like $+5$, or **negative,** like -9. The $+$ and $-$ are called the **signs** of the numbers, and the 5 and 9 are called the **absolute values** of the numbers. A number with a negative sign is the opposite of the same number with a positive sign. (If the sign is not shown, the number is understood to be positive.) The **absolute value** of a positive number is that number, while the absolute value of a negative number is the opposite of that number. Thus, the absolute value of $+5$ is 5, but the absolute value of -9 is 9. The number 0 has no sign, and the absolute value of 0 is simply 0.

The relationship between positive and negative numbers is illustrated by the number line in Figure A-1. On the number line, the absolute value of a number is the distance between the number and zero. The distance between two points is always obtained by measuring from left to right (the positive direction), and this distance is a positive number. Thus, the distance between 0 and $+5$ is 5; the distance between -5 and 0 is also 5. This

FIGURE A-1　The Standard Number Line

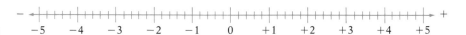

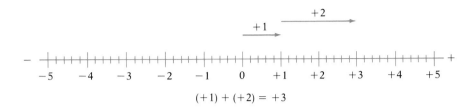

FIGURE A-2 Adding 1 and 2 on the Number Line

$(+1) + (+2) = +3$

means that the absolute value of $+5$ is 5, and the absolute value of -5 is also 5. The number line also illustrates that $-n$ is the opposite of $+n$, where n can be any whole number or decimal number.

EXAMPLE A.1 What is the absolute value of -4?

Solution The number line illustrates that the distance between -4 and 0 is 4. Thus, the absolute value of -4 is 4.

EXERCISE A.1 Give the absolute value of each number.

a. $+6$ **b.** -17 **c.** 0   **d.** $+3$ ■

Answers

a. 6   **b.** 17   **c.** 0   **d.** 3

Addition **Addition** is the operation in which two or more numbers are combined. It can be represented by one or more movements on the number line. Thus, adding 1 and 2, as illustrated in Figure A-2, is the same as starting at $+1$ and moving 2 units in the *positive* direction. The movement brings us to 3, the sum of 1 and 2. Adding -1 and -2 is illustrated in Figure A-3. Here, starting at -1, we move 2 units in the *negative* direction and arrive at -3, the sum of -1 and -2.

As Figures A-2 and A-3 show, if all the numbers to be added have the same sign, then the sum has the same sign as the original numbers. If such an addition involves more than two numbers, adding any two of the numbers corresponds to one movement on the number line. The numbers may be added in any order, and thus the movements may be made in any order.

In adding decimal numbers, special care must be taken to line up the decimal points vertically before adding.

EXAMPLE A.2 (Do not use a calculator.) Add the following numbers: 3.62, 17.6, 14.87, 319.

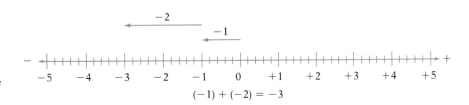

FIGURE A-3 Adding -1 and -2 on the Number Line

$(-1) + (-2) = -3$

Solution   The numbers may be added in any order, but the decimal points must be lined up vertically before beginning. Since the numbers are all positive, the answers will be positive.

$$
\begin{array}{rrr}
3.62 & 3.62 & 319. \\
17.6 & 14.87 & 14.87 \\
14.87 & 319. & 3.62 \\
\underline{319.} & \underline{17.6} & \underline{17.6} \\
355.09 & 355.09 & 355.09
\end{array}
$$
and so forth

EXERCISE A.2   (Do not use a calculator.) Add the following numbers: $-6.0, -459, -74.2, -0.37$. ∎

Answer -539.57

The addition of two numbers having *different* signs is illustrated in Figure A-4. In this case, $+5$ and -3 are added. To perform this addition on the number line, we start at $+5$ and move 3 units in the *negative* direction, arriving at $+2$. The sum of $+5$ and -3 is positive because the absolute value of $+5$ is greater than the absolute value of -3. In adding two numbers of different signs, three steps are necessary:

1. Find the absolute values of the numbers.
2. Take the smaller absolute value from the larger absolute value.
3. Give the answer the sign of the original number having the larger absolute value.

EXAMPLE A.3   (Do not use a calculator.) Add 23.6 and -17.5.

Solution   Step 1: The absolute values of the numbers are 23.6 and 17.5.

Step 2: Taking 17.5 from 23.6 gives 6.1, the absolute value of the answer.

Step 3: Since the original number having the larger absolute value, 23.6, is positive, the answer must also be positive.

$$
\begin{array}{r}
23.6 \\
-17.5 \\
\hline
6.1
\end{array}
$$

EXERCISE A.3   (Do not use a calculator.) Add 47.1 and -117. ∎

Answer -69.9

If three or more numbers of different signs are to be added, the operation can be broken down into three steps:

1. Add together all numbers of positive sign.

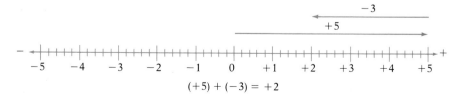

FIGURE A-4 Adding $+5$ and -3 on the Number Line

2. Add together all numbers of negative sign.
3. Treat the problem now as the addition of two numbers of different signs. The positive number is the sum of all original positive numbers, and the negative number is the sum of all original negative numbers.

EXAMPLE A.4   (Do not use a calculator.) Add the following numbers: $-10.86, 65.2, -17.3$.

Solution   Step 1:   Add together all numbers of positive sign. In this case, there is only one positive number, 65.2.

Step 2:   Add together all numbers of negative sign.

$$\begin{array}{r} -10.86 \\ \underline{-17.3} \\ -28.16 \end{array}$$

Step 3:   Now add 65.2 and -28.16.

$$\begin{array}{r} 65.2 \\ \underline{-28.16} \\ 37.04 \end{array}$$

EXERCISE A.4   (Do not use a calculator.) Add the following numbers: $79.2, -14.86, 21.31, -49.7$.   ■

Answer   35.95

Subtraction   **Subtraction** is the operation of finding the difference between two numbers. For example, $4 - 1$ can be illustrated on the number line as in Figure A-5, where reducing 4 by 1 is the same as moving from 4 one unit in the *negative* direction. The difference between 4 and 1 is thus 3. We can break down the process of subtracting one number from another into two steps:

1. Reverse the sign of the number to be subtracted.
2. Then add the numbers.

Thus, *subtraction is a form of addition in which the sign of the number to be subtracted is reversed.* In subtraction, just as in addition, decimal points must be lined up vertically before starting.

EXAMPLE A.5   (Do not use a calculator.) Complete the following calculation: $25.79 - 14.32$.

Solution   Step 1:   Reverse the sign of the number to be subtracted. Change 14.32 to -14.32.

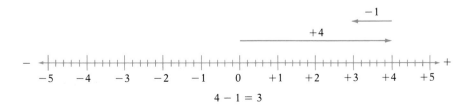

FIGURE A-5 Subtracting 1 from 4 on the Number Line

Step 2:   Add the numbers.

$$
\begin{array}{r}
25.79 \\
-14.32 \\
\hline
11.47
\end{array}
$$

EXERCISE A.5   (Do not use a calculator.) Complete the following calculations:

a. $49.63 - 65.33$ c. $7.56 - 9.62$
b. $17.42 - 12.09$ d. $23.66 - 10.19$ ■

Answers

a. -15.70 b. 5.33 c. -2.06 d. 13.47

Multiplication   **Multiplication** is a form of addition in which the same number is added two or more times. Thus, 2×3 means 2 added three times, or $2 + 2 + 2$; 2×3 also means 3 added twice, or $3 + 3$. The answer of $2 + 2 + 2$ is 6, as is the answer of $3 + 3$. In another example, 2×2832 means both 2 added two thousand eight hundred thirty-two times and 2832 added twice. The answer in both cases is 5664. Multiplication is indicated in various ways as follows:

$$2 \times 3 = 2 \cdot 3 = 2(3) = (2)(3)$$

The result of a multiplication is called the **product.**
   Signs must be treated carefully in multiplication; use the following rules.

RULE 1

If all numbers multiplied have positive signs, the product has a positive sign. Thus, $(2)(3) = 6$ means $2 + 2 + 2 = 6$, or $3 + 3 = 6$.

EXAMPLE A.6   Carry out the multiplication $(15)(65)(48)$.

Solution   Since all the numbers to be multiplied have positive signs, the product will have a positive sign:

$$(15)(65)(48) = 46,800$$

EXERCISE A.6   Carry out the following multiplications:

a. $(17)(36)$ c. $(2)(7)(15)(22)$
b. $(4)(27)(13)$ d. $(6)(75)(3)(14)(42)$ ■

Answers

a. 612   b. 1404   c. 4620   d. 793,800

RULE 2

If an even number of multipliers (the numbers being multiplied) have negative signs, the product has a positive sign. Thus, $(-2)(3)(-4) = 24$. The basis for this rule is that a number with a negative sign is to the left of zero on the number line, but multiplication of one negative number by another changes the direction of movement to the positive direction. Thus, an even number of negative multipliers results in a positive product.

EXAMPLE A.7   Carry out the following multiplication: $(-7)(4)(-36)$.

Solution   Since an even number of the multipliers (two) have negative signs, the product has a positive sign. Thus,

$$(-7)(4)(-36) = 1008$$

EXERCISE A.7   Carry out the following multiplications:

a. $(-7)(-3)$ c. $(86)(4)(-6)(-3)$
b. $(-2)(-6)(17)(8)$ d. $(-3)(7)(-8)(12)$ ■

Answers

a. 21   b. 1632   c. 6192   d. 2016

RULE 3

If an odd number of the multipliers have negative signs, the product has a negative sign. Thus, $(-2)(3)(4) = -24$. The basis for this rule is that multiplying one negative number by another results in a positive product (Rule 2), but multiplication of the positive product by a third negative number *reverses* the direction on the number line. Hence, multiplying an odd number of negative numbers together brings about an odd number of direction reversals on the number line, always ending in the negative direction.

EXAMPLE A.8   Carry out the following multiplication: $(-2)(3)(-18)(-7)$

Solution   Since there is an odd number of negative multipliers, the product has a negative sign:

$$(-2)(3)(-18)(-7) = -756$$

EXERCISE A.8   Carry out the following multiplications:

a. $(-6)(-75)(-5)$ c. $(-23)(4)(-99)(-2)$
b. $(16)(-2)$ d. $(-17)(4)(-18)(-7)(-2)(-7)$ ■

Answers

a. -2250 b. -32 c. $-18,216$ d. $-119,952$

Roots and Exponents   A **root** is a number that, when multiplied by itself n times, gives another number y. If n is 2, the root is called the **square root** of y, and if n is 3, the root is called the **cube root** of y. The square root of y is expressed as

$$\sqrt{y}$$

and the cube root of y is expressed as

$$\sqrt[3]{y}$$

The **square root** of a number is one of the number's two equal factors. Thus, the square root of 4 is 2, or

$$\sqrt{4} = 2$$

This means that $(2)(2) = 4$.

The **cube root** of a number is one of the number's three equal factors. For example, the cube root of 8 is 2, or

$$\sqrt[3]{8} = 2$$

This means simply that $(2)(2)(2) = 8$, or 2 multiplied by itself three times is 8.

Whole-number roots are found by finding the whole number which, when multiplied by itself the specified number of times, gives the original number. Decimal-number roots can be computed on a calculator or found in tables. If these are not available, decimal-number roots can be approximated by educated guessing.

EXAMPLE A.9 Find the following whole-number roots:

a. $\sqrt{16}$ b. $\sqrt[3]{27}$

Solution a. By considering small, whole-number factors of 16, we find that 4 multiplied by itself twice gives 16. Thus,

$$\sqrt{16} = 4$$

b. By considering small, whole-number factors of 27, we find that 3 multiplied by itself three times gives 27. Thus,

$$\sqrt[3]{27} = 3$$

EXERCISE A.9 Find the following whole-number roots:

a. $\sqrt{36}$ b. $\sqrt{81}$ c. $\sqrt[3]{125}$ d. $\sqrt[3]{64}$

Answers

a. 6   b. 9   c. 5   d. 4

A number n multiplied by itself 2 times can be expressed as n^2, where 2 is called the **power** or the **exponent** of n. Thus, $2^2 = (2)(2) = 4$ and $2^3 = (2)(2)(2) = 8$. In general terms, then, n^x means a number n multiplied by itself x times. Other examples are

$$3^2 = (3)(3) = 9 \quad \text{(three ''squared'')}$$
$$4^3 = (4)(4)(4) = 64 \quad \text{(four ''cubed'')}$$
$$2^5 = (2)(2)(2)(2)(2) = 32 \quad \text{(two raised to the fifth power)}$$

The concept of powers (exponents) applies to units of measurement as well as to pure numbers. Thus, in^2 means inches ''squared'' or square inches, and cm^3 means centimeters ''cubed,'' or cubic centimeters.

In the preceding multiplication examples, it was assumed that numbers without units were exact; that is, they contained an infinite number of zeros to the right of the decimal point. Most numbers used in chemistry, however, represent measurements and thus are not exact. In these cases, answers arrived at by multiplication must be expressed to the correct number of significant figures; this topic is discussed in sections 2.2 and 2.3 of the textbook. Numbers representing measurements also have units associated with them, and the units must be included in calculations. Thus, the area of a square that has sides 1.0 cm (centimeters) long is calculated as follows:

$$A = 1^2$$
$$= (1.0 \text{ cm})(1.0 \text{ cm})$$

Since numbers and units can be multiplied in any order,

$$A = (1.0)(1.0)(\text{cm})(\text{cm})$$
$$= (1.0)^2 \text{ cm}^2$$
$$= 1.0 \text{ cm}^2$$

EXAMPLE A.10

Find the area of a circle having a radius of 7.2 cm. ($A = \pi r^2$; $\pi = 3.14$).

Solution

$$A = \pi r^2$$

Substituting values for π and r,

$$A = (3.14)(7.2 \text{ cm})^2$$
$$= (3.14)(7.2)(7.2)(\text{cm})(\text{cm})$$

Now we can multiply the numbers together and round off to the correct number of significant figures:

$$= 160(\text{cm})(\text{cm})$$

Finally, we express (cm)(cm) as cm^2,

$$A = 160 \text{ cm}^2$$

EXERCISE A.10

Find the volume of a rectangular solid whose dimensions are as follows: length (l) = 4.3 cm, width (w) = 1.2 cm, height (h) = 2.4 cm. ($V = lwh$). ▒

Answer   12 cm^3

Division   **Division** is simply the repeated subtraction of one number from another. If we subtract one number from another as many times as is necessary to get to zero, the number of subtractions is the answer (the **quotient**) of the division. As an illustration, let's subtract 4 from 20 until we get to zero.

$$20 - 4 = 16$$
$$16 - 4 = 12$$
$$12 - 4 = 8$$
$$8 - 4 = 4$$
$$4 - 4 = 0$$

Since it took five subtractions, we conclude that 20 contained five fours, or that $20 \div 4 = 5$. Another way of writing $20 \div 4$ is $^{20}\!/_4$. Thus, $^{20}\!/_4 = 5$. In division, the number that is divided (20 in this case) is called the **dividend,** and the number that does the dividing (4 in this case) is called the **divisor.** If the dividend and divisor have different signs, the quotient will have a negative sign. If the dividend and divisor have the same sign, the quotient will have a positive sign.

EXAMPLE A.11

Carry out the following operation: $\dfrac{(-5)(4)}{2}$

Solution

The dividend, the product of $(-5)(4)$, has a negative sign. Since the signs of the dividend and divisor are different, the quotient will have a negative sign.

$$\frac{(-5)(4)}{2} = \frac{-20}{2} = -10$$

EXERCISE A.11

Carry out the following operations:

a. $\dfrac{(16)(3)(-2)}{4}$ c. $\dfrac{(4)(5)(15)}{-10}$

b. $\dfrac{(5)(-6)(-3)}{30}$ d. $\dfrac{(-2)(7)(-6)}{-3}$

Answers

a. -24 b. 3 c. -30 d. -28

When the same number appears in the dividend and divisor, the calculation can be simplified by dividing the number into itself. This process is known as cancellation. In the division

$$\frac{(2)(3)(4)}{(2)(5)}$$

the 2 of the divisor divides into the 2 of the dividend one time, or $\frac{2}{2} = 1$. We can show the cancellation as

$$\frac{(\cancel{2})(3)(4)}{(\cancel{2})(5)} = \frac{(3)(4)}{5}$$

When numbers represent measurements, the numbers have units associated with them. For example, 2 cm and 4 cm represent measurements. When one measurement is divided by another measurement, as in 4 cm/2 cm, the units that are the same can be cancelled. Thus, in our example, cm/cm = 1, and

$$\frac{4 \cancel{\text{ cm}}}{2 \cancel{\text{ cm}}} = \frac{(4)(1)}{(2)(1)} = 2$$

Let's look at another example where identical units are raised to different powers:

$$\frac{10 \text{ cm}^2}{2 \text{ cm}} = \frac{(10)(\text{cm})(\cancel{\text{cm}})}{(2)(\cancel{\text{cm}})} = \frac{(10)(\text{cm})(1)}{(2)(1)} = 5 \text{ cm}$$

If, on the other hand, the units are not the same, they cannot be cancelled, and they appear as the units of the answer:

$$\frac{10 \text{ cm}}{2 \text{ sec}} = 5 \text{ cm/sec} \quad (5 \text{ centimeters per second})$$

EXAMPLE A.12 Carry out the following operation: $\dfrac{30 \text{ cm}^3}{5 \text{ cm}}$

Solution
$$\frac{30 \text{ cm}^3}{5 \text{ cm}} = \frac{(30)(\text{cm})(\text{cm})(\text{cm})}{(5)(\text{cm})} = \frac{(30)(\text{cm})(\text{cm})(1)}{(5)(1)} = 6 \text{ cm}^2$$

EXERCISE A.12 Carry out the following operations:

a. $\dfrac{(15 \text{ mL})(20 \text{ g})}{4 \text{ mL}}$

c. $\dfrac{(4.0 \text{ in})(9.0 \text{ cm})}{(2.0 \text{ in})(15 \text{ sec})}$

b. $\dfrac{(10 \text{ lb})(16 \text{ oz})}{1.0 \text{ lb}}$

d. $\dfrac{15 \text{ lb}}{(2.0 \text{ in})(3.0 \text{ in})}$

Answers

a. 75 g   b. 160 oz   c. 1.2 cm/sec   d. 2.5 lb/in²

Fractions A **fraction** is a ratio of two numbers, such as ¼ or 9/2. When a fraction is written, as shown in the preceding sentence, with one number divided by another, the fraction is called a **common fraction.** The number on the top can be referred to as the dividend, but it is more often called the **numerator.** The number on the bottom is actually the divisor, but it is usually called the **denominator** of the fraction.

$$\frac{1}{4} \quad \begin{array}{l} \text{numerator (dividend)} \\ \text{denominator (divisor)} \end{array}$$

The fraction ¼ means one part out of a total of four parts; it also means 1 divided by 4.

A fraction can have a numerator or a denominator of 1, as in ¼ and 4/1. A number that is not written as a fraction is understood to have a denominator of 1; thus the number 4 is understood to be 4/1. A fraction can never have a denominator of 0, because division by 0 would give an answer of infinity. A fraction with a numerator of 0, 0/4 for example, has the value of 0, because 0 divided by any number is always 0.

A fraction whose numerator is greater than its denominator, such as 9/2, is referred to as an **improper fraction.** An improper fraction can be converted to a **mixed fraction** by carrying out the division (9/2 = 4½), but if the improper fraction is to be used in a calculation, it is convenient to leave it in improper form. A fraction whose numerator is greater than its denominator has a value greater than 1, while a fraction whose numerator is less than its denominator has a value less than 1.

When fractions are to be added or subtracted, the fractions must have the same denominator, which is called a **common denominator.** If one fraction has a denominator that is a whole-number multiple of the denominator of another fraction, such as $\frac{1}{3}$ and $\frac{1}{6}$, the larger number (6) is used as the common denominator. The other fraction is converted to a second, equivalent fraction having the common denominator as follows:

1. Divide the denominator of the other fraction ($\frac{1}{3}$) into the common denominator (6).

$$\frac{6}{3} = 2$$

2. Multiply the quotient of step 1 by the original numerator of the other fraction.

$$(2)(1) = 2$$

3. The product of step 2 then becomes the numerator of the new fraction.

$$\frac{1}{3} = \frac{2}{6}$$

The two fractions can now be added. Addition is performed by adding the numerators; the denominator of the answer is the common denominator.

$$\frac{1}{3} + \frac{1}{6} = \frac{2}{6} + \frac{1}{6} = \frac{3}{6}$$

We can simplify the answer, $\frac{3}{6}$, by dividing both numerator and denominator by 3:

$$\frac{3}{6} = \frac{\frac{3}{3}}{\frac{6}{3}} = \frac{1}{2}$$

Dividing or multiplying both the numerator and the denominator by the same quantity does not change the value of the fraction.

EXAMPLE A.13   Carry out the following operations:

a. $\frac{3}{16} + \frac{1}{8}$ **b.** $\frac{2}{3} - \frac{4}{9}$

Solution   **a.** First, find the common denominator:

Step 1: $\frac{16}{8} = 2$

Step 2: $(2)(1) = 2$

Step 3: $\frac{1}{8} = \frac{2}{16}$

Now, add the two fractions:

$$\frac{3}{16} + \frac{1}{8} = \frac{3}{16} + \frac{2}{16} = \frac{5}{16}$$

b. First, find the common denominator:

Step 1: $\frac{9}{3} = 3$

Step 2: $(3)(2) = 6$

Step 3: $\frac{2}{3} = \frac{6}{9}$

Now, subtract the fractions:

$$\frac{2}{3} - \frac{4}{9} = \frac{6}{9} - \frac{4}{9} = \frac{2}{9}$$

EXERCISE A.13 Carry out the following operations:

a. $6/4 + 1/2$ b. $1/3 + 1/6$ c. $7/24 - 1/8$ d. $9/15 - 2/5$ ■

Answers

a. $8/4$ or 2   b. $3/6$ or $1/2$ c. $4/24$ or $1/6$ d. $3/15$ or $1/5$

If one of the fractions to be added or subtracted does not have a denominator that is a whole-number multiple of the other denominator(s), as in

$$1/4 - 1/6$$

the denominators are multiplied together, $(4)(6) = 24$, to find the common denominator. Then each fraction is converted to an equivalent fraction having the common denominator, as described in steps 1 and 2 above,

$$1/4 = 6/24$$
$$1/6 = 4/24$$

and the addition or subtraction is performed.

$$1/4 - 1/6 = 6/24 - 4/24 = 2/24$$

We can simplify the answer, $2/24$, by dividing numerator and denominator by 2

$$2/24 = \frac{2/2}{24/2} = 1/12$$

EXAMPLE A.14 Carry out the following operations:

a. $2/3 + 3/5$ b. $3/4 - 1/3$

Solution   a. First, find the common denominator:

$$(3)(5) = 15$$

Next, convert each fraction to the equivalent fraction having the common denominator:

$$2/3 = 10/15$$
$$3/5 = 9/15$$

Last, add the new fractions:

$$2/3 + 3/5 = 10/15 + 9/15 = 19/15 \text{ or } 1\,4/15$$

b. First, find the common denominator:

$$(4)(3) = 12$$

Next, convert each fraction to the equivalent fraction having the common denominator:

$$3/4 = 9/12$$
$$1/3 = 4/12$$

Last, subtract the new fractions:

$$3/4 - 1/3 = 9/12 - 4/12 = 5/12$$

EXERCISE A.14   Carry out the following operations:

a. $\frac{1}{6} + \frac{2}{5}$ b. $\frac{1}{2} + \frac{4}{3}$ c. $\frac{3}{4} - \frac{1}{6}$ d. $\frac{4}{5} - \frac{2}{3}$ ■

Answers

a. $\frac{17}{30}$ b. $\frac{11}{6}$ or $1\frac{5}{6}$ c. $\frac{14}{24}$ or $\frac{7}{12}$ d. $\frac{2}{15}$

Fractions need not have a common denominator to be multiplied or divided. In multiplication, the numerators are multiplied to give the numerator of the product, and denominators are multiplied to give the denominator of the product.

$$\frac{2}{3} \times \frac{1}{6} = \frac{(2)(1)}{(3)(6)} = \frac{2}{18} = \frac{1}{9}$$

$$\frac{3}{4} \times \frac{1}{6} = \frac{(3)(1)}{(4)(6)} = \frac{3}{24} = \frac{1}{8}$$

In dividing one fraction by another, the denominator fraction is inverted (turned upside down) and then multiplied by the numerator fraction.

$$\frac{1}{3} \div \frac{1}{6} = \frac{\frac{1}{3}}{\frac{1}{6}} = \left(\frac{1}{3}\right)\left(\frac{6}{1}\right) = \frac{(1)(6)}{(3)(1)} = \frac{6}{3} = 2$$

$$\frac{1}{4} \div \frac{1}{6} = \frac{\frac{1}{4}}{\frac{1}{6}} = \left(\frac{1}{4}\right)\left(\frac{6}{1}\right) = \frac{(1)(6)}{(4)(1)} = \frac{6}{4} = \frac{3}{2}$$

EXAMPLE A.15   Carry out the following operations:

a. $\frac{3}{4} \times \frac{2}{5}$ b. $\frac{1}{7} \div \frac{2}{3}$

Solution

a. $\frac{3}{4} \times \frac{2}{5} = \frac{(3)(2)}{(4)(5)} = \frac{6}{20} = \frac{3}{10}$

b. $\frac{1}{7} \div \frac{2}{3} = \frac{\frac{1}{7}}{\frac{2}{3}} = \left(\frac{1}{7}\right)\left(\frac{3}{2}\right) = \frac{3}{14}$

EXERCISE A.15   Carry out the following operations:

a. $\frac{3}{5} \times \frac{1}{2}$ b. $\frac{7}{16} \times \frac{2}{3}$ c. $\frac{4}{11} \div \frac{3}{4}$ d. $\frac{5}{8} \div \frac{1}{9}$ ■

Answers

a. $\frac{3}{10}$ b. $\frac{14}{48}$ or $\frac{7}{24}$ c. $\frac{16}{33}$ d. $\frac{45}{8}$ or $5\frac{5}{8}$

TABLE A.1   Decimal Numbers Corresponding to Common Fractions

Fraction	Decimal Number
$\frac{1}{5}$	0.200
$\frac{1}{4}$	0.250
$\frac{1}{3}$	0.333
$\frac{3}{8}$	0.375
$\frac{2}{5}$	0.400
$\frac{1}{2}$	0.500

Decimal Numbers and Percent   Since every fraction can be represented as a/b, or $a \div b$, the division of the numerator by the denominator produces a **decimal number** (sometimes called a **decimal fraction**). A **decimal number** is simply a fraction written in decimal form. Table A.1 lists examples of decimal numbers corresponding to some common fractions. Note that a zero is usually placed before a decimal number whose value is less than 1 to emphasize the location of the decimal point.

A decimal number can be thought of as a fraction having a multiple of ten as its denominator. Thus, $0.2 = \frac{2}{10}$, and $0.50 = \frac{50}{100}$. A decimal number can be expressed as a percentage by multiplying the decimal number by 100%. The term **percent** means parts per 100. Thus, 0.50 expressed as a percentage is simply $(0.50)(100\%) = 50\%$, meaning "50 parts out of 100 parts." Percent is obtained as follows:

$$a/b \times 100\% = \text{\_\_\_\_}\%$$

To convert % to a decimal number, simply divide % by 100%:

$$50\%/100\% = 0.50$$

EXAMPLE A.16   Convert the following numbers to percent:

   a. ¾   **b.** 0.67

Solution   **a.** ¾ $\times$ 100% = 0.75 $\times$ 100% = 75%   **b.** 0.67 $\times$ 100% = 67%

EXERCISE A.16   Convert the following numbers to percent:

   a. ³⁄₁₆   **b.** ⁴⁄₉   **c.** 0.33   **d.** 0.147   ■

Answers

   a. 18.8%   **b.** 44.4%   **c.** 33%   **d.** 14.7%

A.3 DIAGNOSTIC TEST FOR BASIC ALGEBRA

Following this paragraph is a short diagnostic test on basic algebra. First, work out answers to each problem, and then check your work using the answers given at the end of the test. If your score was 100%, then your background in basic algebra is probably sufficient for introductory chemistry, and you can go on to the next diagnostic test. If you got at least 80% of the answers correct, go back through your work and find your mistakes. If you made only minor errors and can rework the problems correctly, there should be no need for further review. However, if you cannot correct your work or if you scored less than 80%, you should work through Section A.4, Review of Basic Algebra.

DIAGNOSTIC TEST FOR BASIC ALGEBRA

A. Solve each equation for x.

   1. $2x - 6 = 10$ **4.** $3(2x - 1) + 2x = 4(x - 3)$

   2. $5x - 7 = 2x + 2$ **5.** $\dfrac{x}{10} = \dfrac{5}{6}$

   3. $2(3x - 1) = 5x - (x + 2)$ **6.** $\dfrac{25}{(3x)} + \dfrac{1}{3} = \dfrac{10}{x}$

B. Solve each problem.

   1. The difference between two numbers is 3. If five times the smaller number is divided by the larger number, the quotient is 4. What are the numbers?

   2. A collection of quarters and half dollars has a value of $8.75. If there are 8 fewer half dollars than quarters, how many quarters and half dollars are in the collection?

   3. For any rectangle, the perimeter is given by the formula $P =$

$2(l + w)$, where l is the length and w is the width of the rectangle. The length of a particular rectangle is 6 feet greater than its width. If its perimeter is 44 feet, find the length and width of the rectangle.

4. How many ounces of a 30% acid solution should be added to 20 ounces of a 60% acid solution in order to obtain a 50% acid solution?

5. The pressure on an object submerged in a liquid is directly proportional to the depth of the object's submersion. In an experiment, the pressure on the object at a depth of 4 m is 9 kg/cm². What pressure will be exerted at 10 m?

6. Boyle's law states that the volume (V) of a gas is inversely proportional to the pressure (P) at fixed temperature. If $V = 500$ L when $P = 25$ torr, find V when $P = 40$ torr.

C. Complete each of the following.
1. log 500 =   3. antilog 2.43 =
2. log 0.39 =   4. antilog -1.65 =

Answers

Section A. 1. 8   2. 3   3. 0   4. $-\frac{9}{4}$ 5. $\frac{25}{3}$ 6. 5
Section B. 1. 12, 15   2. 17 quarters, 9 half dollars   3. 14 ft, 8 ft
    4. 10 ounces   5. 22.5 kg/cm²   6. 312.5 L
Section C. 1. 2.7   2. -0.41 3. 2.7×10^2 4. 2.2×10^{-2}

A.4   REVIEW OF BASIC ALGEBRA

Solving Algebraic Equations   One of the most important uses of algebra in chemistry is formulating and solving equations. An **equation** is a statement that two quantities are equal. One of the simplest kinds of equations is a **linear equation,** an equation in which an unknown quantity, often designated as x, has an exponent of 1. Although linear equations can have many different forms, each can be written in the following standard form:

$$ax = b$$

where a and b are numbers and x is the unknown quantity. A linear equation is solved by isolating the unknown quantity on one side of the equation and the other quantities on the other side. Thus, the solution of $ax = b$ is $x = b/a$. To solve a linear equation, such as

$$3(2x + 6) = 30$$

use the following steps.

1. Simplify both sides of the equation, if necessary, by carrying out any indicated multiplication or division that is possible. In our example, we can multiply 3 by $2x + 6$. Thus,

$$3(2x + 6) = 30$$
$$3(2x) + 3(6) = 30$$
$$6x + 18 = 30$$

2. Then, rearrange the equation to the standard form $ax = b$ by using the following principles.

a. A quantity may be added to or subtracted from both sides of an equation without changing the meaning of the equation.

b. Both sides of an equation can be multiplied or divided by the same quantity without changing the meaning of the equation.

c. Both sides of an equation can be inverted (turned upside down) without changing the meaning of the equation.

d. Quantities equal to the same quantity are equal to each other.

In our example, if we subtract 18 from both sides of the equation, (principle a),

$$6x + 18 - 18 = 30 - 18$$

or

$$6x = 12$$

we can remove 18 from the left side. Now the equation is of the standard form $ax = b$.

3. Last, solve for the unknown quantity by dividing both sides of the equation by a. In this case, a is 6.

$$\frac{(6x)}{6} = \frac{12}{6}$$

$$x = 2$$

The solution can be checked easily by substituting 2 for x in the original equation:

$$3(2x + 6) = 30$$
$$3[(2)(2) + 6] = 30$$
$$3[4 + 6] = 30$$
$$3(10) = 30$$
$$30 = 30$$

Example A.17 shows the solution of linear equations.

EXAMPLE A.17

Solve each equation for the unknown quantity.

a. $x + 3 = 8$ **d.** $2(3x - 5) = 7(2x + 5) - 5$

b. $y = 12 - y$ **e.** $\dfrac{t}{6} + 2 = \dfrac{t}{4}$

c. $\dfrac{2z}{3} = 8$ **f.** $\dfrac{(3x - 3)}{6} = \dfrac{(4x + 1)}{15} + 2$

Solution

a. Step 1: Simplify the equation. In this case, step 1 is not necessary.

$$x + 3 = 8$$

Step 2: Rearrange the equation to the standard form $ax = b$. In this case, we must subtract 3 from each side.

$$x + 3 - 3 = 8 - 3$$
$$x = 5$$

Step 3:   Solve for x by dividing both sides of the equation by a. In this case, a is 1, and the equation is solved. To check, substitute 5 for x in the original equation.

$$x + 3 = 8$$
$$5 + 3 = 8$$
$$8 = 8$$

b. Step 1:   Simplify the equation. In this case, step 1 is not necessary.

$$y = 12 - y$$

Step 2:   Rearrange the equation to the standard form $ax = b$. In this case, we must add y to each side.

$$y + y = 12 - y + y$$

Then, we get

$$2y = 12$$

Step 3:   Solve for the unknown quantity. In this case, we must divide each side by 2.

$$\frac{2y}{2} = \frac{12}{2}$$
$$y = 6$$

Checking:

$$y = 12 - y$$
$$6 = 12 - 6$$
$$6 = 6$$

c. Step 1:   Simplify the equation. In this case, step 1 is not necessary.

$$\frac{2z}{3} = 8$$

Step 2:   Rearrange the equation to the standard form $ax = b$. In this case, we must multiply both sides by 3.

$$\frac{2z}{3} = 8$$
$$(3)\left(\frac{2z}{3}\right) = (3)(8)$$
$$2z = 24$$

Step 3:   Solve for the unknown quantity. Now we must divide both sides by 2.

$$\frac{2z}{2} = \frac{24}{2}$$
$$z = 12$$

Checking:

$$\frac{2z}{3} = 8$$

$$\frac{(2)(12)}{3} = 8$$

$$\frac{24}{3} = 8$$

$$8 = 8$$

d. Step 1:   Simplify the equation. In this case, we must carry out the indicated multiplication

$$2(3x - 5) = 7(2x + 5) - 5$$
$$2(3x) - 2(5) = 7(2x) + 7(5) - 5$$
$$6x - 10 = 14x + 35 - 5$$

and then the subtraction $(35 - 5)$ on the right side.

$$6x - 10 = 14x + 30$$

Step 2:   Rearrange the equation to the standard form $ax = b$. To do this, we must add 10 to both sides,

$$6x - 10 + 10 = 14x + 30 + 10$$
$$6x = 14x + 40$$

and subtract $14x$ from both sides.

$$6x - 14x = 14x + 40 - 14x$$
$$-8x = 40$$

Step 3:   Solve for the unknown quantity. Now we must divide both sides by -8.

$$\frac{-8x}{-8} = \frac{40}{-8}$$

$$x = -5$$

Checking:

$$2(3x - 5) = 7(2x + 5) - 5$$
$$2[(3)(-5) - 5] = 7[(2)(-5) + 5] - 5$$
$$2[-15 - 5] = 7[-10 + 5] - 5$$
$$2(-20) = 7(-5) - 5$$
$$-40 = -35 - 5$$
$$-40 = -40$$

e. Step 1:   Simplify the equation. In this case, the equation

$$\frac{t}{6} + 2 = \frac{t}{4}$$

contains fractions, and we must clear it of the fractions before proceeding. To do this we first determine the common denominator of all the fractions in the equation. The lowest common denominator (the lowest whole number into which 6 and 4 divide evenly) is 12, so we must multiply both sides of the equation by 12 to clear the fractions.

$$12\left(\frac{t}{6}+2\right)=12\left(\frac{t}{4}\right)$$

$$12\left(\frac{t}{6}\right)+12(2)=12\left(\frac{t}{4}\right)$$

$$2t+24=3t$$

Now we can proceed to step 2.

Step 2: Rearrange the equation to the standard form $ax=b$. In this case, we must subtract $2t$ from both sides:

$$2t+24-2t=3t-2t$$

$$24=t$$

Step 3: Solve for the unknown quantity. In this case, a is 1 and the equation is already solved:

$$24=t$$

Checking:

$$\frac{t}{6}+2=\frac{t}{4}$$

$$\frac{24}{6}+2=\frac{24}{4}$$

$$4+2=6$$

$$6=6$$

f. Step 1: Simplify the equation. In this case, as in part e, we must clear the equation of fractions by multiplying both sides by the common denominator.

$$\frac{(3x-3)}{6}=\frac{(4x+1)}{15}+2$$

The lowest common denominator (the lowest whole number into which 6 and 16 divide evenly) is 30. Thus,

$$30\left(\frac{3x-3}{6}\right)=30\left[\frac{(4x+1)}{15}+2\right]$$

$$5(3x-3)=2(4x+1)+60$$

To simplify further, we carry out any multiplication or division that is possible.

$$5(3x-3)=2(4x+1)+60$$

$$5(3x)-5(3)=2(4x)+2(1)+60$$

$$15x-15=8x+2+60$$

$$15x-15=8x+62$$

Step 2: Rearrange the equation to the standard form $ax = b$. In this case, we must add 15 to both sides,

$$15x - 15 + 15 = 8x + 62 + 15$$
$$15x = 8x + 77$$

and subtract $8x$ from both sides.

$$15x - 8x = 8x + 77 - 8x$$
$$7x = 77$$

Step 3: Solve for the unknown quantity. Now we must divide both sides by 7.

$$\frac{7x}{7} = \frac{77}{7}$$
$$x = 11$$

Checking:

$$\frac{(3x - 3)}{6} = \frac{(4x + 1)}{15} + 2$$
$$\frac{[(3)(11) - 3]}{6} = \frac{[(4)(11) + 1]}{15} + 2$$
$$\frac{(33 - 3)}{6} = \frac{(44 + 1)}{15} + 2$$
$$\frac{30}{6} = \frac{45}{15} + 2$$
$$5 = 3 + 2$$
$$5 = 5$$

EXERCISE A.17   Solve each equation for the unknown quantity.

a. $\dfrac{3m}{7} = 21$ d. $(4k - 3) - 2(k + 4) = 3(k + 7)$

b. $2t + 3 = t$ e. $\dfrac{2w}{7} = 1 - \dfrac{(2w + 1)}{3}$

c. $5v = 6 + 3v$ f. $\dfrac{v}{3} - \dfrac{1}{6} = \dfrac{(v + 1)}{9}$ ■

Answers

a. $m = 49$ c. $v = 3$ e. $w = \frac{7}{10}$
b. $t = -3$ d. $k = -32$ f. $v = \frac{5}{4}$

A second type of algebraic equation used in chemistry is the **quadratic equation,** an equation in which an unknown quantity appears with an exponent of 2 in one term and an exponent of 1 in another term. (A **term** is a part of an equation separated from other parts by $+$ or $-$ signs.) The standard form of a quadratic equation is

$$ax^2 + bx + c = 0$$

where a, b, and c are numbers, and x is the unknown quantity. The only restriction is that a cannot be 0. (If a were 0, then the equation would become $bx + c = 0$, a linear equation.) Certain kinds of quadratic equations can be solved by specialized methods, but most quadratic equations used in chemistry are solved using the **quadratic formula:**

$$x = \frac{-b \pm \sqrt{b^2 - 4ac}}{2a}$$

where a, b, and c have the same values as in the specific quadratic equation being solved. The $\pm$ sign means that there are two values of the unknown quantity:

$$x = \frac{-b + \sqrt{b^2 - 4ac}}{2a}$$

and

$$x = \frac{-b - \sqrt{b^2 - 4ac}}{2a}$$

To solve a quadratic equation using the quadratic formula, you must first write the quadratic equation in standard form ($ax^2 + bx + c = 0$), and substitute values of a, b, and c into the quadratic formula. Then perform the indicated operations to solve for x. Example A.18 illustrates the process.

EXAMPLE A.18   Solve the following quadratic equations.

  a. $x^2 - 5x - 6 = 0$ **b.** $10y^2 - 11y - 6 = 0$ **c.** $3x^2 = 8 - 10x$

Solution   **a.** $x^2 - 5x - 6 = 0$

$$x = \frac{-b \pm \sqrt{b^2 - 4ac}}{2a}$$

$$x = \frac{-(-5) \pm \sqrt{(-5)^2 - 4(1)(-6)}}{2(1)}$$

$$= \frac{5 \pm \sqrt{25 + 24}}{2}$$

$$= \frac{5 \pm \sqrt{49}}{2}$$

$$= \frac{5 \pm 7}{2}$$

$$x = \frac{5 + 7}{2} = \frac{12}{2} = 6$$

and

$$x = \frac{5 - 7}{2} = \frac{-2}{2} = -1$$

Checking $x = 6$:

$$x^2 - 5x - 6 = 0$$
$$6^2 - 5(6) - 6 = 0$$
$$36 - 30 - 6 = 0$$
$$0 = 0$$

Checking $x = -1$:

$$x^2 - 5x - 6 = 0$$
$$(-1)^2 - 5(-1) - 6 = 0$$
$$1 + 5 - 6 = 0$$
$$0 = 0$$

b. $\qquad 10y^2 - 11y - 6 = 0$

$$y = \frac{-b \pm \sqrt{b^2 - 4ac}}{2a}$$

$$y = \frac{-(-11) \pm \sqrt{(-11)^2 - 4(10)(-6)}}{2(10)}$$

$$= \frac{11 \pm \sqrt{121 + 240}}{20}$$

$$= \frac{11 \pm \sqrt{361}}{20}$$

$$= \frac{11 \pm 19}{20}$$

$$x = \frac{11 + 19}{20} = \frac{30}{20} = \frac{3}{2}$$

and

$$x = \frac{11 - 19}{20} = \frac{-8}{20} = -\frac{2}{5}$$

Checking $y = \frac{3}{2}$:

$$10y^2 - 11y - 6 = 0$$

$$10\left(\frac{3}{2}\right)^2 - 11\left(\frac{3}{2}\right) - 6 = 0$$

$$10\left(\frac{9}{4}\right) - 11\left(\frac{3}{2}\right) - 6 = 0$$

$$\frac{90}{4} - \frac{33}{2} - 6 = 0$$

$$\frac{90}{4} - \frac{66}{4} - \frac{24}{4} = 0$$

$$0 = 0$$

Checking $y = -\tfrac{2}{5}$:

$$10y^2 - 11y - 6 = 0$$

$$10\left(-\frac{2}{5}\right)^2 - 11\left(-\frac{2}{5}\right) - 6 = 0$$

$$10\left(\frac{4}{25}\right) - 11\left(-\frac{2}{5}\right) - 6 = 0$$

$$\frac{40}{25} + \frac{22}{5} - 6 = 0$$

$$\frac{40}{25} + \frac{110}{25} - \frac{150}{25} = 0$$

$$0 = 0$$

c. In this case, we must first rearrange the quadratic equation to standard form.

$$3x^2 = 8 - 10x$$
$$3x^2 + 10x = 8 - 10x + 10x$$
$$3x^2 + 10x = 8$$
$$3x^2 + 10x - 8 = 8 - 8$$
$$3x^2 + 10x - 8 = 0$$

Now we apply the quadratic formula.

$$x = \frac{-b \pm \sqrt{b^2 - 4ac}}{2a}$$

$$= \frac{-10 \pm \sqrt{(10)^2 - 4(3)(-8)}}{2(3)}$$

$$= \frac{-10 \pm \sqrt{100 + 96}}{6}$$

$$= \frac{-10 \pm \sqrt{196}}{6}$$

$$= \frac{-10 \pm 14}{6}$$

$$x = \frac{-10 + 14}{6} = \frac{4}{6} = \frac{2}{3}$$

and

$$x = \frac{-10 - 14}{6} = \frac{-24}{6} = -4$$

Checking $x = \tfrac{2}{3}$:

$$3x^2 = 8 - 10x$$

$$3\left(\frac{2}{3}\right)^2 = 8 - 10\left(\frac{2}{3}\right)$$

$$3\left(\frac{4}{9}\right) = 8 - 10\left(\frac{2}{3}\right)$$

$$\frac{12}{9} = 8 - \frac{20}{3}$$

$$\frac{12}{9} = \frac{72}{9} - \frac{60}{9}$$

$$\frac{12}{9} = \frac{12}{9}$$

Checking $x = -4$:

$$3x^2 = 8 - 10x$$

$$3(-4)^2 = 8 - 10(-4)$$

$$3(16) = 8 - 10(-4)$$

$$48 = 8 + 40$$

$$48 = 48$$

Word Problems Most mathematical problems in chemistry are in the form of word problems, in which information is stated and a question is asked. In these problems it is necessary to read carefully, identify the given information, and use it to construct an algebraic equation. The equation should contain the unknown quantity, often designated as x, and at least some of the given information. Solving the equation for the unknown quantity then provides the answer to the question.

The key to solving word problems is the formulation of mathematical expressions that describe verbal information. Examples of such verbal information and their corresponding mathematical expressions are given below.

Verbal Information: The sum of a number and 10

Mathematical Expression: $x + 10$

Verbal Information: Four more than twice a number

Mathematical Expression: $4 + 2x$

Verbal Information: The sum of a number and four times itself

Mathematical Expression: $x + 4x$

Verbal Information: One number, x, is larger than another number, y, by 4.

Mathematical Expression: $x = y + 4$ or $x - y = 4$

Verbal Information: One number, x, has the same value as four times another number, y.

Mathematical Expression: $x = 4y$

Verbal Information: The square of a number, w, is 25% of another number, z.

Mathematical Expression: $w^2 = 25\% z$ or, changing percent to a decimal number, $w^2 = 0.25z$

In solving word problems, the four general steps listed on the top of page 496 are used. These steps are illustrated in the examples that follow.

1. Read the problem carefully; write down the given information and the question asked.
2. Write mathematical expressions for the verbal information, and use these expressions to formulate an equation that contains the unknown quantity.
3. Solve the equation for the unknown quantity.
4. Check your answer whenever possible; if checking is not possible, evaluate your answer to see if it is a reasonable answer for the question asked.

EXAMPLE A.19   The sum of a certain number and 25 is equal to six times the number. What is the number?

Solution   Step 1:   Given information: A certain number (let x be the number) is added to 25; this sum is equal to $6x$.

Question: What is x?

Step 2:

$$x + 25 = 6x$$

Step 3:

$$x + 25 - 25 = 6x - 25$$
$$-x + x + 25 = 6x - x$$
$$25 = 5x$$
$$5 = x$$

Step 4:

$$x + 25 = 6x$$
$$5 + 25 = 6(5)$$
$$30 = 30$$

EXERCISE A.19   If a certain number is increased by 6, it will be 2 less than twice its original value. What is the number?   ■

Answer   The number is 8.

EXAMPLE A.20   A collection of nickels and dimes has a value of $7.25. How many nickels and dimes are in the collection if there are 38 more dimes than nickels?

Solution   Step 1:   Given information: The total value of nickels and dimes is $7.25. The number of dimes exceeds the number of nickels by 38.

Question: How many nickels and dimes are present?

Step 2:   Let x be the number of nickels. Then the number of dimes is $x + 38$. If we think of the values of nickels and dimes in terms of cents, then 1 nickel = 5 cents, 1 dime = 10 cents, $7.25 = 725 cents, and value of nickels in cents + value of dimes in cents = value of collection in cents or

$$5x + 10(x + 38) = 725$$

Step 3:

$$5x + 10(x + 38) = 725$$
$$5x + 10x + 380 = 725$$
$$15x + 380 = 725$$
$$15x + 380 - 380 = 725 - 380$$
$$15x = 345$$
$$x = 23$$

Thus, there are 23 nickels and $23 + 38 = 61$ dimes in the collection.

Step 4:

$$5x + 10(x + 38) = 725$$
$$5(23) + 10(23 + 38) = 725$$
$$115 + 230 + 380 = 725$$
$$725 = 725$$

EXERCISE A.20 A student has \$4.55 to spend for lunch. The money is all in nickels and quarters, and the student has five more quarters than nickels. How many nickels and quarters does the student have? ▧

Answer The student has 11 nickels and 16 quarters.

EXAMPLE A.21 An 84-foot rope is cut into three pieces. The length of the first piece is twice the second, and the length of the third piece is three times that of the second piece. What is the length of each piece of rope?

Solution Step 1: Given information: Three pieces of rope have a combined length of 84 ft. The lengths of the first and third pieces are related to the length of the second piece:

$$\text{length of piece } 1 = 2 \times \text{length of piece } 2$$
$$\text{length of piece } 3 = 3 \times \text{length of piece } 2$$

Step 2: length of piece 1 + length of piece 2 + length of piece 3 = 84 ft

Let $x =$ length of piece 2; then

$$\text{length of piece } 1 = 2x$$

and

$$\text{length of piece } 3 = 3x$$

Thus,

$$2x + x + 3x = 84 \text{ ft}$$

Step 3:

$$2x + x + 3x = 84 \text{ ft}$$
$$6x = 84 \text{ ft}$$
$$x = 14 \text{ ft}$$

Thus, the length of piece 1 ($2x$) is $2(14\text{ ft}) = 28$ ft; the length of piece 2 (x) is 14 ft; and the length of piece 3 ($3x$) is $3(14\text{ ft}) = 42$ ft.

Step 4:

$$2x + x + 3x = 84 \text{ ft}$$
$$2(14 \text{ ft}) + 14 \text{ ft} + 3(14 \text{ ft}) = 84 \text{ ft}$$
$$28 \text{ ft} + 14 \text{ ft} + 42 \text{ ft} = 84 \text{ ft}$$
$$84 \text{ ft} = 84 \text{ ft}$$

EXERCISE A.21   The perimeter of any rectangle is the sum of the lengths of its four sides. A rectangle having a perimeter of 56 ft has a length that is 4 ft greater than its width. Find the length and width of the rectangle.   ■

Answer   The length is 16 ft and the width is 12 ft.

EXAMPLE A.22   How many liters of a 20% salt solution should be added to 30 liters of a 60% salt solution to obtain a 50% salt solution?

Solution   Step 1:   Given information: An *unknown volume* of a 20% salt solution is to be mixed with *30 liters of a 60% salt solution* to make a 50% salt solution.

Question: How many *liters of the 20% salt solution* are needed?

Step 2:   Since the total amount of salt in the two solutions that are mixed must be the same as the amount of salt in the final solution, then

amt of salt in 20% solution + amt of salt in 60% solution
$$= \text{amt of salt in 50\% solution}$$

Let $x =$ liters of 20% solution; then

amt of salt in 20% solution $= 20\% \, x = 0.20x$
amt of salt in 60% solution $= 60\%(30) = 0.60(30)$
amt of salt in 50% solution $= 50\%(x + 30) = 0.50(x + 30)$

and

$$0.20x + 0.60(30) = 0.50(x + 30)$$

Step 3:

$$0.20x + 0.60(30) = 0.50(x + 30)$$
$$0.20x + 0.60(30) = 0.50x + 0.50(30)$$
$$0.20x + 18 = 0.50x + 15$$
$$0.20x + 18 - 18 = 0.50x + 15 - 18$$
$$0.20x = 0.50x - 3$$
$$0.20x - 0.50x = 0.50x - 3 - 0.50x$$
$$-0.30x = -3$$
$$x = 10$$

Thus, 10 liters of the 20% solution are needed.

Step 4:

$$0.20x + 0.60(30) = 0.50(x + 30)$$
$$0.20(10) + 0.60(30) = 0.50(10 + 30)$$
$$2 + 18 = 20$$
$$20 = 20$$

EXERCISE A.22   How many grams of an alloy containing 40% iron must be mixed with an alloy containing 70% iron to obtain 20 grams of an alloy containing 45% iron?   ■

Answer   The needed amount of alloy containing 40% iron is 16⅔ grams.

Proportionalities   There are many relationships in chemistry in which a change in one quantity produces a change in a related quantity. One example is the relationship between the absolute temperature of a gas and its volume at constant pressure: if the absolute temperature of a gas is increased, the volume of the gas increases by the same proportion if the pressure is constant. This relationship, known as Charles' law, is a **direct proportion.** If a relationship is a **direct proportion,** an increase in one quantity causes a proportionate increase in another quantity; conversely, a decrease in one quantity causes a proportionate decrease in another quantity.

A direct proportionality is written as

$$A \propto B$$

where $\propto$ is the proportionality symbol. The proportionality is read as "A is directly proportional to B." A direct proportionality is converted to an equation by replacing the proportionality sign with an equals sign followed by a constant, k:

$$A = kB$$

The constant k has a fixed numerical value and

$$A/B = k$$

Thus, k is always equal to A/B. When either A or B changes, the other changes proportionately *in the same direction*.

In the relationship between the volume of a gas and its absolute temperature,

$$V \propto T$$

and

$$V = kT$$

The constant of proportionality has a value that is determined experimentally. Thus, experimental results allow calculation of the value of k, and this value remains the same regardless of how the other factors may change. In solving proportionality problems, the value of k is often calculated from a set of experimental results, then the value of k is used to predict other experimental results. Example A.23 illustrates this procedure.

EXAMPLE A.23   The distance (d) a spring stretches is directly proportional to the weight (w) of an object hung from the spring. In an experiment, it was found that a 5-kilogram (kg) object stretches a spring 6 centimeters (cm). How many centimeters will the spring stretch when an 8-kilogram object is hung from the spring?

Solution   Step 1:   Given information: d is directly proportional to w;

$$d = 6 \text{ cm}$$
$$w = 5 \text{ kg}$$

Question: What is d when w is 8 kg?

Step 2:

$$d \propto w$$
$$d = kw$$

Step 3:

$$d = kw$$
$$k = \frac{d}{w}$$

From the experimental results,

$$k = \frac{6 \text{ cm}}{5 \text{ kg}} = \frac{1.2 \text{ cm}}{\text{kg}}$$

For the prediction of d with the new w,

$$d = kw$$
$$d = \left(\frac{1.2 \text{ cm}}{\text{kg}}\right)(8 \text{ kg})$$
$$d = 9.6 \text{ cm}$$

Thus, the new distance will be 9.6 cm.

Step 4:

$$d = kw$$
$$9.6 \text{ cm} = \left(\frac{1.2 \text{ cm}}{\text{kg}}\right)(8 \text{ kg})$$
$$9.6 \text{ cm} = 9.6 \text{ cm}$$

EXERCISE A.23   The volume of a gas is directly proportional to the absolute temperature of the gas at constant pressure. (Absolute temperature is measured in kelvins, K.) If the volume of a gas is 60 liters (L) at 200 K, what will be the volume of the gas at 300 K if the pressure remains constant? ■

Answer   The new volume will be 90 L.

In some scientific relationships, an increase in one quantity causes a proportionate decrease in another quantity, and vice versa; this kind of relationship is called an **inverse proportionality.** An inverse proportionality is written as

$$A \propto \frac{1}{B}$$

and read as "*A* is inversely proportional to *B*." Converting to an equation, we get

$$A = k\left(\frac{1}{B}\right)$$

or

$$A = \frac{k}{B}$$

and

$$k = AB$$

One of the inverse proportionalities of chemistry is Boyle's law; it states that the volume of a gas is inversely proportional to its pressure at constant temperature:

$$V \propto \frac{1}{P}$$

or

$$V = \frac{k}{P}$$

and

$$k = PV$$

Thus, if *V* increases, *P* decreases proportionately; conversely, if *V* decreases, *P* increases proportionately. Problems involving inverse proportionalities are solved by the same method used for problems involving direct proportionalities.

EXAMPLE A.24   The current, *I*, in an electrical circuit is inversely proportional to the resistance, *R*. If the current in a circuit is 40 amp when the resistance is 8 ohm, find the current when the resistance is 12 ohm.

Solution   Step 1: Given information:

$$I \propto \frac{1}{R}$$
$$I = 40 \text{ amp}$$
$$R = 8 \text{ ohm}$$

Question: What is *I* when *R* is 12 ohm?

Step 2:

$$I \propto \frac{1}{R}$$

$$I = \frac{k}{R}$$

Step 3:

$$I = \frac{k}{R}$$

$$k = IR$$

For the first set of values,

$$k = (40 \text{ amp})(8 \text{ ohm})$$

$$k = 320 \text{ amp ohm}$$

To predict the new current,

$$I = \frac{k}{R}$$

$$I = \frac{(320 \text{ amp } \cancel{\text{ohm}})}{12 \cancel{\text{ohm}}}$$

$$I = 27 \text{ amp}$$

Thus, the new current will be 27 amp.

Step 4:

$$I = \frac{k}{R}$$

$$27 \text{ amp} = \frac{(320 \text{ amp } \cancel{\text{ohm}})}{12 \cancel{\text{ohm}}}$$

$$27 \text{ amp} = 27 \text{ amp}$$

EXERCISE A.24   A gas has a volume of 25 cm³ and exerts a pressure of 785 torr. If the gas is compressed to a volume of 20 cm³, without changing its temperature, what will be the pressure of the gas?   ■

Answer   The new pressure will be 981 torr.

Logarithms   A **logarithm** is the exponent of a number. Most calculations in chemistry use **common logarithms** (abbreviated as log), in which the base 10 is used. Thus, the log of 10^exponent is the exponent:

$$\log 10^{\text{exponent}} = \text{exponent}$$

The logarithm of 10 raised to any power is simply the power of 10, and the log can be determined by inspection. Remember that 10^0 is always 1, and thus the log of 1 is 0.

   Logarithms of numbers between 1 and 10 cannot be determined by inspection, but they can be obtained from the table of logarithms at the end of this section or from some pocket calculators. The numbers in the table do not have decimal points, but each of the numbers listed under N is assumed to have a decimal point following the first digit, and each logarithm listed is assumed to have a decimal point immediately preceding the value given. Thus, the logarithm of 2.720 is 0.4346. (Verify this for yourself by looking in the table.)

The logarithm of a number greater than 10 or less than 1 is obtained by the following steps:

1. Express the number in scientific notation. For example, 1750 expressed in scientific notation is 1.75×10^3 (to three significant figures).
2. Since logarithms are exponents, and since exponents are added when numbers are multiplied, then

$$\log 1.75 \times 10^3 = \log 1.75 + \log 10^3$$

Now determine the logarithm of each number and add the logarithms together. (The number of significant figures in each number determines the number of significant figures in each logarithm; whole-number exponents are considered to be exact numbers.) Thus,

$$\log 1.75 \times 10^3 = 0.243 + 3.000 = 3.243$$

EXAMPLE A.25 Find the logarithm of 0.0257.

Solution
$$0.0257 = 2.57 \times 10^{-2}$$
$$\log 2.57 \times 10^{-2} = \log 2.57 + \log 10^{-2}$$
$$= 0.410 + (-2.000)$$
$$= -1.590$$

EXERCISE A.25 Find the logarithm of each number.

a. 7.29×10^4 c. 85,100 (three significant figures)
b. 2.36×10^{-3} d. 0.000295

Answers

a. 4.863   b. −2.627   c. 4.930   d. −3.530

Sometimes it is necessary to find an **antilogarithm,** the number that corresponds to a given logarithm. The given logarithm is composed of two parts: a decimal number (the **mantissa**) and a positive or negative whole number (the **characteristic**). For example, in the logarithm 2.959, .959 is the mantissa and +2 is the characteristic. The antilogarithm (abbreviated as antilog) is found by reversing the procedure for finding a logarithm.

1. Using the table of logarithms at the end of this section (or a pocket calculator), find the antilogarithm of the mantissa (0.959).

$$\text{antilog } 0.959 = 9.10$$

2. Next, find the antilogarithm of the characteristic (+2) by inspection.

$$\text{antilog } 2 = 10^2$$

3. Last, since logs are added when numbers are multiplied, show the total antilog in scientific notation.

$$\text{antilog } 2.959 = \text{antilog } 0.959 + \text{antilog } 2$$
$$= 9.10 \times 10^2$$

EXAMPLE A.26　Find the antilog of 3.841.

Solution
$$\text{antilog } 3.841 = \text{antilog } 0.841 + \text{antilog } 3$$
$$= 6.93 \times 10^3$$

EXERCISE A.26　Find the antilog of each of the following logarithms.

a. 0.845　　**b.** 2.480　　**c.** 5.781　　**d.** 1.891 ▪

Answers

a. 7.00　　**b.** 3.02×10^2　　**c.** 6.04×10^5　　**d.** 7.8×10^1

All of the mantissas in a table of logarithms are positive, and this fact must be taken into account when an antilog of a negative number is found. For example, to take the antilog of -5.802, we must rewrite the logarithm in such a way that the mantissa is positive. Thus,

$$\text{antilog}(-5.802) = \text{antilog}(+0.198 - 6.000)$$
$$= 1.58 \times 10^{-6}$$

or

$$-5.802 = \log(1.58 \times 10^{-6})$$

EXAMPLE A.27　Find the antilog of -3.850.

Solution
$$\text{antilog}(-3.850) = \text{antilog}(+0.150 - 4.000)$$
$$= 1.41 \times 10^{-4}$$

EXERCISE A.27　Find the antilog of each of the following logarithms.

a. -2.493　　**b.** -4.959　　**c.** -0.225　　**d.** -10.873 ▪

Answers

a. 3.21×10^{-3}　　**b.** 1.10×10^{-5}　　**c.** 5.96×10^{-1}　　**d.** 1.34×10^{-11}

Table of Logarithms

N	0	1	2	3	4	5	6	7	8	9
10	0000	0043	0086	0128	0170	0212	0253	0294	0334	0374
11	0414	0453	0492	0531	0569	0607	0645	0682	0719	0755
12	0792	0828	0864	0899	0934	0969	1004	1038	1072	1106
13	1139	1173	1206	1239	1271	1303	1335	1367	1399	1430
14	1461	1492	1523	1553	1584	1614	1644	1673	1703	1732
15	1761	1790	1818	1847	1875	1903	1931	1959	1987	2014
16	2041	2068	2095	2122	2148	2175	2201	2227	2253	2279
17	2304	2330	2355	2380	2405	2430	2455	2480	2504	2529
18	2553	2577	2601	2625	2648	2672	2695	2718	2742	2765
19	2788	2810	2833	2856	2878	2900	2923	2945	2967	2989
20	3010	3032	3054	3075	3096	3118	3139	3160	3181	3201
21	3222	3243	3263	3284	3304	3324	3345	3365	3385	3404
22	3424	3444	3464	3483	3502	3522	3541	3560	3579	3598
23	3617	3636	3655	3674	3692	3711	3729	3747	3766	3784
24	3802	3820	3838	3856	3874	3892	3909	3927	3945	3962
25	3979	3997	4014	4031	4048	4065	4082	4099	4116	4133
26	4150	4166	4183	4200	4216	4232	4249	4265	4281	4298
27	4314	4330	4346	4362	4378	4393	4409	4425	4440	4456
28	4472	4487	4502	4518	4533	4548	4564	4579	4594	4609
29	4624	4639	4654	4669	4683	4698	4713	4728	4742	4757
30	4771	4786	4800	4814	4829	4843	4857	4871	4886	4900
31	4914	4928	4942	4955	4969	4983	4997	5011	5024	5038
32	5051	5065	5079	5092	5105	5119	5132	5145	5159	5172
33	5185	5198	5211	5224	5237	5250	5263	5276	5289	5302
34	5315	5328	5340	5353	5366	5378	5391	5403	5416	5428
35	5441	5453	5465	5478	5490	5502	5514	5527	5539	5551
36	5563	5575	5587	5599	5611	5623	5635	5647	5658	5670
37	5682	5694	5705	5717	5729	5740	5752	5763	5775	5786
38	5798	5809	5821	5832	5843	5855	5866	5877	5888	5899
39	5911	5922	5933	5944	5955	5966	5977	5988	5999	6010
40	6021	6031	6042	6053	6064	6075	6085	6096	6107	6117
41	6128	6138	6149	6160	6170	6180	6191	6201	6212	6222
42	6232	6243	6253	6263	6274	6284	6294	6304	6314	6325
43	6335	6345	6355	6365	6375	6385	6395	6405	6415	6425
44	6435	6444	6454	6464	6474	6484	6493	6503	6513	6522
45	6532	6542	6551	6561	6571	6580	6590	6599	6609	6618
46	6628	6637	6646	6656	6665	6675	6684	6693	6702	6712
47	6721	6730	6739	6749	6758	6767	6776	6785	6794	6803
48	6812	6821	6830	6839	6848	6857	6866	6875	6884	6893
49	6902	6911	6920	6928	6937	6946	6955	6964	6972	6981
50	6990	6998	7007	7016	7024	7033	7042	7050	7059	7067
51	7076	7084	7093	7101	7110	7118	7126	7135	7143	7152
52	7160	7168	7177	7185	7193	7202	7210	7218	7226	7235
53	7243	7251	7259	7267	7275	7284	7292	7300	7308	7316
54	7324	7332	7340	7348	7356	7364	7372	7380	7388	7396

To find the logarithm of a number between 1 and 10, first assume that each number listed under N has a decimal point following the first digit. Locate the first two digits of the number in the left-hand column and then move your finger across the table to the column under the third digit of the original number at the top of the table. You will then be pointing to the logarithm of the number. The number of digits to be included to the right of the decimal point in the logarithm is the same as the number of significant figures in the original number. For example, the logarithm of 3.49 is 0.543 (down to 34 and across to 9).

(continued on next page)

Table of Logarithms (continued)

N	0	1	2	3	4	5	6	7	8	9
55	7404	7412	7419	7427	7435	7443	7451	7459	7466	7474
56	7482	7490	7497	7505	7513	7520	7528	7536	7543	7551
57	7559	7566	7574	7582	7589	7597	7604	7612	7619	7627
58	7634	7642	7649	7657	7664	7672	7679	7686	7694	7701
59	7709	7716	7723	7731	7738	7745	7752	7760	7767	7774
60	7782	7789	7796	7803	7810	7818	7825	7832	7839	7846
61	7853	7860	7868	7875	7882	7889	7896	7903	7910	7917
62	7924	7931	7938	7945	7952	7959	7966	7973	7980	7987
63	7993	8000	8007	8014	8021	8028	8035	8041	8048	8055
64	8062	8069	8075	8082	8089	8096	8102	8109	8116	8122
65	8129	8136	8142	8149	8156	8162	8169	8176	8182	8189
66	8195	8202	8209	8215	8222	8228	8235	8241	8248	8254
67	8261	8267	8274	8280	8287	8293	8299	8306	8312	8319
68	8325	8331	8338	8344	8351	8357	8363	8370	8376	8382
69	8388	8395	8401	8407	8414	8420	8426	8432	8439	8445
70	8451	8457	8463	8470	8476	8482	8488	8494	8500	8506
71	8513	8519	8525	8531	8537	8543	8549	8555	8561	8567
72	8573	8579	8585	8591	8597	8603	8609	8615	8621	8627
73	8633	8639	8645	8651	8657	8663	8669	8675	8681	8686
74	8692	8698	8704	8710	8716	8722	8727	8733	8739	8745
75	8751	8756	8762	8768	8774	8779	8785	8791	8797	8802
76	8808	8814	8820	8825	8831	8837	8842	8848	8854	8859
77	8865	8871	8876	8882	8887	8893	8899	8904	8910	8915
78	8921	8927	8932	8938	8943	8949	8954	8960	8965	8971
79	8976	8982	8987	8993	8998	9004	9009	9015	9020	9025
80	9031	9036	9042	9047	9053	9058	9063	9069	9074	9079
81	9085	9090	9096	9101	9106	9112	9117	9122	9128	9133
82	9138	9143	9149	9154	9159	9165	9170	9175	9180	9186
83	9191	9196	9201	9206	9212	9217	9222	9227	9232	9238
84	9243	9248	9253	9258	9263	9269	9274	9279	9284	9289
85	9294	9299	9304	9309	9315	9320	9325	9330	9335	9340
86	9345	9350	9355	9360	9365	9370	9375	9380	9385	9390
87	9395	9400	9405	9410	9415	9420	9425	9430	9435	9440
88	9445	9450	9455	9460	9465	9469	9474	9479	9484	9489
89	9494	9499	9504	9509	9513	9518	9523	9528	9533	9538
90	9542	9547	9552	9557	9562	9566	9571	9576	9581	9586
91	9590	9595	9600	9605	9609	9614	9619	9624	9628	9633
92	9638	9643	9647	9652	9657	9661	9666	9671	9675	9680
93	9685	9689	9694	9699	9703	9708	9713	9717	9722	9727
94	9731	9736	9741	9745	9750	9754	9759	9763	9768	9773
95	9777	9782	9786	9791	9795	9800	9805	9809	9814	9818
96	9823	9827	9832	9836	9841	9845	9850	9854	9859	9863
97	9868	9872	9877	9881	9886	9890	9894	9899	9903	9908
98	9912	9917	9921	9926	9930	9934	9939	9943	9948	9952
99	9956	9961	9965	9969	9974	9978	9983	9987	9991	9996

DECIMAL pH VALUES

Decimal pH values are calculated using logarithms. (Logarithms are discussed in the last part of Appendix A.) For example, let's calculate the pH of a solution containing $2.3 \times 10^{-4} M$ H^+.

$$\begin{aligned}
pH &= -\log[H^+] \\
&= -\log(2.3 \times 10^{-4}) \\
&= -[\log 2.3 + \log 10^{-4}] \\
&= -[0.36 - 4.00] \\
&= -(-3.64) \\
&= 3.64
\end{aligned}$$

EXAMPLE B.1 Calculate the pH of a solution containing $4.5 \times 10^{-8} M$ H^+.

Solution

$$\begin{aligned}
pH &= -\log[H^+] \\
&= -\log(4.5 \times 10^{-8}) \\
&= -[\log 4.5 + \log 10^{-8}] \\
&= -[0.65 + (-8.00)] \\
&= -[0.65 - 8.00] \\
&= -(-7.35) \\
&= 7.35
\end{aligned}$$

EXERCISE B.1 Calculate the pH corresponding to each of the following H^+ concentrations.

a. $7.93 \times 10^{-2} M$ c. $5.38 \times 10^{-7} M$
b. $3.79 \times 10^{-10} M$ d. $9.85 \times 10^{-3} M$ ■

Answers

a. 1.101   b. 9.421   c. 6.269   d. 2.007

Hydrogen ion concentrations are calculated from decimal pH values by finding antilogs. (Antilogs are discussed in the last part of Appendix A.) For example, let's calculate $[H^+]$ for a solution having a pH of 7.45.

$$\begin{aligned}
pH &= 7.45 \\
-\log[H^+] &= 7.45 \\
\log[H^+] &= -7.45 \\
[H^+] &= \text{antilog}(-7.45) \\
&= \text{antilog}(0.55) + \text{antilog}(-8.00) \\
&= 3.5 \times 10^{-8} M
\end{aligned}$$

EXAMPLE B.2   Calculate [H^+] for a solution having a pH of 3.38.

Solution

$$pH = 3.38$$
$$-\log[H^+] = 3.38$$
$$\log[H^+] = -3.38$$
$$[H^+] = \text{antilog}(-3.38)$$
$$= \text{antilog}(0.62) + \text{antilog}(-4.00)$$
$$= 4.2 \times 10^{-4} M$$

EXERCISE B.2   Calculate [H^+] for each of the following pH values.

a. 6.29   **b.** 9.35   **c.** 7.05   **d.** 4.47

Answers

a. $5.1 \times 10^{-7} M$ **c.** $8.9 \times 10^{-8} M$
b. $4.5 \times 10^{-10} M$ **d.** $3.4 \times 10^{-5} M$

SOLUTIONS TO EXERCISES AND ANSWERS TO SELECTED STUDY QUESTIONS AND PROBLEMS

APPENDIX C

CHAPTER 1

EXERCISES

1.1 **a.** Since water boils at 212° F, it is a gas at 450° F.   **b.** Stainless steel is used in cooking utensils and thus has a very high melting point. It is a solid at 450° F.   **c.** Candle wax melts at relatively low temperatures and is a liquid at 450° F.   **d.** Cheese melts at relatively low temperatures and is a liquid at 450° F.

1.2 **a.** Orange juice with pulp is heterogeneous because it is a liquid containing suspended bits of solid material.   **b.** Sea sand is heterogeneous because it consists of many sizes of small, well-defined solid particles.   **c.** Vinegar is homogeneous because it consists of one phase.   **d.** Whipped cream is heterogeneous because it consists of two phases: pockets of air (a gas) distributed through cream (a liquid).

1.3 **a.** Smoke is a heterogeneous mixture because it consists of different substances in separate phases: small, solid, soot particles suspended in air (a gas).   **b.** Gasoline is a homogeneous mixture because it contains a mixture of different substances (various additives and petroleum fractions) in one phase.   **c.** Swiss cheese is a heterogeneous mixture because it is a mixture of different substances in separate phases: air (the holes) distributed through a mixture of milk solids.   **d.** Oxygen is a pure substance; it consists of one substance in one phase.

1.4 **a.** Fermentation is a chemical change; in this case, an identity change occurs when grape juice changes to wine.   **b.** The mixing of cake batter is a physical change because the ingredients maintain their identities.   **c.** Wood being sawed is a physical change because only a change in the size of the wood particles occurs.   **d.** The formation of yogurt is another example of fermentation, a chemical change. In this case, an identity change occurs when milk changes to yogurt.

1.5 **a.** Being corrodible is a chemical property.   **b.** Hardness and physical appearance are physical properties.   **c.** Physical state is a physical property.   **d.** Taste is a physical property.

1.6 **a.** Since the gopher is moving, it has kinetic energy.   **b.** A stretched rubber band has the potential of doing work when it is allowed to contract, and thus it has potential energy.   **c.** A ripe apple in a tree is immobile but has the potential of doing work when it falls to the ground; thus, it has potential energy.   **d.** A dormant volcano has the potential of doing work when it erupts; thus, it has potential energy.

STUDY QUESTIONS AND PROBLEMS

1. **a.** liquid   **b.** solid   **c.** gas   **d.** solid   **e.** solid

4. **a.** condensation   **b.** freezing   **c.** evaporation   **d.** freezing   **e.** boiling   **f.** freezing

7. **a.** heterogeneous   **b.** homogeneous   **c.** heterogeneous   **d.** homogeneous   **e.** heterogeneous   **f.** heterogeneous

8. There are two phases of water in fog: water vapor and tiny droplets of liquid water suspended in air.

14. **a.** homogeneous mixture   **b.** heterogeneous mixture   **c.** heterogeneous mixture   **d.** heterogeneous mixture   **e.** heterogeneous mixture   **f.** heterogeneous mixture

18. **a.** chemical change   **b.** physical change   **c.** physical change   **d.** chemical change   **e.** physical change   **f.** physical change

22. **a.** Physical appearance and physical state are physical properties.   **b.** Electrical conductivity is a physical property.   **c.** Being corrodible is a chemical property.   **d.** Flammability is a chemical property.   **e.** Taste is a physical property.   **f.** Burning is a chemical property.

25. When coal burns, it is changed from a solid (carbon) to a gas (carbon dioxide). (A few ashes may be left.) Energy is released in the form of heat and light during the burning. Since this is a chemical reaction, no appreciable amount of mass is lost, but a very small amount of mass is converted to the energy that is released by the reaction.

27. **a.** kinetic energy   **b.** potential energy   **c.** kinetic energy   **d.** potential energy   **e.** potential energy
f. potential energy

30. **a.** endothermic   **b.** exothermic   **c.** exothermic   **d.** endothermic

CHAPTER 2

EXERCISES

2.1 **a.** 4   **b.** 2   **c.** 3   **d.** 3

2.2 **a.** 76   **b.** 0.0078   **c.** 1.2   **d.** 580,000

2.3 **a.** 1.38×10^5 **b.** 5.7×10^3 **c.** 9.549×10^6 **d.** 1.4×10^5

2.4 **a.** 6.0×10^{-3} **b.** 3.71×10^{-4} **c.** 9.63×10^{-2} **d.** 3.6×10^{-7}

2.5 **a.** $(3.3 \text{ m})\left(\dfrac{1000 \text{ mm}}{1 \text{ m}}\right) = 3,300 \text{ mm} = 3.3 \times 10^3 \text{ mm}$ **b.** $(16.5 \text{ m}^3)\left(\dfrac{1057 \text{ qt}}{1 \text{ m}^3}\right) = 17,400 \text{ qt} = 1.74 \times 10^4 \text{ qt}$

c. $(25.0 \text{ lb})\left(\dfrac{1 \text{ kg}}{2.2 \text{ lb}}\right) = 11 \text{ kg}$ **d.** $(17 \text{ mL})\left(\dfrac{1 \text{ L}}{1000 \text{ mL}}\right) = 0.017 \text{ L}$

2.6 **a.** $(1750 \text{ mg})\left(\dfrac{1 \text{ g}}{1000 \text{ mg}}\right)\left(\dfrac{1 \text{ kg}}{1000 \text{ g}}\right) = 1.75 \times 10^{-3} \text{ kg}$ **b.** $(4230 \text{ ft})\left(\dfrac{12 \text{ in}}{1 \text{ ft}}\right)\left(\dfrac{1 \text{ m}}{39.37 \text{ in}}\right)\left(\dfrac{1 \text{ km}}{1000 \text{ m}}\right) = 1.29 \text{ km}$

c. $(3.65 \text{ kg})\left(\dfrac{1000 \text{ g}}{1 \text{ kg}}\right)\left(\dfrac{10^6 \text{ μg}}{1 \text{ g}}\right) = 3.65 \times 10^9 \text{ μg}$ **d.** $(4.30 \text{ m})\left(\dfrac{1000 \text{ mm}}{1 \text{ m}}\right) = 4.30 \times 10^3 \text{ mm}$

2.7 $d = \dfrac{m}{v} = \dfrac{82.7 \text{ g}}{10.7 \text{ mL}} = 7.73 \text{ g/mL}$

2.8 $d = \dfrac{m}{v}$

$v = \dfrac{m}{d} = \dfrac{11.23 \text{ g}}{2.32 \text{ g/mL}} = 4.84 \text{ mL}$

2.9 $d = \dfrac{m}{v}$

$m = dv = (2.70 \text{ g/mL})(45 \text{ mL}) = 120 \text{ g} = 1.2 \times 10^2 \text{ g}$

2.10 specific gravity $= \dfrac{1.02 \text{ g/mL}}{1.0000 \text{ g/mL}} = 1.02$

2.11 **a.** $°C = \frac{5}{9}(°F - 32) = \frac{5}{9}(72 - 32) = \frac{5}{9}(40) = 22$ **b.** $°F = \frac{9}{5}°C + 32 = \frac{9}{5}(54) + 32 = 97 + 32 = 129$
c. $°C = \frac{5}{9}(°F - 32) = \frac{5}{9}(0 - 32) = -18$ **d.** $°F = \frac{9}{5}°C + 32 = \frac{9}{5}(-45) + 32 = -81 + 32 = -49$

2.12 specific heat $= \dfrac{\text{cal}}{\text{g } \Delta T} = \dfrac{2,178 \text{ cal}}{125 \text{ g}(87° \text{ C} - 20° \text{ C})} = \dfrac{2,178 \text{ cal}}{125 \text{ g}(67° \text{ C})} = 0.26 \text{ cal/g }°C$

2.13 specific heat $= \dfrac{\text{cal}}{\text{g } \Delta T}$

$\text{cal} = (\text{specific heat})(g)(\Delta T) = (0.038 \text{ cal/g }°C)(15.0 \text{ g})(25° \text{ C}) = 14 \text{ cal}$

2.14

$$\text{specific heat} = \frac{\text{cal}}{\text{g } \Delta T}$$

$$(\text{specific heat})(\Delta T) = \frac{\text{cal}}{\text{g}}$$

$$\Delta T = \frac{\text{cal}}{(\text{g})(\text{specific heat})} = \frac{1{,}500 \text{ cal}}{(11.4 \text{ g})(0.108 \text{ cal/g} \cdot {}^\circ\text{C})} = 1200^\circ \text{ C}$$

$$\Delta T = T_2 - T_1$$
$$T_2 = \Delta T + T_1 = 1200^\circ \text{ C} + 10^\circ \text{ C} = 1210^\circ \text{ C}$$

STUDY QUESTIONS AND PROBLEMS

2. Yes. A series of measurements could give values that are very close together, yet far from the value that is accepted as correct.

4. In most cases, accuracy is more dependent on human error, since precision usually depends more on the calibration of the measuring device than on human ability.

6. **a.** 4 **b.** 3 **c.** 2 **d.** 4 **e.** 3 **f.** 4 **g.** 1 **h.** 4

8. **a.** 4700 **b.** 0.0037 **c.** 2.6 **d.** 0.17 **e.** 550 **f.** 36

10. **a.** 1145 **b.** 3.9 **c.** 4.7 **d.** 73 **e.** 130 **f.** 0.220

13. **a.** 7.326×10^2 **b.** 1.004×10^2 **c.** 7×10^6 **d.** 5.38×10^{-2} **e.** 4.37×10^{-11} **f.** 1.0573×10^4

15. **a.** 568.9 **b.** 23,670 **c.** 0.0000001499 **d.** 0.00365 **e.** 4,761,000 **f.** 0.0000000072517

17. The metric system is easy to use because it is based on powers of ten.

19. **a.** 6.73×10^{-8} m **b.** 8.492×10^{-2} kg **c.** 4.57×10^{-4} L **d.** 5.23 cm **e.** 5.9×10^7 ng **f.** 3.94×10^2 mm

21. **a.** 6.67×10^3 g **b.** 0.4399 m **c.** 23.49 L **d.** 11.9 m **e.** 103 mi **f.** 15 in

23. **a.** 381 m **b.** 3.81×10^{-1} km **c.** 3.81×10^4 cm **d.** 3.81×10^5 mm

25. 55 cm³

27. 3.85×10^5 km

29. 14.2 km/hr

32. The bowling ball weighs more in Death Valley, where it is closer to the center of the earth than it is on Mount Whitney.

34. 0.9981 g/mL

36. 22.5 g/mL

38. 1.47 mL

40. 86 g

43. 8.29 mL

45. aluminum

47. **a.** 648.9° C, assuming four significant figures in 1200 **b.** 0° C **c.** 99° F **d.** 113° F **e.** 26° C **f.** 1090° F, assuming three significant figures in 590

49. 2.1° C

51. 7.50×10^3 cal

53. 1.5×10^3 cal

55. 2.4×10^3 cal

CHAPTER 3

EXERCISES

3.1 Since the mass number is the sum of protons and neutrons, the mass number for potassium is 39.

3.2 number of neutrons = mass number − number of protons = 119 − 50 = 69
Since the atom contains 50 protons, 50 is also its atomic number. The name of element 50 is tin.

3.3   number of protons = mass number − number of neutrons = 55 − 30 = 25
Since there are 25 protons in the atom, its atomic number must also be 25. Element 25 is manganese.

3.4   Since the atomic number of Cd, cadmium, is 48, an atom of cadmium contains 48 protons and 48 electrons. The difference between the mass number and atomic number, 112 − 48, gives the number of neutrons, 64.

3.5   The ion must have the same number of protons and neutrons as its parent atom; thus $^{27}_{13}\text{Al}^{3+}$ contains 13 protons and 14 neutrons. Its three positive charges mean that it contains 3 more protons than electrons; thus, it contains 10 electrons.

3.6 $^{80}_{35}\text{Br}^-$ has 35 protons, 45 neutrons, and 36 electrons (one more than the number of protons).

3.7   Assume a mass of 1.0 amu for each neutron and proton. Then the mass of $^{24}_{12}\text{Mg}$ is 24.0 amu, the mass of $^{25}_{12}\text{Mg}$ is 25.0 amu, and the mass of $^{26}_{12}\text{Mg}$ is 26.0 amu. Then,

$$^{24}_{12}\text{Mg:} \quad (0.7899)(24.0 \text{ amu}) = 19.0 \text{ amu}$$
$$^{25}_{12}\text{Mg:} \quad (0.1000)(25.0 \text{ amu}) = 2.50 \text{ amu}$$
$$^{26}_{12}\text{Mg:} \quad (0.1101)(26.0 \text{ amu}) = \underline{2.86 \text{ amu}}$$
$$24.36 \text{ amu}$$
$$\text{or } 24.4 \text{ amu}$$

3.8   **a.**   FW of CH_4 = (1 atom C)(atomic weight of C) + (4 atoms H)(atomic weight of H) = $(1 \text{ atom C})\left(\dfrac{12.0 \text{ amu}}{\text{atom C}}\right) +$ $(4 \text{ atoms H})\left(\dfrac{1.0 \text{ amu}}{\text{atom H}}\right)$ = 12.0 amu + 4.0 amu = 16.0 amu   **b.**   FW of $C_6H_{12}O_6$ = (6 atoms C)(atomic weight of C) + (12 atoms H)(atomic weight of H) + (6 atoms O)(atomic weight of O) = $(6 \text{ atoms C})\left(\dfrac{12.0 \text{ amu}}{\text{atom C}}\right) +$ $(12 \text{ atoms H})\left(\dfrac{1.0 \text{ amu}}{\text{atom H}}\right) + (6 \text{ atoms O})\left(\dfrac{16.0 \text{ amu}}{\text{atom O}}\right)$ = 72.0 amu + 12.0 amu + 96.0 amu = 180.0 amu   **c.**   FW of Na_2SO_4 = (2 atoms Na)(atomic weight of Na) + (1 atom S)(atomic weight of S) + (4 atoms O)(atomic weight of O) = $(2 \text{ atoms Na})\left(\dfrac{23.0 \text{ amu}}{\text{atom Na}}\right) + (1 \text{ atom S})\left(\dfrac{32.1 \text{ amu}}{\text{atom S}}\right) + (4 \text{ atoms O})\left(\dfrac{16.0 \text{ amu}}{\text{atom O}}\right)$ = 46.0 amu + 32.1 amu + 64.0 amu = 142.1 amu

3.9   FW of $NaHCO_3$ = $(1 \text{ atom Na})\left(\dfrac{23.0 \text{ amu}}{\text{atom Na}}\right) + (1 \text{ atom H})\left(\dfrac{1.0 \text{ amu}}{\text{atom H}}\right) + (1 \text{ atom C})\left(\dfrac{12.0 \text{ amu}}{\text{atom C}}\right) + (3 \text{ atoms O})$ $\left(\dfrac{16.0 \text{ amu}}{\text{atom O}}\right)$ = 23.0 amu + 1.0 amu + 12.0 amu + 48.0 amu = 84.0 amu

$$\% \text{ Na} = \left(\frac{\text{mass of Na in NaHCO}_3}{\text{FW of NaHCO}_3}\right)(100\%) = \left(\frac{23.0 \text{ amu}}{84.0 \text{ amu}}\right)(100\%) = 27.4\%$$

$$\% \text{ H} = \left(\frac{\text{mass of H in NaHCO}_3}{\text{FW of NaHCO}_3}\right)(100\%) = \left(\frac{1.0 \text{ amu}}{84.0 \text{ amu}}\right)(100\%) = 1.2\%$$

$$\% \text{ C} = \left(\frac{\text{mass of C in NaHCO}_3}{\text{FW of NaHCO}_3}\right)(100\%) = \left(\frac{12.0 \text{ amu}}{84.0 \text{ amu}}\right)(100\%) = 14.3\%$$

$$\% \text{ O} = \left(\frac{\text{mass of O in NaHCO}_3}{\text{FW of NaHCO}_3}\right)(100\%) = \left(\frac{48.0 \text{ amu}}{84.0 \text{ amu}}\right)(100\%) = 57.1\%$$

STUDY QUESTIONS AND PROBLEMS

13.   **a.** Anion; it contains one more electron than protons.   **b.** Cation; it contains two fewer electrons than protons.   **c.** Anion; it contains two more electrons than protons.   **d.** Cation; it contains three fewer electrons than protons.

15. **a.** 82   **b.** 93   **c.** 67   **d.** 55   **e.** 43   **f.** 42
18. **a.** iron   **b.** mercury   **c.** gold   **d.** uranium   **e.** krypton   **f.** lead

21.

	Protons	Neutrons	Electrons
a.	56	81	54
b.	1	0	2
c.	13	14	10
d.	52	76	54
e.	50	69	48
f.	37	48	36

22. a and c; b, e, and f
28. 20.2 amu
34. 19.6 g
37. **a.** 2 atoms Na; 1 atom C; 3 atoms O   **b.** 3 atoms Mg; 2 atoms P; 8 atoms O   **c.** 1 atom Ca; 2 atoms H; 2 atoms S; 8 atoms O   **d.** 1 atom Ba; 4 atoms C; 6 atoms H; 4 atoms O   **e.** 1 atom Al; 3 atoms Br   **f.** 1 atom Sr; 2 atoms N; 6 atoms O
38. **a.** Na_2O **b.** HNO_3 **c.** $NaC_2H_3O_2$ **d.** Al_2O_3 **e.** CCl_4 **f.** Fe_2S_3
40. **a.** 410.7 amu   **b.** 189.3 amu   **c.** 208.1 amu   **d.** 392.3 amu
42. 69.9% Fe; 30.1% O
44. 42.1% C; 6.4% H; 51.5% O
46. 3×10^{22} protons; 3×10^{22} neutrons; 5×10^{25} electrons
48. 19.6 g
50. 1.1 g of hydrogen atoms

CHAPTER 4

EXERCISES

4.1 From the list of elements inside the front cover, we find the atomic number of each element. The atomic number gives us the number of electrons for each atom. Then we place the electrons in atomic orbitals using Figure 4-17. As a check, we count electrons in each electron configuration to be sure that we have accounted for all electrons in each atom.   **a.** $1s^22s^22p^63s^23p^64s^23d^{10}4p^65s^1$ or $[Kr]5s^1$ **b.** $1s^22s^22p^63s^23p^5$ or $[Ne]3s^23p^5$ **c.** $1s^22s^22p^63s^23p^64s^23d^{10}4p^65s^24d^{10}5p^66s^2$ or $[Xe]6s^2$ **d.** $1s^22s^22p^63s^23p^64s^23d^{10}4p^65s^24d^{10}5p^6$

4.2 Using the list of elements and the charge on each ion, we determine the number of electrons in each ion. Then we place the electrons in atomic orbitals according to Figure 4-17. As a check, we count electrons in each electron configuration to be sure that we have accounted for all electrons in each ion.
   a. $1s^22s^22p^63s^23p^6$ or $[Ar]$ **b.** $1s^22s^22p^63s^23p^64s^23d^{10}4p^6$ or $[Kr]$ **c.** $1s^22s^22p^63s^23p^64s^23d^{10}4p^6$ or $[Kr]$
   d. $1s^22s^22p^63s^23p^6$ or $[Ar]$

4.3 Since $l = 0, 1, 2, 3, \ldots, n-1$, then the values allowed are $l = 0$, $l = 1$, $l = 2$, $l = 3$, and $l = 4$. The five values of l signify five atomic orbitals of increasing energy; therefore, the five orbital types must be s, p, d, f, and g.

4.4 For $n = 4$, $l = 0$, $l = 1$, $l = 2$, and $l = 3$. The four values of l indicate four orbital types: s, p, d, and f.

For $l = 0$,

$$m_l = 0$$

For $l = 1$,

$$m_l = +l = 1$$
$$m_l = +(l-1) = +(1-1) = 0$$
$$m_l = -l = -1$$

For $l = 2$,

$$m_l = +l = 2$$
$$m_l = +(l-1) = +(2-1) = 1$$
$$m_l = +(l-2) = +(2-2) = 0$$
$$m_l = -(l-1) = -(2-1) = -1$$
$$m_l = -l = -2$$

For $l = 3$,

$$m_l = +l = 3$$
$$m_l = +(l-1) = +(3-1) = 2$$
$$m_l = +(l-2) = +(3-2) = 1$$
$$m_l = +(l-3) = +(3-3) = 0$$
$$m_l = -(l-2) = -(3-2) = -1$$
$$m_l = -(l-1) = -(3-1) = -2$$
$$m_l = -l = -3$$

For $l = 0$, the single value for m_l indicates one s orbital. For $l = 1$, the three values for m_l indicate three p orbitals. For $l = 2$, the five values for m_l indicate five d orbitals. For $l = 3$, the seven values for m_l indicate seven f orbitals.

STUDY QUESTIONS AND PROBLEMS

25. Each shell has an electron capacity of $2n^2$, where n is the principal energy level.   **a.** 2   **b.** 8   **c.** 18   **d.** 32
30. The first principal energy level does not have any p orbitals. All other principal energy levels have three p orbitals each.
32. A $3p$ orbital contains more energy than a $2p$ orbital, and a $3p$ orbital is farther from the nucleus than a $2p$ orbital.
38. **a.** $1s^22s^22p^63s^23p^64s^23d^{10}4p^65s^24d^{10}5p^6$ or [Xe]   **b.** $1s^22s^22p^63s^23p^64s^23d^{10}4p^65s^24d^{10}5p^66s^24f^{14}5d^{10}6p^6$ or [Rn]   **c.** $1s^22s^22p^63s^23p^64s^23d^{10}4p^6$ or [Kr]   **d.** $1s^22s^22p^6$ or [Ne]
42. $m_l = 2, 1, 0, -1, -2$; there are five d orbitals.
45. **a.** $3d$ **b.** $4d$ **c.** $5p$ **d.** $4d$

CHAPTER 5

EXERCISES

5.1 S and Te are in the same family (Group VIA). Since first ionization energies decrease within a group, the element higher in the group, in this case S, would be expected to have the higher first ionization energy.
5.2 Na and S are in the same period (Period 3). Since electron affinity tends to increase across a period, the element farther to the right, in this case S, would be expected to have the higher electron affinity.
5.3 Cl and I are in the same family (Group VIIA). Since metals have a greater tendency to lose electrons than do nonmetals, the more metallic element will be the one with the greater tendency to lose electrons. Since first ionization energies decrease going down a group, the element lower in the group will have the greater tendency to lose an electron. Although neither Cl nor I is actually considered to be a metal, I would be expected to be more nearly metallic than Cl.

STUDY QUESTIONS AND PROBLEMS

6. **a.** 18   **b.** 20   **c.** 8   **d.** 2
8. **a.** 4   **b.** 2   **c.** 3   **d.** 2   **e.** 1   **f.** 5
10. Each period is numbered according to the principal energy level whose filling is begun by the first element in the period.

11. **a.** $2s^2 2p^2$ **b.** $6s^2$ **c.** $3s^2 3p^6$ **d.** $4s^2 4p^3$ **e.** $6s^1$ **f.** $3s^2 3p^4$

22. **a.** representative **b.** transition **c.** representative **d.** transition **e.** representative **f.** representative **g.** representative **h.** representative **i.** transition **j.** representative **k.** representative **l.** representative

28. hydrogen, H_2; nitrogen, N_2; oxygen, O_2; fluorine, F_2; chlorine, Cl_2; bromine, Br_2; iodine, I_2

31. Group 0

37. hydrogen; francium

41. **a.** Cl would have the higher first ionization energy, because it is farther to the right and in the same period as Al. **b.** As would have the higher first ionization energy, because in the periodic table it is higher and farther to the right than Rb. **c.** B would have the higher first ionization energy, because it is farther to the right and in the same period as Li. **d.** F would have the higher first ionization energy, because in the periodic table it is higher and farther to the right than Cs.

45. Values of ionization energies for ions such as Na^+, Mg^{2+}, and Al^{3+} illustrate that it is very difficult to remove an electron from a filled octet. This indicates the stability of a filled octet.

49. **a.** Fe would have the higher electron affinity, because in the periodic table it is higher and farther to the right than K. **b.** Se would have the higher electron affinity, because it is farther to the right and in the same period as Ca. **c.** N would have the higher electron affinity, because it is higher and in the same group as As. **d.** Cl would have the higher electron affinity, because in the periodic table it is higher and farther to the right than Po.

53. **a.** Ca would be more metallic, because in the periodic table it is lower and farther to the left than B. **b.** Ba would be more metallic, because in the periodic table it is lower and farther to the left than N. **c.** Fr would be more metallic, because it is lower and in the same group as Li. **d.** Al would be more metallic, because it is farther to the left and in the same period as S.

59. **a.** Al^{4+} is not likely to exist, because the fourth electron would have to be removed from the filled second principal energy level of Al^{3+}. **b.** Mg^{3+} is not likely to exist, because the third electron would have to be removed from the filled second principal energy level of Mg^{2+}. **c.** Rb^+ is likely to exist, because it contains a full octet of outer shell electrons. **d.** F^+ is not likely to exist, because F has an extremely high first ionization energy and an extremely high electron affinity. **e.** Sr^+ is not likely to exist, because its outer shell would contain only one electron; that electron is likely to be lost, forming Sr^{2+}, which contains a full octet of outer shell electrons. **f.** K^{2+} is not likely to exist, because its outer shell would contain seven electrons. K^+ is much more likely to exist, because it contains a full octet of outer shell electrons.

CHAPTER 6

EXERCISES

6.1 We can recognize an ionic compound from its formula by the presence of both a metal and one or more nonmetals. Thus, compounds a and c are ionic. In contrast, molecular compounds usually contain nonmetals only; thus, b and d are molecular compounds.

6.2 **a.** An atom of gallium, a Group IIIA element, has three valence electrons. A gallium atom forms a cation by losing its three valence electrons and taking on a charge of 3+. Therefore, the electron dot formula of the gallium cation is Ga^{3+}. **b.** An atom of sulfur, a Group VIA element, has six valence electrons; it will form an anion by gaining enough electrons (two) to complete its valence shell octet. Therefore, the electron dot formula of the sulfur anion is $\ddot{\underset{..}{S}}:^{2-}$. **c.** An atom of bromine, a Group VIIA element, has seven valence electrons; it will form an anion by gaining enough electrons (one) to complete its valence shell octet. Therefore, the electron dot formula of the bromine anion is $:\ddot{\underset{..}{Br}}:^-$.

6.3 **a.** An atom of potassium, Group IA, has one valence electron. Since potassium is a metal, it will form a cation by losing its valence electron. Thus, the electron dot formula of the cation is K^+. An atom of sulfur, Group VIA, has six valence electrons. Since sulfur is a nonmetal, it will form an anion by gaining two electrons to complete its valence shell octet. Therefore, the electron dot formula of the sulfur anion is $:\ddot{\underset{..}{S}}:^{2-}$. Now, write the electron dot formulas of the cation and anion together, and choose the appropriate subscript(s) to make total positive charge equal to total negative charge: $(K^+)_2:\ddot{\underset{..}{S}}:^{2-}$. **b.** Through reasoning similar to that used in part a, we find that the electron dot formula for the calcium cation is Ca^{2+} and the electron dot formula for the oxygen anion is $:\ddot{\underset{..}{O}}:^{2-}$. Now, write the electron dot formulas of the cation and anion together. In this case, total positive charge is already equal to total negative charge: $Ca^{2+} : \overset{..}{\underset{..}{O}} :^{2-}$.   **c.** Through reasoning similar to that used in part a, we find that the electron dot formula for the aluminum cation is Al^{3+} and the electron dot formula for the fluorine anion is $: \overset{..}{\underset{..}{F}} :^{-}$. Now, write the electron dot formulas of the cation and anion together, and choose the appropriate subscript(s) to make total positive charge equal to total negative charge: $Al^{3+} (: \overset{..}{\underset{..}{F}} :^{-})_3$.

6.4   Write down the formulas of the cation and anion in each case. Then use the number of the charge on each ion as a guide for choosing the subscript(s).

    a. $Ba^{②-} \quad O^{②-} = Ba_2O_2 = Ba_{2/2}O_{2/2} = BaO$

    b. $Ga^{③+} \quad Br^{①-} = GaBr_3$

    c. $Sr^{②+} \quad N^{③-} = Sr_3N_2$

6.5   Since the oxidation numbers of atoms and monatomic ions are the same as their charges (but written in reverse order), the oxidation numbers are as follows: **a.** $+1$ **b.** 0 **c.** -1 **d.** -2

6.6   **a. 1.** Draw the atomic skeleton to show C as the central atom, and join the atoms with single bonds, making sure that carbon has a valence of 4 and hydrogen a valence of 1.

$$\begin{array}{c} H \\ | \\ H - C - H \\ | \\ H \end{array}$$

    2. Count total valence electrons.

$$\begin{array}{r} 4\,H = 4(1) = 4 \\ C = \underline{4} \\ 8 \text{ total valence electrons} \end{array}$$

    3. Subtract two valence electrons for each single bond and distribute the remaining valence electrons as nonbonding electrons.

$$8 - 4(2) = 8 - 8 = 0 \text{ remaining valence electrons to be distributed as nonbonding electrons}$$

This gives

$$\begin{array}{c} H \\ | \\ H - C - H \\ | \\ H \end{array}$$

    4. Check to see that each atom has a noble gas valence shell configuration. They do, so our electron dot formula of CH_4 is correct. Using the same four steps as in part a, we get

    b.

$$\begin{array}{c} :\overset{..}{Cl}: \\ | \\ :\overset{..}{\underset{..}{Cl}} - \overset{..}{N} - \overset{..}{\underset{..}{Cl}}: \end{array}$$

    c.

$$H - \overset{..}{\underset{..}{S}} - H$$

6.7   **a. 1.** Draw the atomic skeleton to show C bonded to O: $C-O$ **2.** Count total valence electrons.

$$\begin{array}{r} C = 4 \\ O = \underline{6} \\ 10 \text{ total valence electrons} \end{array}$$

    3. Subtract two electrons for each single bond and distribute the remaining electrons as nonbonding electrons.

$$10 - 2 = 8 \text{ remaining electrons to be distributed as nonbonding electrons}$$

This gives

$$:\ddot{C}-\ddot{O}:$$

4. If necessary, shift nonbonding electrons to create multiple bonds and give each atom a noble gas configuration.

$$:\ddot{C}\overset{\oplus}{\underset{}{\frown}}\overset{\oplus}{\ddot{O}}: \longrightarrow :C\equiv O:$$

b. 1. Draw the atomic skeleton. Since no atom appears only once in the formula, there is no central atom. Furthermore, it is likely that two or more identical nonmetal atoms are bonded together. Since there are only two nitrogen atoms, let's try bonding them together, as a first approach. Thus, our trial atomic skeleton is

2. Count total valence electrons.

$$\begin{aligned} 4\,O &= 4(6) = 24 \\ 2\,N &= 2(5) = \underline{10} \\ &\quad\;\; 34 \text{ total valence electrons} \end{aligned}$$

3. Subtract two electrons for each single bond and distribute remaining electrons as nonbonding electrons.

$$34 - 5(2) = 34 - 10 = 24 \text{ remaining electrons to be distributed as nonbonding electrons}$$

This gives

4. If necessary, shift nonbonding electrons to create multiple bonds and give each atom a noble gas configuration.

(Note that nitrogen has a valence of 4 instead of the usual 3.)

c. 1. Draw the atomic skeleton with C as the central atom: $O-C-O$

2. Count total valence electrons.

$$\begin{aligned} 2\,O &= 2(6) = 12 \\ C &= \quad\;\; \underline{4} \\ &\quad\;\; 16 \text{ total valence electrons} \end{aligned}$$

3. Subtract two electrons for each single bond and distribute remaining electrons as nonbonding electrons.

$$16 - 2(2) = 16 - 4 = 12 \text{ remaining electrons to be distributed as nonbonding electrons}$$

This gives

$$:\ddot{O}-C-\ddot{O}:$$

4. If necessary, shift nonbonding electrons to create multiple bonds and give each atom a noble gas configuration.

$$:\overset{\oplus}{\ddot{O}}\overset{}{\frown}C\overset{}{\frown}\overset{\oplus}{\ddot{O}}: \longrightarrow \ddot{O}=C=\ddot{O}$$

6.8 **1.** Draw the atomic skeleton with C as the central atom. In this case, oxygen apparently does not have a valence of 2, because this valence is not compatible with C as the central atom.

$$\begin{bmatrix} & O & \\ & | & \\ O & -\!\!\overset{}{C}\!\!- & O \end{bmatrix}^{2-}$$

2. Count total valence electrons.

$$3 O = 3(6) = 18$$
$$C = 4$$
$$\text{charge} = \underline{2}$$
$$24 \text{ total valence electrons}$$

3. Find the number of nonbonding electrons and distribute them.

$$24 - 3(2) = 24 - 6 = 18 \text{ nonbonding electrons}$$

$$\left[\begin{array}{c} :\overset{..}{O}: \\ | \\ :\overset{..}{\underset{..}{O}}-C-\overset{..}{\underset{..}{O}}: \end{array} \right]^{2-}$$

4. Create multiple bonds if necessary.

$$\left[\begin{array}{c} :\overset{..}{O}: \\ | \\ :\overset{..}{\underset{..}{O}}-C-\overset{\oplus}{O}: \end{array} \right]^{2-} \longrightarrow \left[\begin{array}{c} :\overset{..}{O}: \\ | \\ :\overset{..}{\underset{..}{O}}-C=\overset{..}{O} \end{array} \right]^{2-}$$

6.9 First, derive the electron dot formula: $:\overset{..}{\underset{..}{F}}-\overset{..}{\underset{..}{O}}-\overset{..}{\underset{..}{F}}:$. We now see that the oxygen atom has four pairs of valence electrons, with four electrons in two nonbonded pairs. This means that there will be tetrahedral geometry about the oxygen atom, and each pair of valence electrons on oxygen will occupy the corner of a tetrahedron. This gives the OF_2 molecule a bent shape, very similar to that of H_2O.

$$\overset{..}{\underset{..}{F}}\diagdown \diagup \overset{..}{\underset{..}{F}} \\ \overset{..}{O}$$

6.10 First, derive the electron dot formula:

$$\begin{array}{c} :O: \\ \| \\ :\overset{..}{\underset{..}{O}}-S-\overset{..}{\underset{..}{O}}: \end{array}$$

(Note that two oxygen atoms have valences of 1 instead of the usual 2.) We now have the sulfur atom bonded to two oxygen atoms by single bonds and to a third oxygen atom by a double bond. Since the double bond behaves like a single bond, minimum bond repulsion will exist when the bond angles about S are 120°. Thus, we would predict the SO_3 molecule to have a trigonal shape.

6.11 **a.** Since the two atoms of O_2 are identical, the difference in electronegativity is 0, and the bond is nonpolar covalent. **b.** First we find the electronegativity difference in the Cl—F bond.

$$\text{electronegativity difference} = 4.0 - 3.0 = 1.0$$

The value of 1.0 indicates that the ClF bond is probably polar covalent. The bond polarity is

$$\overset{\delta+}{Cl}-\overset{\delta-}{F}$$

c. First we find the electronegativity difference in the C—O bond.

$$\text{electronegativity difference} = 3.5 - 2.5 = 1.0$$

The value of 1.0 indicates that the CO bond is probably polar covalent. The bond polarity is

$$\overset{\delta+}{C}\equiv\overset{\delta-}{O}$$

6.12 To answer the question, we must decide whether the electrical charge is distributed symmetrically through the NH_3 molecule. We should first determine whether there are any polar covalent bonds in NH_3. Since all bonds are identical (the three N—H bonds), we only need to find the electronegativity difference between N and H.

$$\text{electronegativity difference} = 3.0 - 2.1 = 0.9$$

The value of 0.9 indicates that the N—H bond is probably polar covalent. Now we must sketch the molecule, indicating electron-withdrawing effects, and determine whether these effects are symmetrical.

$$\ddot{\underset{\delta^-}{N}}$$

$$\underset{\delta^+}{H} \quad \underset{\underset{\delta^+}{H}}{\overset{}{|}} \quad \underset{\delta^+}{H}$$

The sketch suggests that the electron-withdrawing effect of the nitrogen atom is not symmetrically distributed through the NH_3 molecule; thus, we would expect the NH_3 molecule to be polar.

STUDY QUESTIONS AND PROBLEMS

4. Ionic: a, d, e, f, g, h, j, k, l Molecular: b, c, i

9. **a.** $:\ddot{Se}:^{2-}$ **b.** Cs^+ **c.** $:\ddot{Br}:^-$ **d.** Sr^{2+} **e.** $:\ddot{S}:^{2-}$ **f.** In^{3+}

11. **a.** $C\ddot{s}_2O$ **b.** Al_2S_3 **c.** $CaAt_2$ **d.** $BaSe$ **e.** $\ddot{N}a_3N$ **f.** K_2Se

14. **a.** 0 **b.** +3 **c.** 0 **d.** +2 **e.** −1 **f.** −3

16.

	Oxidized	Reduced
a.	Fe	H^+
b.	Ni	Cu^{2+}
c.	Pb	Zn^{2+}
d.	Cu	Ag^+
e.	Pb	Sn^{2+}
f.	Zn	AgCl

27. $:N≡N:$

32.

a. $H—\ddot{S}—H$

b. $H—\underset{\underset{H}{|}}{\overset{\overset{H}{|}}{C}}—\ddot{O}—H$

c. $:\ddot{C}l—\underset{\underset{:\ddot{C}l:}{|}}{\ddot{P}}—\ddot{C}l:$

d. $:\ddot{O}—\ddot{O}=\ddot{O}$

e. $:\ddot{O}—\underset{\underset{:O:}{\|}}{\overset{\overset{:O:}{\|}}{P}}—\ddot{O}—\underset{\underset{:O:}{\|}}{\overset{\overset{:O:}{\|}}{P}}—\ddot{O}:$

f. $H—\underset{\underset{:O:}{\|}}{C}—H$

34. **a.** $Ca^{2+} \left[:\ddot{O}—\underset{\underset{:\ddot{O}:}{|}}{\overset{\overset{:\ddot{O}:}{\|}}{S}}—\ddot{O}: \right]^{2-}$

b. $Na^+ \left[:\ddot{O}—\underset{\underset{:\ddot{O}:}{|}}{N}=\ddot{O} \right]^-$

c. $(K^+)_3 \left[:\ddot{O}—\underset{\underset{:\ddot{O}:}{|}}{\overset{\overset{:O:}{\|}}{P}}—\ddot{O}: \right]^{3-}$

d. $(Na^+)_2 \left[:\ddot{O}—\underset{\underset{:\ddot{O}:}{|}}{\overset{\overset{:\ddot{O}:}{\|}}{S}}—\ddot{O}: \right]^{2-}$

e. $Mg^{2+} \left[H—\underset{\underset{H}{|}}{\overset{\overset{H}{|}}{C}}—\underset{\underset{}{}}{\overset{\overset{:O:}{\|}}{C}}—\ddot{O}: \right]^-$

f. $Sr^{2+} \left[:\ddot{O}—\underset{\underset{:\ddot{O}:}{|}}{\overset{\overset{:\ddot{O}:}{\|}}{Cl}}—\ddot{O}: \right]^-_2$

36. **a.** tetrahedron **b.** pyramid **c.** bent shape **d.** trigonal shape

38. **a.** bent shape; 109.5°

$$\underset{H \quad\quad\quad H}{\overset{S}{\diagup \diagdown}}$$

b. pyramid; 109.5°

$$
\begin{array}{c}
\text{P} \\
\diagup \, | \, \diagdown \\
\text{Cl} \quad \text{Cl} \\
\text{Cl}
\end{array}
$$

c. tetrahedron; 109.5°

$$
\begin{array}{c}
\text{F} \\
| \\
\text{C} \\
\diagup \, | \, \diagdown \\
\text{F} \; \text{F} \; \text{F}
\end{array}
$$

44. $\overset{\delta+\;\;\delta-}{\text{N—Cl}}$ (nonpolar), $\overset{\delta+\;\;\delta-}{\text{S—O}}$, $\overset{\delta+\;\;\delta-}{\text{H—O}}$, $\overset{\delta+\;\;\delta-}{\text{P—F}}$

47. **a.** Nonpolar; tetrahedral shape gives symmetrical distribution of electron-withdrawing effects of Br atoms. **b.** Polar; bent shape gives unsymmetrical distribution of electron-withdrawing effect of S atom. **c.** Nonpolar; Br—Br bond is nonpolar.

50. **a.** 24 **b.** 32 **c.** 32 **d.** 32 **e.** 32 **f.** 24

52.

	Electron Dot Formula	Structural Formula	Shape	Bond Angles
a.	:N≡N:	N≡N	linear	—
b.	H—C≡N:	H—C≡N	linear	180°
c.	H—C—H with ‖ O below	H, H bonded to C with ‖ O below	trigonal	120°
d.	H—C≡C—H	H—C≡C—H	linear	180°
e.	:O—S=O:	O—S=O	bent	120°
f.	S=C=S	S=C=S	linear	180°

54. **a.** Nonpolar; all bonds are nonpolar. **b.** Nonpolar; the trigonal shape gives symmetrical distribution of the electron-withdrawing effect of the O atom. **c.** Nonpolar; both bonds are nonpolar.

CHAPTER 7

EXERCISES

7.1 Aluminum oxide is a binary compound composed of Al^{3+} and O^{2-}, and the oxidation number of each ion is the same as its charge: $\underset{+3\;-2}{Al_2O_3}$. Notice that the sum of oxidation numbers is zero: $2(+3) + 3(-2) = 6 - 6 = 0$

7.2 The sum of oxidation numbers in a polyatomic ion must equal the charge on the ion. Since we know the oxidation number of oxygen, -2, we can deduce the oxidation number of nitrogen.

$$\text{oxidation number of N} + 3(-2) = -1 \quad \text{(charge on polyatomic ion)}$$

$$\text{oxidation number of N} + (-6) = -1$$

$$\text{oxidation number of N} = -1 + 6 = +5$$

Thus, the oxidation numbers are $\underset{+5\;-2}{N\,O_3^-}$.

7.3 Since the sum of oxidation numbers in a compound is zero, and since we know the oxidation number of hydrogen $(+1)$ and oxygen (-2), we can deduce the oxidation number of phosphorus.

$$3(+1) + \text{oxidation number of P} + 4(-2) = 0$$

$$3 + \text{oxidation number of P} + (-8) = 0$$

$$\text{oxidation number of P} = 0 - 3 + 8 = +5$$

Then, the oxidation numbers are $\underset{+1\;+5\;-2}{H_3\,P\,O_4}$.

7.4　**a.** calcium selenide　**b.** manganese(IV) oxide

7.5.　**a.** $Pb^{2+}\ S^{2-} \longrightarrow Pb_2S_2 \longrightarrow Pb_{2/2}S_{2/2} \longrightarrow PbS$

　　b. $Al^{3+}\ O^{2-} \longrightarrow Al_2O_3$

7.6　**a.** mercuric oxide　**b.** ferrous chloride

7.7.　**a.** $Hg^{2+}\ Br^{1-} \longrightarrow HgBr_2$

　　b. $Sn^{4+}\ Cl^{1-} \longrightarrow SnCl_4$

7.8　**a.** sodium cyanide　**b.** potassium permanganate

7.9.　**a.** $Cs^{1+}\ PO_4^{3-} \longrightarrow Cs_3PO_4$

　　b. $Al^{3+}\ OH^{1-} \longrightarrow Al(OH)_3$

7.10　**a.** phosphorus pentachloride　**b.** dinitrogen oxide
7.11　**a.** HBr　**b.** CS_2
7.12　**a.** Oxyanion name: chromate　**b.** Oxyanion name: carbonate
　　　 Acid name: chromic acid　　　 Acid name: carbonic acid
7.13　**a.** Since the name has no *hydro-* prefix, the compound must be an oxyacid. Also, since phosphor*ous* acid must contain the phosph*ite* anion, the formula must be H_3PO_3.　**b.** Since the name has no *hydro-* prefix, the compound must be an oxyacid. Also, since permangan*ic* acid must contain the permangan*ate* anion, the formula must be $HMnO_4$.
7.14　**a.** barium hydroxide octahydrate　**b.** ammonium oxalate hydrate
7.15　**a.** $CoCl_2 \cdot 6\ H_2O$　**b.** $Al(BrO_3)_3 \cdot 9\ H_2O$

STUDY QUESTIONS AND PROBLEMS

3.　**a.** $\underset{+1\ +4\ -2}{H\ C\ O_3^-}$　**b.** $\underset{+2\ -2}{S_2\ O_3^{2-}}$　**c.** $\underset{+3\ -2}{C_2\ O_4^{2-}}$　**d.** $\underset{-2\ +1}{O\ H^-}$　**e.** $\underset{-3\ +1}{N\ H_4^+}$

5.　**a.** gallium iodide　**b.** iron(II) sulfide　**c.** mercury(I) oxide　**d.** platinum(II) chloride

7.　**a.** ferrous fluoride　**b.** cuprous sulfide　**c.** plumbic chloride　**d.** mercurous iodide　**e.** stannic oxide

10.　**a.** aluminum hydrogen sulfate or aluminum bisulfate　**b.** calcium cyanide　**c.** iron(III) hydrogen phosphate
　　d. strontium iodate

12.　**a.** diboron tetrabromide　**b.** bromine trifluoride　**c.** calcium diselenide　**d.** hydrogen fluoride

14.　**a.** hydrocyanic acid　**b.** sulfuric acid　**c.** acetic acid　**d.** nitric acid　**e.** hydrofluoric acid

18.　**a.** $Cd(NO_3)_2 \cdot 4\ H_2O$　**b.** $Au(CN)_3 \cdot 3\ H_2O$　**c.** $Pb(ClO_4)_4 \cdot 3\ H_2O$　**d.** $K_2CO_3 \cdot 2\ H_2O$　**e.** $Na_2HPO_4 \cdot 7\ H_2O$

20.　**a.** MnF_3　**b.** $Cr_2(SO_4)_3$　**c.** $KClO$　**d.** VBr_4　**e.** $Sr(BrO_3)_2$

22.　**a.** dichlorine heptoxide　**b.** chromium(III) acetate hydrate　**c.** mercury(I) carbonate　**d.** phosphoric acid
　　e. ammonium hydrogen sulfite or ammonium bisulfite

CHAPTER 8

EXERCISES

8.1　**a.** $FW = (2\ \text{atoms H}) \left(\dfrac{1.01\ \text{amu}}{\text{atom H}} \right) = 2.02$ amu. The formula weight expressed in grams, 2.02 g, is the weight of

one mole of H_2.　**b.** $FW = (1\ \text{atom C}) \left(\dfrac{12.0\ \text{amu}}{\text{atom C}} \right) + (2\ \text{atoms O}) \left(\dfrac{16.0\ \text{amu}}{\text{atom O}} \right) = 12.0$ amu $+ 32.0$ amu $=$

44.0 amu. The weight of one mole of CO_2 is 44.0 g.　**c.** $FW = (1\ \text{atom Mg}) \left(\dfrac{24.3\ \text{amu}}{\text{atom Mg}} \right) +$

$$(1 \text{ atom S}) \left(\frac{32.1 \text{ amu}}{\text{atom S}} \right) + (4 \text{ atoms O}) \left(\frac{16.0 \text{ amu}}{\text{atom O}} \right) = 24.3 \text{ amu} + 32.1 \text{ amu} + 64.0 \text{ amu} = 120.4 \text{ amu.}$$ The weight

of one mole of $MgSO_4$ is 120.4 g.   **d.** $FW = (1 \text{ ion K}^+) \left(\frac{39.1 \text{ amu}}{\text{ion K}^+} \right) = 39.1 \text{ amu.}$ The weight of one mole of K^+

is 39.1 g.

8.2   First we must find the formula weight of CO_2.

$$FW = (1 \text{ atom C}) \left(\frac{12.0 \text{ amu}}{\text{atom C}} \right) + (2 \text{ atoms O}) \left(\frac{16.0 \text{ amu}}{\text{atom O}} \right)$$

$$= 12.0 \text{ amu} + 32.0 \text{ amu} = 44.0 \text{ amu}$$

The formula weight of CO_2 expressed in grams, 44.0 g, is the weight of one mole of CO_2. We can use this relationship along with Avogadro's number to find the number of molecules in 26.3 g of CO_2.

$$1 \text{ mol } CO_2 = 44.0 \text{ g } CO_2$$

$$1 \text{ mol } CO_2 = 6.02 \times 10^{23} \text{ molecules of } CO_2$$

$$\text{Number of molecules} = (26.3 \text{ g } CO_2) \left(\frac{1 \text{ mol } CO_2}{44.0 \text{ g } CO_2} \right) \left(\frac{6.02 \times 10^{23} \text{ molecules } CO_2}{1 \text{ mol } CO_2} \right)$$

$$= 3.60 \times 10^{23} \text{ molecules } CO_2$$

8.3   First, use the unit-factor method to find moles of H_2O molecules:

$$1 \text{ mol } H_2O \text{ molecules} = 6.02 \times 10^{23} \text{ } H_2O \text{ molecules}$$

$$\text{moles } H_2O \text{ molecules} = (7.63 \times 10^{20} \text{ } H_2O \text{ molecules}) \left(\frac{1 \text{ mol } H_2O \text{ molecules}}{6.02 \times 10^{23} \text{ } H_2O \text{ molecules}} \right)$$

$$= 1.27 \times 10^{-3} \text{ mol } H_2O \text{ molecules}$$

Then convert moles of H_2O to mass:

$$1 \text{ mol } H_2O \text{ molecules} = 18.0 \text{ g } H_2O$$

$$\text{mass } H_2O = (1.27 \times 10^{-3} \text{ mol } H_2O \text{ molecules}) \left(\frac{18.0 \text{ g } H_2O}{\text{mol } H_2O \text{ molecules}} \right) = 2.29 \times 10^{-2} \text{ g } H_2O$$

Notice that we could have combined the two preceding steps into one expression:

$$\text{mass } H_2O = (7.63 \times 10^{20} \text{ } H_2O \text{ molecules}) \left(\frac{1 \text{ mol } H_2O \text{ molecules}}{6.02 \times 10^{23} \text{ } H_2O \text{ molecules}} \right) \left(\frac{18.0 \text{ g } H_2O}{\text{mol } H_2O \text{ molecules}} \right)$$

$$= 2.28 \times 10^{-2} \text{ g } H_2O$$

(Note that the discrepancy between answers, 2.29 g H_2O vs. 2.28 g H_2O, arises from rounding off 1.27×10^{-3} mol H_2O molecules in the first step of the first solution.)

8.4   First, find the molar mass of acetic acid:

$$FW = (4 \text{ atoms H}) \left(\frac{1.0 \text{ amu}}{\text{atom H}} \right) + (2 \text{ atoms C}) \left(\frac{12.0 \text{ amu}}{\text{atom C}} \right) + (2 \text{ atoms O}) \left(\frac{16.0 \text{ amu}}{\text{atom O}} \right)$$

$$= 4.0 \text{ amu} + 24.0 \text{ amu} + 32.0 \text{ amu} = 60.0 \text{ amu}$$

Therefore, the molar mass of $HC_2H_3O_2$ is 60.0 g. Then, use the unit-factor method to find moles of $HC_2H_3O_2$:

$$1 \text{ mol } HC_2H_3O_2 = 60.0 \text{ g } HC_2H_3O_2$$

$$\text{moles } HC_2H_3O_2 = (17.9 \text{ g } HC_2H_3O_2) \left(\frac{1 \text{ mol } HC_2H_3O_2}{60.0 \text{ g } HC_2H_3O_2} \right) = 0.298 \text{ mol } HC_2H_3O_2$$

8.5 First, find the molar mass of $Al(OH)_3$:

$$FW = (1 \text{ atom Al}) \left(\frac{27.0 \text{ amu}}{\text{atom Al}} \right) + (3 \text{ atoms O}) \left(\frac{16.0 \text{ amu}}{\text{atom O}} \right) + (3 \text{ atoms H}) \left(\frac{1.0 \text{ amu}}{\text{atom H}} \right)$$

$$= 27.0 \text{ amu} + 48.0 \text{ amu} + 3.0 \text{ amu} = 78.0 \text{ amu}$$

Therefore, the molar mass of $Al(OH)_3$ is 78.0 g. Now use the unit-factor method to find mass of $Al(OH)_3$:

$$\text{mass } Al(OH)_3 = (2.83 \text{ mol } Al(OH)_3) \left(\frac{78.0 \text{ g } Al(OH)_3}{\text{mol } Al(OH)_3} \right) = 221 \text{ g } Al(OH)_3$$

8.6 Step 1: Assume 100 g of the compound and calculate the actual weight of each element.

$$\text{wt N} = (0.304)(100 \text{ g}) = 30.4 \text{ g}$$
$$\text{wt O} = (0.696)(100 \text{ g}) = 69.6 \text{ g}$$

Step 2: Calculate moles of each element.

$$\text{moles N} = (30.4 \text{ g N}) \left(\frac{1 \text{ mol N}}{14.0 \text{ g N}} \right) = 2.17 \text{ mol N}$$

$$\text{moles O} = (69.6 \text{ g O}) \left(\frac{1 \text{ mol O}}{16.0 \text{ g O}} \right) = 4.35 \text{ mol O}$$

Step 3: Find the smallest whole-number ratio of elements.

$$\text{N: } \frac{2.17}{2.17} = 1.00$$

$$\text{O: } \frac{4.35}{2.17} = 2.00$$

Therefore, the simplest whole-number ratio of N:O is 1:2, and the empirical formula is NO_2.

8.7 Step 1:
$$\text{wt C} = (0.400)(100 \text{ g}) = 40.0 \text{ g}$$
$$\text{wt H} = (0.0667)(100 \text{ g}) = 6.67 \text{ g}$$
$$\text{wt O} = (0.533)(100 \text{ g}) = 53.3 \text{ g}$$

Step 2:
$$\text{moles C} = (40.0 \text{ g C}) \left(\frac{1 \text{ mol C}}{12.0 \text{ g C}} \right) = 3.33 \text{ mol C}$$

$$\text{moles H} = (6.67 \text{ g H}) \left(\frac{1 \text{ mol H}}{1.01 \text{ g H}} \right) = 6.60 \text{ mol H}$$

$$\text{moles O} = (53.3 \text{ g O}) \left(\frac{1 \text{ mol O}}{16.0 \text{ g O}} \right) = 3.33 \text{ mol O}$$

Step 3:
$$\text{C: } \frac{3.33}{3.33} = 1.00$$

$$\text{H: } \frac{6.60}{3.33} = 1.98$$

$$\text{O: } \frac{3.33}{3.33} = 1.00$$

Therefore, the simplest whole-number ratio of C:H:O is 1:2:1, and the empirical formula is CH_2O.

8.8 Step 1: Since the weight of each element is given, we can skip step 1 and go on to step 2.

Step 2:

$$\text{moles O} = (15.5 \text{ g O}) \left(\frac{1 \text{ mol O}}{16.0 \text{ g O}} \right) = 0.969 \text{ mol O}$$

$$\text{moles F} = (12.3 \text{ g F}) \left(\frac{1 \text{ mol F}}{19.0 \text{ g F}} \right) = 0.647 \text{ mol F}$$

Step 3:

$$\text{O:} \quad \frac{0.969}{0.647} = 1.50$$

$$\text{F:} \quad \frac{0.647}{0.647} = 1.00$$

We have not yet reached a whole-number ratio of elements, so we must multiply both values by the smallest number that will convert both to whole numbers. This number is 2.

$$\text{O:} \quad (1.50)(2) = 3.00$$

$$\text{F:} \quad (1.00)(2) = 2.00$$

Thus, the smallest whole-number ratio of O:F is 3:2, and the empirical formula is O_3F_2.

8.9 First, find the apparent molar mass:

$$\text{Apparent FW} = (1 \text{ atom H}) \left(\frac{1.01 \text{ amu}}{\text{atom H}} \right) + (1 \text{ atom O}) \left(\frac{16.0 \text{ amu}}{\text{atom O}} \right)$$

$$= 1.01 \text{ amu} + 16.0 \text{ amu} = 17.0 \text{ amu}$$

Thus, the apparent molar mass is 17.0 g/mol. Next, find n:

$$n = \frac{\text{actual molar mass}}{\text{apparent molar mass}} = \frac{34.0 \text{ g/mol}}{17.0 \text{ g/mol}} = 2.00$$

Finally, multiply the empirical formula by n:

$$\text{true formula} = H_{(1 \times 2)} O_{(1 \times 2)} = H_2O_2$$

8.10 First, find the empirical formula:

$$\text{wt C} = (0.400)(100 \text{ g}) = 40.0 \text{ g}$$

$$\text{wt H} = (0.0667)(100 \text{ g}) = 6.67 \text{ g}$$

$$\text{wt O} = (0.0533)(100 \text{ g}) = 53.3 \text{ g}$$

$$\text{moles C} = (40.0 \text{ g C}) \left(\frac{1 \text{ mol C}}{12.0 \text{ g C}} \right) = 3.33 \text{ mol C}$$

$$\text{moles H} = (6.67 \text{ g H}) \left(\frac{1 \text{ mol H}}{1.01 \text{ g H}} \right) = 6.67 \text{ mol H}$$

$$\text{moles O} = (53.3 \text{ g O}) \left(\frac{1 \text{ mol O}}{16.0 \text{ g O}} \right) = 3.33 \text{ mol O}$$

$$\text{C:} \quad \frac{3.33}{3.33} = 1.00$$

$$\text{H:} \quad \frac{6.67}{3.33} = 2.00$$

$$\text{O:} \quad \frac{3.33}{3.33} = 1.00$$

Therefore, the empirical formula is CH_2O. Next, find the apparent molar mass:

$$\text{Apparent FW} = (1 \text{ atom C}) \left(\frac{12.0 \text{ amu}}{\text{atom C}} \right) + (2 \text{ atoms H}) \left(\frac{1.01 \text{ amu}}{\text{atom H}} \right) + (1 \text{ atom O}) \left(\frac{16.0 \text{ amu}}{\text{atom O}} \right)$$

$$= 12.0 \text{ amu} + 2.02 \text{ amu} + 16.0 \text{ amu} = 30.0 \text{ amu}$$

Thus the apparent molar mass is 30.0 g/mol. Then, find n:

$$n = \frac{\text{actual molar mass}}{\text{apparent molar mass}} = \frac{180.0 \text{ g/mol}}{30.0 \text{ g/mol}} = 6.00$$

Finally, multiply the empirical formula by n:

$$\text{true formula} = C_{(1\times6)}H_{(2\times6)}O_{(1\times6)} = C_6H_{12}O_6$$

STUDY QUESTIONS AND PROBLEMS

3. **a.** 5.79×10^{24} **b.** 2.45×10^{20} **c.** 1.16×10^{24} **d.** 1.5×10^{25}
7. **a.** 0.174 moles; 1.05×10^{23} atoms **b.** 1.80×10^{-2} g; 2.14×10^{-4} moles **c.** 17.2 g; 7.40×10^{23} atoms
   d. 3.53×10^{17} g; 6.78×10^{15} moles **e.** 1.26 moles; 7.59×10^{23} atoms **f.** 754 g; 4.73×10^{24} atoms
9. **a.** 4.21×10^{24} Na$^+$; 2.11×10^{24} Cr$_2$O$_7{}^{2-}$ **b.** 2.11×10^{24} Hg^{2+}; 2.11×10^{24} C$_2$O$_4{}^{2-}$ **c.** 4.21×10^{24} K$^+$;
   2.11×10^{24} SO$_4{}^{2-}$
11. **a.** 262.0 g **b.** 288.6 g **c.** 174.3 g
13. 2.90×10^{19} Na$^+$; 2.90×10^{19} OH$^-$
15. 8.13×10^{-5} g
21. C_4H_9
23. $BiPO_4$
25. MgO
27. N_2H_4
29. **a.** 2.56×10^{24} molecules; 2.56×10^{24} Xe atoms; 2.56×10^{24} O atoms; 1.02×10^{25} F atoms **b.** 2.56×10^{24} mole-
    cules; 5.12×10^{24} C atoms; 1.54×10^{25} H atoms **c.** 2.56×10^{24} molecules; 2.56×10^{24} P atoms; 2.56×10^{24} S
    atoms; 5.12×10^{24} Br atoms; 2.56×10^{24} Cl atoms
31. 3.60×10^{30} g
33. CH_4

CHAPTER 9

EXERCISES

9.1 In this reaction, carbon and oxygen are the reactants, and carbon dioxide is the product. Since the total mass of
     reactants is 18.3 g, the mass of the product must also be 18.3 g:

$$\text{total mass of reactants} = \text{total mass of products}$$
$$5.00 \text{ g} + 13.3 \text{ g} = 18.3 \text{ g}$$

9.2 **a.** Step 1: Initial equation:

$$Na(s) + Cl_2(g) \longrightarrow NaCl(s)$$

Step 2: Count atoms:

Element	Reactants	Products
Na	1	1
Cl	2	1

Step 3: Since sodium is already balanced, insert a coefficient for Cl in front of NaCl:

$$Na(s) + Cl_2(g) \longrightarrow 2 \text{ NaCl}(s)$$

Step 4:   Recount atoms:

Element	Reactants	Products
Na	1	2
Cl	2	2

Step 5:   Insert the coefficient 2 for Na(s) to rebalance Na:

$$2\,Na(s) + Cl_2(g) \longrightarrow 2\,NaCl(s)$$

b. Step 1:   Initial equation:

$$Mg(s) + O_2(g) \longrightarrow MgO(s)$$

Step 2:   Count atoms:

Element	Reactants	Products
Mg	1	1
O	2	1

Step 3:   Since magnesium is already balanced, insert a coefficient for O in front of MgO:

$$Mg(s) + O_2(g) \longrightarrow 2\,MgO(s)$$

Step 4:   Recount atoms:

Element	Reactants	Products
Mg	1	2
O	2	2

Step 5:   Insert the coefficient 2 for Mg(s) to rebalance Mg:

$$2\,Mg(s) + O_2(g) \longrightarrow 2\,MgO(s)$$

9.3   **a.** Step 1:   Initial equation:

$$S_8(s) + O_2(g) \longrightarrow SO_2(g)$$

Step 2:   Count atoms:

Element	Reactants	Products
S	8	1
O	2	2

Step 3:   Insert coefficients:
Oxygen is balanced, but we need to insert coefficients for sulfur. Insertion of 8 before SO_2 will balance sulfur atoms:

$$S_8(s) + O_2(g) \longrightarrow 8\,SO_2(g)$$

Step 4: Recount atoms:

Element	Reactants	Products
S	8	8
O	2	16

Step 5: Change coefficients if necessary. In this case, we need to insert 8 before O_2:

$$S_8(s) + 8\ O_2(g) \longrightarrow 8\ SO_2(g)$$

b. Step 1: Initial equation:

$$C_2H_6(g) + O_2(g) \longrightarrow CO_2(g) + H_2O(g)$$

Step 2: Count atoms:

Element	Reactants	Products
C	2	1
H	6	2
O	2	3

Step 3: Insert coefficients:
First balance carbon atoms.

$$C_2H_6(g) + O_2(g) \longrightarrow 2\ CO_2(g) + H_2O(g)$$

Next, balance hydrogen atoms.

$$C_2H_6(g) + O_2(g) \longrightarrow 2\ CO_2(g) + 3\ H_2O(g)$$

Last, balance oxygen atoms.

$$C_2H_6(g) + \tfrac{7}{2}\ O_2(g) \longrightarrow 2\ CO_2(g) + 3\ H_2O(g)$$

Now multiply the equation by 2.

$$2\ C_2H_6(g) + 7\ O_2(g) \longrightarrow 4\ CO_2(g) + 6\ H_2O(g)$$

Step 4: Recount atoms:

Element	Reactants	Products
C	4	4
H	12	12
O	14	14

Step 5: Change coefficients if necessary. This step is not necessary in this case.

9.4 **a.** Step 1: Initial equation:

$$CaCl_2(aq) + Na_3PO_4(aq) \longrightarrow Ca_3(PO_4)_2(s) + NaCl(aq)$$

Step 2:   Count atoms:

Element	Reactants	Products
Ca	1	3
Cl	2	1
Na	3	1
P	1	2
O	4	8

Step 3:   Insert coefficients:
We start by inserting coefficients for the metals: 3 in front of $CaCl_2$ and 3 in front of NaCl.

$$3\ CaCl_2(aq) + Na_3PO_4(aq) \longrightarrow Ca_3(PO_4)_2(s) + 3\ NaCl(aq)$$

Next, we balance the nonmetals. For Cl we must change the coefficient of NaCl from 3 to 6. (This "unbalances" Na, but we can adjust Na later if necessary.)

$$3\ CaCl_2(aq) + Na_3PO_4(aq) \longrightarrow Ca_3(PO_4)_2(s) + 6\ NaCl(aq)$$

Since the polyatomic ion PO_4^{3-} appears on both sides of the equation, we can treat it as a unit. Thus, we must insert a coefficient of 2 in front of Na_3PO_4 to balance PO_4^{3-} ions.

$$3\ CaCl_2(aq) + 2\ Na_3PO_4(aq) \longrightarrow Ca_3(PO_4)_2(s) + 6\ NaCl(aq)$$

Step 4:   Recount atoms:

Element	Reactants	Products
Ca	3	3
Cl	6	6
Na	6	6
P	2	2
O	8	8

Step 5:   Change coefficients if necessary. This step is not necessary in this case, since balancing PO_4^{3-} ions also balanced Na.

b. Step 1:   Initial equation:

$$H_2SO_4(aq) + NaOH(aq) \longrightarrow Na_2SO_4(aq) + H_2O(l)$$

Step 2:   Count atoms:

Element	Reactants	Products
H	3	2
S	1	1
O	5	5
Na	1	2

Step 3:   Insert coefficients:
Starting with the metal Na,

$$H_2SO_4(aq) + 2\ NaOH(aq) \longrightarrow Na_2SO_4(aq) + H_2O(l)$$

Next, we examine coefficients for the polyatomic ion SO_4^{2-}, and we find that it is balanced. Then we balance H.

$$H_2SO_4(aq) + 2\ NaOH(aq) \longrightarrow Na_2SO_4(aq) + 2\ H_2O(l)$$

Last, we examine the coefficients of the O that is not a part of SO_4^{2-} and find that it is balanced.

Step 4: Recount atoms:

Element	Reactants	Products
H	4	4
S	1	1
O	6	6
Na	2	2

Step 5: Change coefficients if necessary. This step is not necessary in this case.

9.5 **a.** This reaction follows the pattern $A + B \rightarrow AB$, where A is P_4, B is O_2, and AB is P_2O_5. Thus, this is a combination reaction.   **b.** This reaction follows the pattern $A + BC \rightarrow AC + B$, where A is H_2, BC is Fe_2O_3, AC is H_2O, and B is Fe. Thus, this is a single replacement reaction.   **c.** This reaction follows the pattern $AB \rightarrow A + B$, where AB is NaCl, A is NA, and B is Cl_2. Thus, this is a decomposition reaction.   **d.** This reaction follows the pattern $AB + CD \rightarrow AD + CB$, where AB is H_2SO_4, CO is NaOH, AD is H_2O, and CB is Na_2SO_4. Thus, this is a double replacement reaction.

9.6 **a.** According to the general pattern for double replacement, CaF_2 corresponds to AB and H_2SO_4 to CD:

$$AB + \quad CD \quad \longrightarrow AD + CB$$
$$CaF_2 + H_2SO_4 \longrightarrow$$

Thus, $A = Ca^{2+}$, $B = F^-$, $C = H^+$, and $D = SO_4^{2-}$. When the reactants trade ionic partners, the products are shown below.

$$AB + \quad CD \quad \longrightarrow \quad AD \quad + CB$$
$$CaF_2 + H_2SO_4 \longrightarrow CaSO_4 + HF$$

To balance the equation, we must place a 2 in front of HF:

$$CaF_2 + H_2SO_4 \longrightarrow CaSO_4 + 2\ HF$$

b. According to the general pattern for double replacement, $NaNO_3$ corresponds to AB and H_2SO_4 to CD:

$$AB \quad + \quad CD \quad \longrightarrow AD + CB$$
$$NaNO_3 + H_2SO_4 \longrightarrow$$

Thus, $A = Na^+$, $B = NO_3^-$, $C = H^+$, and $D = SO_4^{2-}$. When the reactants trade ionic partners, the products are as shown below.

$$AB \quad + \quad CD \quad \longrightarrow \quad AD \quad + \quad CD$$
$$NaNO_3 + H_2SO_4 \longrightarrow Na_2SO_4 + HNO_3$$

To balance the equation, we must place a 2 in front of $NaNO_3$ and HNO_3.

$$2\ NaNO_3 + H_2SO_4 \longrightarrow Na_2SO_4 + 2\ HNO_3$$

9.7 There will be two mole ratios for each different pair of substances. Since there are four substances, there are six different pairs of substances and a total of 12 mole ratios.

$$\frac{3\text{ mol NaOH}}{1\text{ mol }H_3PO_4}, \frac{1\text{ mol }H_3PO_4}{3\text{ mol NaOH}}$$

$$\frac{3\text{ mol NaOH}}{1\text{ mol }Na_3PO_4}, \frac{1\text{ mol }Na_3PO_4}{3\text{ mol NaOH}}$$

$$\frac{3 \text{ mol NaOH}}{3 \text{ mol H}_2\text{O}}, \frac{3 \text{ mol H}_2\text{O}}{3 \text{ mol NaOH}}$$

$$\frac{1 \text{ mol H}_3\text{PO}_4}{1 \text{ mol Na}_3\text{PO}_4}, \frac{1 \text{ mol Na}_3\text{PO}_4}{1 \text{ mol H}_3\text{PO}_4}$$

$$\frac{1 \text{ mol H}_3\text{PO}_4}{3 \text{ mol H}_2\text{O}}, \frac{3 \text{ mol H}_2\text{O}}{1 \text{ mol H}_3\text{PO}_4}$$

$$\frac{1 \text{ mol Na}_3\text{PO}_4}{3 \text{ mol H}_2\text{O}}, \frac{3 \text{ mol H}_2\text{O}}{1 \text{ mol Na}_3\text{PO}_4}$$

9.8 Step 1: Write down the given number of moles: 0.283 mol K

Step 2: Multiply the given number of moles by the mole ratio that relates K_2O and K and allows cancellation of K:

$$(0.283 \text{ mol K})\left(\frac{2 \text{ mol K}_2\text{O}}{4 \text{ mol K}}\right) = 0.142 \text{ mol K}_2\text{O}$$

9.9 Step 1:

$$5.62 \text{ mol KClO}_3$$

Step 2:

$$(5.62 \text{ mol KClO}_3)\left(\frac{3 \text{ mol O}_2}{2 \text{ mol KClO}_3}\right) = 8.43 \text{ mol O}_2$$

9.10 Step 1: Write down the given number of moles: 1.50 mol NH_3

Step 2: Convert moles of NH_3 to moles of NO by using the appropriate mole ratio:

$$(1.50 \text{ mol NH}_3)\left(\frac{4 \text{ mol NO}}{4 \text{ mol NH}_3}\right) = 1.50 \text{ mol NO}$$

Step 3: Convert moles of NO to weight of NO using the molar mass:

$$\text{FW of NO} = (1 \text{ atom N})\left(\frac{14.0 \text{ amu}}{\text{atom N}}\right) + (1 \text{ atom O})\left(\frac{16.0 \text{ amu}}{\text{atom O}}\right)$$
$$= 14.0 \text{ amu} + 16.0 \text{ amu} = 30.0 \text{ amu}$$

Therefore the molar mass of NO is 30.0 g.

$$(1.50 \text{ mol NO})\left(\frac{30.0 \text{ g NO}}{\text{mol NO}}\right) = 45.0 \text{ g NO}$$

9.11 Step 1:

$$1.75 \text{ mol CaO}$$

Step 2:

$$(1.75 \text{ mol CaO})\left(\frac{1 \text{ mol H}_2\text{O}}{1 \text{ mol CaO}}\right) = 1.75 \text{ mol H}_2\text{O}$$

Step 3:

$$\text{FW of H}_2\text{O} = (2 \text{ atoms H})\left(\frac{1.0 \text{ amu}}{\text{atom H}}\right) + (1 \text{ atom O})\left(\frac{16.0 \text{ amu}}{\text{atom O}}\right)$$
$$= 2.0 \text{ amu} + 16.0 \text{ amu} = 18.0 \text{ amu}$$
$$(1.75 \text{ mol H}_2\text{O})\left(\frac{18.0 \text{ g H}_2\text{O}}{\text{mol H}_2\text{O}}\right) = 31.5 \text{ g H}_2\text{O}$$

9.12 Step 1: Write down the given weight: 46.5 g N_2

Step 2: Convert weight of N_2 to moles of N_2 by using the molar mass:

$$(46.5 \text{ g N}_2)\left(\frac{1 \text{ mol N}_2}{28.0 \text{ g N}_2}\right) = 1.66 \text{ mol N}_2$$

Step 3:   Convert moles of N_2 to moles of NH_3 by using the appropriate mole ratio:

$$(1.66 \text{ mol } N_2)\left(\frac{2 \text{ mol } NH_3}{1 \text{ mol } N_2}\right) = 3.32 \text{ mol } NH_3$$

9.13 Step 1:   17.2 g Hg

Step 2:
$$(17.2 \text{ g Hg})\left(\frac{1 \text{ mol Hg}}{200.6 \text{ g Hg}}\right) = 8.57 \times 10^{-2} \text{ mol Hg}$$

Step 3:
$$(8.57 \times 10^{-2} \text{ mol Hg})\left(\frac{1 \text{ mol } O_2}{2 \text{ mol Hg}}\right) = 4.28 \times 10^{-2} \text{ mol } O_2$$

9.14 Step 1: Write down the given weight: 14.5 g C_2H_2

Step 2: Convert the given weight to moles by using the molar mass:

$$(14.5 \text{ g } C_2H_2)\left(\frac{1 \text{ mol } C_2H_2}{26.0 \text{ g } C_2H_2}\right) = 0.558 \text{ mol } C_2H_2$$

Step 3: Convert moles of C_2H_2 to moles of CO_2 using the appropriate mole ratio:

$$(0.558 \text{ mol } C_2H_2)\left(\frac{4 \text{ mol } CO_2}{2 \text{ mol } C_2H_2}\right) = 1.12 \text{ mol } CO_2$$

Step 4: Convert moles of CO_2 to weight of CO_2 using the molar mass:

$$(1.12 \text{ mol } CO_2)\left(\frac{28.0 \text{ g } CO_2}{\text{mol } CO_2}\right) = 31.4 \text{ g } CO_2$$

or, combining steps,

$$(14.5 \text{ g } C_2H_2)\left(\frac{1 \text{ mol } C_2H_2}{26.0 \text{ g } C_2H_2}\right)\left(\frac{4 \text{ mol } CO_2}{2 \text{ mol } C_2H_2}\right)\left(\frac{28.0 \text{ g } CO_2}{\text{mol } CO_2}\right) = 31.2 \text{ g } CO_2$$

(The small difference in the two answers is due to rounding off in the four individual steps.)

9.15 Step 1:   14.5 g Zn

Step 2:
$$(14.5 \text{ g Zn})\left(\frac{1 \text{ mol Zn}}{65.4 \text{ g Zn}}\right) = 0.222 \text{ mol Zn}$$

Step 3:
$$(0.222 \text{ mol Zn})\left(\frac{2 \text{ mol HCl}}{1 \text{ mol Zn}}\right) = 0.444 \text{ mol HCl}$$

Step 4:
$$(0.444 \text{ mol HCl})\left(\frac{36.5 \text{ g HCl}}{\text{mol HCl}}\right) = 16.2 \text{ g HCl}$$

or, combining steps,

$$(14.5 \text{ g Zn})\left(\frac{1 \text{ mol Zn}}{65.4 \text{ g Zn}}\right)\left(\frac{2 \text{ mol HCl}}{1 \text{ mol Zn}}\right)\left(\frac{36.5 \text{ g HCl}}{\text{mol HCl}}\right) = 16.2 \text{ g HCl}$$

9.16 Step 1: Since the actual yield of Cr is given in units of grams, we must calculate theoretical yield in units of grams. This is a weight-weight stoichiometric calculation.

$$(0.500 \text{ g C})\left(\frac{1 \text{ mol C}}{12.0 \text{ g C}}\right)\left(\frac{2 \text{ mol Cr}}{3 \text{ mol C}}\right)\left(\frac{52.0 \text{ g Cr}}{\text{mol Cr}}\right) = 14.4 \text{ g Cr}$$

Step 2:   Calculate the percentage yield using the actual yield and the theoretical yield:

$$\% \text{ Yield} = \frac{\text{actual yield}}{\text{theoretical yield}} \times 100\% = \frac{12.0 \text{ g Cr}}{14.4 \text{ g Cr}} \times 100\% = 83.3\%$$

9.17 Step 1:

$$(0.0200 \text{ mol Pb(NO}_3)_2) \left(\frac{1 \text{ mol PbCl}_2}{1 \text{ mol Pb(NO}_3)_2} \right) \left(\frac{278.2 \text{ g PbCl}_2}{\text{mol PbCl}_2} \right) = 5.56 \text{ g PbCl}_2$$

Step 2:

$$\% \text{ Yield} = \frac{4.34 \text{ g PbCl}_2}{5.56 \text{ g PbCl}_2} \times 100\% = 78.1\%$$

9.18 Step 1:   Calculate the number of moles of each reactant.

$$\text{Al:} \quad (22.3 \text{ g Al}) \left(\frac{1 \text{ mol Al}}{27.0 \text{ g Al}} \right) = 0.826 \text{ mol Al}$$

$$\text{O}_2\text{:} \quad (47.6 \text{ g O}_2) \left(\frac{1 \text{ mol O}_2}{32.0 \text{ g O}_2} \right) = 1.49 \text{ mol O}_2$$

Step 2:   Calculate the number of moles of Al_2O_3 that would be produced by each reactant.

$$\text{Al:} \quad (0.826 \text{ mol Al}) \left(\frac{2 \text{ mol Al}_2\text{O}_3}{4 \text{ mol Al}} \right) = 0.413 \text{ mol Al}_2\text{O}_3$$

$$\text{O}_2\text{:} \quad (1.49 \text{ mol O}_2) \left(\frac{2 \text{ mol Al}_2\text{O}_3}{3 \text{ mol O}_2} \right) = 0.993 \text{ mol Al}_2\text{O}_3$$

Since Al produces the lower theoretical yield, Al is the limiting reactant.

9.19 Step 1:

$$\text{AgNO}_3\text{:} \quad (15.0 \text{ g AgNO}_3) \left(\frac{1 \text{ mol AgNO}_3}{169.9 \text{ g AgNO}_3} \right) = 0.0883 \text{ mol AgNO}_3$$

$$\text{CaCl}_2\text{:} \quad (12.0 \text{ g CaCl}_2) \left(\frac{1 \text{ mol CaCl}_2}{111.1 \text{ g CaCl}_2} \right) = 0.108 \text{ mol CaCl}_2$$

Step 2:

$$\text{AgNO}_3\text{:} \quad (0.0883 \text{ mol AgNO}_3) \left(\frac{2 \text{ mol AgCl}}{2 \text{ mol AgNO}_3} \right) = 0.0883 \text{ mol AgCl}$$

$$\text{CaCl}_2\text{:} \quad (0.108 \text{ mol CaCl}_2) \left(\frac{2 \text{ mol AgCl}}{1 \text{ mol CaCl}_2} \right) = 0.216 \text{ mol AgCl}$$

Since $AgNO_3$ produces the lower theoretical yield, $AgNO_3$ is the limiting reactant.

9.20 Step 1:   Calculate the number of moles of each reactant.

$$\text{CaCO}_3\text{:} \quad (20.0 \text{ g CaCO}_3) \left(\frac{1 \text{ mol CaCO}_3}{100.1 \text{ g CaCO}_3} \right) = 0.200 \text{ mol CaCO}_3$$

$$\text{H}_3\text{PO}_4\text{:} \quad (15.0 \text{ g H}_3\text{PO}_4) \left(\frac{1 \text{ mol H}_3\text{PO}_4}{98.0 \text{ g H}_3\text{PO}_4} \right) = 0.153 \text{ mol H}_3\text{PO}_4$$

Step 2:   Find the limiting reactant on the basis of its theoretical yield of product. In this case, use the $Ca_3(PO_4)_2$ product, since its actual yield is given.

$$\text{CaCO}_3\text{:} \quad (0.200 \text{ mol CaCO}_3) \left(\frac{1 \text{ mol Ca}_3(\text{PO}_4)_2}{3 \text{ mol CaCO}_3} \right) = 0.0667 \text{ mol Ca}_3(\text{PO}_4)_2$$

$$\text{H}_3\text{PO}_4\text{:} \quad (0.153 \text{ mol H}_3\text{PO}_4) \left(\frac{1 \text{ mol Ca}_3(\text{PO}_4)_2}{2 \text{ mol H}_3\text{PO}_4} \right) = 0.0765 \text{ mol Ca}_3(\text{PO}_4)_2$$

Therefore $CaCO_3$ is the limiting reactant.

Step 3: Calculate the percentage yield based on the theoretical yield of $Ca_3(PO_4)_2$ by $CaCO_3$.

$$\text{Theoretical yield (in g)} = (0.0667 \text{ mol } Ca_3(PO_4)_2) \left(\frac{310.3 \text{ g } Ca_3(PO_4)_2}{\text{mol } Ca_3(PO_4)_2} \right)$$

$$= 20.7 \text{ g } Ca_3(PO_4)_2$$

$$\% \text{ Yield} = \frac{18.5 \text{ g } Ca_3(PO_4)_2}{20.7 \text{ g } Ca_3(PO_4)_2} \times 100\% = 89.4\%$$

STUDY QUESTIONS AND PROBLEMS

3. **a.** One atom of solid magnesium reacts with two molecules of gaseous hydrogen chloride to produce one formula unit of solid magnesium chloride and one molecule of gaseous hydrogen. **b.** One molecule of gaseous hydrogen chloride reacts with one molecule of gaseous ammonia to produce one formula unit of solid ammonium chloride. **c.** Two formula units of aqueous potassium bromide react with one molecule of gaseous chlorine to produce two formula units of aqueous potassium chloride and one molecule of liquid bromine. **d.** One formula unit of aqueous sulfuric acid reacts with two formula units of solid sodium cyanide to produce two molecules of gaseous hydrogen cyanide and one formula unit of aqueous sodium sulfate.

5. 8.5 g

7. 24.6

9. **a.** $2 B_2O_3(s) + 7 C(s) \rightarrow B_4C(s) + 6 CO(g)$ **b.** $CaC_2(s) + 2 H_2O(l) \rightarrow Ca(OH)_2(s) + C_2H_2(g)$
   c. $Ba(NO_3)_2(aq) + H_2SO_4(aq) \rightarrow BaSO_4(s) + 2 HNO_3(aq)$ **d.** $4 Bi(s) + 3 O_2(g) \rightarrow 2 Bi_2O_3(s)$
   e. $(NH_4)_2Cr_2O_7(s) \xrightarrow{\Delta} N_2(g) + 4 H_2O(g) + Cr_2O_3(s)$

10. **a.** double replacement **b.** decomposition **c.** double replacement **f.** double replacement

13. **a.** Pb **b.** KI **c.** CuO **d.** SO_3 **e.** $CuSO_4$ **f.** HCl

15. **a.** $\dfrac{1 \text{ mol } Fe_3O_4}{3 \text{ mol Fe}}$ **b.** $\dfrac{4 \text{ mol } H_2O}{3 \text{ mol Fe}}$ **c.** $\dfrac{4 \text{ mol } H_2}{1 \text{ mol } Fe_3O_4}$ **d.** $\dfrac{1 \text{ mol } Fe_3O_4}{4 \text{ mol } H_2O}$ **e.** $\dfrac{1 \text{ mol } Fe_3O_4}{4 \text{ mol } H_2}$ **f.** $\dfrac{4 \text{ mol } H_2}{3 \text{ mol Fe}}$

17. 4.00 g

19. 146 g

21. 22.3 g

23. $2 Na + 2 H_2O \rightarrow 2 NaOH + H_2$; 0.659 g

25. 215.6 g

27. 1751 g

29. 49.8%

31. O_2

33. 61.9%

35. 92.1%

37. $CaO + H_2O \rightarrow Ca(OH)_2$; 132 g

39. 0.0161 mol

41. limiting reactant: HNO_3; grams Al_2O_3 left: 8.18

43. 3.79 g

CHAPTER 10

EXERCISES

10.1 Step 1: State the problem:

$$V_i = 75.0 \text{ mL}$$
$$P_i = 720.0 \text{ torr}$$
$$P_f = 700.0 \text{ torr}$$
$$V_f = ? \text{ mL}$$

Step 2:   Set up the solution and do the calculations.

$$V_f = (75.0 \text{ mL}) \left(\frac{720.0 \text{ torr}}{700.0 \text{ torr}} \right) = 77.1 \text{ mL}$$

Step 3:   Check to see that the volume increased, as it should have in response to a pressure decrease.

10.2   Step 1:

$$V_i = 140.0 \text{ mL}$$
$$P_i = 1.00 \text{ atm}$$
$$V_f = 200.0 \text{ mL}$$
$$P_f = ? \text{ torr}$$

Step 2:

$$P_f = (1.00 \text{ atm}) \left(\frac{760 \text{ torr}}{1 \text{ atm}} \right) \left(\frac{140.0 \text{ mL}}{200.0 \text{ mL}} \right) = 532 \text{ torr}$$

Step 3:   When we compare the final pressure (532 torr) to the initial pressure (1.00 atm = 760 torr), we see that there was a pressure decrease, as there should be when volume increases.

10.3   Step 1:

$$V_i = 10.0 \text{ mL}$$
$$T_i = 100.0° \text{ C}$$
$$V_f = 8.00 \text{ mL}$$
$$T_f = ? \text{ °C}$$

Step 2:

$$T_i = 100.0 + 273 = 373 \text{ K}$$
$$T_f = (373 \text{ K}) \left(\frac{8.00 \text{ mL}}{10.0 \text{ mL}} \right) = 298 \text{ K}$$
$$T_f = \text{K} - 273 = 298 - 273 = 25° \text{ C}$$

Step 3:   A check of the final answer shows a decrease in temperature, as should occur in response to a decrease in volume.

10.4   Step 1:

$$V_i = 1.00 \text{ L}$$
$$T_i = 25.0° \text{ C}$$
$$T_f = 50.0° \text{ C}$$
$$V_f = ? \text{ L}$$

Step 2:

$$T_i = 25.0 + 273 = 298 \text{ K}$$
$$T_f = 50.0 + 273 = 323 \text{ K}$$
$$V_f = (1.00 \text{ L}) \left(\frac{323 \text{ K}}{298 \text{ K}} \right) = 1.08 \text{ L}$$

Step 3:   We check our final answer to see that it represents a volume increase, as should occur in response to a temperature increase.

10.5   Step 1:

$$P_i = 2.50 \text{ atm}$$
$$T_i = 100.0° \text{ C}$$
$$T_f = 50.0° \text{ C}$$
$$P_f = ? \text{ torr}$$

Step 2:

$$T_i = 100.0 + 273 = 373 \text{ K}$$
$$T_f = 50.0 + 273 = 323 \text{ K}$$
$$P_f = (2.50 \text{ atm}) \left(\frac{323 \text{ K}}{373 \text{ K}} \right) \left(\frac{760 \text{ torr}}{1 \text{ atm}} \right) = 1650 \text{ torr}$$

Step 3: A check of the final answer shows a decrease in pressure (from 2.50 atm = 1900 torr to 1650 torr), as there should be in response to a temperature decrease.

10.6

$$V_i = 1.50 \text{ L}$$
$$P_i = 800.0 \text{ torr}$$
$$T_i = 150.0° \text{ C} = 423 \text{ K}$$
$$P_f = 700.0 \text{ torr}$$
$$T_f = 50.0° \text{ C} = 323 \text{ K}$$
$$V_f = ? \text{ L}$$

From equation (2), $V_f = V_i \times P$ ratio $\times T$ ratio

$$V_f = (1.50 \text{ L}) \left(\frac{800.0 \text{ torr}}{700.0 \text{ torr}}\right) \left(\frac{323 \text{ K}}{423 \text{ K}}\right) = 1.31 \text{ L}$$

10.7

$$V_i = 250.0 \text{ mL}$$
$$T_i = 25.0° \text{ C} = 298 \text{ K}$$
$$P_i = 760.0 \text{ torr}$$
$$V_f = 200.0 \text{ mL}$$
$$T_f = 50.0° \text{ C} = 323 \text{ K}$$
$$P_f = ? \text{ torr}$$

From equation (3), $P_f = P_i \times V$ ratio $\times T$ ratio

$$P_f = (760.0 \text{ torr}) \left(\frac{250.0 \text{ mL}}{200.0 \text{ mL}}\right) \left(\frac{323 \text{ K}}{298 \text{ K}}\right) = 1.03 \times 10^3 \text{ torr}$$

10.8

$$V_i = 7.85 \text{ L}$$
$$T_i = 24° \text{ C} = 297 \text{ K}$$
$$P_i = 545 \text{ torr}$$
$$V_f = 9.60 \text{ L}$$
$$P_f = 435 \text{ torr}$$
$$T_f = ?° \text{ C}$$

From equation (4), $T_f = T_i \times V$ ratio $\times P$ ratio

$$T_f = (297 \text{ K}) \left(\frac{9.60 \text{ L}}{7.85 \text{ L}}\right) \left(\frac{435 \text{ torr}}{545 \text{ torr}}\right) = 2.90 \times 10^2 \text{ K}$$

$$T_f = \text{K} - 273 = 290 - 273 = 17° \text{ C}$$

10.9

$$n = \frac{PV}{RT} = \frac{(2.00 \text{ atm})(1.25 \text{ L})}{(0.0821 \text{ L atm/mol K})(303 \text{ K})} = 0.100 \text{ mol}$$

10.10

$$n = \frac{PV}{RT} = \frac{(1.50 \text{ atm})(80.1 \text{ L})}{(0.0821 \text{ L atm/mol K})(293 \text{ K})} = 4.99 \text{ mol}$$

$$\text{Molar mass} = \frac{220.0 \text{ g}}{4.99 \text{ mol}} = 44.1 \text{ g/mol}$$

10.11

$$T = \frac{PV}{nR} = \frac{(1.20 \text{ atm})(140.0 \text{ L})}{(7.25 \text{ mol})(0.0821 \text{ L atm/mol K})} = 282 \text{ K}$$

$$T = \text{K} - 273 = 282 - 273 = 9° \text{ C}$$

10.12

$$\frac{P_i V_i}{T_i} = \frac{P_f V_f}{T_f}$$

$$T_f = \frac{P_f V_f T_i}{P_i V_i} = \frac{(1.50 \text{ atm})(2.00 \text{ L})(0.0 + 273 \text{ K})}{(2.00 \text{ atm})(1.35 \text{ L})} = 303 \text{ K}$$

$$T_f = 303 \text{ K} - 273 = 30° \text{ C}$$

10.13

$$P_{\text{total}} = P_{\text{hydrogen chloride}} - P_{\text{water vapor}}$$

$$P_{\text{hydrogen chloride}} = P_{\text{total}} - P_{\text{water vapor}} = 750 \text{ torr} - 24 \text{ torr} = 726 \text{ torr}$$

10.14

$$V = \frac{nRT}{P} = \frac{(1.00 \text{ mol})(0.0821 \text{ L atm/mol K})(303 \text{ K})}{2.00 \text{ atm}} = 12.4 \text{ L}$$

$$d = \frac{m}{v} = \frac{17.0 \text{ g}}{12.4 \text{ L}} = 13.7 \text{ g/L}$$

10.15   Step 1:   2.35 mol NaCl

       Step 2:

$$(2.35 \text{ mol NaCl}) \left(\frac{2 \text{ mol HCl}}{2 \text{ mol NaCl}} \right) = 2.35 \text{ mol HCl}$$

       Step 3:

$$V = \frac{nRT}{P} = \frac{(2.35 \text{ mol})(0.0821 \text{ L atm/mol K})(273 \text{ K})}{1 \text{ atm}} = 52.7 \text{ L}$$

       or,

$$\left(\frac{2 \text{ mol}}{2 \text{ mol}} \right) \frac{(2.35 \text{ mol})(0.0821 \text{ L atm/mol K})(273 \text{ K})}{(1 \text{ atm})} = 52.7 \text{ L}$$

10.16   Step 1:   10.5 g N_2

       Step 2:

$$(10.5 \text{ g N}_2) \left(\frac{1 \text{ mol N}_2}{28.0 \text{ g N}_2} \right) = 0.375 \text{ mol N}_2$$

       Step 3:

$$(0.375 \text{ mol N}_2) \left(\frac{3 \text{ mol H}_2}{1 \text{ mol N}_2} \right) = 1.12 \text{ mol H}_2$$

       Step 4:

$$V \text{ (of H}_2) = \frac{nRT}{P} = \frac{(1.12 \text{ mol})(0.0821 \text{ L atm/mol K})(325 \text{ K})}{2.35 \text{ atm}} = 12.7 \text{ L}$$

       or,

$$(10.5 \text{ g N}_2) \left(\frac{1 \text{ mol N}_2}{28.0 \text{ g N}_2} \right) \left(\frac{3 \text{ mol H}_2}{1 \text{ mol N}_2} \right) \frac{(0.0821 \text{ L atm/mol K})(325 \text{ K})}{(2.35 \text{ atm})} = 12.8 \text{ L}$$

(Note that the small difference in the two answers is a result of rounding off the four steps individually.)

10.17   Step 1:   44.3 L

       Step 2:

$$(44.3 \text{ L}) \left(\frac{4 \text{ mol}}{1 \text{ mol}} \right) = 177 \text{ L}$$

STUDY QUESTIONS AND PROBLEMS

10. 270 mL
12. **a.** 0.800 atm   **b.** 2.00 atm   **c.** 0.200 atm
14. 10.0 atm
18. 68.2 mL
20. 5.67×10^3 mL

22. 106° C
24. 2640 torr
26. 0.361 atm
28. 357° C
32. 1.55×10^{-2} L
34. 0.894 atm
36. 2.88 mol N_2; 80.6 g N_2
38. 3.18 L
46. 190 torr or 0.25 atm
48. The partial pressure of oxygen will decrease to 80.0 torr.
50. **a.** 1.96 g/L **b.** 3.17 g/L **c.** 3.58 g/L
52. 6.04×10^3 mL HCN; 3.02×10^3 mL NO_2
54. 0.291 L
57. 180 weather balloons
59. **a.** 822° C **b.** −127° C **c.** 161° C
61. 0.507 atm
63. 60.1 g/mol
65. 5.00 mL O_2

CHAPTER 11

EXERCISES

11.1 To arrive at our answer, we must determine the type(s) of intermolecular attraction (dipolar attractions, hydrogen bonds, London dispersion forces) exerted by each molecule on the basis of its structure. If the molecule is not polar, it can only exert London dispersion forces. Molecule c (H_2) fits this category. If the molecule is polar but does *not* have hydrogen attached to nitrogen, oxygen, or fluorine, the molecule will exert both London dispersion forces and dipolar attractions. Molecules a (CO) and b (H_2S) are in this category. Finally, if the molecule is polar, with polar bonds between hydrogen and nitrogen, oxygen, or fluorine, the molecule will exert London dispersion forces and will form hydrogen bonds. Molecule d (HF) is such a molecule.

11.2 First, find the number of moles in 2.50 g of Hg:

$$(2.50 \text{ g Hg}) \left(\frac{1 \text{ mol Hg}}{200.6 \text{ g Hg}} \right) = 0.0125 \text{ mol Hg}$$

Then calculate kilojoules of heat needed to vaporize 0.0125 mol Hg at its boiling point, using the heat of vaporization for liquid mercury (56.9 kJ/mol) given in Table 11-7:

$$\text{heat required} = (0.0125 \text{ mol Hg}) \left(\frac{56.9 \text{ kJ}}{\text{mol Hg}} \right) = 0.711 \text{ kJ}$$

Or,

$$(2.50 \text{ g Hg}) \left(\frac{1 \text{ mol Hg}}{200.6 \text{ g Hg}} \right) \left(\frac{56.9 \text{ kJ}}{\text{mol Hg}} \right) = 0.709 \text{ kJ}$$

(The small discrepancy in the answers is caused by rounding off the two steps in the first solution separately.)

11.3 Use the molar heat of fusion of benzene (9.83 kJ/mol) given in Table 11-8 to solve the problems as follows:

$$\text{heat required} = (15.0 \text{ g benzene}) \left(\frac{1 \text{ mol benzene}}{78.0 \text{ g benzene}} \right) \left(\frac{9.83 \text{ kJ}}{\text{mol benzene}} \right) = 1.89 \text{ kJ}$$

11.4 Step 1: Calculate the heat in kilojoules required to heat the ice from −5.0° C to 0° C:

$$\text{heat required} = \left(\frac{0.492 \text{ cal}}{\text{g } H_2O \text{ °C}} \right) (10.0 \text{ g } H_2O)(5.0° \text{ C}) \left(\frac{4.185 \text{ J}}{\text{cal}} \right) \left(\frac{1 \text{ kJ}}{1000 \text{ J}} \right) = 0.10 \text{ kJ}$$

Step 2: Calculate the heat required to melt the ice at 0° C:

$$\text{heat required} = (10.0 \text{ g H}_2\text{O})\left(\frac{1 \text{ mol H}_2\text{O}}{18.0 \text{ g H}_2\text{O}}\right)\left(\frac{5.98 \text{ kJ}}{\text{mol H}_2\text{O}}\right) = 3.32 \text{ kJ}$$

Step 3: Calculate the heat in kilojoules required to raise the temperature of liquid water from 0° C to 100.0° C:

$$\text{heat required} = \left(\frac{1.00 \text{ cal}}{\text{g H}_2\text{O} \, °\text{C}}\right)(10.0 \text{ g H}_2\text{O})(100.0° \text{C})\left(\frac{4.185 \text{ J}}{\text{cal}}\right)\left(\frac{1 \text{ kJ}}{1000 \text{ J}}\right) = 4.18 \text{ kJ}$$

Step 4: Calculate the heat required to convert the liquid water to steam at 100.0° C:

$$\text{heat required} = (10.0 \text{ g H}_2\text{O})\left(\frac{1 \text{ mol H}_2\text{O}}{18.0 \text{ g H}_2\text{O}}\right)\left(\frac{40.6 \text{ kJ}}{\text{mol H}_2\text{O}}\right) = 22.6 \text{ kJ}$$

Then, add together all of the heat requirements:

Step 1	0.10 kJ
Step 2	3.32 kJ
Step 3	4.18 kJ
Step 4	22.6 kJ
	30.2 kJ

STUDY QUESTIONS AND PROBLEMS

3. **a.** London dispersion forces   **b.** London dispersion forces and hydrogen bonds   **c.** London dispersion forces
    d. London dispersion forces   **e.** London dispersion forces   **f.** London dispersion forces and hydrogen bonds
31. 8.46 kJ
51. 16.6 kJ
57. 1.82×10^4 cal
59. 43.6 kJ
65. 23.3 kJ/mol
68. 2.82 kJ

CHAPTER 12

EXERCISES

12.1
$$(1.00 \text{ L})\left(\frac{0.50 \text{ mol C}_6\text{H}_{12}\text{O}_6}{\text{L}}\right)\left(\frac{180.0 \text{ g C}_6\text{H}_{12}\text{O}_6}{\text{mol C}_6\text{H}_{12}\text{O}_6}\right) = 90 \text{ g C}_6\text{H}_{12}\text{O}_6$$

12.2
$$(250.0 \text{ mL})\left(\frac{\text{L}}{1000 \text{ mL}}\right)\left(\frac{1.50 \text{ mol Mg}_3(\text{PO}_4)_2}{\text{L}}\right)\left(\frac{263 \text{ g Mg}_3(\text{PO}_4)_2}{\text{mol Mg}_3(\text{PO}_4)_2}\right) = 98.6 \text{ g Mg}_3(\text{PO}_4)_2$$

12.3
$$\text{Volume} = (26.2 \text{ g LiNO}_3)\left(\frac{1 \text{ mol LiNO}_3}{68.9 \text{ g LiNO}_3}\right)\left(\frac{\text{L}}{3.70 \text{ mol LiNO}_3}\right) = 0.103 \text{ L}$$

12.4
$$\frac{25.0 \text{ g solute}}{200.0 \text{ mL solution}} \times 100\% = 12.5\% \text{ (w/v)}$$

12.5
$$\frac{4.50 \text{ mL solute}}{55.0 \text{ mL solution}} \times 100\% = 8.18\% \text{ (v/v)}$$

12.6
$$\frac{17.6 \text{ g solute}}{50.0 \text{ g solution}} \times 100\% = 35.2\% \text{ (w/w)}$$

12.7
$$C_d = C_c \left(\frac{V_c}{V_d}\right) = 2.50\ M \left(\frac{250.0\ \text{mL}}{800.0\ \text{mL}}\right) = 0.781\ M$$

12.8
$$V_d = V_c \left(\frac{C_c}{C_d}\right) = 100.0\ \text{mL} \left(\frac{2.00\ M}{0.100\ M}\right) = 2000\ \text{mL or } 2.00 \times 10^3\ \text{mL}$$

12.9
$$(1.000\ \text{g HC}_2\text{H}_3\text{O}_2) \left(\frac{1\ \text{mol HC}_2\text{H}_3\text{O}_2}{60.0\ \text{g HC}_2\text{H}_3\text{O}_2}\right) \left(\frac{1\ \text{mol NaOH}}{1\ \text{mol HC}_2\text{H}_3\text{O}_2}\right) \left(\frac{L}{0.4600\ \text{mol NaOH}}\right) = 0.0362\ L$$

12.10
$$M = (0.02705\ L) \left(\frac{0.1800\ \text{mol NaOH}}{L}\right) \left(\frac{1\ \text{mol H}_2\text{C}_2\text{O}_4}{2\ \text{mol NaOH}}\right) \left(\frac{1}{0.02500\ L}\right) = 0.09738\ \text{mol H}_2\text{C}_2\text{O}_4/L$$

12.11 First, find moles of ethylene glycol in the solution:

$$(9.55\ \text{g C}_2\text{H}_6\text{O}_2) \left(\frac{1\ \text{mol C}_2\text{H}_6\text{O}_2}{62.0\ \text{g C}_2\text{H}_6\text{O}_2}\right) = 0.154\ \text{mol C}_2\text{H}_6\text{O}_2$$

Then, convert g of water to kg of water and calculate molality of the solution:

$$(100.0\ \text{g}) \left(\frac{1\ \text{kg}}{1000\ \text{g}}\right) = 0.1000\ \text{kg}$$

$$\frac{0.154\ \text{mol solute}}{0.1000\ \text{kg solvent}} = 1.54\ m$$

12.12 First, calculate Δt_f:

$$\Delta t_f = \text{normal freezing point} - \text{freezing point of solution}$$
$$= 0.0°\ C - (-10.0°\ C) = 10.0°\ C$$

Next, convert g of solvent to kg of solvent:

$$(150.0\ \text{g}) \left(\frac{1\ \text{kg}}{1000\ \text{g}}\right) = 0.150\ \text{kg}$$

Last, find k_f for water in Table 12-6 and calculate molar mass from equation (5):

$$\text{molar mass} = k_f \left(\frac{\text{g solute}}{\Delta t_f}\right) \left(\frac{1}{\text{kg solvent}}\right)$$

$$= 1.86\ (°C)(\text{kg solvent})(\text{mol solute})^{-1} \left(\frac{50.0\ \text{g solute}}{10.0°\ C}\right) \left(\frac{1}{0.150\ \text{kg solvent}}\right)$$

$$= 62.0\ \text{g solute (mol solute)}^{-1} \text{ or } 62.0\ \text{g solute/mol solute}$$

STUDY QUESTIONS AND PROBLEMS

12. **a.** $MgCl_2(s) \xrightarrow{H_2O} Mg^{2+}(aq) + 2\ Cl^-(aq)$ **b.** $NaOH(s) \xrightarrow{H_2O} Na^+(aq) + OH^-(aq)$
c. $Al_2(SO_4)_3(s) \xrightarrow{H_2O} 2\ Al^{3+}(aq) + 3\ SO_4{}^{2-}(aq)$ **d.** $LiNO_3(s) \xrightarrow{H_2O} Li^+(aq) + NO_3{}^-(aq)$

18. Usually, the more heat released during dissolving, the greater the solubility; hence, NaOH is likely to be most soluble, followed by NaI and then NaCl.

25. **a.** 0.171 mol/L   **b.** 2.68 mol/L   **c.** 0.346 mol/L   **d.** 0.259 mol/L   **e.** 0.254 mol/L

27. **a.** 0.0150 mol   **b.** 0.0500 mol   **c.** 0.720 mol   **d.** 0.300 mol   **e.** 2.78 mol   **f.** 0.0450 mol

29. 4 g

31. **a.** 11.3% (v/v)   **b.** 9.88% (v/v)   **c.** 6.24% (v/v)   **d.** 4.99% (v/v)   **e.** 22.7% (v/v)   **f.** 0.573% (v/v)

34. **a.** 3.00×10^2 mL   **b.** 1.00×10^3 mL   **c.** 1.44×10^4 mL   **d.** 6.00×10^3 mL   **e.** 5.55×10^4 mL
f. 9.00×10^2 mL

36. **a.** 5.00×10^2 mL   **b.** 2.00×10^2 mL   **c.** 41.7 mL   **d.** 66.7 mL   **e.** 13.5 mL   **f.** 5.00×10^2 mL

38. 396 mL

40. 2.72×10^{-3} mol/L
45. **a.** 0.125 m **b.** 2.50 m **c.** 0.250 m **d.** 0.0975 m
53. $Na_2SO_4 > MgSO_4 > Al_2(SO_4)_3 \cdot 6\,H_2O$
55. **a.** 146 mL **b.** 902 mL **c.** 85.5 mL **d.** 382 mL **e.** 185 mL **f.** 24.4 mL
57. 4.5×10^2 mL
59. 1.80×10^2 g/mol
61. Whole milk contains fat droplets suspended as colloidal particles, causing whole milk to be more opaque than nonfat milk.

CHAPTER 13

EXERCISES

13.1 **a.** rate = k[A] = k[CH$_3$CHO] **b.** Assume that the original concentration of CH$_3$CHO was 1 mol/L; then the original rate was

$$\text{rate} = k[\text{CH}_3\text{CHO}] = k(1 \text{ mol/L}) = 1\,k \text{ mol/L}$$

If the concentration of CH$_3$CHO is decreased to one-half its original value, then

$$\text{rate} = k[\text{CH}_3\text{CHO}] = \left(\frac{1}{2} \text{ mol/L}\right) = \frac{1}{2}\,k \text{ mol/L}$$

Thus, halving the concentration of CH$_3$CHO gives a new rate that is one-half the first rate.

13.2
$$K_{eq} = \frac{[\text{NH}_3]^2}{[\text{N}_2][\text{H}_2]^3}$$

13.3
$$K_{eq} = \frac{[\text{NH}_3]^2}{[\text{N}_2][\text{H}_2]^3}$$
$$= \frac{(0.0434 \text{ mol/L})^2}{(0.9783 \text{ mol/L})(0.935 \text{ mol/L})^3}$$
$$= \frac{(0.0434 \text{ mol/L})(0.0434 \text{ mol/L})}{(0.9783 \text{ mol/L})(0.935 \text{ mol/L})(0.935 \text{ mol/L})(0.935 \text{ mol/L})}$$
$$= 2.36 \times 10^{-3}(\text{mol/L})^{-2} \text{ or } 2.36 \times 10^{-3} \text{ mol}^{-2}\,\text{L}^2$$

13.4 **a.** $K_{eq} = \dfrac{[\text{NO}_2]^2}{[\text{N}_2\text{O}_4]}$ An increase in [NO$_2$] would cause an increase in [N$_2$O$_4$] and a shift of the equilibrium to the left.

b. $K_{eq} = \dfrac{[\text{CO}]^2[\text{O}_2]}{[\text{CO}_2]^2}$ An increase in [CO$_2$] would cause an increase in [CO], an increase in [O$_2$], and a shift of the equilibrium to the right. **c.** $K_{eq} = \dfrac{[\text{H}_2\text{O}]^2}{[\text{H}_2]^2[\text{O}_2]}$ A decrease in [H$_2$O] would cause a decrease in [H$_2$], a decrease in

[O$_2$], and a shift of the equilibrium to the right.

13.5 **a.** The equilibrium will shift to the right, because five moles of products occupies more volume than three moles of reactants. **b.** The equilibrium will shift to the right, because two moles of products occupies more volume than one mole of reactant. **c.** The equilibrium will remain unchanged, because the number of moles of reactants is the same as the number of moles of product in the balanced equation.

STUDY QUESTIONS AND PROBLEMS

3. a.

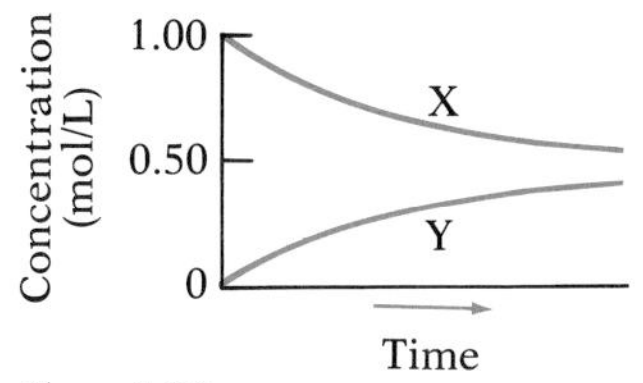

b.

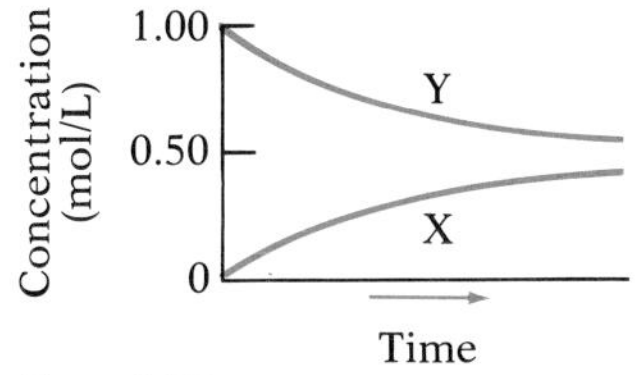

c.

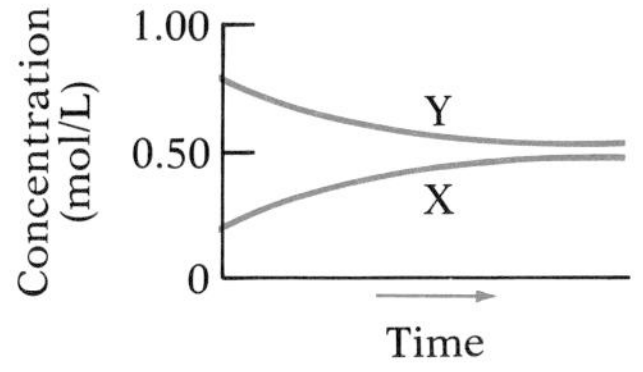

8. Reaction 2 has the higher activation energy, because it has the lower rate.

11. a. The rate will be doubled. **b.** The rate will be doubled. **c.** The rate will be quadrupled.

13. a. The rate of the reverse reaction will be increased by a factor of 1000. **b.** The equilibrium constant will not be affected, because the rate of the reverse reaction will also be increased by a factor of 500.

15. a. $K_{eq} = \dfrac{[NO]^4[H_2O]^6}{[NH_3]^4[O_2]^5}$ **b.** $K_{eq} = \dfrac{[N_2O_4]}{[NO_2]^2}$ **c.** $K_{eq} = \dfrac{[SO_2]^2[O_2]}{[SO_3]^2}$ **d.** $K_{eq} = \dfrac{[PCl_3][Cl_2]}{[PCl_5]}$ **e.** $K_{eq} = \dfrac{[HCl]^2}{[H_2][Cl_2]}$

17. 1.0

21. a. to the right **b.** to the left **c.** to the right **d.** to the right **e.** to the left

25. a. $[N_2]$ will decrease, $[NO]$ will increase, and the equilibrium will shift to the right. **b.** $[N_2]$ and $[H_2]$ will increase, and the equilibrium will shift to the left. **c.** $[CO]$ and $[H_2O]$ will decrease, $[CO_2]$ will increase, and the equilibrium will shift to the right. **d.** $[PCl_3]$ and $[Cl_2]$ will decrease, and the equilibrium will shift to the left.

28. a. increase **b.** increase **c.** increase

34. The rate will be increased by a factor of 1024.

36. a. The enzyme lowers the activation energy, because the enzyme increases the reaction rate. **b.** The enzyme has no effect on the equilibrium constant, because catalysts affect the rates of the forward reaction and the reverse reaction to the same extent.

38. 4.2×10^{-2} mol/L

40. a. The equilibrium will be shifted to the left. **b.** The equilibrium will be shifted to the right. **c.** The equilibrium will be shifted to the right.

CHAPTER 14

EXERCISES

14.1 The conjugate base of each acid is the ion or molecule formed when the acid loses a proton. Thus, the conjugate bases are:

a.

$$HCO_3^- \longrightarrow H^+ + CO_3^{2-}$$
acid   conjugate base

b.

$$H_2SO_4 \longrightarrow H^+ + \underset{\text{conjugate base}}{HSO_4^-}$$
$$\underset{\text{acid}}{}$$

c.

$$HF \longrightarrow H^+ + \underset{\text{conjugate base}}{F^-}$$
$$\underset{\text{acid}}{}$$

The conjugate acid of each base is the ion or molecule formed when the base gains a proton. Thus, the conjugate acids are:

d.

$$\underset{\text{base}}{ClO_4^-} + H^+ \longrightarrow \underset{\text{conjugate acid}}{HClO_4}$$

e.

$$\underset{\text{base}}{HCO_3^-} + H^+ \longrightarrow \underset{\text{conjugate acid}}{H_2CO_3}$$

f.

$$\underset{\text{base}}{NH_3} + H^+ \longrightarrow \underset{\text{conjugate acid}}{NH_4^+}$$

14.2

$$HA(aq) + H_2O(l) \rightleftharpoons H_3O^+(aq) + A^-(aq)$$

$$K_a = \frac{[H_3O^+][A^-]}{[HA]} = \frac{(2.0 \times 10^{-3})(2.0 \times 10^{-3})}{0.10} = 4.0 \times 10^{-5}$$

14.3

$$HCN(aq) + H_2O(l) \rightleftharpoons H_3O^+(aq) + CN^-(aq)$$

At equilibrium, $[H_3O^+] = [CN^-]$

$$K_a = \frac{[H_3O^+][CN^-]}{[HCN]} = \frac{[H_3O^+]^2}{[HCN]}$$

$$[H_3O^+]^2 = K_a[HCN]$$

$$[H_3O^+] = \sqrt{K_a[HCN]} = \sqrt{(4.9 \times 10^{-10})(0.15)} = \sqrt{0.74 \times 10^{-10}}$$

$$= 0.86 \times 10^{-5}\ M = 8.6 \times 10^{-6}\ M$$

14.4

$$K_b = \frac{[BH^+][OH^-]}{[B]} = \frac{(7.7 \times 10^{-8})(7.7 \times 10^{-8})}{0.40} = 1.5 \times 10^{-14}$$

14.5 First, find the oxidation number of Cl in $HClO_4$:

Atoms	Individual Oxidation Numbers	Sum of Oxidation Numbers = 0
H	+1	+1
4 O	−2	−8
Cl	+7	+7
		0

Then, find the formula of the oxide of chlorine in which Cl has an oxidation number of +7.

Atoms	Individual Oxidation Numbers	Sum of Oxidation Numbers = 0
2 Cl	+7	+14
7 O	−2	−14
		0

Therefore, the acid anhydride of $HClO_4$ must be Cl_2O_7.

14.6 First, find the charge of Fe, using the charge of OH^- as a guide. Since one Fe ion is combined with two OH^-, the charge on the Fe ion must be 2+. Then, find the formula of the oxide of Fe, remembering that it contains Fe^{2+} and O^{2-}:

$$\overset{\text{②+}}{Fe} \diagdown \overset{\text{②−}}{O} = Fe_2O_2 = Fe_{2/2}O_{2/2} = FeO$$

14.7
$$H_2SO_4(aq) + 2\,NaOH(aq) \longrightarrow Na_2SO_4(aq) + 2\,H_2O(l)$$

$$M = \frac{(18.27\text{ mL})\left(\dfrac{1\text{ L}}{1000\text{ mL}}\right)\left(\dfrac{0.1586\text{ mol NaOH}}{L}\right)\left(\dfrac{1\text{ mol }H_2SO_4}{2\text{ mol NaOH}}\right)}{(20.00\text{ mL})\left(\dfrac{1\text{ mL}}{1000\text{ mL}}\right)} = 0.0724\text{ mol }H_2SO_4/L$$

14.8
$$N = \frac{(125\text{ g NaOH})\left(\dfrac{1\text{ eq NaOH}}{40.0\text{ g NaOH}}\right)}{1.50\text{ L}} = 2.08\text{ eq NaOH/L}$$

14.9
$$1\text{ mol }H_3PO_4 = 3\text{ eq }H_3PO_4$$

$$(500.0\text{ mL})\left(\dfrac{1\text{ L}}{1000\text{ mL}}\right)\left(\dfrac{6.00\text{ eq }H_3PO_4}{1\text{ L }H_3PO_4}\right)\left(\dfrac{1\text{ mol }H_3PO_4}{3\text{ eq }H_3PO_4}\right)\left(\dfrac{98.0\text{ g }H_3PO_4}{\text{mol }H_3PO_4}\right) = 98.0\text{ g }H_3PO_4$$

14.10
$$\left(\dfrac{0.1000\text{ eq}}{L}\right)\left(\dfrac{1\text{ L}}{1000\text{ mL}}\right)(20.00\text{ mL})\left(\dfrac{\text{meq}}{0.001\text{ eq}}\right) = 2.00\text{ meq}$$

14.11
$$N = \left(\dfrac{0.0200\text{ mol }Ca(OH)_2}{L}\right)\left(\dfrac{2\text{ eq }Ca(OH)_2}{\text{mol }Ca(OH)_2}\right) = 0.0400\text{ eq }Ca(OH)_2/L$$

14.12
$$N_a = N_b\left(\dfrac{V_b}{V_a}\right) = (0.1048\text{ eq/L})\left(\dfrac{26.21\text{ mL}}{20.00\text{ mL}}\right) = 0.1373\text{ eq/L}$$

14.13 Total ionic equation:
$$Ba^{2+}(aq) + 2\,Cl^-(aq) + 2\,Na^+(aq) + SO_4^{2-}(aq) \longrightarrow BaSO_4(s) + 2\,Na^+(aq) + 2\,Cl^-(aq)$$

Net ionic equation:
$$Ba^{2+}(aq) + SO_4^{2-}(aq) \longrightarrow BaSO_4(s)$$

14.14 The possible products from a double replacement reaction are HNO_3 and $MgSO_4$. Since both HNO_3 and $MgSO_4$ are water-soluble, there will not be a reaction.

14.15
$$Mn(OH)_2(s) \rightleftharpoons Mn^{2+}(aq) + 2\,OH^-(aq); \quad K_{sp} = [Mn^{2+}][OH^-]^2$$

14.16 $BaSO_4(s) \rightleftharpoons Ba^{2+}(aq) + SO_4^{2-}(aq)$. Compare the product of the proposed concentrations of dissolved ions to the K_{sp} value:

$$[Ba^{2+}][SO_4^{2-}] = (3.5 \times 10^{-5})(7.4 \times 10^{-2}) = 2.6 \times 10^{-6}$$
$$K_{sp} = 1.1 \times 10^{-10}$$

Since the product of $[Ba^{2+}]$ and $[SO_4^{2-}]$ in the proposed solution would be higher than the K_{sp} value, $BaSO_4$ would precipitate.

14.17
$$CaF_2(s) \rightleftharpoons Ca^{2+}(aq) + 2\,F^-(aq)$$

$$K_{sp} = [Ca^{2+}][F^-]^2$$
$$[F^-]^2 = \frac{K_{sp}}{[Ca^{2+}]}$$
$$[F^-] = \sqrt{\frac{K_{sp}}{[Ca^{2+}]}} = \sqrt{\frac{4.0 \times 10^{-11}}{0.10}} = \sqrt{40 \times 10^{-11}} = \sqrt{4.0 \times 10^{-10}} = 2.0 \times 10^{-5}\ M$$

STUDY QUESTIONS AND PROBLEMS

	Conjugate Bases		Conjugate Acids
4. a.	I^-	**e.**	H_2SO_4
b.	ClO_3^-	**f.**	H_2O
c.	HPO_4^{2-}	**g.**	NH_4^+
d.	NH_3	**h.**	HCO_3^-

10. 6.0 M

12. 4.9×10^{-10}

14. a. strong   **b.** weak   **c.** strong   **d.** weak

21. 2.0 M

25. a. CaO   **b.** Cl_2O **c.** SiO_2 **d.** ZnO   **e.** B_2O_3

27. Release of NO_2 into the atmosphere results in formation of the strong acid HNO_3 according to the following reaction.

$$3 \ NO_2(g) + H_2O(l) \longrightarrow 2 \ HNO_3(aq) + NO(g)$$

(The source of H_2O is moisture in the air.) Reduction of NO_2 to N_2 before release into the atmosphere prevents formation of HNO_3.

29. 0.3858 M

31. 0.1125 M

34. a. diprotic   **b.** triprotic   **c.** monoprotic   **d.** diprotic

36. 2.00 N

38. 1.000 meq

40. 6.00 N

42. 6.00 N

44. 37.50 mL

46. a. $\cancel{2 \ Na^+(aq)} + 2 \ OH^-(aq) + 2 \ H^+(aq) + \cancel{SO_4^{2-}(aq)} \rightarrow \cancel{2 \ Na^+(aq)} + \cancel{SO_4^{2-}(aq)} + 2 \ H_2O(l)$
   b. $2 \ OH^-(aq) + 2 \ H^+(aq) \rightarrow 2 \ H_2O(l)$

49. b, c, e

52. a. $Al^{3+}(aq) + 3 \ OH^-(aq) \rightarrow Al(OH)_3(s)$ **b.** $Ba^{2+}(aq) + SO_4^{2-}(aq) \rightarrow BaSO_4(s)$
   c. $Ag^+(aq) + ClO_3^-(aq) \rightarrow AgClO_3(s)$ **d.** $Fe^{2+}(aq) + 2 \ OH^-(aq) \rightarrow Fe(OH)_2(s)$

54. 2.0×10^{-14}

56. $CuCO_3$

59. $5.8 \times 10^{-3} \ M$

61. 1.5×10^{-14}

63. SO_2 forms H_2SO_3 on reaction with moisture in the atmosphere, but SO_3 forms H_2SO_4, a much stronger acid.

65. 76.79% (w/w)

67. The resulting solution was acidic, because 5.00×10^{-3} eq of NaOH (0.0500 L $\times$ 0.100 eq/L) was added to 6.00×10^{-3} eq of HCl (0.0150 L $\times$ 0.400 eq/L). Thus, 5.00×10^{-3} eq of acid was neutralized, leaving 1.00×10^{-3} eq of acid unreacted.

69. 2.2×10^{-4} mol/L

CHAPTER 15

EXERCISES

15.1 $pH = -\log [H_3O^+] = -\log (1.0 \times 10^{-9}) = -(-9) = 9$. The solution is basic, because the pH is greater than 7.

15.2

$$pH = -\log [H_3O^+]$$
$$\log [H_3O^+] = -pH = -6.0$$
$$[H_3O^+] = 1 \times 10^{-6} \ M$$

15.3 $pH + pOH = 14$; $pOH = 14 - pH = 14 - 8 = 6$

15.4 $pH + pOH = 14$; $pOH = 14 - pH = 14 - 5 = 9 = -\log[OH^-]$

$$\log[OH^-] = -9$$
$$[OH^-] = 1 \times 10^{-9}\ M$$

15.5   Since hydrochloric acid is a strong acid, it ionizes completely, producing a solution in which $[H_3O^+]$ is 0.001 M. Thus,

$$pH = -\log[H_3O^+] = -\log 0.001 = -\log(1 \times 10^{-3}) = -(-3) = 3$$

15.6   Since KOH is a strong base, it ionizes completely in water, producing a solution in which $[OH^-]$ is 0.01 M. Thus,

$$[OH^-] = 0.01\ M = 1 \times 10^{-2}\ M$$
$$pOH = -\log[OH^-] = -\log(1 \times 10^{-2}) = 2$$
$$pH + pOH = 14$$
$$pH = 14 - pOH = 14 - 2 = 12$$

15.7 $pOH = -\log[OH^-] = -\log(6.2 \times 10^{-3}) = -(\log 6.2 + \log 10^{-3}) = -[0.79 + (-3)] = -[-2.21] = 2.21$

15.8 $pH + pOH = 14$; $pOH = 14 - pH = 14 - 7.85 = 6.15$. (Note that 14 as used here is an exact number.)

$$pOH = -\log[OH^-]$$
$$\log[OH^-] = -pOH = -6.15$$
$$[OH^-] = \text{antilog}(-6.15)$$
$$= \text{antilog}(0.85 - 7)$$
$$= [\text{antilog } 0.85][\text{antilog}(-7)]$$
$$= 7.1 \times 10^{-7}\ M$$

15.9   First, write the equation for ionization in water:

$$HA(aq) + H_2O(l) \rightleftharpoons H_3O^+(aq) + A^-(aq)$$

Then, convert pH to $[H_3O^+]$:

$$pH = -\log[H_3O^+]$$
$$\log[H_3O^+] = -pH = -2.16$$
$$[H_3O^+] = \text{antilog}(-2.16)$$
$$= \text{antilog}(0.84 - 3)$$
$$= [\text{antilog } 0.84][\text{antilog}(-3)]$$
$$= 6.9 \times 10^{-3}\ M$$

Next, calculate [HA] at equilibrium, remembering that $[HA]_{dissociated} = [H_3O^+]$, because each mole of HA that dissociates produces one mole of H_3O^+:

$$[HA]_{equilibrium} = [HA]_{initial} - [HA]_{dissociated} = 0.25\ M - 6.9 \times 10^{-3}\ M = 0.24\ M$$

Last, calculate K_a, remembering that $[H_3O^+] = [A^-]$:

$$K_a = \frac{[H_3O^+][A^-]}{[HA]} = \frac{(6.9 \times 10^{-3})^2}{0.24} = 2.0 \times 10^{-4}$$

15.10   First, write the equation for ionization of the weak acid in water:

$$HA(aq) + H_2O(l) \rightleftharpoons H_3O^+(aq) + A^-(aq)$$

Then, calculate $[H_3O^+]$:

$$[HA]_{dissociated} = (0.15\ M)(0.0203) = 3.0 \times 10^{-3}\ M = [H_3O^+]$$

Next, calculate pH:

$$pH = -\log[H_3O^+] = -\log(3.0 \times 10^{-3}) = -(\log 3.0 + \log 10^{-3}) = -(0.48 - 3) = 2.52$$

Last, calculate K_a:

$$[HA]_{equilibrium} = [HA]_{initial} - [HA]_{dissociated} = 0.15\,M - 3.0 \times 10^{-3}\,M = 0.15\,M$$

$$K_a = \frac{[H_3O^+][A^-]}{[HA]} = \frac{(3.0 \times 10^{-3})^2}{0.15} = 6.0 \times 10^{-5}$$

(Note that pH is not needed for the calculation of K_a; thus, K_a could have been calculated before pH.)

15.11 Consider both the cation and the anion of the salt. If one is the conjugate of a strong acid or base and the other is not, the solution will not be neutral. However, in this case, the salt is composed of a cation (K^+) from a strong base (KOH) and an anion (NO_3^-) from a strong acid (HNO_3). Thus, hydrolysis cannot occur, and the solution will be neutral.

STUDY QUESTIONS AND PROBLEMS

12. 1.00×10^{-11}

14. **a.** 13 **b.** 2 **c.** 10 **d.** 5 **e.** 14 **f.** 8

16. **a.** $1.00 \times 10^{-3}\,M$ **b.** $1.00 \times 10^{-11}\,M$ **c.** 11 **d.** 3

18. When CO_2 dissolves in water, H_2CO_3 is formed:

$$CO_2(g) + H_2O(l) \rightleftharpoons H_2CO_3(aq)$$

The weak acid H_2CO_3 dissociates to make the water slightly acidic:

$$H_2CO_3(aq) + H_2O(l) \rightleftharpoons H_3O^+(aq) + HCO_3^-(aq)$$

22. **a.** 0.60 **b.** 13.54 **c.** 8.64 **d.** 2.58

24. **a.** $8.9 \times 10^{-4}\,M$ **b.** $2.1 \times 10^{-7}\,M$ **c.** $7.6 \times 10^{-7}\,M$ **d.** $2.8 \times 10^{-3}\,M$

26. 3.3×10^{-7}

28. $pH = 2.22$; $K_a = 1.5 \times 10^{-4}$

30. **a.** basic; $NO_2^-(aq) + H_2O(l) \rightleftharpoons HNO_2(aq) + OH^-(aq)$
b. acidic; $NH_4^+(aq) + H_2O(l) \rightleftharpoons NH_3(aq) + H_3O^+(aq)$
c. basic; $C_2H_3O_2^-(aq) + H_2O(l) \rightleftharpoons HC_2H_3O_2(aq) + OH^-(aq)$ **d.** neutral

37. $Ba(OH)_2(aq) + H_2SO_4(aq) \rightarrow BaSO_4(s) + 2\,H_2O(l)$

39. $[H_3O^+] = [OH^-] = 1.56 \times 10^{-7}\,M$

41. 2.0×10^{-5}

43. $pH = 2.47$; 1.7%

45. **a.** Yes; the solution would contain a weak acid (HNO_2) and its salt (KNO_2). **b.** No; the solution would contain neither a weak acid and its salt nor a weak base and its salt. **c.** No; the solution contains a strong acid (HNO_3) and its salt (NH_4NO_3), instead of a weak acid and its salt. **d.** Yes; the solution contains a weak acid ($H_2PO_4^-$) and its salt (K_2HPO_4).

CHAPTER 16

EXERCISES

16.1

$$\underset{0}{H_2} + \underset{0}{Br_2} \longrightarrow 2\underset{+1\ -1}{H\,Br}$$

The substance that is oxidized is the one whose oxidation number increases; that substance is hydrogen. It is also the reducing agent. The substance that is reduced is the one whose oxidation number decreases; that substance is bromine. It is also the oxidizing agent.

16.2 Balancing oxidation with reduction, we get

$$\text{2 HNO}_3 + \text{3 H}_2\text{S} \longrightarrow \text{3 S} + \text{2 NO} + \text{H}_2\text{O}$$

Then, balancing the rest of the equation we get

$$\text{2 HNO}_3 + \text{3 H}_2\text{S} \longrightarrow \text{3 S} + \text{2 NO} + \text{4 H}_2\text{O}$$

16.3 Step 1:

$$\underset{+2}{\text{Mn}^{2+}} \longrightarrow \underset{+7}{\text{MnO}_4^-} \qquad \text{(oxidation)}$$

$$\underset{+5}{\text{BiO}_3^-} \longrightarrow \underset{+3}{\text{Bi}^{3+}} \qquad \text{(reduction)}$$

Step 2:

$$\text{4 H}_2\text{O} + \text{Mn}^{2+} \longrightarrow \text{MnO}_4^- \qquad \text{(oxidation)}$$

$$\text{BiO}_3^- \longrightarrow \text{Bi}^{3+} + \text{3 H}_2\text{O} \qquad \text{(reduction)}$$

Step 3:

$$\text{4 H}_2\text{O} + \text{Mn}^{2+} \longrightarrow \text{MnO}_4^- + \text{8 H}^+ \qquad \text{(oxidation)}$$

$$\text{6 H}^+ + \text{BiO}_3^- \longrightarrow \text{Bi}^{3+} + \text{3 H}_2\text{O} \qquad \text{(reduction)}$$

Step 4:

$$\text{4 H}_2\text{O} + \text{Mn}^{2+} \longrightarrow \text{MnO}_4^- + \text{8 H}^+ + 5\,e^- \qquad \text{(oxidation)}$$

$$\text{2}\,e^- + \text{6 H}^+ + \text{BiO}_3^- \longrightarrow \text{Bi}^{3+} + \text{3 H}_2\text{O} \qquad \text{(reduction)}$$

Step 5:

$$2[\text{4 H}_2\text{O} + \text{Mn}^{2+} \longrightarrow \text{MnO}_4^- + \text{8 H}^+ + 5\,e^-] \qquad \text{(oxidation)}$$

$$5[\text{2}\,e^- + \text{6 H}^+ + \text{BiO}_3^- \longrightarrow \text{Bi}^{3+} + \text{3 H}_2\text{O}] \qquad \text{(reduction)}$$

Step 6:

$$\text{8 H}_2\text{O} + \text{2 Mn}^{2+} \longrightarrow \text{2 MnO}_4^- + \text{16 H}^+ + 10\,e^- \qquad \text{(oxidation)}$$

$$\underline{10\,e^- + \text{30 H}^+ + \text{5 BiO}_3^- \longrightarrow \text{5 Bi}^{3+} + \text{15 H}_2\text{O}} \qquad \text{(reduction)}$$

$$\cancel{\text{8 H}_2\text{O}} + \text{2 Mn}^{2+} + \cancel{10\,e^-} + \overset{14}{\cancel{\text{30 H}^+}} + \text{5 BiO}_3^- \longrightarrow \text{2 MnO}_4^- + \cancel{\text{16 H}^+} + \cancel{10\,e^-} + \text{5 Bi}^3 + \overset{7}{\cancel{\text{15 H}_2\text{O}}}$$

or,

$$\text{2 Mn}^{2+} + \text{14 H}^+ + \text{5 BiO}_3^- \longrightarrow \text{2 MnO}_4^- + \text{5 Bi}^{3+} + \text{7 H}_2\text{O}$$

16.4 Step 1:

$$\underset{-2}{\text{S}^{2-}} \longrightarrow \underset{+6}{\text{SO}_4^{2-}} \qquad \text{(oxidation)}$$

$$\underset{0}{\text{I}_2} \longrightarrow \underset{-1}{\text{2 I}^-} \qquad \text{(reduction)}$$

Step 2:

$$\text{8 OH}^- + \text{S}^{2-} \longrightarrow \text{SO}_4^{2-} \qquad \text{(oxidation)}$$

$$\text{I}_2 \longrightarrow \text{2 I}^- \qquad \text{(reduction)}$$

Step 3:

$$\text{8 OH}^- + \text{S}^{2-} \longrightarrow \text{SO}_4^{2-} + \text{4 H}_2\text{O} \qquad \text{(oxidation)}$$

$$\text{I}_2 \longrightarrow \text{2 I}^- \qquad \text{(reduction)}$$

Step 4:

$$\text{8 OH}^- + \text{S}^{2-} \longrightarrow \text{SO}_4^{2-} + \text{4 H}_2\text{O} + 8\,e^- \qquad \text{(oxidation)}$$

$$\text{2}\,e^- + \text{I}_2 \longrightarrow \text{2 I}^- \qquad \text{(reduction)}$$

Step 5:

$$\text{8 OH}^- + \text{S}^{2-} \longrightarrow \text{SO}_4^{2-} + \text{4 H}_2\text{O} + 8\,e^- \qquad \text{(oxidation)}$$

$$4[\text{2}\,e^- + \text{I}_2 \longrightarrow \text{2 I}^-] \qquad \text{(reduction)}$$

Step 6:

$$\text{8 OH}^- + \text{S}^{2-} \longrightarrow \text{SO}_4^{2-} + \text{4 H}_2\text{O} + 8\,e^- \qquad \text{(oxidation)}$$

$$\underline{\text{8}\,e^- + \text{4 I}_2 \longrightarrow \text{8 I}^-} \qquad \text{(reduction)}$$

$$\text{8 OH}^- + \text{S}^{2-} + \cancel{8\,e^-} + \text{4 I}_2 \longrightarrow \text{SO}_4^{2-} + \text{4 H}_2\text{O} + \cancel{8\,e^-} + \text{8 I}^-$$

or,

$$8\,OH^- + S^{2-} + 4\,I_2 \longrightarrow SO_4{}^{2-} + 4\,H_2O + 8\,I^-$$

16.5 Fe is listed above Pb in the activity series, and thus there will be a reaction in which Fe replaces Pb in solution:

$$Fe + Pb^{2+} + \cancel{2\,NO_3{}^-} \longrightarrow Fe^{2+} + Pb + \cancel{2\,NO_3{}^-}$$
$$Fe + Pb^{2+} \longrightarrow Fe^{2+} + Pb$$

STUDY QUESTIONS AND PROBLEMS

2. Only reaction b is a redox reaction.

4. **a.** $2\,KMnO_4 + 10\,KCl + 8\,H_2SO_4 \rightarrow 2\,MnSO_4 + 6\,K_2SO_4 + 8\,H_2O + 5\,Cl_2$
    b. $6\,H_2O + P_4 + 10\,HClO \rightarrow 4\,H_3PO_4 + 10\,HCl$ **c.** $4\,Sb + 4\,HNO_3 \rightarrow Sb_4O_6 + 4\,NO + 2\,H_2O$
    d. $PbO_2 + 4\,HI \rightarrow PbI_2 + I_2 + 2\,H_2O$

7. **a.** $6\,H_2O(l) + 4\,AsH_3(g) + 24\,Ag^+(aq) \rightarrow As_4O_6(s) + 24\,H^+(aq) + 24\,Ag(s)$
    b. $3\,Zn(s) + 12\,H^+(aq) + 2\,H_2MoO_4(aq) \rightarrow 3\,Zn^{2+}(aq) + 2\,Mo^{3+}(aq) + 8\,H_2O(l)$
    c. $3\,H_2O(l) + 3\,Se(s) + 2\,BrO_3{}^-(aq) \rightarrow 3\,H_2SeO_3(aq) + 2\,Br^-(aq)$
    d. $5\,H_3AsO_3(aq) + 6\,H^+(aq) + 2\,MnO_4{}^-(aq) \rightarrow 5\,H_3AsO_4(aq) + 2\,Mn^{2+}(aq) + 3\,H_2O(l)$

9. **a.** $8\,OH^-(aq) + S^{2-}(aq) + 4\,I_2(s) \rightarrow SO_4{}^{2-}(aq) + 4\,H_2O(l) + 8\,I^-(aq)$
    b. $Si(s) + H_2O(l) + 2\,OH^-(aq) \rightarrow SiO_3{}^{2-}(aq) + 2\,H_2(g)$
    c. $4\,OH^-(aq) + 2\,Cr(OH)_3(s) + 3\,BrO^-(aq) \rightarrow 2\,CrO_4{}^{2-}(aq) + 5\,H_2O(l) + 3\,Br^-(aq)$
    d. $2\,OH^-(aq) + S_2O_3{}^{2-}(aq) + 4\,ClO^-(aq) \rightarrow 2\,SO_4{}^{2-}(aq) + H_2O(l) + 4\,Cl^-(aq)$

11. **a.** $Ba + 2\,H^+ \rightarrow Ba^{2+} + H_2$ **b.** No reaction   **c.** $2\,Cr + 6\,H^+ \rightarrow 2\,Cr^{3+} + 3\,H_2$ **d.** $Pb + 2\,H^+ \rightarrow Pb^{2+} + H_2$

14. The reaction will not occur because Al is above As in the activity series; thus, As will not replace Al^{3+} in solution. However, the reverse reaction, replacement of As^{3+} by Al, would occur.

18. $Zn + Cu^{2+} \rightarrow Zn^{2+} + Cu$
    anode: $Zn \rightarrow Zn^{2+} + 2\,e^-$ (oxidation)
    cathode: $Cu^{2+} + 2\,e^- \rightarrow Cu$ (reduction)

21. Pb(s) is oxidized and thus is the reducing agent. The Pb in $PbO_2(s)$ is reduced and thus $PbO_2(s)$ is the oxidizing agent.

24. Zn(s) is oxidized and thus is the reducing agent. The Mn in $MnO_2(s)$ is reduced and thus $MnO_2(s)$ is the oxidizing agent.

29. $2\,H_2O(l) \rightarrow 2\,H_2(g) + O_2(g)$

31. **a.** $2\,Fe(s) + 6\,H_2O(l) + 3\,NiO_2(s) \rightarrow 2\,Fe(OH)_3(s) + 3\,Ni(OH)_2(s)$
    b. $8\,OH^-(aq) + I^-(aq) + 8\,MnO_4{}^-(aq) \rightarrow IO_4{}^-(aq) + 4\,H_2O(l) + 8\,MnO_4{}^{2-}(aq)$
    c. $10\,OH^-(aq) + S_2O_3{}^{2-}(aq) + 4\,I_2(s) \rightarrow 2\,SO_4{}^{2-}(aq) + 5\,H_2O(l) + 8\,I^-(aq)$
    d. $2\,As(s) + 6\,OH^-(aq) \rightarrow 2\,AsO_3{}^{3-}(aq) + 3\,H_2(g)$

CHAPTER 17

EXERCISES

17.1 By looking up the atomic number of nitrogen, we can write the following incomplete equation:

$$^{13}_{7}N \longrightarrow {}^{0}_{1}\beta + ?$$

Since the sum of atomic numbers on each side of the equation must be the same (7 in this case), then the atomic product must have an atomic number of 6 ($1 + 6 = 7$). Similarly, we deduce the mass number of the atomic product to be 13. Since the element with atomic number 6 is carbon, the complete equation is

$$^{13}_{7}N \longrightarrow {}^{0}_{1}\beta + {}^{13}_{6}C$$

17.2 The incomplete equation is

$$? \longrightarrow {}^{4}_{2}\alpha + {}^{186}_{76}Os$$

Since the atomic number of the starting radioisotope must be 78 ($2 + 76 = 78$) and the mass number must be

190 (4 + 186), the complete equation is

$$^{190}_{78}\text{Pt} \longrightarrow {}^{4}_{2}\alpha + {}^{186}_{76}\text{Os}$$

17.3 The incomplete equation

$$^{99}_{46}\text{Pd} \longrightarrow {}^{99}_{45}\text{Rh} + ?$$

indicates that the nuclear decay produces a particle of atomic number 1 ($46 - 45 = 1$) and mass number 0 ($99 - 99 = 0$). This corresponds to a positron, so the complete equation is

$$^{99}_{46}\text{Pd} \longrightarrow {}^{99}_{45}\text{Rh} + {}^{0}_{1}\beta$$

STUDY QUESTIONS AND PROBLEMS

16. **a.** $^{199}_{84}\text{Po} \rightarrow {}^{4}_{2}\alpha + {}^{194}_{82}\text{Pb}$ **b.** $^{210}_{82}\text{Pb} \rightarrow {}^{0}_{-1}\beta + {}^{210}_{83}\text{Bi}$ **c.** $^{192}_{78}\text{Pt} \rightarrow {}^{188}_{76}\text{Os} + {}^{4}_{2}\alpha$ **d.** $^{205}_{84}\text{Po} \rightarrow {}^{4}_{2}\alpha + {}^{201}_{82}\text{Pb}$

19. **a.** $^{239}_{94}\text{Pu} + {}^{1}_{0}n \rightarrow {}^{240}_{95}\text{Am} + {}^{0}_{-1}\beta$ **b.** $^{239}_{94}\text{Pu} + {}^{4}_{2}\alpha \rightarrow {}^{242}_{96}\text{Cm} + {}^{1}_{0}n$ **c.** $^{241}_{95}\text{Am} + {}^{4}_{2}\alpha \rightarrow {}^{243}_{97}\text{Bk} + 2\,{}^{1}_{0}n$
   d. $^{242}_{96}\text{Cm} + {}^{4}_{2}\alpha \rightarrow {}^{245}_{98}\text{Cf} + {}^{1}_{0}n$

27. A single large dose of radiation is more likely to damage cellular DNA so extensively that it cannot be repaired.

47. **a.** $^{56}_{27}\text{Co} \rightarrow {}^{0}_{1}\beta + \gamma + {}^{56}_{26}\text{Fe}$ **b.** $^{223}_{88}\text{Ra} \rightarrow {}^{219}_{86}\text{Rn} + {}^{4}_{2}\alpha$ **c.** $^{10}_{6}\text{C} \rightarrow {}^{0}_{1}\beta + {}^{10}_{5}\text{B}$ **d.** $^{18}_{7}\text{N} \rightarrow {}^{0}_{-1}\beta + {}^{18}_{8}\text{O}$
   e. $^{190}_{77}\text{Ir} + {}^{0}_{-1}\beta \rightarrow {}^{190}_{76}\text{Os}$

49. 6.25%

CHAPTER 18

EXERCISES

18.1 First, locate the main chain. In this case, it has five carbon atoms; thus, the parent compound is pentane.

$$\boxed{\text{CH}_3-\text{CH}_2-\text{CH}}\!-\!\text{CH}_3$$
$$|$$
$$\text{CH}_2$$
$$|$$
$$\text{CH}_3$$

Next, number the carbon atoms in the main chain. In this case, we may start at either end, since the branch is the same distance from each end.

$$\boxed{\underset{1}{\text{CH}_3}-\underset{2}{\text{CH}_2}-\underset{3}{\text{CH}}}\!-\!\text{CH}_3$$
$$\underset{4}{\text{CH}_2}$$
$$\underset{5}{\text{CH}_3}$$

There is one methyl substituent; hence, the name is 3-methylpentane.

18.2 First, locate the main chain. In this case it is heptane.

$$\text{CH}_3\boxed{\!-\!\text{CH}-\text{CH}_2-\text{CH}-\text{CH}_2-\text{CH}_3}$$
$$|\qquad\qquad\quad|$$
$$\text{CH}_2\qquad\quad\text{CH}_3$$
$$|$$
$$\text{CH}_3$$

Next, number the main chain. In this case we can start at either end, since each is the same distance from a branch,

$$\text{CH}_3\boxed{\!-\!\underset{3}{\text{CH}}-\underset{4}{\text{CH}_2}-\underset{5}{\text{CH}}-\underset{6}{\text{CH}_2}-\underset{7}{\text{CH}_3}}$$
$$\underset{2}{\text{CH}_2}\qquad\qquad\text{CH}_3$$
$$\underset{1}{\text{CH}_3}$$

Since both substituents are methyl groups, the name is 3,5-dimethylheptane.

18.3 This alkene is linear, with four carbon atoms and one double bond; hence, its name must end in *-butene*. We can number the chain from either end, since the double bond is the same distance from each end.

$$CH_3 \underset{1}{—} CH \underset{2}{=} CH \underset{3}{—} CH_3 \atop \quad\quad\quad 4$$

Thus, the correct name is 2-butene.

18.4 We first select the main chain and number it:

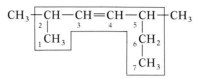

Since both substituents are methyl groups, the correct name is 2,5-dimethyl-3-heptene.

18.5 We number the chain as follows:

$$\boxed{CH_3 \underset{5}{—} CH_2 \underset{4}{—} CH \underset{3}{—} C \underset{2}{\equiv} CH \atop 1}$$
$$\underset{CH_3}{|}$$

and the correct name is 3-methyl-1-pentyne.

STUDY QUESTIONS AND PROBLEMS

2. **a.** single, double, and triple **b.** single **c.** single **d.** single **e.** single **f.** single, double, and triple
g. single, double

8. C_nH_{2n+2}. The open-chain alkanes are b and e.

12. **a.** methylpropane **b.** 2,4,4,6-tetramethyloctane **c.** 2-methylpentane **d.** 3-methyl-5-propyloctane
e. 2,2,4-trimethylpentane

14. **a.**
$$\underset{\underset{\displaystyle CH_3}{|}}{CH_3—CH—CH_2—CH_2—CH_3}$$
b.
$$\underset{\underset{\displaystyle CH_3}{|}}{\overset{\overset{\displaystyle CH_3}{|}}{CH_3—C—CH_2—CH_3}}$$

c.
$$\underset{\underset{\underset{\displaystyle CH_3}{|}}{\underset{\displaystyle CH_2}{|}}}{\overset{\overset{\displaystyle CH_3}{|}}{CH_3—CH_2—C—CH_2—CH—CH_2—CH_2—CH_2—CH_2—CH_3}}$$

d.
$$\underset{\underset{\displaystyle CH_3\ \ CH_3}{|\quad\ |}}{CH_3—CH_2—CH—CH—CH—CH_2—CH_2—CH_2—CH_3}$$
$$CH_2\ CH_2\ CH_3$$

e.
$$CH_3—CH—CH_2—CH—CH_2—CH_2—CH_2—CH_3$$
$$\underset{\underset{\underset{\displaystyle CH_3}{|}}{\underset{\displaystyle CH_2}{|}}}{CH_2}\qquad\underset{\underset{\underset{\displaystyle CH_3}{|}}{\underset{\displaystyle CH_2}{|}}}{CH_2}$$

17. C_nH_{2n}. The open-chain alkenes are a and d.

19. **a.** cis **b.** neither **c.** trans **d.** cis

21. C_nH_{2n-2}. The open-chain alkynes are c and f.

24. **a.** propene **b.** 4-methyl-1-hexyne **c.** 4-methyl-1-pentene **d.** 3-methyl-2-pentene
e. 5-ethyl-3-methyl-1-heptyne **f.** 3-heptene **g.** 6-methyl-3-heptene

34. **a.**

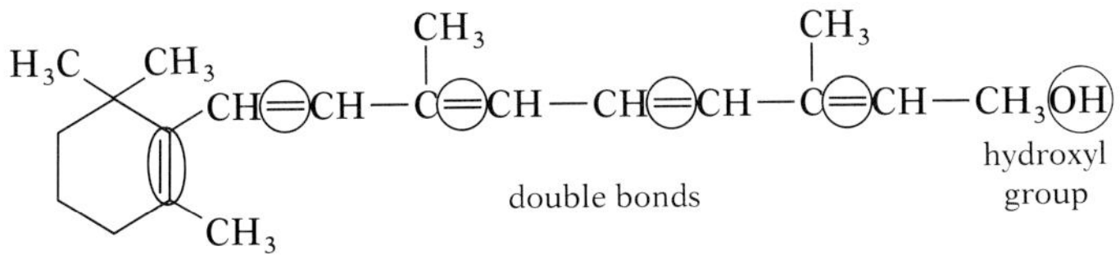

double bonds · hydroxyl group

b.

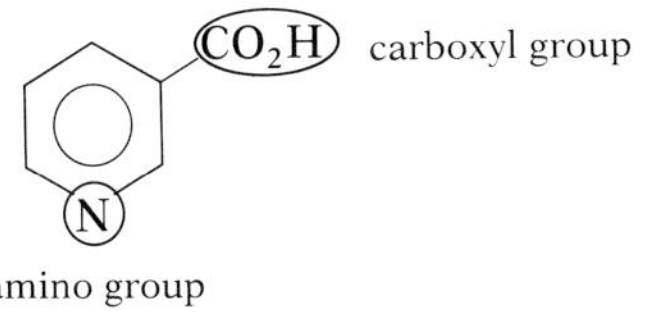

carboxyl group · amino group

c.

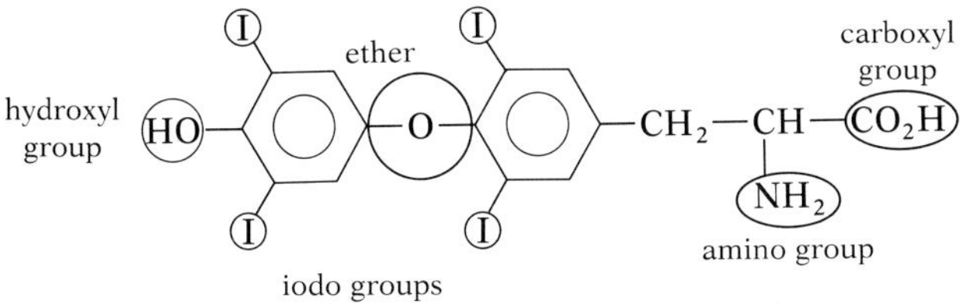

hydroxyl group · ether · iodo groups · carboxyl group · amino group

d.

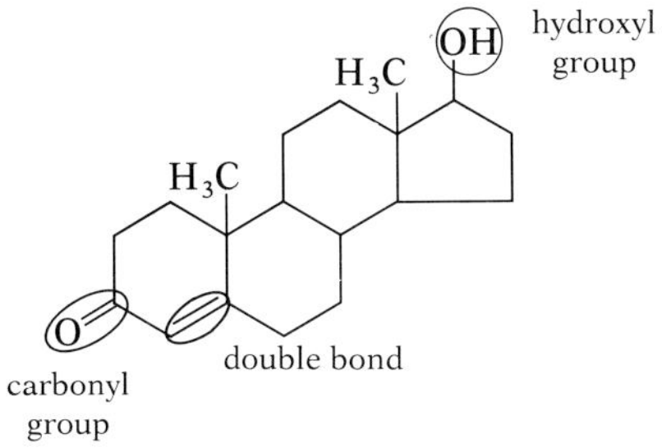

hydroxyl group · carbonyl group · double bond

e.

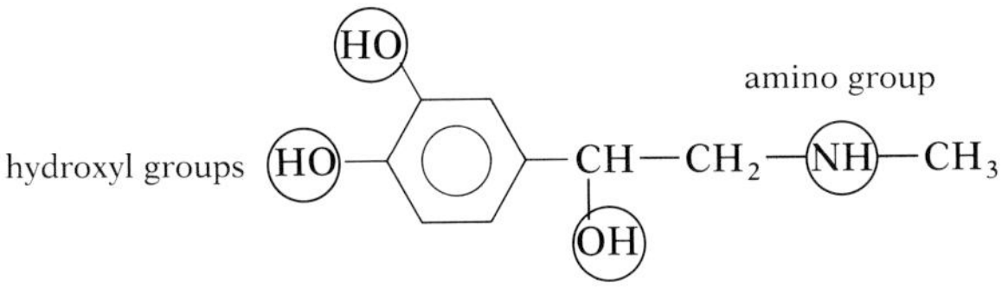

hydroxyl groups · amino group

f. 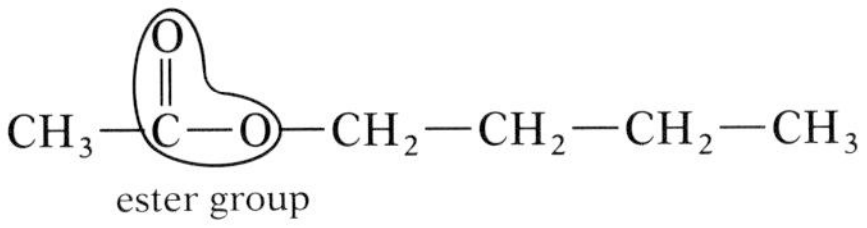 CH_3—C—O—CH_2—CH_2—CH_2—CH_3

ester group

43. **a.** sigma molecular orbital   **b.** one sigma molecular orbital and one pi molecular orbital   **c.** one sigma molecular orbital and two pi molecular orbitals

46. There are so few carbon atoms in C_3H_8 that there is only one possible arrangement for them: CH_3—CH_2—CH_3.

47.

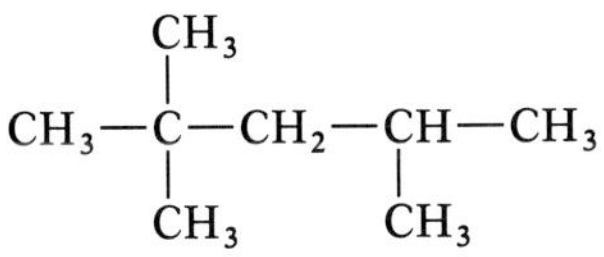

2,2,4-trimethylpentane CH_3—CH_2—CH_2—CH_2—CH_2—CH_2—CH_3 heptane

49. **a.** CH_3—CH_2—CH=CH—CH_2—CH_2—CH_3 **b.** 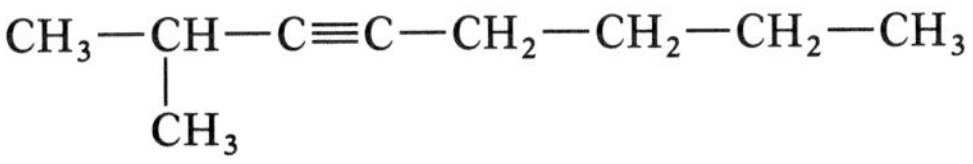 CH_3—CH—C≡C—CH_2—CH_2—CH_2—CH_3

c. $CH_3-\underset{\underset{\displaystyle CH_3}{|}}{C}=CH-CH_2-CH_2-CH_3$ **d.** $CH_3-CH=\underset{\underset{\displaystyle CH_3}{\underset{|}{\underset{\displaystyle CH_2}{\underset{|}{\underset{\displaystyle CH_2}{|}}}}}}{C}-CH_2-CH_2-CH_2-CH_2-CH_3$

e. $CH_3-\underset{\underset{\displaystyle CH_3}{\underset{|}{\underset{\displaystyle CH_2}{|}}}}{\overset{\overset{\displaystyle CH_2-CH_3}{|}}{C}}-CH_2-\underset{\underset{\displaystyle CH_3}{|}}{CH}-CH=CH_2$ **f.** $CH_3-\underset{\underset{\displaystyle CH_3}{|}}{CH}-CH=\underset{\underset{\displaystyle CH_3}{|}}{C}-CH_2-CH_2-CH_2-CH_2-CH_3$

g. $CH_3-CH=CH-CH=CH-CH=CH-CH_2-CH_2-CH_3$

GLOSSARY

Absolute temperature scale: *see* Kelvin temperature scale.

Accuracy: how well a measurement or multiple measurements agree with the true value.

Acid: a substance that produces hydrogen ions (H^+) in water; alternatively, a proton donor.

Acid anhydride: an oxide of a nonmetal that reacts with water to form an acid.

Acid-base indicator: a substance that is one color in acid solution and another color in basic solution.

Activated complex: the structure that is midway between the structures of the reactants and products of a chemical reaction; also known as the *transition state;* it possesses the energy of activation (activation energy).

Activation energy (E_a): the amount of energy needed to cross the highest energy barrier in a reaction pathway.

Activity: in radiation studies, the number of radioactive disintegrations in a given period of time.

Activity series of metals: a list of metals arranged according to their decreasing ability to displace hydrogen from acids.

Alchemy: the medieval art whose principal goals were finding an elixir to prolong life indefinitely and changing base metals into gold.

Aliphatic: referring to all nonaromatic organic compounds.

Alkali metals: elements in Group IA.

Alkaline earth metals: elements in Group IIA.

Alkane: a type of hydrocarbon containing only single bonds.

Alkene: a type of hydrocarbon whose molecules each have one or more double bonds between carbon atoms; also called an *olefin.*

Alkyne: a type of hydrocarbon whose molecules each have one or more triple bonds between carbon atoms.

Allotropes: different forms of the same phase of an element, such as graphite and diamond, which are both solid forms of carbon.

Alloy: a homogeneous blend of two or more metals.

Alpha decay: the emission of an alpha particle by a radioisotope.

Alpha particle: a particle containing two protons and two neutrons; a helium-4 nucleus.

Amorphous: lacking definite shape or form.

Amorphous solid: a solid composed of disordered particles having no particular geometric pattern.

Anion: a negatively charged ion.

Anode: the electrode in an electrochemical cell where oxidation occurs and to which anions migrate.

Apparent molar mass: the molar mass calculated from an empirical formula.

Aqueous solution: a solution in which water is the solvent.

Aromatic: referring to organic compounds that have special ring structures containing multiple covalent bonds; alternatively, any compound that is structurally related to benzene.

Aromatic hydrocarbon: a cyclic hydrocarbon having one or more double bonds spread over all the carbon atoms in a planar ring.

Artificial radioactivity: radioactive decay created by bombardment of stable atomic nuclei with subatomic particles.

Atom: the smallest unit of an element that has properties of the element.

Atomic bomb: an explosive device employing nuclear fission.

Atomic crystals: crystals composed of single nonmetallic atoms held at lattice points by London dispersion forces.

Atomic emission spectrum: the light emitted by a hot, gaseous element.

Atomic mass unit (amu): a relative unit used to express masses of atoms, ions, and molecules; an amu is $1/12$ the mass of an atom of the carbon isotope with mass number 12.

Atomic nucleus: the dense core of an atom where protons and neutrons are found.

Atomic number: the number of protons (also the number of electrons) in an atom.

Atomic orbital: the region of space near an atomic nucleus where an electron is most likely to be found.

Atomic weight: the average mass of all isotopes of an element found in nature.

Average binding energy: the nuclear binding energy of a particular nucleus divided by the number of neutrons and protons in the nucleus.

Avogadro's hypothesis: the theory that at the same conditions of temperature and pressure, equal volumes of gases contain equal numbers of molecules.

Base: a substance that produces hydroxide ions (OH^-) in water; alternatively, a proton acceptor.

Basic anhydride: a metal oxide that reacts with water to form a base.

Battery: a commercial voltaic cell used as a source of electricity.

Beta decay: a type of radioactive decay involving electron emission, positron emission, or electron capture by an atomic nucleus.

Beta particle: an electron emitted by an atomic nucleus.

Bimolecular reaction: a reaction whose rate is determined by the collision of two formula units of reactants.

Binary: composed of two.

Bohr model: the scientific model that describes an atom as having a nucleus at the center and concentric paths (orbits) about the nucleus in which electrons move.

Boiling point: the temperature at which the vapor pressure of a liquid is equal to atmospheric pressure.

Boyle's law: the statement that the pressure exerted on a fixed mass of gas is inversely proportional to the volume of the gas at constant temperature.

Brownian motion: the constant, irregular motion of small particles.

Buffer: a solution that resists pH change when moderate amounts of acid or base are added.

Buffer pair: the components of a buffer: a weak acid and its salt or a weak base and its salt.

Calorie (cal): an amount of energy that is equivalent to 4.184 Joules.

Catalysis: the action of a catalyst.

Catalyst: a substance that does not undergo permanent change but affects the rate of a chemical reaction by its presence in the reaction mixture; most catalysts increase the rates of specific chemical reactions.

Cathode: the electrode in an electrochemical cell where reduction occurs and to which cations migrate.

Cation: a positively charged ion.

Celsius temperature scale: the temperature scale that sets the freezing point of water at 0° and the boiling point of water at 100°; formerly called the centigrade scale.

Centigrade temperature scale: see Celsius temperature scale.

Change of state: a change from one physical state to another, as from liquid to solid or from gas to liquid.

Charles' law: the statement that the volume of a gas varies directly with its absolute temperature if its pressure and mass are kept constant; also, the statement that the pressure exerted by a gas at constant volume is directly proportional to absolute temperature.

Chemical bonds: the attractive forces between the elements in the simplest unit of a chemical compound.

Chemical change: a change that alters the identity of a substance; also called a *chemical reaction*.

Chemical equation: a symbolic description of a chemical reaction.

Chemical family: see chemical group.

Chemical formula: a representation of a compound that uses symbols for elements and subscripts to indicate the proportions in which atoms or ions of elements are combined.

Chemical group: a vertical column of elements in the periodic table; also known as a *chemical family*.

Chemical kinetics: the study of rates of chemical reactions.

Chemical period: a horizontal row of elements in the periodic table.

Chemical periodicity: the periodic recurrence of similar properties when elements are arranged in order of increasing atomic number.

Chemical reaction: see chemical change.

Chemistry: the study of matter and its transformations.

Coefficient: a number placed in front of a quantity to indicate multiplication.

Colligative properties: properties of solutions that depend only on the number of solute particles and are completely independent of the identity of the solute.

Colloid: a homogeneous mixture of dispersed particles that are larger than most molecules.

Combination reaction: a chemical reaction in which two or more atoms or simpler compounds combine to form a compound.

Combustion: a chemical reaction in which oxygen is one reactant, characterized by the production of heat and light; burning.

Common ion effect: the process of increasing the concentration of one of the ions of a slightly soluble ionic compound, causing the saturation equilibrium to shift to the left and thus decreasing the concentration of the other ion.

Compound: a pure substance composed of two or more elements combined in definite proportions by weight.

Compressibility: the ability to be compressed.

Concentration: the amount of solute in a given quantity of solution.

Condensation: see liquefaction.

Condensed structural formula: a chemical formula that gives the major structural features of a molecule.

Conjugate acid: the ion or molecule formed when one formula unit of a base accepts a proton.

Conjugate base: the ion or molecule formed when one formula unit of an acid loses a proton.

Continuous spectrum: a continuous display of wavelengths of the entire range of visible light.

Coordinate covalent bond: a covalent bond composed of a pair of electrons provided by only one of the connected atoms.

Cosmic rays: streams of subatomic particles coming into the earth's atmosphere from the sun and outer space.

Covalent bond: a chemical bond composed of two, four, or six electrons shared between two atoms.

Covalent crystal: a crystal composed of a network of atoms held at lattice points by covalent bonds.

Critical mass: the smallest amount of fissionable isotope required for a self-sustaining nuclear fission reaction.

Crystal lattice: the orderly array of atoms, molecules, or ions in a crystal.

Crystal lattice forces: the forces that hold crystals together.

Crystalline solids: solids that exist in the form of crystals.

Dalton's law of partial pressures: *see* law of partial pressures.

Decomposition reaction: a chemical reaction in which a single reactant is broken down into two or more simpler products.

Density: the amount of mass (m) in a volume (v) of one unit; density (d) $= m/v$.

Derivative: a compound made (derived) from another compound.

Diffusion: the ability of one substance to move and intermingle with another.

Dilution: the addition of solvent to a concentrated stock solution to lower the concentration of the stock solution to a more desirable concentration.

Dimensional analysis: the use of the dimensions (units) associated with a quantity as an aid in setting up the solution to a problem.

Dipolar attractions: attractions between oppositely charged ends of dipoles.

Discrete spectrum: a spectrum composed of only a few discrete wavelengths; also known as a *line spectrum*.

Dosage: in radiation studies, the amount of radiation delivered by a substance.

Double bond: a covalent bond composed of four electrons shared between two atoms.

Double replacement reactions: a chemical reaction in which two parts of two different reactants replace each other.

Dynamic equilibrium: a state of equilibrium in which there is continuous movement of particles in opposite directions in equal numbers per unit time.

Electrochemical cell: the apparatus in which an electrochemical reaction is run.

Electrochemical reaction: a chemical reaction capable of producing electrical energy.

Electrode: a conductor through which an electric current enters or leaves a system.

Electrolysis: the process of forcing nonspontaneous reactions to occur by providing electrical energy from an outside source.

Electrolyte: a substance that conducts electricity when melted or dissolved in water.

Electrolytic cell: an electrochemical cell in which electrolysis is carried out.

Electron: a subatomic particle found outside the nucleus of an atom; it has a charge of -1 and negligible mass.

Electron affinity: the energy change, usually an energy release, that is brought about by adding an electron to a gaseous ground-state atom.

Electron capture: a type of beta decay in which an atomic nucleus captures an electron from outside.

Electron configuration: the distribution of electrons in the atomic orbitals of an atom.

Electron dot formula: a chemical formula that shows all bonds and nonbonding electrons.

Electronegativity: the ability of an atom to attract bonding electrons toward itself.

Electron emission: a type of beta decay in which an electron is emitted by an atomic nucleus.

Electron shell: a region of space in an atom where the probability of finding an electron is high; also known as a *principal energy level*.

Element: a pure substance that cannot be separated into simpler substances by ordinary processes.

Elemental analysis: an analysis of a compound that finds the percentage of each constituent element.

Emission spectrum: *see* atomic emission spectrum.

Empirical formula: the simplest whole-number ratio of ions or atoms in a compound.

Endothermic: heat-absorbing.

Energy: the capacity to do work.

Energy of activation: *see* activation energy.

Energy level: in radiation studies, the energy of radiation emitted by radioactive materials or X-ray generators.

Enthalpy change (ΔH): the heat absorbed or released by a chemical reaction.

Equilibrium: a state of rest or balance.

Equilibrium constant: the constant ratio of concentrations in an equilibrium mixture formed by a reversible reaction at a specific temperature.

Equivalent: the weight of an acid that neutralizes one mole of hydroxide ions or the weight of a base that neutralizes one mole of hydrogen ions.

Evaporation: vaporization that significantly diminishes the volume of a liquid.

Exact number: a number that has no uncertainty associated with it.

Excited state: an unstable, high energy state of an atom, resulting from absorption of energy and elevation of one or more electrons to higher shells.

Exothermic: heat-releasing.

Exponent: the power to which a number is raised.

Fahrenheit temperature scale: the temperature scale that sets the freezing point of water at 32° and the boiling point of water at 212°.

Flammability: the ability to burn.

Fluid: a substance that changes its shape easily and is capable of flowing.

Formula unit: the simplest unit of a compound as described by the formula of the compound.

Formula weight: the mass, in atomic mass units (amu), of a formula unit of a compound.

Free radical: a highly reactive structure having one or more unpaired electrons.

Freezing point: *see* melting point.

Functional group: a distinctive group of atoms that is a part of the structure of many different organic compounds but always reacts the same.

Fusion: the process in which particles combine at elevated temperatures.

Galvanic cell: *see* voltaic cell.

Gamma decay: the release of gamma rays by a radioisotope.

Gamma rays: a highly penetrating form of energy emitted by a radioisotope.

Gas: the state of matter in which particles are far away from each other and move rapidly and randomly; a gas has neither fixed volume nor shape, and it can expand or contract to fill containers of all shapes and sizes.

Geometric isomers: isomers of alkenes in which identical substituents on the two carbon atoms connected by a double bond can be on the same side of the double bond or on opposite sides of the double bond.

Ground state: the stable, minimum energy state of an atom.

Half-life: in radioactive decay, the time required for half of the atoms of a radioisotope to decay.

Halogens: elements of Group VIIA.

Heat: the form of energy related to the average kinetic energy of the particles in a sample of matter; heat flows from a region of higher temperature to a region of lower temperature.

Heat capacity: the capacity for absorbing heat.

Heat of fusion: *see* molar heat of fusion.

Heat of solution: *see* molar heat of solution.

Heat of vaporization: *see* molar heat of vaporization.

Heisenberg uncertainty principle: the statement that it is not possible to know both the exact energy of an electron and its location in space.

Heterogeneous: nonuniform throughout.

Heterogeneous matter: a mixture of different substances in separate phases or different phases of the same substance.

Homogeneous: uniform throughout.

Homogeneous matter: matter consisting of only one phase.

Homogeneous mixture: a mixture having the same composition throughout and consisting of only one phase.

Homologous series: a series of compounds in which each member differs from the preceding member only by a single, constant group.

Hybrid orbital: an atomic orbital formed by blending two different types of standard atomic orbitals.

Hydrate: *see* inorganic hydrate.

Hydration: the process in which water molecules surround solute particles in a solution.

Hydrocarbons: binary molecular compounds composed of hydrogen and carbon.

Hydrogen bomb: an explosive device employing nuclear fusion, usually of hydrogen atoms.

Hydrogen bonds: polar attractions between molecules in which hydrogen is bonded to nitrogen, oxygen, or fluorine.

Hydrolysis: the reaction in which water is split into hydrogen ions and hydroxide ions by reacting with other substances.

Hydrophilic: having strong attractions for water.

Hydrophobic: not attracted to water and hence not soluble in water.

Hypothesis: a proposition set forth as an explanation for a set of observations.

Ideal gas equation: the equation that relates pressure (P), volume (V), number of moles (n), and absolute temperature (T) of an ideal gas; $PV = nRT$.

Inorganic chemistry: the chemistry of nonliving substances, or the study of all elements and compounds except for most covalent carbon compounds.

Inorganic hydrate: an ionic compound containing a definite amount of water in its crystals.

Intermolecular forces: forces that attract one molecule to another.

Ion: an atom or group of atoms having a net electrical charge other than zero.

Ionic bonds: the electrical attractions between oppositely charged ions in a crystal lattice.

Ionic compounds: compounds composed of ions.

Ionic crystals: crystals composed of ions held at lattice points by ionic forces.

Ionization energy: the energy required to remove the most loosely held electron from a gaseous, ground-state atom or ion.

Ionizing radiation: radiation that causes ionization in substances that it strikes.

Ion-product constant (K_w): the ionization constant of water.

Isoelectronic: having the same number of electrons.

Isomers: compounds having the same molecular formula but different arrangements of atoms.

Isotopes: atoms of the same element that differ only by the number of neutrons they possess.

Kelvin temperature scale: the temperature scale derived from the theoretical behavior of gases; the freezing point of water on the Kelvin scale is 273 and the boiling point of water is 373; also called the *absolute temperature scale.*

Kinetic energy: energy possessed by a moving object.

Kinetic molecular theory: the explanation for gaseous behavior that suggests that gas particles move freely and rapidly along straight lines, with frequent collisions causing variation in velocity and direction.

Law of conservation of energy: energy cannot be created or destroyed, but it can be changed from one form to another.

Law of conservation of mass: matter is neither created nor destroyed during a chemical reaction.

Law of definite proportions: a given compound always contains the same elements combined in the same proportions by weight.

Law of partial pressures: each component in a gas mixture exerts its own pressure independent of other gases, and the total pressure of a gas mixture is the sum of the partial pressures of the individual gases; also known as *Dalton's law of partial pressures.*

LeChatelier's principle: if a chemical system in equilibrium is disturbed, the system will readjust in such a way as to offset the disturbance partially and restore equilibrium.

Light: radiant energy having wavelengths in the range of $10^{-8} - 10^{-6}$ m.

Limiting reactant: a reactant that is present in smaller quantities than indicated by the mole ratios of the balanced equation for a reaction.

Line spectrum: *see* discrete spectrum.

Liquefaction: the process in which a gas condenses into a liquid; also known as *condensation.*

Liquid: the condensed state of matter in which particles are in contact with each other but moving freely throughout the sample of matter; a liquid has a definite volume but assumes the shape of its container.

London dispersion forces: attractions between the instantaneous dipoles of atoms and molecules.

Lone pairs: *see* nonbonding electrons.

Magnetic orbital quantum number: the quantum number symbolized by m_l that describes the number of spatial orientations a particular type of atomic orbital may have; also known as the *orientation quantum number.*

Mass: the amount of matter contained in a substance.

Mass defect: the calculated mass deficiency in an atomic nucleus.

Mass number: the sum of the number of protons and neutrons in the nucleus of an atom.

Matter: anything that takes up space and has mass.

Melting: disruption of a crystal lattice so that the particles move freely about and the solid becomes a liquid.

Melting point: the temperature at which a liquid and its solid are in equilibrium; also known as the *freezing point.*

Metal displacement: a reaction in which atoms of a metal spontaneously lose electrons to cations of another metal (or hydrogen) listed below the first metal in the activity series of metals.

Metallic crystals: crystals composed of metal atoms held at lattice points by electrical forces.

Metalloids: elements having some properties of metals and others of nonmetals; in the periodic table, metalloids border the zigzag line that separates metals from nonmetals.

Metals: elements that generally have hard, lustrous surfaces and can be pounded into sheets and drawn into wires; metals are usually good conductors of heat and electricity.

Miscible: mutually soluble (with reference to liquids).

Mixture: two or more intermingled substances or phases.

Molality: solution concentration expressed as moles of solute per kilogram of solvent.

Molar heat of fusion (ΔH_{fus}): the amount of heat needed to melt one mole of a crystalline solid at its melting point.

Molar heat of solution ($\Delta H_{solution}$): the heat change that occurs when one mole of a solute dissolves in a solvent.

Molar heat of vaporization (ΔH_{vap}): the amount of heat required to vaporize one mole of a liquid at its boiling point under one atmosphere of pressure.

Molarity: solution concentration expressed as moles of solute per liter of solution.

Molar mass: the mass of 6.02×10^{23} particles; this mass corresponds to the formula weight of a substance expressed in grams.

Molar volume: 22.4 liters, the volume of an ideal gas at one atmosphere pressure and 273 K.

Mole: 6.02×10^{23} objects or particles.

Molecular compounds: compounds composed of molecules.

Molecule: a group of two or more atoms held together by the attraction of individual nuclei for electrons belonging to other atoms of the same molecule.

Molecular crystals: crystals composed of molecules held at lattice points by intermolecular forces.

Molecular formula: a chemical formula that shows the number and kinds of atoms present in a molecule.

Molecular orbital: an electron orbital formed when atomic orbitals in a molecule merge.

Multiple covalent bond: a chemical bond composed of two or three pairs of electrons shared between two atoms.

Neutralization reaction: the reaction of an acid with a base.

Neutron: a subatomic particle found in the nucleus of an atom; it has a mass of 1 amu and no electrical charge.

Noble gases: elements of Group O.

Nonbonding electrons: valence electrons not involved in a covalent bond; also called *unshared pairs* or *lone pairs*.

Nonelectrolyte: a substance that produces no ions when dissolved in water; thus the solution is a nonconductor of electricity.

Nonmetals: elements that generally crumble easily (if they are solids) and are poor conductors of heat and electricity.

Normality: solution concentration expressed as the number of equivalents of solute per liter of solution.

Nuclear binding energy: the energy equivalent of the mass defect.

Nuclear fission: the process of splitting a heavy atomic nucleus into lighter nuclei.

Nuclear fusion: the process of combining two light atomic nuclei to produce a heavier nucleus.

Nuclear radiation: radiation emitted specifically by radioactive matter.

Nuclear reaction: a change in the nucleus of an atom.

Nuclear reactor: the equipment in which nuclear fission is carried out.

Nucleon: a subatomic particle inside an atomic nucleus.

Nucleus: *see* atomic nucleus.

Octet rule: an atom or monatomic ion is least reactive when it contains eight electrons in its valence shell.

Olefins: *see* alkene.

Orbital: *see* atomic orbital.

Orbital type quantum number: the quantum number symbolized by *l* that indicates the orbital type of an electron.

Organic chemistry: the study of most covalent carbon compounds.

Orientation quantum number: *see* magnetic orbital quantum number.

Osmosis: the movement of water across an osmotic membrane from a region of high water concentration to one of lower water concentration.

Osmotic membrane: a special class of semipermeable membrane that allows only water to pass across.

Osmotic pressure: the amount of external pressure that must be applied to the surface of a solution to prevent osmosis.

Oxidant: *see* oxidizing agent.

Oxidation: loss of one or more electrons, or an increase in oxidation number.

Oxidation number: a number assigned to an element (uncombined or in a compound) to keep track of bond formation and electron transfer between atoms.

Oxidizing agent: a substance that causes oxidation and becomes reduced; also known as an *oxidant*.

Oxyacid: an acid that contains oxygen.

Oxyanion: a polyatomic anion containing oxygen.

Partial pressure: the pressure exerted by an individual component of a gas mixture.

Percentage composition (of a compound): the weight percentage of every element in a compound.

Percentage yield: the actual amount of a product obtained from a chemical reaction, divided by the theoretical yield and multiplied by 100%.

Periodic law: statement that the physical and chemical properties of the elements are functions of their atomic numbers.

Periodic table: the table that contains all the elements listed in order of atomic number and arranged in horizontal rows called periods and vertical columns called groups or families.

Peroxide: a compound in which an oxygen atom is bonded to another oxygen atom, as in hydrogen peroxide, $H-O-O-H$.

pH: the negative logarithm of the hydronium ion concentration.

Phase: a state of matter having clearly defined and distinguishable boundaries.

pH scale: the scale of range 0–14 used for expressing levels of acidity in aqueous solutions.

Physical change: a change in the form of a substance; a change that does not alter the identity of a substance.

Physical states: *see* states of matter.

Pi bond: *see* pi molecular orbital.

Pi molecular orbital: an electron orbital formed in a molecule by side-to-side overlap of two *p* atomic orbitals from two different atoms; it is also called a *pi bond*.

Plum pudding model: *see* Thomson plum pudding model.

pOH: the negative logarithm of the hydroxide ion concentration.

Polar covalent bond: a covalent bond between atoms of different electronegativities in which the bonding electrons are shifted toward the atom of higher electronegativity, resulting in an electric dipole.

Polar molecule: a molecule with an unsymmetrical distribution of electric charge.

Polyatomic ion: an ion composed of two or more covalently bonded atoms.

Polycyclic aromatic hydrocarbon: a hydrocarbon consisting of two or more aromatic rings fused together.

Polymer: a large molecule composed of many repeating units.

Polyprotic acid: an acid capable of donating more than one proton from each of its formula units.

Position of equilibrium: the relative concentrations of products and reactants at equilibrium.

Positron: a particle similar to an electron but with a positive charge.

Positron emission: a type of beta decay in which a positron is emitted by an atomic nucleus.

Potential energy: energy stored in an object because of the object's position, condition, or composition.

Precipitate: an insoluble solid that forms in a solution.

Precision: the extent to which repeated measurements of the same quantity agree with each other.

Pressure: a force that acts on a surface.

Primary cosmic rays: the cosmic rays composed of mostly protons, with some electrons, alpha particles, and atomic nuclei.

Principal energy level: *see* electron shell.

Principal quantum number: a positive whole number, symbolized by n, corresponding to a principal energy level of an atom.

Properties of matter: the characteristics and features that distinguish one kind of matter from all other kinds.

Proton: a subatomic particle found in an atomic nucleus; it has a $+1$ charge and a mass of 1 amu; a proton is the same as a hydrogen ion (H^+).

Pure substance: a sample of matter containing only one substance and only one phase.

Quantum: a definite quantity of energy.

Quantum numbers: the four numbers that indicate principal energy level, orbital type, orbital orientation, and electron spin for each electron in an atom.

Radiant energy: energy given off by an object.

Radiation: energy emitted by matter.

Radioactive decay: the change in the nucleus of a radioisotope as it attempts to gain stability.

Radioactive disintegration series: a succession of radioactive disintegrations that ultimately converts a radioisotope into a stable isotope.

Radioactivity: the process in which unstable atomic nuclei are transformed into more stable nuclei.

Rate constant: the constant of proportionality in a rate equation.

Rate equation: the mathematical relationship between the rate of a chemical reaction and the concentrations of the reactants; also known as a *rate law*.

Rate law: *see* rate equation.

Rate of reaction: the number of formula units of a reaction product formed in a given period of time.

Reaction pathway: the hypothetical route in which reactants are converted to products in a chemical reaction.

Reagent: a reactive chemical agent.

Redox reaction: a reaction that involves changes in oxidation states.

Reducing agent: a substance that causes reduction and becomes oxidized; also known as a *reductant*.

Reductant: *see* reducing agent.

Reduction: gain of one or more electrons, or decrease in oxidation number.

Representative elements: elements in the A groups and Group O.

Resonance structures: molecular structures that show alternative placements of electrons.

Reversible reaction: a reaction that proceeds in both forward and reverse directions at the same time.

Rutherford model: the early scientific model that described an atom as a nucleus surrounded by electrons traveling in orbits much like the planetary orbits of our solar system; also known as the *solar system model*.

Salt: the product of acid-base neutralization, contains cations from the base and anions from the acid.

Saturated hydrocarbons: hydrocarbons having only single bonds.

Saturated solution: a solution in which undissolved solute is in equilibrium with the solution.

Scientific model: a mental image that describes the behavior of a scientific system.

Secondary cosmic rays: the cosmic rays composed of all known elementary particles and formed by collisions of primary cosmic rays with atoms and molecules of atmospheric gases.

Semipermeable membrane: a membrane that allows the passage of only water and certain solute particles.

Shell: *see* electron shell.

Sigma bond: *see* sigma molecular orbital.

Sigma molecular orbital: a molecular orbital that is symmetrical about its axis; it corresponds to a single covalent bond; also known as a *sigma bond*.

Significant figures: the digits in a number that have actual physical meaning.

Single replacement reaction: a chemical reaction in which an ion, atom, or group of atoms replaces an ion, atom, or group of atoms in a compound.

Solar system model: *see* Rutherford model.

Solid: the condensed state of matter in which particles are in contact with each other and occupy fixed positions in space; a solid has a definite volume and fixed shape.

Solubility: the maximum amount of a solute that will dissolve in a given quantity of a solvent at a particular temperature.

Solubility product constant (K_{sp}): for a slightly soluble ionic compound, the product of the dissolved ion concentrations raised to powers corresponding to the coefficients in the balanced equation for the saturation equilibrium.

Solute: a minor component of a solution.

Solution: a mixture of two or more components uniformly distributed through each other.

Solvation: the process in which solvent molecules surround solute particles in a solution.

Solvent: the major component of a solution.

Specific gravity: the density of a material relative to that of water at 4° C.

Specific heat: the amount of heat, measured in calories, required to raise the temperature of one gram of a substance by 1° C.

Spectator ion: an ion in a reaction mixture that does not participate in the reaction.

Spectrum: an array of light waves ordered by wavelength.

Spin quantum number: the quantum number symbolized by m_s that indicates the direction ($+\frac{1}{2}$ or $-\frac{1}{2}$) of spin of an electron.

Standard temperature and pressure (STP): 1 atm and 273 K.

States of matter: the three physical forms of matter: solid, liquid, and gas; also called *physical states*.

Stoichiometry: the quantitative relationships among reactants and products that can be derived from balanced chemical equations.

Strong acid: an acid that ionizes almost totally, having an ionization constant of 1 or greater.

Strong base: a base that ionizes completely in water.

Strong electrolyte: a compound that dissociates completely into ions when dissolved in water and thus causes the solution to be a good conductor of electricity.

Structural formula: a chemical formula that indicates the geometry of a molecule or polyatomic ion by providing as many structural features as is possible in two dimensions.

Subatomic particle: a particle that exists within an atom.

Sublimation: the process in which solid matter is converted directly to gaseous matter without passing through the liquid state.

Surface tension: the force that acts on the surface of a liquid and tends to minimize the surface area.

Suspension: in a fluid, a temporary scattering of particles larger than colloidal particles; a suspension is a heterogeneous mixture whose suspended particles will settle by gravity.

Temperature: a measure of the relative coldness or warmth of an object.

Tetrahedron: a four-sided geometric solid.

Theoretical yield: the calculated amount of product that would be produced by a completely efficient chemical reaction.

Thomson plum pudding model: the early scientific model that described an atom as a sphere of uniformly distributed positive charge with electrons stuck in it like raisins in a pudding.

Titration: a procedure in which one measures the volume of a solution required for complete reaction with another solution in order to find the concentration of the second solution.

Transition elements: elements in the B groups.

Transition state: *see* activated complex.

Transmutation: the conversion of common, inexpensive metals such as lead and iron into gold.

Transuranium elements: the elements beyond uranium in the periodic table.

Trigonal: having the shape of a triangle.

Triple bond: a chemical bond composed of three pairs of electrons shared between two atoms.

True formula: a chemical formula that represents the total number of ions or atoms present in a formula unit of a compound.

Tyndall effect: the visible scattering of light as it passes through a colloid.

Unimolecular reaction: a reaction whose rate depends on only one reactant concentration.

Unit factor: a fraction, composed of equivalent units, whose value is 1 and which is used as a conversion factor.

Unsaturated hydrocarbons: hydrocarbons whose molecules have one or more multiple bonds.

Unshared pairs: *see* nonbonding electrons.

Valence: the number of covalent bonds an atom can form.

Valence electrons: the outer shell electrons of an atom; also referred to as the *valence shell*.

Valence shell: *see* valence electrons.

Valence shell electron pair repulsion (VSEPR) theory: the theory that valence shell electron pairs of an atom will stay as far away from each other as possible and will thus determine the shape of a molecule.

Vaporization: the process in which molecules pass from the liquid to the gaseous state.

Vapor pressure: the pressure exerted by a vapor in equilibrium with its liquid.

Viscosity: the resistance to flow exerted by a fluid.

Volatile: having a high vapor pressure and a low boiling point; such a substance vaporizes easily.

Voltaic cell: an electrochemical cell that provides electricity; also called a *galvanic cell*.

Volume: the space occupied by an object.

Volume-to-volume percentage concentration (%(v/v)): percentage concentration based on the number of milliliters of solute per milliliter of solution.

Wavelength: the distance a wave moves before it starts to repeat itself; the distance between two equivalent points in the path traveled by a wave.

Weak acid: an acid whose ionization constant has a value of less than 1.

Weak base: a base whose ionization constant has a value of less than 1.

Weak electrolyte: a compound that dissociates incompletely into ions when dissolved in water, thus making the solution a weak conductor of electricity.

Weight: the force exerted on an object's mass by the earth's gravitational attraction.

Weight-to-volume percentage concentration (%(w/v)): percentage concentration based on the number of grams of solute per milliliter of solution.

Weight-to-weight percentage concentration (%(w/w)): percentage concentration based on the number of grams of solute per gram of solution.

X rays: high-energy ultraviolet light given off due to electron transitions within atoms.

INDEX

absolute temperature scale, 38, 226–227
absolute value, 472
absolute zero, 227
acetic acid, 337
acid anhydrides, 345, 359
acid-base conjugate pairs, 335, 340
acid-base indicators, 336, 337, 373, 381
acid rain, 346
acidosis, 380
acids, 169, 332, 359
    Arrhenius definition, 332, 359
    binary, 166, 169
    Bronsted-Lowry definition, 334, 359
    commercial, concentrations, 338
    commercial production, 345
    common, 337, 338
    conjugate, 335, 359
    as electrolytes, 366
    equivalents, 349, 359
    ionization constants, 338
    names, 166
    polyprotic, 341–342, 349, 359
    properties, 336, 337
    strength, 338–341
    strong, 338, 359
    weak, 339, 359
accuracy, 20, 42
actinides, 103, 104, 110
activated complex, 312
activation energy, 312, 325
activity series of metals, 393–395
actual yield, 208, 213
addition, 473–475
alchemy, 2, 3
alcohols, 454–456
aldehydes, 455
aliphatic hydrocarbons, 439
alkali metals, 104–105, 117
alkaline earth metals, 105, 117
alkaline solutions, 104; *also see* bases
alkalosis, 380
alkanes, 439–447, 463
    general formula, 440
    isomers, 443
    names, 442, 444
    properties, 463
alkenes, 447–449, 454–455, 463
    bond angles, 447, 463
    geometric isomers, 447–448, 463
    molecular orbitals, 458–460
    names, 449–450
    production, 448

uses, 448–449
alkyl groups, 444
alkynes, 449, 454–455, 463
    bond angles, 449, 463
    molecular orbitals, 460, 463
    names, 451
allotropes, 106
alloys, 105
alpha decay, 409–410, 431
alpha particles, 75, 409
    penetrating ability, 418, 419
    shielding, 420
aluminum, 106
amines, 455, 456
amino acids, 456
amino group, 455, 456
ammonia, 337
    commercial production, 315
ammonium hydroxide, 337 .
ammonium ion, 137
-ane, 444
anhydrous, 167
anion, 55, 67
anode, 397, 402
anthracene, 454
antifreeze, 297
antilogarithms, 503–504
antimony, 107
(aq), 190
archeological dating, 422, 431
argon, 103
Aristotle, 53
aromatic hydrocarbons, 439, 452–454,
    463–464
    molecular orbitals, 460
    polycyclic, 453–454, 464
Arrhenius, Svante, 332
Arrhenius theory of acids and bases, 332,
    333, 334, 359
arrows, in chemical equations, 190
arsenic, 107
astatine, 108
-ate, 164, 166, 169
atmosphere, definition, 221
atmosphere, Earth's, 229
atom
    definition, 54
    excited state, 79
    formation of ions, 55
    ground state, 79
    nucleus, 54, 67
    size, 54, 111

atomic bomb, 427, 429–430
atomic emission spectra, 78, 92
    hydrogen, 78, 79–80
atomic mass unit, 59, 67
atomic models, 74–76, 78–81, 92
    Bohr, 78–80, 92
    Kelvin's, 74
    Rutherford, 75–76, 78, 92
    solar system model, 74–75
    Thomson plum pudding, 74, 92
atomic number
    chemical periodicity, 99
    definition, 55, 67
    of isotopes, 57
atomic orbitals, 81, 92
    d, 82, 84, 104
    filling, 87, 104
    f, 82, 104
    p, 82, 83, 104
    relative energies, 87
    s, 81, 82, 83, 104
    shapes, 82–84
atomic size, 111, 117
atomic theory, 53–54
atomic weight, 60, 67
    of radioactive elements, 61, 408
average binding energy, 425
Avogadro, Amedeo, 174, 235, 247
Avogadro's number, 174
Avogadro's hypothesis, 235, 239, 247, 248

baking soda, 344
balancing chemical equations, 191–196
balancing redox equations, 387–393, 401
    oxidation state method, 387–389, 401
    ion-equation method, 389–393, 401
barometer, 221
base (of exponentiation), 24
bases, 332, 359
    Arrhenius definition, 332, 359
    Bronsted-Lowry definition, 334, 359
    commercial, concentrations, 338
    common, 337, 338
    conjugate, 335, 359
    as electrolytes, 336
    equivalent, 349, 359
    ionization constant, 343
    properties, 336–337
    strength, 343–344
    strong, 344, 359
    weak, 344, 359
basic anhydrides, 346, 359

batteries, 397–399, 402
Becquerel, Henri, 408, 411
benzene, 452, 463
    derivatives, 453, 463
    molecular orbitals, 460, 461
3,4-benzpyrene, 454
beryllium, 105
beta decay, 410, 431
    detecting, 416
beta particles, 410
    penetrating ability, 418, 419
    shielding against, 420
binary acids, 166, 169
binary ionic compounds, 131, 151
    names, 161–162
binary molecular compounds, 164–165
bismuth, 107
blood
    acid-base balance, 380
    buffers, 380–381, 382
Bohr, Niels, 78, 430
boiling point, 262–263
    and atomic number, 100
    definition, 6
    elevation, 296
    ionic compounds, 131, 151
    and intermolecular forces, 255–258
    molecular compounds, 135
    water, variation with pressure, 263
bond angles, 142–145, 447, 449, 463
boron, 105
boron trifluoride, 144
boundary surface diagrams, 81, 92
Boyle, Robert, 48, 223, 247
Boyle's law, 223–226, 232, 240, 247
British system, 27
Broglie, Louis de, 80, 89
bromine, 108
Bronsted, Johannes, 333
Bronsted-Lowry theory, 333, 334, 359
Brown, Robert, 300
Brownian motion, 300, 302
buffers, 366, 378–381, 381–382
    common, 374
    in blood, 380–381
butane, 441–442

calcium, 105
calculations, basic, 472–485
calorie, 30, 42
Calorie, 30
calorimeter, 323, 324
Calvin, Melvin, 423
carbon, 106
    electron distribution, 458
    molecular orbitals, 458
    in organic compounds, 438
    valence, 139
carbon-14, 413, 416

in archeological dating, 422, 431
    half-life, 417
    as tracer in photosynthesis, 423
carbon dioxide
    electron dot formula, 135
    molecular shape, 145
    sublimation, 260
carbonic acid, 337, 342
carbonyl group, 455
carboxyl group, 455
carboxylic acids, 455, 456
cathode, 397, 402
cation, 55, 67
catalysis, 314–315
catalysts, 314–315, 325
Celsius scale, 37
centigrade scale, 37
cesium, 103
cesium-137, 423
Chadwick, James, 415
chain reaction, 428
Charles, J. A. C., 226
Charles' law, 226–231, 232, 240, 247
chemical bonds
    covalent, 124, 133–135, 137, 146–148,
        152
    definition, 124
    ionic, 124, 125, 151
    types, 124
chemical change, 9–10, 14; *also see* chem-
        ical reactions
chemical periodicity, 98, 115
    atomic size, 111, 117
    boiling point, 100
    electron affinity, 114–115, 117
    ionization energy, 112–114, 117
chemical properties, 10, 11, 14
chemical reactions, 9–10
    activation energy, 312
    bimolecular, 313
    conservation of mass, 11–12, 14
    endothermic, 326
    energy barriers, 311
    energy changes, 13, 322–323
    exothermic, 323, 326
    rate, 311
    reaction pathway, 311
    reversible, 310–311
    types, 197–198, 213
    unimolecular, 313
chemistry
    definition, 2, 14
    how to study, 3–5
    relation to other sciences, 2
chlorine, 108
cis, 448
cloud density maps, 81, 92
coal gasification, 323
cobalt-60, 423

coefficients, 190, 213
colligative properties, 295–299, 302
colloidal dispersions, 278, 299–300, 302
    types, 301
combination reactions, 197, 213
combined gas law, 321–325
combustion, 191
common ion effect, 358, 359
common names, chemical, 158
compounds
    composition, 62
    definition, 61, 67
    formula unit, 63, 68
    formula weight, 63–65, 68
    formulas, 62–63
    ionic, 61
    molecular, 61
    percentage composition, 65–67
concentration
    change, effect on equilibrium, 319–321
    definition, 286, 302
    effect on reaction rate, 313–314, 325
    molality, 296–302
    molarity, 287–288, 302
    percent, 290–291
    ppb, 291
    ppm, 291
    of solutions, 286–291
condensation, 254
conjugate acids, 335
conjugate bases, 335
conservation of energy, 13, 15
conservation of mass, 11–12, 14
    in chemical reactions, 191, 192, 213
    in nuclear reactions, 410
coordinate covalent bonds, 137, 152
copper, 110
cosmic rays, 413, 431
covalent bonds, 124, 133–135, 146–148,
        152
    coordinate, 137, 152
    definition, 133
    molecular orbitals, 457–463, 464
    multiple, 135–136, 152, 447, 449, 458–
        460, 463, 464
    orbital hybridization, 457–463, 464
    pi, 459, 464
    polar, 147–148
    sigma, 457, 464
crystals
    atomic, 267
    covalent, 266
    defects, 265
    formation, 265–266
    graphite, 266
    ice, 266
    ionic, 129–130, 266
    lattice forces, 265
    liquid, 270

metallic, 266–267
molecular, 135, 266
types, 266–267
cube root, 477, 478
Curie, Marie, 108, 411
Curie, Pierre, 411
cyclic aliphatic hydrocarbons, 451–452
cyclo-, 451

Dalton, John, 49, 241
   atomic theory, 49, 53–54
   law of partial pressures, 241, 248
Daniell, John F., 397
Daniell cell, 397
decimal numbers, 484–485
decomposition reactions, 197, 213
Democritus, 53
density, 33–34, 42
   gases, 242–243, 248
deuterium, 57
diamond, 106, 266
diatomic molecules, 102, 108, 134
dilutions, 291–293
dimensional analysis, 29
dipolar attractions, 255, 258, 272
dipolar bonds; *see* polar covalent bonds
disintegration series, 414
dispersed substances, 300
dispersing medium, 300
dissociation constant, 338; *also see* ionization constant
distillation, 281
   fractional, 446
disulfide, 456
division, 479–480
DNA, radiation damage, 419–420, 431
Dobereiner, Johann Wolfgang, 98
dosimeter, 416
double bonds, 135–136, 152
   in alkenes, 447
   molecular orbitals, 458–460, 464
double replacement reactions, 198, 213
dry cell, 398–399
dynamic equilibrium, 261, 310

Einstein, Albert, 13–14, 430
Einstein equation, 14, 15, 408, 424–425, 432
electric dipole, 147
electrochemical reactions, 396, 401
electrochemical cells, 396–397, 401
electrodes, 396–397
electrolysis, 399–401, 402
electrolytes, 366–367, 381
   in body, 366
   examples, 367
   strong, 367, 381
   weak, 367, 381
electrolytic cells, 399–401, 402

electron
   atomic orbitals, 81–84
   capture, 410
   charge, 54, 60, 67
   configurations, 85–88, 92, back endpiece
   discovery, 74, 408
   distribution in molecules, 256–267
   emission, 410
   energy sublevels, 81–82, 92
   mass, 54, 59, 60
   principal energy levels, 81, 82, 92
   shells, 81, 82, 92
   spin, 83, 84
electron affinity, 114, 117
electron dot formulas, 126–127, 138–139
electronegativity, 146–147, 152
electrons
   lone pairs, 135
   movement during electrolysis, 399
   nonbonding, 135
   shared; *see* covalent bonds
   unshared pairs, 135
   valence, 100, 135
electroplating, 401
elemental analysis, 178–179
elements
   common, 51
   definition, 49
   of earth, 50
   Greek, 49
   historical ideas, 48–49, 67
   isotopes, 57–58
   naming new, 110
   neutron-to-proton ratio, 409
   new, 59
   radioactive, 408
   representative, 104–109
   in solar system, 50
   symbols, 52–53
   transition, 104, 109–110, 116–117
   transuranium, 414
empirical formula, 179, 180, 181, 184, 186
   calculation, 181
endothermic processes, 13, 15
energy
   changes in chemical reactions, 13
   conservation of, 13, 15
   definition, 12
   forms, 13
   kinetic, 12, 15
   and mass, 14
   potential, 12, 15
   units, 27, 36
energy use, 323
enthalpy change, 322–323, 326
enzymes, 315, 325
equations, algebraic, 486–495
   linear, 486–491

quadratic, 491–495
equations, chemical
   balancing, 191–196
   coefficients, 190
   definition, 190
   notation, 190
equations, redox
   balancing, 387–393
   general form, 387
equilibrium constants, 315–319, 325, 326
   and reaction rate, 324–325
equilibrium, position of, 318
   factors affecting, 319–322
equivalent, definition, 349, 350, 359
esters, 455
ethane, 440, 441, 459
ethene, 458–460, 461
ethers, 455
ethylene glycol, 297
ethyne, 449, 460–462, 463
evaporation, 260–261
exact numbers, 22
exothermic processes, 13, 15
exponents, 24–25, 42, 478

Fahrenheit, Daniel Gabriel, 39
Fahrenheit scale, 37
families, 52, 67, 100, 116
Faraday, Michael, 74
fermentation, 190
Fermi, Enrico, 430
film badges, 416, 431
fission, 426–430, 432
fluid, 6
fluorine, 108, 115
   hydrogen bonding, 255
   molecule, 134
formula unit, 63, 68, 130
formula weight, 63–65, 68, 175
formulas, 62–63, 67–68
   condensed structural, 441
   empirical, 179, 180, 181, 184, 186
   ionic compounds, 127–128
   molecular, 179, 441
   structural, 441
   true, 179–180, 181, 186
fractions, 481–484
francium, 103, 105, 115
Franklin, Benjamin, 74
free radicals, 418
freeze-drying, 272
freezing point, 6
   depression constant, 297
   lowering, 296–298, 302
Frisch, Otto, 430
fuels, 442
functional groups, 454–457, 464
fusion, 268, 430–431, 432

(g), 190
Galileo, 39
gallium, 106
Galvani, Luigi, 397
galvanic cells, 396–397, 402
gamma decay, 411, 431
gamma rays, 411
    penetrating ability, 418, 419
    in radiation therapy, 423–424
    shielding, 420–421
gases
    combined gas law, 231–235
    density, 242–243
    ideal behavior, 236, 240
    ideal gas equation, 236–239
    kinetic molecular theory, 239–241
    molar volume, 235
    pressure-temperature relationships, 230–231
    properties, 6, 220–223, 240, 247
    stoichiometric calculations, 243–247
    volume-pressure relationships, 223–226
    volume-temperature relationships, 226–230
Gay-Lussac, Joseph L., 226
Geiger-Muller counter, 415, 416, 431
germanium, 106
glass, 267–268
gold, 110
gram mole, 174
graphite, 106, 266
greases, 442
Group IA, 104–105, 117, 126
Group IIA, 105, 117, 126
Group IIIA, 105–106, 117, 126
Group IVA, 106, 117, 126
Group VA, 106–107, 117, 126
Group VIA, 107–108, 117, 126–127
Group VIIA, 108–109, 117, 127
Group O, 109, 117
groups, 52, 67, 100, 104, 116
von Guericke, Otto, 221

Haber process, 315
half-life, 416–417, 431
half-reaction, 389
halides, 456
halogens, 108, 117
heat, 36, 42
heat capacity, 39
heat of fusion, 268–269, 273
heat of solution, 284–285
heat of vaporization, 263–265, 273
heating curves, 270–272
Heisenberg, Werner, 80
    uncertainty principle, 80–81, 92
helium, 102, 109
heterogeneous matter, 7, 8, 9, 14
homogeneous matter, 7–8, 14

homogeneous mixtures, 8
homologous series, 440
hybrid orbitals, 458–463
hydrates, inorganic, 167–168, 169
hydration, 283, 301
hydro-, 166, 169
hydrocarbons, 438–439, 463
    physical properties, 439
    saturated, 439
    unsaturated, 447
    uses, 439
hydrogen, 105, 109
    atomic emission spectrum, 78, 79–80
    isotopes, 57
    molecule, 133, 457, 458
    oxidation number, 159
    properties, 102
    valence, 133
hydrogen bomb, 430
hydrogen bonds, 255–256, 258, 272
hydrogen cyanide, 145
hydrologic cycle, 6, 7
hydrolysis, 377–378, 381
hydronium ions, 334
hydrophilic substances, 286, 301
hydrophobic substances, 286, 301–302
hydroxide ion, 137
hydroxyl group, 454–457
hypo-, 164, 169
hypothesis, 49

-ic, 162, 166, 169
ice
    crystal structure, 266, 279
    density, 279, 301
    hydrogen bonds, 279
-ide, 165
ideal gas equation, 236–239, 248
indicators, acid-base, 336, 337, 373, 381
indium, 106
inert gases, 109
inorganic chemistry, 137
inorganic compounds, 159
insolubility, 284
instantaneous dipoles, 256–257
intermolecular forces, 254–258
International System of Measurements (SI), 27–28, 42
International Union of Pure and Applied Chemistry (IUPAC), 159, 444, 449, 451
iodine, 108
iodine-131, 424
ion-product constant, 368, 381
ionic bonds, 124, 125, 146, 151
ionic compounds, 61, 124, 129–131, 151
    binary, 131, 151, 161–162
    boiling points, 131, 151
    common, 138
    conducting electricity, 124

dissolving in water, 282–283
    formulas, 127–128
    melting points, 131, 151
    naming, 161–164
    with polyatomic ions, 161–164
ionic equations, 353–354
ionization constant
    acids, 338–341
    bases, 343
    calculating from pH, 376–377
    water, 368, 381
ionization energy, 112–114, 117
ion-electron method, 389–393, 401
    acidic solutions, 389–393
    basic solutions, 392–393
ions
    definition, 55
    formation from atoms, 55, 67, 125, 126–127
    movement during electrolysis, 399
    polyatomic, 136–137, 138–139
iron, 110, 190
isoelectronic, 58
isomers, 441
    geometric, 447–448, 463
isotopes, 57–58, 67; *also see* radioisotopes
    mass, 59
    symbols, 57
    as tracers, 423
-ite, 164, 166, 169

joule, 30, 42
Joule, James Prescott, 30

Kekulé, Friedrich August, 452
    structures for benzene, 452–453
Kelvin, Lord, 74, 226
Kelvin temperature scale, 38, 226–227
ketones, 455
kinetic energy, 12, 15, 254, 255, 259, 260
kinetic molecular theory, 239–241, 248, 254
kinetics, chemical, 311
krypton, 103, 109

(l), 190
lanthanides, 103, 104, 110
Lavoisier, Antoine-Laurent, 11–12, 48, 167, 192
list of elements, 48
laws
    Boyle's, 223–226, 232, 240
    Charles', 226–231, 232, 240
    combined gas, 231–235
    conservation of energy, 13, 15
    conservation of mass, 11–12, 13, 14
    conservation of mass and energy, 14, 15
    definite proportions, 62, 67
    partial pressures, 241, 248

lead, 106
lead storage battery, 397–398
LeChatelier, Henri Louis, 319
LeChatelier's principle, 319–322, 325, 358
Leclanche, Georges, 398
length, units, 27–28
light, 76–77, 92
   energy, 77, 92
   speed, 14
   wavelength, 76, 92
limiting reactant, 210–212, 213
liquid crystals, 270
liquids, 258–265, 272
   boiling point, 262–263
   density, 258–259
   miscible, 284, 301
   properties, 5–6, 254
   surface tension, 259
   vapor pressure, 261, 273
   vaporization, 260–261, 272–273
   viscosity, 259–260
liquefaction, 254
lithium, 102, 105
litmus, 336, 337
logarithms, 502–504
   table, 505–506
London, Fritz, 257
London dispersion forces, 256–258, 272
Lowry, Thomas, 333
lubricants, 442

Magdeburg hemispheres, 221
magnesium, 105
Manhattan project, 430
mass, 32, 42
   in chemical equations, 200
   conservation of, 11–12
   converting to number of particles, 176
   critical, 428, 429
   definition, 12
   and energy, 14
   measuring, 32
   subcritical, 428
   supercritical, 428
   units, 28
mass defect, 425, 432
mass number, 55–56, 59
   definition, 55, 67
   of isotopes, 57
math tips
   antilogarithms, 375
   cancelling units, 224, 244, 298, 352
   equations, working with, 29, 34, 41, 228, 232, 297
   exponents, 314
   exponents, in multiplication and division, 28
   exponents, negative, 27, 313, 358
   fractions, dividing by, 237

K_a, units, 339
K_b, units, 344
Kelvin temperature, 228
logarithms, 374
molar volume, 242
multiplying units, 23
multistep problems, 207
percentages and decimals, 60, 378
proportionality, 77
scientific notation, on calculator, 26
significant figures, 272
square roots, 340, 369
squaring, on calculator, 318
units in stoichiometric calculations, 201
matter 2
   changes of state, 6, 9, 14
   classification, 7–9, 14
   definition, 2, 14
   heterogeneous, 7, 9, 14
   homogeneous, 8, 9, 14
   physical and chemical changes, 9–10
   physical states, 5–6, 14
   properties, 10
measurements
   systems, 27–28
   units, 27
Meitner, Lise, 430
melting, 268, 273
melting point, 268–269
   definition, 6, 268
   ionic compounds, 51, 131
   molecular compounds, 135
Mendeleyev, Dmitry Ivanovich, 98, 116
   periodic table, 98–99
mercury, 110
metal displacements, 395, 401
metalloids, 101, 116
metals, 101, 109, 116
   activity series, 393–395, 401
   chemical behavior, 115
   ion formation, 109, 110, 114
   oxides, 346, 359
   properties, 109
   reaction with acids, 336
methane, 142
   molecular orbitals, 457–458, 459
   structure, 441, 442
   uses, 442
metric system, 27–28, 42
Meyer, Lother, 98
milk of magnesia, 343
milliequivalents, 351
Millikan, Robert, 74
mixture
   definition, 7
   homogenous, 8
molality, 296, 302
molar gas constant (R), 236
molar heat of fusion, 268–269, 273

molar heat of solution, 284–285, 301
   examples, 285
   sign, 284–285
molar heat of vaporization, 263–265, 273
molar mass, 174, 186
   actual, 179, 184
   apparent, 179, 184, 186
   calculating, 176
   in chemical equations, 200
molar volume, 235, 247
molarity, 287–288, 302
   preparing solutions of specified, 289
mole, 174–175, 186
   in chemical equations, 199–200
   definition, 147, 186
   gram, 174
   "maps", 205, 247, 294
   ratio, 200, 213
mole-mole calculations, 201–20
mole-volume calculations, 243–244
mole-weight calculations, 202–203
molecular compounds, 61, 124, 133, 152
   binary, naming, 164–165
   boiling points, 135
   crystals, 135
   electron dot formulas, 138–139
   melting points, 135
   shapes, 141–145
molecular formulas, 179
molecular orbitals, 457–463
   hybrid, 458–459
   pi, 459, 464
   sigma, 457, 464
   sp, 460, 462, 464
   sp^2, 459, 460, 464
   sp^3, 458, 459, 460, 464
molecular shapes, 141–145
molecular weights, 175
molecules
   definition, 61, 67
   diatomic, 102
monatomic, 58
mothballs, 260, 453–454
multiplication, 476

names, chemical
   acids, 166
   alkanes, 442, 444–447
   alkenes, 449–450
   alkynes, 451
   binary ionic compounds, 161–162
   binary molecular compounds, 164–165
   common, 158
   history, 167
   monatomic anions, 161
   new elements, 110
   polyatomic ions, 163
naphthalene, 260, 453–454
neon, 103, 109

neutralization, 332, 336
and hydrolysis, 377–378
neutron
charge, 59, 60, 67
discovery, 415
mass, 54, 59, 60, 67
number in nucleus, 55
Newlands, John Alexander R., 98
nickel, 110
nickel cadmium battery, 399
nitrate ion, 137, 141
nitro, 456
nitrogen, 107
hydrogen bonds, 255
molecule, 136
valence, 139
nitrogen trifluoride, 144
noble gases, 109, 117
noble gas configuration, 125
nomenclature, systematic, 159
nonelectrolytes, 367, 381
nonmetals, 101, 116
chemical behavior, 115
oxides, 345, 359
properties, 101
nonpolar molecules, 257, 258
normality, 349–352, 359
nuclear binding energy, 425, 432
nuclear energy, 424–426
nuclear stability, 425–426
nuclear fission, 426–430, 432
nuclear fusion, 430–431, 432
nuclear medicine, 424
nuclear power plants, 426–429
nuclear reactions, 14, 411–413, 426–427
nuclear waste, 428–429
nucleons, 54, 67
nucleus, 54, 67
number line, 472
numbers
decimal, 472, 484–485
exact, 22
negative, 472
positive, 472
rounding off, 23–24

octet rule, 109, 113–114, 117
olefins, 449
orbital hybridization, 457–463
organic chemistry, 438, 464
osmotic pressure, 298–299, 302
osmotic membranes, 299
osmosis, 299
-ous, 162, 166
oxidant, 386
oxidation, 131–133, 152
definition, 132, 386, 401
in electrochemical cells, 397, 402
oxidation numbers, 132, 151–152, 169

rules for assigning, 159
oxidation-reduction reactions, 132
examples, 386
general equation, 387
oxidation state; see oxidation number
oxidation-state method, 387–389, 401
oxidizing agent, 386, 401
oxyacids, 166, 169
oxyanions, 163
oxygen, 107
hydrogen bonding, 255
oxidation number, 159
valence, 139
ozone, 107

Paracelsus, 3
paraffin, 442
parentheses in formulas, 63
particles, number of
calculating from mass, 176
in mole, 174
pascal, 221
Pascal, Blaise, 221
Pauling electronegativity values, 146
Pauling, Linus, 146, 150
per-, 164, 169
percent, 484–485
percent concentration, 290–291
percentage composition, 65–67, 68, 178–179
empirical formula from, 181
experimental determination, 179, 180
percentage yield, 208–210, 213
periodic law, 98, 99, 116
periodic table, 52–53, 67, 110, 116, front endpiece
history, 98–99
Period 1, 102
Period 2, 102–103
Period 3, 103
Period 4, 103
Period 5, 103
Period 6, 103
Period 7, 103
periods, 52, 67, 100, 101–103, 116
long, 102
number of elements, 102
short, 102
peroxides, 159
petroleum refining, 446
pH, 366, 368–377, 381
calculating, 374–377
calculating decimal values, 507–508
and concentration, 370
definition, 368
of familiar solution, 372
measurement, 373–374
meter, 374, 381
scale, 371–372

phase, 7
phase changes, 270–272, 273
phenanthrene, 454
phosphate ion, 137
phosphoric acid, 337, 342
phosphorous, 107
photosynthesis, 423
physical changes, 9, 14
physical properties, 10, 11, 14
pi molecular orbitals, 459, 464
Planck, Max, 77, 89
Planck's constant, 77
plastics, 268
platinum, 110
pOH
calculating, 374–377
and concentration, 370
definition, 370
polar covalent bonds, 147–148, 152
polar covalent compounds, 283
polar molecules, 147, 149–151, 255, 258
polonium, 108
polyatomic ions, 136–137, 152
electron dot formulas, 138–139
naming compounds, 163–164
polyethylene, 268
polymers, 268
positron emission, 410
potassium, 103, 105
potassium-40, 422
potential energy, 12, 15
powers (exponents), 478
ppb, 291
ppm, 291
precipitate, 355
precision, 20, 41–42
prefixes, Greek, 165, 444
pressure
definition, 221
and solubility, 285–286
and temperature, 230–231
units, 221–222
and volume, 223–226
product (reaction), 190
product (multiplication), 476
sign, 476–477
propane, 441–442
proportionalities, 499–502
direct, 499–500
inverse, 501–502
protium, 57
proton
charge, 54, 60, 67
mass, 54, 59, 60, 67
number in nucleus, 55
pure substances, 8, 9

quantum, 79
quantum, mechanics, 89

quantum numbers, 89–92, 93
   magnetic orbital (m_l), 89
   orbital type (l), 89
   orientation (m_l), 89–90
   principal (n), 89
   spin (m_s), 91
quick-freezing, 265–266

R (molar gas constant), 236
radiation
   activity, 417, 418
   definition, 415
   detection, 415–416, 431
   dosage, 417, 418
   energy level, 417, 418
   exposure, 419–420
   ionizing, 415, 419
   measuring, 416–418
   safety, 418–421
   therapy, 423–424, 431
radioactive decay
   definition, 409
   disintegration series, 414
   types, 409
radioactivity
   artificial, 414–415
   definition, 408, 431
   natural, 413–414
radiochemistry, 421–424, 431–432
radioisotopes
   definition, 408
   in medicine, 408
   nuclear composition, 409
radon, 103, 109
rare gases, 109
rate constant, 313, 325
rate equation, 313
rate law, 313
ratios, 182
reactant
   definition, 190
   limiting, 210–212, 213
   and reaction rate, 314
reaction pathway, 311, 325
reaction rate, 311, 325
   and equilibrium constant, 324–325
   factors affecting, 312–315
reagent, 291
redox reactions, 132, 386–387, 401
reducing agent, 386, 401
reductant, 386
reduction, 131–133, 152
   definition, 132, 386, 401
   in electrochemical cells, 397, 402
representative elements, 104–109, 116
   atomic radii, 111
resonance structures, 453
reversible reactions, 310–311, 325
Rey, Jean, 39

roots, 477
rubidium, 103
rubidium-87, 422
rust, 190
Rutherford, Ernest, 74–75, 409, 414
   gold foil experiment, 75

(s), 190
salts, 332, 352–354, 357
   as electrolytes, 366
   hydrolysis, 377–378, 381
   solubilities, 354–356
scientific model, 74
scientific notation, 24–26, 42
scintillation camera, 424
scintillation counter, 415, 424, 431
seawater, 281
selenium, 108
semipermeable membranes, 298–299
SI units, 27–28, 42
sigma molecular orbitals, 457, 464
significant figures, 21–22, 42
   in calculations, 23
   rules for zeros, 21–22
signs (of numbers), 472
silicates, 106
silicon, 106
silicon dioxide crystals, 267
silver, 110
single replacement reactions, 197, 213
sodium, 103, 105
sodium bicarbonate, 344
sodium chloride
   crystals, 129–130
   electrolysis, 400–401
sodium hydroxide, 337
solids, 265–268, 273
   amorphous, 267–268, 273
   characteristics, 5, 6
   crystalline, 265–267, 273
   properties, 254
solubility, 284–286, 301
   and chemical structure, 286, 301–302
   and pressure, 285–286, 301
   of salts, 354–356
   and temperature, 286, 301
solubility product constant, 356–358, 359
soluble, definition, 284
solutes
   definition, 280, 301
   rate of dissolving, 283
solutions
   aqueous, 280
   colligative properties, 295–299
   definition, 280, 301
   formation, 282–283
   preparing, 289
   properties, 280–281
   rate of formation, 283, 301

saturated, 284, 301
   stoichiometric calculations, 294–295
   types, 280
solvation, 283, 301
solvents, 280, 301
sp molecular orbitals, 460, 462, 464
sp^2 molecular orbitals, 459, 460, 464
sp^3 molecular orbitals, 458, 459, 460, 469
specific gravity, 36, 42
specific heat, 39–40, 42
spectator ions, 353
spectrum
   atomic emission, 78, 92
   continuous, 77
   line, 78, 92
   visible, 76
speed of light, 14
square root, 477, 478
standard temperature and pressure (STP), 235
states of matter, 5–6, 14
   changes, 6, 9, 14
stoichiometry, 199–201, 213
   definition, 200
stoichiometric calculations, 201–207
   fundamental relationships, 205
   gases, 243–247
   solutions, 294–295
STP, 235
strontium-85, 424
structural formulas, 145, 146
subatomic particles, 54–55, 60
   also see electrons, protons, neutrons
sublimation, 260
subscripts (in formulas), 62–63
subtraction, 475–476
sucrose, 174
sulfate ion, 137
sulfur, 107–108
sulfur dioxide, 140
sulfuric acid, 341
sun, nuclear reactions, 430
surface tension, 259
suspensions, 278, 300, 302
symbols
   elements, 52–53
   isotopes, 57, 67
Systeme International (SI), 27–28, 42

tellurium, 108
temperature, 254
   change, effect on equilibrium, 322
   and pressure, 230–231
   and reaction rate, 312–313, 314, 325
   and solubility, 285–286
   scales, 37–38, 226
   units, 27
   and volume, 226–230

tests
  basic algebra, 485–486
  basic calculations, 471–472
thallium, 106
theoretical yield, 208, 213
thermometer, development of, 39
thiols, 456, 457
Thomson, J. J., 74, 408
Thomson plum pudding model
thyroid gland, 424
tin, 106
titrations, 347–348
torr, 221
Torricelli, Evangelista, 221
  barometer, 222
trans, 448
transition elements, 104, 109–110, 116–117
transition state, 312
transmutation, 2, 408, 414, 431
transuranium elements, 414
triads, 98
trigonal, 144
triple bonds, 136, 152
  in alkynes, 449
  molecular orbitals, 460, 463, 464
tritium, 57, 413
true formula, 179–180, 181, 186
  calculation from empirical formula, 184
Tyndall effect, 300, 302

unit factors, 30, 42
unit-factor method, 30–31, 42
units of measurement, 27
  calculations with, 480–481
  conversion, 29–31
  prefixes, 28

uranium-238
  disintegration series, 414
  dating rocks, 422
urea, preparation, 438

valence
  of common elements, 139
  definition, 138
valence electrons, 100, 116
  nonbonding, 135
valence shell, 100, 116
valence shell electron pair repulsion (VSEPR) theory, 142, 152
vapor pressure, 261–263, 272
vapor-pressure lowering, 295–296
vaporization, 260–262, 272–273
viscosity, 259–260
Volta, Alessandro, 397
voltaic cells, 396–397, 402
volume
  change, effect on equilibrium
  measurement, 29
  and pressure, 223–226
  and temperature, 226–230
  units, 28
volume-volume calculations, 245–247
volume-to-volume percent concentration, 290

water
  arrangement of molecules, 26, 278–279
  boiling point, 256, 263
  density, 279
  electron dot formula, 140
  of hydration, 167–168
  hydrogen bonds, 256, 259, 279
  ion-product constant, 368, 381

ionization, 367–368
ionization constant, 368, 381
molar heat of vaporization, 264
molecular shape, 142, 278
physical properties, 279, 280, 301
specific heat, 40
as solvent, 282–283
viscosity, 259
wave mechanics, 89
weight, 33, 42
weight-mole calculations, 203–204
weight-to-volume percent concentration, 290
weight-to-weight percent concentration, 291
weight-volume calculations, 244–245
weight-weight calculations, 205–207
Wöhler, Friedrich, 438
word problems, 495–499
work, 12

X rays, 415
  diagnostic, 421
  penetrating ability, 418, 419
  in radiation therapy, 423–424, 431
  shielding, 420–421
xenon, 103, 109

yield
  actual, 208, 213
  percentage, 208–210, 213
  theoretical, 208, 213
-ylene, 449

zeros
  significant figures, 21–22, 26

Photo Credits *(continued)*

Units in Measurement Systems

Property	Metric	British	Unit Relationships	Property	Metric	British	Unit Relationships
length	kilometer (km)	mile (mi)	1.61 km = 1 mi	mass	**kilogram** (kg)	pound (lb)	1 kg = 2.2 lb
	meter (m)	yard (yd)	1 m = 39.37 in		gram (g)		453.6 g = 1 lb
			1000 m = 1 km				1000 g = 1 kg
	centimeter (cm)	foot (ft)	30.48 cm = 1 ft		milligram (mg)		1000 mg = 1 g
		inch (in)	2.54 cm = 1 in		microgram (μg)		10^6 μg = 1 g
			100 cm = 1 m		nanogram (ng)		10^9 ng = 1 g
	millimeter (mm)		1000 mm = 1 m	energy	**joule** (J)		1 J = 10^7 erg
	micrometer (μm)		10^6 μm = 1 m		kilocalorie (kcal)		1 kcal = 4184 J
	nanometer (nm)		10^9 nm = 1 m		calorie (cal)		1000 cal = 1 kcal
volume	**cubic meter** (m³)	quart (qt)	1 m³ = 1057 qt				1 cal = 4.184 J
	liter (L)		1 L = 1.057 qt		erg (erg)		4.184×10^7 erg = 1 cal
			1 L = 0.001 m³	temperature	**Kelvin** (K)	Fahrenheit (°F)	K = °C + 273
	deciliter (dL)		10 dL = 1 L		Celsius (°C)		°F = $\frac{9}{5}$ °C + 32
	milliliter (mL)		1000 mL = 1 L				°C = (°F − 32)($\frac{5}{9}$)
	cubic centimeter (cm³)		1 cm³ = 1 mL				
	microliter (μL)		10^6 μL = 1 L				

(NOTE: Boldface indicates SI standard units.)

Names and Formulas of Common Ions

Positive Ions		Negative Ions	
Ammonium	NH_4^+	Acetate	$C_2H_3O_2^-$
Copper(I) (Cuprous)	Cu^+	Bromate	BrO_3^-
Hydrogen	H^+	Bromide	Br^-
Potassium	K^+	Chlorate	ClO_3^-
Silver	Ag^+	Chloride	Cl^-
Sodium	Na^+	Chlorite	ClO_2^-
Barium	Ba^{2+}	Cyanide	CN^-
Cadmium	Cd^{2+}	Fluoride	F^-
Calcium	Ca^{2+}	Hydride	H^-
Cobalt(II)	Co^{2+}	Bicarbonate (Hydrogen carbonate)	HCO_3^-
Copper(II) (Cupric)	Cu^{2+}	Bisulfate (Hydrogen sulfate)	HSO_4^-
Iron(II) (Ferrous)	Fe^{2+}	Bisulfite (Hydrogen sulfite)	HSO_3^-
Lead(II)	Pb^{2+}	Hydroxide	OH^-
Magnesium	Mg^{2+}	Hypochlorite	ClO^-
Manganese(II)	Mn^{2+}	Iodate	IO_3^-
Mercury(II) (Mercuric)	Hg^{2+}	Iodide	I^-
Nickel(II)	Ni^{2+}	Nitrate	NO_3^-
Tin(II) (Stannous)	Sn^{2+}	Nitrite	NO_2^-
Zinc	Zn^{2+}	Perchlorate	ClO_4^-
Aluminum	Al^{3+}	Permanganate	MnO_4^-
Antimony(III)	Sb^{3+}	Thiocyanate	SCN^-
Arsenic(III)	As^{3+}	Carbonate	CO_3^{2-}
Bismuth(III)	Bi^{3+}	Chromate	CrO_4^{2-}
Chromium(III)	Cr^{3+}	Dichromate	$Cr_2O_7^{2-}$
Iron(III) (Ferric)	Fe^{3+}	Oxalate	$C_2O_4^{2-}$
Titanium(III) (Titanous)	Ti^{3+}	Oxide	O^{2-}
Manganese(IV)	Mn^{4+}	Peroxide	O_2^{2-}
Tin(IV) (Stannic)	Sn^{4+}	Silicate	SiO_2^{2-}
Titanium(IV) (Titanic)	Ti^{4+}	Sulfate	SO_4^{2-}
Antimony(V)	Sb^{5+}	Sulfide	S^{2-}
Arsenic(V)	As^{5+}	Sulfite	SO_3^{2-}
		Arsenate	AsO_4^{3-}
		Borate	BO_3^{3-}
		Phosphate	PO_4^{3-}
		Phosphite	PO_3^{3-}